# DISEASES

OF

# FOREST

AND

# ORNAMENTAL

# TREES

To
T. Small, OBE, DSc (1894–1976)
*Mycologist to the States of Jersey*, 1931–1954
and
C. W. T. Young (1920–1981)
*Pathology Branch, The Forestry Commission*, 1955–1981

# DISEASES

OF

# FOREST

AND

# ORNAMENTAL

# TREES

SECOND EDITION

D H PHILLIPS &
D A BURDEKIN

**M**

First edition published 1982
Second edition published 1992 by
THE MACMILLAN PRESS LTD
London and Basingstoke

Associated companies in Auckland, Delhi, Dublin, Gaborone,
Hamburg, Harare, Hong Kong, Johannesburg, Kuala Lumpur,
Lagos, Manzini, Melbourne, Mexico City, Nairobi, New York,
Singapore, Tokyo

A CIP catalogue record for this book is available from The
British Library.

ISBN 978-1-349-10955-5     ISBN 978-1-349-10953-1 (eBook)
DOI 10.1007/978-1-349-10953-1

# Contents

# Preface

In the ten years or so since the publication of the first edition of this book, much new information has become available. We have revised and rewritten a large part of the text to incorporate the most important new material, and we have also widened the scope of the book to cover the whole of Europe (with particular reference to the more northerly areas). Many of the diseases described occur elsewhere, especially in North America, and in our account we have therefore drawn on all available literature, and the benefits over the years of our contact with fellow tree pathologists throughout the world. To keep the book to a convenient size we have had to be selective, and our interpretations are inevitably coloured by the fact that our experience of tree diseases has been mainly in the British Isles.

As before, we have described the diseases of fruit trees such as the apple, pear and cherry only in so far as they affect ornamental forms of the trees concerned. These diseases are treated at greater length in texts on diseases of orchard trees.

The depth and length of the treatment given to individual diseases varies, and, in general, reflects both their importance and the amount of information available about them. Where this can usefully be done, we have divided the diseases into those occurring mainly in the nursery and those important chiefly after the nursery stage.

This is not a book on the principles of tree pathology, though many of these principles can be deduced from the accounts of the various diseases described (and we give references to some of the other books which deal with the principles of plant pathology). We have attempted rather to provide information useful in the identification and control of tree diseases, dealing with symptoms, factors affecting disease expression, host range and resistance, and giving brief descriptions of many of the causal organisms. We have not provided a lengthy introduction to mycology, bacteriology and virology, but in the case of fungal and some bacterial diseases have usually supplied enough descriptive detail to enable the professional pathologist to make a diagnosis based on this and the other material given. Further information must be sought in the more extensive texts on fungi, bacteria, viruses, etc., cited in the References.

Name changes of both disease organisms and their hosts cause problems in practical fields such as forestry, arboriculture and plant pathology. We have generally attempted to apply current names to pathogens and to trees. In the case of pathogens we give the better-known synonyms when we use what to some may be unfamiliar names. In the case of trees we usually supply the common names now used in Britain as well as the current scientific names.

Control measures are suggested for many diseases; in some cases, however, no control measures are as yet available. A good many diseases are too minor in their effects for control measures to be required. If it appears necessary to attempt the control of any of these lesser diseases, the general measures discussed in chapter 1 may be applied. Mention is made of many fungicides and bactericides. New chemicals are continually being introduced and old ones may no longer be available or approved or their use may be withdrawn, and the reader wishing to use such materials should consult further, and use only approved products, taking care to observe the recommendations made by the manufacturers. Information on almost all products approved in Britain is given in the *UK Pesticide Guide*, which is revised annually and published by CAB International and the British Crop Protection Council. Similar guides are published in or for most other countries.

The introductory chapter on symptoms, diagnosis and control of diseases is intended mainly for the student, the beginner and the general reader. In the rest of the book the diseases are grouped chiefly under their host genera. Most of the diseases with a wide host range (including the disorders caused by non-living agents such as frost or air pollution), however, are described in separate chapters. In a few cases organisms affect several hosts but are far more important on one than on any of the others; they are then described under the main host. The decay fungi (many of which also have a wide host range) are also drawn together in one chapter.

It is difficult to arrange the trees in a completely logical order reflecting both their importance as hosts and their botanical relationships. In the present edition we have rearranged the order of the chapters somewhat in order to achieve a slightly better compromise. We have also separated out the trees in the family Rosaceae from the formerly very long chapter on 'other broadleaved trees'. We have consolidated the references, formerly at the end of each chapter, into a single bibliography. References to the *Distribution Maps of Plant Diseases* prepared by the International (formerly Commonwealth) Mycological Institute (IMI, formerly CMI), Kew, and published by CAB International, Slough, are referred to in the text (and not in the Bibliography) as CMI Map . . . followed by their numbers in the series. Similarly, the *Descriptions of Plant Pathogenic Bacteria* and *Fungi*, also prepared at the IMI, and the *Descriptions of Plant Viruses*, prepared jointly by the IMI and the Association of Applied Biologists, and also published by CAB International, are referred to in the text in full or as CMI Descr. B. or F. and CMI/AAB Descr. V., followed by their numbers.

Many of our colleagues have made valuable comments on the text. Among them we must especially mention the late Dr S. Batko (who helped particularly by translating literature from many languages), the late Dr W.O. Binns, Dr C.M. Brasier, Dr J.N. Gibbs, Mr B.J.W. Greig, Dr D. Lonsdale, Mr J.D. Low, Mr W.J. McCavish, Dr D.B. Redfern, Mr D.R. Rose, Mr R.G. Strouts, Mr D. Williamson and the late Mr C.W.T. Young. Miss Mary Trusler and Mr G.L. Gate greatly helped by printing photographs, and Mr J. Williams drew figure 32. We are also much indebted to Mrs Catherine

Burdekin, who put most of the text onto disk for us.

The book owes much to the decision of the publishers to illustrate many of the diseases in colour. Many of the photographs are from the collection of the Forestry Commission, to whom special thanks are therefore due. These and the other illustrations are separately listed and acknowledged elsewhere. Fuller acknowledgement of black and white illustrations from sources outside our own collections and those of the Forestry Commission are as follows: Figure 66, C.M. Brasier; Figures 3, 4, 25, 106, B.J.W. Greig; Figure 85, W.B. Grove (1913), *British Rust Fungi*, Cambridge University Press, Cambridge; Figures 48, 53, 56, 64, 93, R. Hartig (1894), *Text-book of the Diseases of Trees*, Macmillan, London; Figure 65, D.B. Redfern; Figure 5, G.A. Salt (1974), *Trans. Br. mycol. Soc.*, **63**, 339–51; Figures 7, 8, 75, 80, 81, 84, 94, K.F. Tubeuf (1895), *Pflanzenkrankheiten*, Springer, Berlin; Figure 10, J.H. Western (1971), *Diseases of Crop Plants*, Macmillan, London; Figure 50, M. Wilson and D.M. Henderson (1966), *British Rust Fungi*, Cambridge University Press, Cambridge; Figures 9, 88, 96, H. Wormald (1955), *Diseases of Fruits and Hops*, Crosby Lockwood, London.

Wrecclesham, 1991

D.H.P.
D.A.B.

# The colour plates

*These appear between pages 238 and 239.*

# 1 General introduction

Disease is a disorganisation of the normal physiological activity of the tree caused by some adverse agent which gives rise to characteristic symptoms. Hence, a diseased tree differs from a healthy one in recognisable ways, and the pathologist must first of all become familiar with the appearance and characteristics of healthy trees. Thus, in cold weather the foliage of nursery plants of western red cedar (*Thuja plicata*) may turn from green to bronze (plate 39). This colour change is normal and reversible, and the affected plants are not diseased. Nevertheless, if the temperature becomes sufficiently low, the foliage of these plants is killed, but they then become an irreversible glossy black. Similarly, the fungus *Didymascella thujina* may attack and kill the scale leaves of *Thuja* seedlings. The affected plants show the small fruit bodies of the fungus (plate 40) and are diseased, and their growth is reduced; if enough scale leaves are destroyed, the plants die. Again, the bark of London plane (*Platanus* × *hispanica*) peels off the tree in a typical pattern, but this is a normal phenomenon, whereas the breaking up of the bark in the late stages of beech bark disease (figure 58) is an indication of severe and irreparable damage by *Nectria coccinea* and other associated fungi.

## Causes of disease

The agents that cause diseases may be non-living or living. Among the non-living agents are climatic factors such as high or low temperatures, drought and lightning, soil factors such as poor soil aeration and the lack of essential nutrients, and injurious chemicals such as atmospheric pollutants and mis-placed herbicides. Damage by these agents is associated with some discrete event or series of events, such as an emission of some polluting chemical, or with the absence of some necessary element (as in the case of nutrient deficiency disorders). Disorders caused by non-living agents are not infectious, so they cannot spread from tree to tree.

The living pathogens include the fungi (the most important), the bacteria and the semiparasitic mistletoes. The viruses, which are ultramicroscopic and can reproduce only in their host cells, are also included here. Some disorders formerly regarded as virus diseases now appear to be caused by mycoplasma-like organisms (MLOs). These are related to bacteria, with which they are included in the Procaryotae (Holliday, 1989). Those affecting trees are still relatively little known, and as the cause of some of these disorders is still uncertain, in this book all the virus and virus-like diseases are included under the same heading. The classification of viruses is discussed by Matthews

(1982). The diseases caused by these living pathogens are infectious and can spread from tree to tree by means of their own spores or other disseminules either alone or with the aid of one or more vectors.

Many fungi cause diseases of trees (and of other plants). The true fungi (excluding the slime moulds) are divided into five main subdivisions. The first two, the Mastigomycotina and the Zygomycotina, were formerly grouped together as the Phycomycetes. These fungi are microscopic, one-celled or formed of mainly non-septate threads (*hyphae*) which in aggregate are described as *mycelium*. Most, if not all, the fungi in these groups that attack trees are classified in the Peronosporales, of the order Oomycetes, which form a class of the Mastigomycotina. Important among the fungi included here are the species of *Phytophthora*.

In the third and fourth subdivisions the hyphae that make up both the vegetative body and the fructifications are septate (that is, they are divided into segments by cross walls). The fruit bodies are generally relatively large, and readily visible to the naked eye or with the aid of a hand lens. In the first of these two groups, the Ascomycotina (Ascomycetes), the sexual spores are borne inside sac-like or tubular *asci*. In a few the sexual or *teleomorphic* (perfect) fruit body containing the asci is very simple, consisting of little or nothing more than an ascal layer (*hymenium*), as in the case of *Taphrina populina*, the cause of poplar leaf blister (plate 28). In others (if we confine ourselves to those causing diseases in trees) the asci are enclosed in more or less globular, flask-shaped, saucer-shaped or boat-shaped fruit bodies. Included here are, for example, the powdery mildews (such as the oak mildew, *Microsphaera alphitoides*), with small, globular fruit bodies (*cleistothecia*) (plate 29), the *Nectria* species (including *N. cinnabarina*, with small, bright red, flask-shaped *perithecia*) (plate 14), *Lachnellula willkommii*, the cause of larch canker (with small, disc-shaped *apothecia* with a white outer wall and a bright orange hymenial disc) (plate 23), and the *Lophodermium* species associated with needle cast in pines, with a boat-shaped *hysterothecium* opening by a slit (figure 36). Many Ascomycetes are active as pathogens in their asexual, *anamorphic* (imperfect) stages, which produce asexual spores in fructifications of various kinds which are followed later by the sexual fruit bodies on the dead host tissues.

In the second of these groups, the Basidiomycotina (Basidiomycetes), the sexual spores are borne outside usually club-shaped *basidia*. Of the three classes of the Basidiomycotina, only two contain pathogens.

The first, the Hemibasidiomycetes, includes the Uredinales (the rusts), such as species of *Cronartium*, *Coleosporium* and *Melampsora*. The rusts often exhibit complex life histories, which in their most complete form involve the production of a series of spore types. Some of these spores arise on one host, while others form on a second, alternate host, which may be necessary for the completion of the life cycle. These spore forms and their associated fructifications are described below.

(1) *Spermatia*: small sex-cells produced in minute *spermagonia*, embedded in the leaves or stems.

(2) *Aeciospores* (*aecidiospores*): produced in *aecia* (*aecidia*), which are often cup-shaped or cylindrical (figure 37), like those of *Coleosporium tussilaginis* on needles of pines, or blister-like, like those of *Peridermium pini* on pine stems (plate 12); sometimes they lack an outer wall, when they are pustular, and known as *caeomata*.

(3) *Urediniospores* (*uredospores*): produced in *uredinia* (*uredosori*), small, often yellow or brown pustules; the uredinia of *C. tussilaginis* occur on *Senecio*, *Tussilago* and other members of the Compositae.

(4) *Teliospores* (*teleutospores*): produced in *telia* (*teleutosori*) which often resemble the uredinia and follow them on the same leaves or other parts of the plant (as they do in the case of *C. tussilaginis*). In some rusts the spores in these sori are massed together into small columns that stand up above the leaf surface like minute bristles. This is so in *Cronartium ribicola*, the white pine blister rust, the telia of which occur on the leaves of the black currant and other species of *Ribes* (figure 44). In species of *Gymnosporangium*, the telia form as large, gelatinous, cushion-shaped or horn-like bodies (figure 49).

(5) The teliospores germinate to produce haploid *basidiospores* (*sporidia*).

The second of these groups among the Basidiomycotina, the Hymenomycetes, includes the Agaricales (plate 17) (the toadstools, such as the *Armillaria* species and similar forms) and the Aphyllophorales, important among which are the pore-bearing bracket fungi, with many wood-decay organisms such as *Heterobasidion annosum* (*Fomes annosus*) (plate 18), *Phaeolus schweinitzii* (plate 62) and *Meripilus giganteus*.

Finally, the fifth subdivision of the fungi is the Deuteromycotina (Deuteromycetes, the Fungi Imperfecti), an artificial assemblage consisting mainly of asexual forms (anamorphic or imperfect stages) of ascomycetes and of fungi with no known affinity with other groups. The fruit bodies of these fungi often superficially resemble those of ascomycetes, but they contain asexual spores instead of asci with ascospores.

The few bacteria known to cause diseases of trees are mainly species of *Agrobacterium*, *Pseudomonas*, *Erwinia* and *Xanthomonas*.

The literature on tree diseases is summarised in the *Review of Plant Pathology* (formerly the *Review of Applied Mycology*), published by the Commonwealth Agricultural Bureaux, Slough. *Forestry Abstracts*, published by the same Bureaux, is also valuable, especially for abstracts of papers on disorders caused by non-living agents. Among useful general texts are those by Tarr (1972), Manners (1982), Gareth Jones (1987), Chet (1987) and Holliday (1989).

Texts dealing with the fungi include those by Webster (1977), Burnett (1977), Alexopoulos and Mims (1979), Hawksworth *et al.* (1983), Cannon *et al.* (1985), Ellis and Ellis (1985) and Rossman *et al.* (1987).

The bacteria and their identification are covered by those of Fahy and

Persley (1983), Bradbury (1986), Lelliott and Stead (1987) and Schaad (1988).

The virus and virus-like diseases and the methods for their identification have been described by Smith (1972), Cooper (1979), Matthews (1981), Regenmortel (1982); Maramorosch and Raychaudhuri (1982), Hill (1984), Hiruki (1988) and Mandahar (1989).

Standard descriptions of many plant-pathogenic fungi and bacteria (referred to in the text as CMI Descr. F. and CMI Descr. B.) have been issued by the Commonwealth Agricultural Bureaux, which also publish corresponding descriptions of plant viruses prepared by the Commonwealth Mycological Institute and the Association of Applied Biologists (CMI/AAB Descr. V).

Pathogenic organisms do not operate in isolation, however. In the development of disease they interact closely with other elements in the environment. Among these are other organisms, and factors, for example, of weather and climate. These factors are considered by Tarr (1972), Manners (1982), Gareth Jones (1987) and others. Only a few examples will be mentioned here, but many others will be found in the accounts of individual diseases in later chapters.

Various competing fungi, the most effective of which is *Phlebia gigantea* (*Peniophora gigantea*), interfere with the colonisation of conifer stumps by *Heterobasidion annosum*. Oidial suspensions of *P. gigantea* are successfully used to control root and butt rot caused by *H. annosum* in pine crops. Antagonistic fungi such as species of *Trichoderma* have also been used in disease control. The fungus *Hendersonia acicola*, a minor pathogen, may prevent the fruiting on pine needles of the rather more damaging *Lophodermella sulcigena*, and also on pines, the hyperparasite *Tuberculina maxima* may greatly reduce the production of aeciospores of the resin top fungus *Peridermium pini* (*Endocronartium pini*). The fungus *Phomopsis oblonga*, which grows in the phloem of stressed elms, produces metabolites inimical to the bark beetles which spread the Dutch elm disease fungus *Ophiostoma ulmi* (Claydon *et al.*, 1985), while among the bacteria, some *Pseudomonas* spp. have a controlling effect on *O. ulmi* itself (Scheffer *et al.*, 1989). Mycorrhizal fungi have a controlling effect on some root diseases. Thus *Paxillus involutus* has been found to suppress root rot of seedling pines caused by *Fusarium oxysporum* (Duchesne *et al.*, 1989).

Beetles and other insects, mites and sometimes nematode worms are important as carriers of many diseases. In the absence of the bark beetles which carry its causal fungus to such effect, Dutch elm disease would probably be of no more than minor importance. Birds and mammals may also play a part in the spread of diseases. The spores of *Cryphonectria parasitica*, the cause of sweet chestnut blight, are carried by birds and small mammals as well as by insects and mites, and perhaps by slugs. Birds are also well known as carriers of the seeds of the semiparasitic common mistletoe.

Animals, both invertebrate and vertebrate, may also cause wounds which allow the entry of disease organisms. Wounds caused, for example, by extraction machinery, by pruning and by the breaking of branches by wind may provide entry points for decay fungi. Natural points of entry for disease organisms are provided by lenticels and by stipule and leaf scars, which are among the infection sites for *Xanthomonas populi*, the cause of bacterial canker of poplars. Leaves may be entered through stomata, or by direct penetration of the cuticle.

Frost damage may give rise to dead tissue which may then be colonised by the grey mould, *Botrytis cinerea*, and other fungi. Frost may also cause stem cracking and death of bark, which also allow the entry of decay fungi.

Other weather factors, especially rainfall and relative humidity and temperature, are also important in the development of disease. The powdery mildews grow and sporulate most actively in dry weather, and drought, especially on shallow soils, may encourage the spread of the *Armillaria* species which kill and decay many trees. Many diseases flourish in damp conditions. Many fungus spores are carried mainly by air currents, but some are spread by rain splash, which also disseminates bacterial slime. Most spores need free water or a high relative humidity for germination and entry into the host plant. Development of many soil-borne diseases, like those caused by *Phytophthora* species, is favoured by wet soil, in which germination of their resting spores takes place.

Many other factors may affect disease development, among them light conditions and soil pH and oxygen concentration. Air pollution is considered to increase the severity of some diseases, but it is difficult to separate its effects from those of other factors such as weather conditions, insect damage, etc. It is known, however, that relatively high levels of sulphur dioxide in the air prevent the development of tar spot of sycamore caused by *Rhytisma acerinum*.

Also important are the genetic characteristics of both the disease organisms and the host trees. Some fungi, for example, exist as various strains which vary in pathogenicity, while within tree species, provenances and individual trees and clones may also show wide differences in susceptibility to various diseases. This applies not only to diseases caused by living organisms: for example, some trees, including the London plane, the small-leaved lime and the maidenhair tree, are much planted in cities because of their resistance to air pollution.

## Symptoms of disease

Symptoms of disease may be found on any parts of the tree, on the leaves, on the stems and branches and on the roots.

## SYMPTOMS ON LEAVES

### Colour changes
*Browning and death*
The leaves may be partly or completely killed, when they usually turn brown (or sometimes black). This death of leaves may have many causes. It may be due to drought, or to wind damage, or frost, or sometimes to air pollution. Fungi may also sometimes destroy leaves in this way. Thus in severe attacks *Lophodermium seditiosum* may kill the needles of young pines in the nursery and *Meria laricis* may similarly affect nursery beds of European larch. The leaves of broadleaved trees may also be killed and become brown. Thus secondary spread of *Ophiostoma ulmi* (*Ceratocystis ulmi*) (the cause of Dutch elm disease) through root systems may produce rapid browning and death of the crown of the trees. Primary spread by insect vectors in its later stages also leads to leaf death. Sometimes leaf browning may be marginal only, as is often the case with leaves scorched by wind.

*Other colour changes*
In some cases browning and death of leaves may be replaced or preceded by other colour changes. As already noted, severely cold-damaged fronds of *Thuja plicata* die and turn glossy black.

Nursery plants (and sometimes plantation trees) may change colour when affected by mineral deficiencies (plates 5, 6, 7, 8). Thus, plants deficient in nitrogen tend to be pale green, while those short of potash may become pinkish, purplish or sometimes yellow. Magnesium deficiency is indicated by a bright yellow colour which on the leaves of broadleaved trees occurs mainly between the veins. Pines on very alkaline sites often become yellow at about 10–15 years of age, and then gradually die. This is due to a complex deficiency disorder called lime-induced chlorosis. The same disease may also affect other trees (including broadleaved ones), though usually less markedly.

Virus diseases may also cause yellowing of leaves. This yellowing may take the form of a more or less regular mottle or mosaic, or it may follow the leaf veins ('vein clearing'), or appear as lines or rings (figure 79).

Some wilt diseases that block the vessels (and at least some of which produce toxins) may cause the foliage of parts of the crown to become yellow or red. Thus, one of the earliest primary symptoms of Dutch elm disease, caused by *Ophiostoma ulmi*, is a yellowing of the foliage of isolated branches of the tree. Eventually (as noted above) the affected leaves die and become brown. Leaves in parts of the crown of willows affected by watermark disease (caused by the bacterium *Erwinia salicis*) turn bright red.

Some rusts may cause a yellowing of the leaves; an example is provided by *Chrysomyxa* spp. which produce yellow bands on diseased spruce needles (plate 19).

## Leaf distortion and reduction

A more or less general distortion of leaf shape may be caused by hormone weed killers such as 2,4-D.

Some fungi may also cause leaf distortion. Among these are the various species of *Taphrina*, including *T. deformans* which gives rise to leaf curl in peaches and almonds (plate 55), and *T. populina* which causes leaf blister in poplars (plate 28).

Many root disease fungi reduce the size of leaves. Pine trees attacked by *Heterobasidion annosum* (*Fomes annosus*) at first have short, sparse needles, which eventually fall as the tree dies.

Some virus diseases may cause the distortion of leaves and the production of small outgrowths (enations) from their surfaces.

## General mould growth

Fungal attack may show itself as a more or less general mould growth. *Botrytis cinerea*, the common grey mould, appears as a greyish, sparse or more woolly growth of mould (plate 4) (often with small sporing heads and sometimes with rounded black fungal aggregates called *sclerotia*) on the leaf surface. The oidial stages of powdery mildews, such as oak mildew (*Microsphaera alphitoides*), form a powdery white (or at first yellowish or brownish) cover on the leaf surface (e.g. figures 52, 95; plates 25, 52).

## Fungal fructifications

More clearly marked fungal fruit bodies may also be found on leaves (usually on spots or areas of dead or dying leaf tissue). The fungus *Rhytisma acerinum* forms large thickened black stromata ('tar spots') on the leaves of sycamore (plate 26).

Examination of the concave surfaces of the blisters on poplar leaves affected by the leaf blister disease mentioned earlier will reveal a golden-yellow lining, consisting of a layer of the asci of the ascomycete *Taphrina populina* (plate 28).

Other, more distinctive, ascocarps produced on leaves include, for example, the cleistothecia (cleistocarps), perithecia, hysterothecia and apothecia described above.

Among the Basidiomycetes, the rusts give rise to various fructifications on the leaves. These fruit bodies have also been listed above.

Finally, various small, asexual, so-called imperfect fructifications (fruit bodies of the Deuteromycotina, including asexual or anamorphic stages of ascomycetes) may be found on leaves. Many look like small black perithecia (when they are called pycnidia), but they contain asexual spores instead of asci.

## SYMPTOMS ON STEMS AND BRANCHES

### Splits in bark and wood
Splits in the bark and wood are often due to drought or frost damage. In the later stages of beech bark disease (associated with *Nectria coccinea*) the bark of affected trees may break into pieces and eventually fall off (figure 58).

### Watery and gummy exudations
Sap may exude from cracks in tree trunks. Continuing, often foul-smelling, sap flowing from the bark of broadleaved trees may indicate the bacterial disorder wetwood (plate 11). Exudations may also be caused by *Phytophthora* species, as in the case of the bleeding canker sometimes found on the horse chestnut (figure 91) and lime, but the exudate is then gummy, drying later to become hard and brittle.

### Exudations of bacterial slime
Wetwood exudations may often become slimy through the activities of bacteria and yeasts, and are then known as slime fluxes. In some bacterial diseases (notably in the case of bacterial canker of poplar, caused by *Xanthomonas populi*) thick bacterial slime may exude from small cracks in the bark of young stems and branches.

### Resin flows
Resin may sometimes exude through the bark of conifers and flow down the trunk. Such resin flows often indicate root damage by fungi such as *Phytophthora* spp., *Armillaria mellea* or *Heterobasidion annosum* (figure 16). Resin may also exude from the stem cankers produced on pines by *Cronartium* spp. and other fungi.

### Distortions and malformations
In some seasons plant stems (including small branches of trees) may become wide, flattened and strap-like. This abnormality is known as *fasciation*.

Cold damage insufficient to cause death may distort the growth of tree shoots.

Distortion of tree shoots may also be caused by fungi. Thus the pine twisting rust, *Melampsora pinitorqua* (a race of *M. populnea*), causes a twisting of the pine shoots (mainly of Scots pine) on which it grows (figure 39). The ends of the shoots of elms affected by Dutch elm disease often bend to form 'shepherds' crooks' which provide a useful diagnostic feature after leaf fall (figure 70).

Some fungi may cause the production of dense groups of adventitious shoots called 'witches' brooms'. An example is the witches' broom of birch caused by *Taphrina betulina* (figures 62, 63). Some viruses (and some mites) can also give rise to similar malformations on some trees.

Very marked distortions of stems of young trees may be caused by the honeysuckle (*Lonicera periclymenum*). This damage is mechanical, since the honeysuckle is not a parasite.

Chemical distortion, especially of rather soft shoots, may follow applications of hormone weed killers.

## Cankers

Sometimes a disease agent kills part of the cambium and the adjacent bark. The cambium around the lesion then produces new tissue, which grows inwards to cover the dead area. If no further damage occurs, the canker will be occluded and enveloped by the stem tissues. Often, however, further annual damage occurs, accompanied by further growth and a large, swollen canker results. Such cankers are caused on poplar by the bacterium *Xanthomonas populi* (figure 78). Rough-surfaced cankers are produced on ash by another bacterium, *Pseudomonas syringae* ssp. *savastanoi* pv. *fraxini* (plate 31).

Other cankers are caused by fungi. Among these is the canker of larch caused by *Lachnellula willkommii* (figures 46, 47). Rusts of the genus *Cronartium* also give rise to cankers on pine trunks and branches. Among these rusts are *Cronartium ribicola*, the white pine blister rust, and *Peridermium pini* (an aecial form of *C. flaccidum*).

## General fungal growths

Grey mould (*Botrytis cinerea*), already mentioned in connection with leaves, may also produce lesions (often sunken) on twigs, and on stems of nursery plants. On these lesions it produces its mycelium, with masses of sporing heads, and sometimes its black sclerotia.

The honey fungus, *Armillaria mellea sensu lato*, will often be found on dead or diseased trees as a thick sheet of creamy or whitish mycelium under the bark at the base of the trunk. At a later stage the fungus produces a network of flattened, bootlace-like rhizomorphs, also under the bark (figure 15). These rhizomorphs have a black outer surface, but have a white and (unlike roots) structureless tissue within. *A. mellea* and other *Armillaria* spp. also form rhizomorphs in the soil. These are similar to those described above, but are round in section. They grow out from infested stumps and roots, and may attack the roots of surrounding trees and shrubs (and sometimes other plants).

## Fungal fructifications

When these occur, they are very valuable in diagnosis. Examples among the Basidiomycetes are the honey-coloured toadstools of the *Armillaria* spp., which in autumn usually grow in tufts at the base of affected trees or on the ground around them (plate 17).

The bracket-shaped root and butt-rotting fungus *Heterobasidion annosum* also occurs at the bases of diseased trees and on infested stumps (plate 18). Another butt-rotting polypore, *Phaeolus schweinitzii* (plate 62) (the velvet-topped fungus), also usually occurs on the ground or at the stem base (but may also be found higher up the trunk), while another, *Polyporus squamosus* (the Dryad's saddle), causes a top rot and so fructifies near the top of the tree trunk, often on wounds caused by the loss of branches.

Also among the Basidiomycetes, species of *Cronartium* produce large, blister-like, cream or white aecia filled with orange spores on the cankers caused by these rusts on the stems and branches of pines (plate 12, figure 43).

The fruit bodies of ascomycetes may also be found on trunks and branches. Sheets of the small, red, flask-shaped perithecia of *Nectria coccinea* may grow on the trunks of beech trees affected by beech bark disease.

Reddish-brown stromata containing the embedded perithecia of *Cryptodiaporthe castanea* may be found in cracks in the bark of coppice shoots of sweet chestnuts affected by Cryptodiaporthe canker.

Cup-like or disc-like apothecia of other ascomycetes also occur, like those of *Lachnellula willkommii* on the cankers caused by this fungus on larch (plate 23).

Asexual fruit bodies, anamorphic (imperfect) stages of ascomycetes or the fructifications of deuteromycetes (Fungi Imperfecti) may also be found. Thus, the pink, cushion-like anamorphic stage of *Nectria cinnabarina* is common on twigs attacked by this fungus or colonised by it saprophytically (plate 14). Again, the pycnidial stage of *Cryptodiaporthe castanea*, *Discella castanea*, occurs around the edges of the perithecial stroma.

**Other plant growths**
Algae, mosses and liverworts, ferns and lichens often grow on tree trunks and branches, but these growths are purely epiphytic. Plants of mistletoe (*Viscum album*) seen on some trees, however, are semiparasitic.

**Internal symptoms**
Some symptoms may be hidden inside the twigs, branches or trunks, either as stains, or rots, or as characteristic rings.

The bacterium *Erwinia salicis* causes rusty or inky stains inside the stems of the cricket bat willow and other *Salix* spp. (plate 57). Usually after felling, various fungi cause a blue stain of the wood of pines. *Verticillium dahliae* (and the less common *V. albo-atrum*), the cause of Verticillium wilt, may give rise to a brown or greenish stain in the stems of affected trees (plate 27) (as it does in the case of many other hosts, including the tomato).

In the early stages of attack, many root-rot and butt-rot fungi also cause inky or rusty stains. Later the affected wood usually becomes soft and rotten, and may then occupy the centre of the stem, surrounded by a stained zone. Some of these fungi cause white rot, others a brown rot. Typical white rot fungi are *Armillaria mellea* and *Heterobasidion annosum*. The first of these causes a soft, wet, stringy rot in a conical zone usually reaching about 60 cm into the butt (figure 17). The second produces a rot with small white pockets of fungal mycelium, and the rot usually reaches much further up the trunk, to a level of about 300–400 cm (figures 27, 28, 30). *Phaeolus schweinitzii*, on the other hand, is one of the brown rot fungi. It breaks the wood for a long distance up the trunk into brown, cubical blocks (figure 105), often separated by thin, yellow sheets of mycelium, and the affected wood smells of turpentine.

Rings of abnormal tissue produced by damaged cambium may be produced in twigs in the case of some disorders, and can be seen if the damaged twigs are sectioned and examined under a lens or a microscope. Such rings may be caused by frost (figure 1), drought or lightning.

A rather different type of ring may be found in the twigs of elms affected by Dutch elm disease (and occasionally by some other diseases). Here, sections of diseased twigs show a ring of brown dots in their outer wood (figure 71). If, instead of sectioning the twigs, their bark and outer wood are shaved away, these dots appear as brown streaks (figure 72).

### Stem blisters
*Cryptostroma corticale*, the cause of the sooty bark disease of sycamore, raises the bark of affected trees into large blisters, which eventually break open to release the mass of brown, powdery spores that give the disease its common name (plate 32).

## SYMPTOMS ON ROOTS

Symptoms on roots caused by root-rot and butt-rot fungi, such as *Armillaria mellea* and *Heterobasidion annosum*, are broadly similar to those caused in trunks by the same fungi. The roots of conifers attacked by *H. annosum* may exude resin (as happens with the trunks of some trees attacked by this fungus) and internally the roots show characteristic forms of rot.

The ascomycete *Rhizina undulata* (*R. inflata*) is associated with a root disease in conifers, especially in Sitka spruce. Its large, brown, hollow fruit bodies grow on the ground around diseased trees (figure 11), attached to the damaged roots, which also show characteristic lesions (figure 14).

Among the Phycomycetes, *Phytophthora cambivora* and *P. cinnamomi* cause Phytophthora root rot in various broadleaved trees. In sweet chestnut the damaged roots may exude a blue, inky stain which seeps out into the soil. This stain gives rise to the name 'ink disease', but it is not in itself diagnostic.

## Diagnosis of diseases

In the above account the various symptoms have been isolated to draw attention to them. It must be emphasised, however, that the pathologist must usually look for symptoms as patterns or groups rather than individually, and the various symptoms added together give a picture of a given disease. The pathologist may also find clues pointing to the causes of various diseases and disorders by noting, for example, the relationship of affected trees to possible sources of air pollution or their position in frost hollows or areas of waterlogged land. He (or she) may also gain much in some cases from a study of local weather data, which will assist in the diagnosis of frost or drought damage, or damage by other adverse weather conditions. Growth changes (seen in studies of leader growth and of successive annual rings) may also provide valuable information, indicating, for example, when a decline in growth set in; it may then be possible to correlate the decline with some known adverse factor.

When dealing with diseases caused by living agents, laboratory work is usually needed to isolate and identify the organisms concerned. In the case of a new disease it is necessary to re-inoculate the isolated organism into specimens of the host plant to confirm that they do indeed produce the disease. As final confirmation they must then be re-isolated and rechecked.

The techniques used for the identification of bacteria and viruses are described in the texts listed above at the end of the section on 'causes of disease'. They now include the use of enzyme-linked immunosorbent assay (ELISA). ELISA kits for the identification of some of the viruses, including apple mosaic virus, cherry leaf roll virus and poplar mosaic virus, and some bacteria, including *Erwinia amylovora* (the cause of fireblight of the Pomoideae), are now commercially available. Similar methods have been applied to some of the fungi, among them *Ophiostoma ulmi* (Dewey and Brasier, 1988; Jeng *et al.*, 1988; Dewey *et al.*, 1989). Interfertility tests have been used to separate species of *Armillaria* (see chapter 3).

## Losses and damage caused by disease

It is usually impossible to make any precise assessment of the losses caused by tree diseases. Some diseases cause little appreciable damage, but others cause clearly visible loss of some kind or other.

Diseases such as anthracnose (*Marssonina salicicola*) (figure 82) of weeping willow make the tree look unsightly, a serious matter in the case of an ornamental tree of parks and gardens. Similarly, blue-stain fungi may give an unpleasant appearance to the wood of pine and so reduce its value and usefulness for many purposes (figure 45). Watermark disease (*Erwinia salicis*) in the cricket bat willow (plate 57) stains the wood and makes it brittle, rendering it useless for the manufacture of cricket bats. A wood of high value then becomes virtually unmarketable.

In nurseries, species of *Phytophthora* may destroy plants; the number of dead plants may then be counted. Such a count, however, only estimates the minimum loss, as apparently healthy plants which leave an infected nursery may also be diseased. In many nurseries until recently it was impossible to grow plants of *Thuja plicata* because of the depredations of the needle blight fungus *Didymascella thujina*, though this blight may now be comparatively well controlled by spraying with various fungicides.

The stem rust *Cronartium ribicola* makes it virtually impossible to grow the valuable five-needled pine *Pinus strobus* at least on a commercial scale in

Britain. So far this rust cannot be controlled (though much work on the breeding of resistant pines has been done in the USA). The fungus *Ascocalyx abietina* (*Gremmeniella abietina*), on the other hand, does not prevent the growing of Corsican pine in Britain, but confines it to lowland areas below about 275 m.

Again, some diseases restrict the growing of susceptible varieties or clones (or of species within a susceptible group). Thus fireblight (*Erwinia amylovora*) was reduced to a commercially acceptable level in British orchards by the removal of the susceptible pear variety Laxton's Superb. Canker of larch restricts the growing of susceptible forms of European larch, and poplar canker (*Xanthomonas populi*) prevents the wide use of susceptible poplar clones.

Some diseases cause death of trees. *Heterobasidion annosum* kills pines on certain sites, though in other trees it usually causes a root rot or butt rot without killing the tree. Dutch elm disease is exceptional among tree diseases in the losses it has caused in those areas (so far in North America and Europe) in which the causal fungi *Ophiostoma ulmi* and the more aggressive *O. novoulmi* and their associated vectors occur with susceptible elm populations. The latter fungus causes rapid death of susceptible elms, and has destroyed almost all the elms in much of southern Britain, and many in other parts of Europe, as well as large numbers of the even more susceptible elms in North America.

As mentioned above, the decay fungus *Heterobasidion annosum* may kill trees. It also causes loss of wood in timber trees by rotting the interior of the trunk. Many other decay fungi also render trees dangerous by rotting roots (and so leading to windblow) or stems (leading to stem break) or branches (which may then suddenly fall).

Many diseases cause a reduction in growth, and so of increment, which is important to the forester. Thus *Heterobasidion annosum* may cause marked reduction in the growth and increment of pines on alkaline sites, owing to extensive death of the roots (which eventually leads to the death of the trees).

## Control measures

Control measures can first be divided into:

(1) Legislative measures to prevent the entry of alien diseases (and pests).
(2) Measures used to control diseases already present.


PLANT HEALTH FRONTIER LEGISLATION

Broad international guidance for plant health authorities is given under the auspices of the Food and Agriculture Organization of the United Nations (FAO) by the International Convention for the Protection of Plants. This was first issued in 1881, and widened in 1929, and has since been revised (Anon., 1954; Chock, 1979). Under the Convention, regional plant protection organisations were set up. That for Europe is the European Plant Protection Organisation (EPPO), formed in 1951 and based in Paris. EPPO, the work of which has been reviewed by Smith (1979), aims *inter alia* to advise the states within its borders so that they can develop a common strategy, based on sound and agreed scientific principles. Their regulations should be made as effective as possible, but as far as can be consistent with this, they should cause the minimum interference with trade. In particular, EPPO has prepared lists of threatening organisms absent from all countries in its area, and against which all should take legislative action (List A1), and of such organisms occurring in parts of the EPPO area but not throughout (List A2). To assist further it has prepared data sheets on these harmful organisms (EPPO, 1978, 1979a, 1980), and it also issues periodically updated summaries of the plant health legislation of all its member states. In its *EPPO Bulletin* it publishes many papers concerning plant health.

The member states of EPPO base their plant health regulations directly or indirectly on its recommendations. As a group within EPPO, the European Community (EC) has prepared its own Plant Health Directive (CEC, 1977), which is binding on all its member states and is subject to periodical amendment. The principles behind the Plant Health Directive, derived from those formulated by EPPO, are as follows:

(1) The EC must take all reasonable steps to protect its crops from alien organisms.

(2) At the same time the Community depends for its livelihood on trade, especially between its member states.

(3) Hence, all measures used to prevent the entry of plant diseases and pests must be demonstrably justifiable on plant health grounds; it must be clear that they are not used as a means of trade protection.

(4) With only a few exceptions, the pests and diseases of Europe occur more or less throughout.

(5) Hence, within the European area generally, legislation need deal only with those European pests and diseases still limited in their distribution. Only those Member States without a given European disease or pest can justifiably take steps to control its entry. Thus the United Kingdom may take steps to keep out *Cryphonectria parasitica*, the cause of sweet chestnut blight, because it occurs on the European mainland but not in Britain. It may not take steps against bacterial canker of poplar caused by *Xanthomonas populi*, which is already present and well-established there.

(6) In the world at large there are many potentially dangerous diseases and pests which have not yet reached Europe, and many of these are difficult to assess. Strong legislation is justifiable against these, though again it must be soundly based and justifiable on plant health grounds; and it must not be disproportionate to the risks involved.

(7) Some alien pests and diseases clearly pose a threat; these can be listed, and special measures against them can be laid down.

(8) Some crops are of particular importance, and in the case of these, more rigorous measures may be justifiable than in the case of others.

(9) In general, food crops are of special importance, and as they are grown on a wide scale in monoculture, they may be especially vulnerable to epidemics. The same applies to forest trees and a few very widely grown ornamental trees, which are also in some cases in special need of protection.

In theory, pests and diseases of trees may be carried with seeds and fruits, or with living plants, or in wood with the bark (and occasionally perhaps without it). In practice, though some tree diseases may be carried by seeds and fruits (Neergaard, 1979; Rees and Phillips, 1986), tree seeds are far less important in the carriage of these diseases than are plants for planting and wood (especially with bark attached). So far, therefore, importation of seed is not subject to plant health controls in the EC. Some important pests and diseases may be carried with plants, however, and some with wood with the bark attached. Hence, various controls are placed on these in appropriate cases.

In Great Britain, the Orders designed to prevent the entry of alien organisms harmful to plants are made under the Plant Health Act 1967. General controls are laid down in the Import and Export (Plant Health) (Great Britain) Order 1980, and those specifically concerning forest and ornamental trees are listed in the Plant Health (Forestry) (Great Britain) Order 1989. Similar but not identical legislation covers Northern Ireland, and the authorities in Jersey and Guernsey have issued corresponding Orders. Details must be sought in the individual Orders (which are subject to amendment from time to time), but it may be said here that in Britain the tree genera specifically covered are *Abies*, *Larix*, *Picea*, *Pinus*, *Pseudotsuga* and *Tsuga* among the conifers, and *Acer*, *Castanea*, *Platanus*, *Populus*, *Quercus* and *Ulmus* among the broadleaved trees. These genera are known to be susceptible to important identifiable alien pests and diseases. Others are covered in some other ways, however, as all 'plants, planted or intended for planting' must be inspected in the exporting country and covered by plant health certificates. Some, including *Fagus* and *Tilia*, are covered by requirements intended to prevent the entry of San Jose scale, which is primarily a pest of fruit trees; these requirements effectively prevent the entry of many other pests and diseases.

The legislation lays down measures to prevent the entry of about 30 individual diseases and pests of trees, the following being the most important.

(1) On broadleaved trees.

    (a) Oak wilt (*Ceratocystis fagacearum*), which occurs only in the USA.

    (b) Elm phloem necrosis and its carrier, the leaf hopper *Scaphoideus luteolus*, which are also North American

    (c) The poplar canker fungi *Mycosphaerella populorum* (occurring in

North and South America) and *Hypoxylon mammatum* (which occurs in North America and, in an apparently non-aggressive form, in parts of France).

(d) Sweet chestnut blight (*Cryphonectria parasitica*, syn. *Endothia parasitica*), which is present in North America, parts of Asia and Southern Europe.

(e) Sap streak of maple.

(f) Canker stain of plane.

(2) On conifers.

(a) Non-European bark beetles (*Scolytidae*), particularly those of North America.

(b) North American cankers of pine caused by species of *Atropellis*.

(c) Non-European dwarf mistletoes (*Arceuthobium* spp.), mostly occurring in North America.

(d) Various conifer rusts (species of *Chrysomyxa*, *Cronartium* and *Melampsora*) of North America and (in a few cases) South America and Asia.

(e) Shoot blight of larch caused by *Guignardia laricina* (an Asiatic fungus).

(f) The laminated root-rot fungus *Phellinus* (*Poria*) *weirii* from North America.

(g) *Pissodes* beetles from North America.

(h) *Ips* spp. and *Dendroctonus micans* from Europe.

Of these diseases and disease organisms, *Cryphonectria parasitica* and *Hypoxylon mammatum* are described respectively in chapters 12 and 16 of this book. Short accounts of some of the others are given by Burdekin and Phillips (1977) (who include elm phloem necrosis), Gibbs (1984) (oak wilt) and Hansen (1985) (who includes dwarf mistletoe and Atropellis canker on lodgepole pine, and *Phellinus weirii*). The possible importance of *Cronartium comptoniae* has been considered by Pawsey (1974).

The legislation prescribes various measures to deal with these organisms, in addition to the general ones (for example, on the certification of plants planted, or for planting) already mentioned. Some are broad steps to exclude the listed organisms, either in general or as contaminants of named produce (for example, conifer wood with bark). In other cases, the importation of certain plants or plant products from given areas is prohibited, or importation may be only from areas free from specified organisms, or subject to some form of treatment.

This legislation and its background, and the assessment of the potential of alien pests and diseases of trees, are further discussed by Phillips and Bevan (1966), Burdekin and Phillips (1977), Burdekin (1986), Phillips (1978, 1979, 1980a,b, 1981), Gibbs and Wainhouse (1986) and (with regard to oak wilt) Gibbs (1978, 1979). Venn (1986) has discussed the significance of timber and wood chips as carriers of pests and diseases.

Changes in British plant health legislation are recorded in the *Annual Reports of the Forestry Commission of Great Britain* (HMSO, London).

It is likely that in 1992, when the customs barriers to trade are abolished within the European Community, legislation referring to controls on internal trade of plants will be changed. In place of a Phytosanitary Certificate there will be a 'Plant Passport' issued at the place of production certifying that the necessary conditions for import have been complied with.

## MEASURES FOR CONTROL OF ESTABLISHED DISEASES

Only a few general points on disease control are given here. Where appropriate, details of control measures for individual diseases are given in succeeding chapters.

Some diseases are of minor significance and need no special control measures. An example is the leaf blister of poplar caused by *Taphrina populina*. In other cases no control measures can really be taken. Thus there are no measures available to prevent infection of conifers by the decay fungus *Phaeolus schweinitzii*.

In many cases control is by *disease avoidance* of some kind. This is especially so in forestry, for the forester is concerned with populations rather than with single trees, and expenditure on disease control must be kept as low as possible.

Thus, to avoid blue-stain, pines may be felled between September and February, when infection by blue-stain fungi is at a low level, and the logs quickly removed from the forest.

Nurseries may be sited to avoid frost hollows and waterlogged areas. Again, they should not be placed near mature crops which may be sources of infection by, for example, pine needle fungi such as the species of *Lophodermium*, or the larch needle fungus *Meria laricis*. Composite weeds which may act as alternate hosts of the rust *Coleosporium tussilaginis* should be kept down, for they may lead to damage to nursery pines.

Similarly, it may be necessary to destroy aspen in young pine plantations to avoid damage by the pine twisting rust *Melampsora populnea* (*M. pinitorqua*).

On suitable sites, this fungus may also be avoided by growing Corsican pine instead of Scots pine, an example of a use of differing resistance. Genetical resistance may also be used to avoid larch canker caused by *Lachnellula willkommii*, which attacks most provenances of European larch. Both Japanese larch and its hybrid with European larch, on the other hand, are generally more resistant. Again, the different forms of Douglas fir vary in their resistance to the needle cast fungus *Rhabdocline pseudotsugae*, the green form of *Pseudotsuga menziesii* being much more resistant than the others, which, therefore, are no longer grown to any extent in Britain. Among the broadleaved trees, many poplar clones have been bred for resistance to the bacterial canker caused by *Xanthomonas populi*.

The use of clonal material, as in the case of poplars, presents problems of its own. Clones may be chosen *inter alia* for their uniform good growth and wood density, for other characteristics such as their form and freeness of flowering, and for their resistance to disease. When breeding, however, it may be necessary to seek resistance to more than one disease. The history of plant pathology provides many examples of disease organisms which have overcome the resistance bred into clones. The results have often been severe, especially when clones have been closely planted over considerable areas (as in the case of poplars) or placed in large numbers throughout the countryside (as in the case of elms). For this reason elms were destroyed in millions in Europe and North America when a new and more aggressive form of the Dutch elm disease fungus developed and quickly spread throughout the uniform elm populations, which were largely made up of susceptible clones. This emphasises the need to acquire a good knowledge of the resistance spectrum of new clones and seek ways to forecast the possible adaptability of disease organisms. It also suggests that it is desirable when setting up plantations of trees such as poplars to build them up from a mixture of clones with differing resistance characteristics, and in the countryside to avoid the widespread and close planting of elms undiluted with large numbers of other trees. Many ornamental trees are grafted clones, which present similar problems. Thus in Britain, following cold spells, the Japanese flowering cherry, *Prunus serrulata* 'Kanzan', has often been affected by a bark infection (perhaps by *Pseudomonas syringae* pv. *syringae*), followed by wilting, and often by the death of many trees (Strouts *et al.*, 1983; Strouts, personal communication). In the past, broadleaved and coniferous forest plantations have been raised mainly from seed collections from the wild. Forest populations have therefore been very diverse. With time, the use of seed from selected trees and from seed orchards has tended to reduce this diversity. Work on the production of clonal stocks even of the major coniferous species has steadily increased. If large areas of forest are eventually raised from such material, the problems outlined above may need careful consideration.

In some cases disease may be avoided or reduced in its effect by sanitation felling. This may be controlled by internal plant health legislation (as opposed to the frontier controls described above). Such legislation is applied in Britain in the case of watermark disease of willow and of Dutch elm disease. Action under the Watermark Disease (Local Authorities) Order 1974 (as amended) has kept watermark disease at a very low level. Dutch elm disease is covered by two orders, the Dutch Elm Disease (Restriction on Movement of Elms) Order 1984 and the Dutch Elm Disease (Local Authorities) Order 1984, which not only allow for sanitation felling but also control the movement of felled elm wood which is likely to spread the disease and its vector beetles from one part of the country to another. This legislation has slowed the spread of the disease, especially in favoured areas where elms are isolated by their geographical position, and where, towards the north of the country, the climate is less favourable to the vector beetles and the elms are slightly less susceptible than those dominant in the south and also less densely planted.

On sites containing many conifer stumps infested by *Heterobasidion anno-*

*sum* it may in a similar manner be necessary to remove the old stumps before replanting with another susceptible crop.

Sometimes disease may be controlled by pruning out affected parts. Removal of diseased limbs was sometimes effective in controlling the non-aggressive strain of the Dutch elm disease fungus *Ophiostoma ulmi*, though this treatment rarely inhibits the newer aggressive form of the fungus (*O. novo-ulmi*).

Nursery plants may be protected from both frost and sunscorch by covering with overhead lath shading. Again, reasonably shade-bearing trees can be protected from frost in the establishment phase by underplanting in a mature overstorey.

Drainage (or the avoidance of wet areas) can be used to control Phytophthora root rot ('ink disease') of sweet chestnut.

Chemical control, using fungicides, is relatively little used to deal with tree diseases (except those of orchard trees). In forestry, the use of fungicides is generally too expensive except in the nursery.

Many chemicals for disease control are mentioned in the accounts of individual diseases. It must be realised, however, that the use of such materials is strictly controlled, and only those approved by the authorities may be used. Approval in Britain is given under the Control of Pesticides Regulations 1986. The information on fungicides and other pesticides given in this book can be used only for guidance, and current information on approved products and their safe use must therefore be sought from the official advisory services, and careful attention paid to all instructions given on product labels. Much information on pesticides and their uses is given in the pages of the *Pest and Disease Control Handbook* (Scopes and Stables, 1989) and the *UK Pesticide Guide 1990* (Ivens, 1990). These texts are updated from time to time, the first every few years and the second annually. They explain the legislation and list available chemicals and their uses, the *Pesticide Guide* giving accepted common names of materials and the proprietary names used by manufacturers.

Subject to the comments made above, the materials most commonly used in British forest nurseries are captan, thiram and benomyl to control grey mould (*Botrytis cinerea*) and maneb, zineb and benomyl to deal with pine needle disease caused by *Lophodermium seditiosum*. Wettable sulphur, dinocap and benomyl are used against oak mildew (*Microsphaera alphitoides*) and wettable sulphur and zineb against needle cast of European larch caused by *Meria laricis*. Continued and frequent use of benomyl is not recommended, as it may lead to the development of resistant strains of the fungi concerned. A nursery crop affected by damping off may be given an emergency drench of captan. In nursery areas in which damping off has occurred, the soil may be treated between crops with a soil sterilant such as dazomet.

Spraying of established garden trees against various diseases such as anthracnose of weeping willow and leaf curl of almond may sometimes be recommended, though it is usually possible to achieve adequate cover only on fairly small trees. The only routine chemical treatment used in the forest against a tree disease is the painting of conifer stumps with urea solution to prevent their colonisation by the root-rot and butt-rot fungus *Heterobasidion annosum*. In the case of pine stumps, a suspension of oidia of the competing fungus *Phlebia gigantea* (*Peniophora gigantea*) may be substituted for the chemical to provide biological control.

In the case of ornamental trees (as well as those in fruit orchards) various chemical protectants are often applied to wounds to prevent the entry of decay fungi. So far these chemicals have had little effect, though some help to promote healing. Efforts to find better materials are being made. Pruning wound protection is discussed in chapter 20.

Much work has been done on the injection of trees with fungicides (mainly benomyl and related materials) to control diseases, especially Dutch elm disease. Some success has been achieved but only with relatively small trees.

Some diseases are carried by insect vectors, and attempts have been made to control some diseases by dealing with those carriers. Thus sanitation felling of elms to control Dutch elm disease is primarily intended to destroy and prevent the breeding of the elm bark beetles which spread the disease. Experiments on the control of these beetles by the use of insecticides have also been done; the results have not been very encouraging, and there are environmental objections to the widespread use of many insecticides.

# 2 Disorders caused by non-living agents

This part of the book is concerned with damage caused to trees by a diverse range of non-living agents. Some, including climatic and soil factors, are often (though not always) outside the direct influence of man. Others, including toxic chemicals such as atmospheric pollutants and herbicides, are produced largely as a result of man's activities. It is important in times of increasing concern about the environment to remember that plants, including trees, can be damaged by natural forces as well as by pollution in its general sense.

The identification of disorders caused by non-living agents may sometimes be straightforward. A severe drought is likely to cause visible damage to some trees and a hard spring frost may have obvious effects on young developing shoots and foliage. In the majority of cases diagnosis is more difficult and this is particularly so when the causal event occurred some time before damage was observed. The visual symptoms may have been obscured by healthy recovery growth or the tree may be no more than a skeleton. In these circumstances examination of the internal growth patterns of the tree may yield valuable information (for example, through tree ring analysis) about the time or season at which the damage occurred.

Therefore, it is important for the pathologist to have some understanding of the physiological processes that have led to the damage so that appropriate clues may be sought. Some indication of the effects of various climatic and chemical factors have been given in this chapter. However, the interested reader who wishes to study these factors in more depth should seek information in the wide range of literature that exists on many of these subjects. Some important references are included but the list is necessarily selective and many others will surely provide valuable information.

It is perhaps too early to detect any possible signs of the so-called greenhouse effect, but it is likely that one or more of the symptoms described in the following pages may become more prevalent if global warming does occur.

## Climatic factors

### EFFECTS OF LOW TEMPERATURES

The damaging effect of low temperature on trees, often broadly termed frost injury, can be particularly serious in Britain. At first sight this may seem somewhat surprising, bearing in mind the oceanic conditions and the generally mild winters which characterise the British climate. However, it is precisely these conditions which have frequently persuaded foresters and arboriculturists to plant less hardy species and when periods of low temperature do occur serious damage can ensue. On the continent of Europe and in much of North America far lower temperatures are frequently recorded and frost injury may occur on more hardy trees.

Peace (1962) undertook a comprehensive review of the effect of frost on trees and more recent work has served largely to confirm and extend his observations. He pointed out that frost damage could occur both in the growing and in the dormant season and that although there would inevitably be some overlap between the seasons, the effects of autumn and spring frosts should be differentiated from that of winter cold. Not only are the trees in a different physiological condition at these times, but also, as we shall see, the climatic factors that determine the low temperature differ.

**Spring and autumn frosts**
Such frosts occur primarily at night when the air is still and cloud cover is absent. Heat is lost by radiation from the plant and soil surfaces and their temperature falls. The air temperature in their vicinity also decreases and so-called 'inversion conditions' develop where the temperature of the air close to the ground is lower than that a few metres above. Such cooling may continue until the temperature falls to freezing point and below. In areas where these conditions occur the ground may be level but more frequently there are small depressions, slopes or valleys and the cold air, being heavier than the warm air above, tends to flow downhill and accumulate at the lowest points. In this way frost pockets or hollows can develop and their occurrence clearly depends on very local topographical variations. There are many localities in Britain, particularly on the eastern side, where frosts are regularly recorded in most months of the year. Cannell (1985) has analysed the risks of frost damage to forest trees by an examination of meteorological data over a period of years and related these risks to spring temperatures which are likely to be damaging.

Over the past decade a number of severe frosts have been recorded in Britain. Redfern *et al.* (1982) reported a spring frost in April 1981 which caused severe damage to Sitka spruce but only minor damage to other conifers in Scotland. Strouts *et al.* (1985) reported severe frosts in early May 1984 in Wales which damaged newly flushed beech shoots and injured the lower stems of 8-year-old Noble fir. Redfern *et al.* (1987) reported autumn frost injury to young Sitka spruce following a series of frosts in early September 1986. A grass minimum temperature of $-6.2\ °C$ in the Scottish borders was followed by severe foliage browning on Sitka spruce and in some cases lammas growth was also killed. Damage from the same frosts was also reported on Douglas fir seedlings and, exceptionally, on young Norway spruce.

The cooling of the air close to the ground during inversion conditions is influenced by a number of factors. One is the nature of the soil and its vegetation cover. Young tree crops, less than 0.5 m or so in height, are more likely to be damaged when growing in a dense grass mat than over a bare soil. When heat is lost to the atmosphere by radiation from the bare soil surface, it is replaced by heat from below, particularly when the underlying soil is moist. However, a grass cover is a poor conductor and heat lost from its surface is not replaced, and its temperature, together with that of the adjoining air mass, falls.

Overhead cover can also influence the occurrence of ground frosts. Plants established beneath an overstorey of mature trees radiate less heat than those in an open ground situation and in consequence are less liable to frost damage.

The use of artificial shade, especially under nursery conditions, has a similar effect of lessening radiation on cold, still nights and in addition may slow down the rate of thawing in the morning which may be a further advantage.

The above description of the conditions that influence temperature at ground level applies to both autumn and spring frosts. Spring frosts tend to be damaging as the early spring growth of many species is more susceptible to frost than that formed in the autumn.

*Symptoms*
The nature and extent of spring frost damage depends upon the stage of growth at the time as well as on the severity of the frost. If the frost occurs before flushing but after physiological activity has begun within the tree, damage may be limited to the buds. When the buds are killed this leads to the growth of adventitious buds which, in turn, can give rise to the production of bushy plants.

Once the trees have flushed the young growth may be particularly susceptible. The tips of the needles or peripheral parts of leaves may become necrotic or in more severe cases whole needles or leaves and often the shoots to which they are attached may be killed. In such cases the brown foliage remains attached to the shoots, which themselves hang down in a characteristic, permanently bent position (plate 1). As in cases where buds are killed, adventitious growth may subsequently develop to give the tree a bushy habit. Sometimes the leader may survive repeated frosting while the lower branches are killed and the tree then assumes a columnar habit with dead branches around the base.

Small areas of bark may be killed by spring frosts, particularly on smooth-barked species such as *Thuja plicata* and *Nothofagus*. If the stem is girdled, the distal portion will die back, but where damage is local, callus growth will occur. The wound may subsequently heal or, if frost recurs, further damage and perhaps fungal infection may follow.

On trees where external signs of damage are present and in others where they are absent, internal symptoms may be evident. Thus the cambial region is often affected and frost rings may develop (figure 1). Glerum and Farrar (1966) have studied the development of frost rings by subjecting seedlings of various conifer species to freezing temperatures. They concluded that while most cambial initials remained alive, differentiating tracheids and xylem mother cells were killed by the frost. This ring of dead tissues contained underlignified and crushed tracheids. Outside the ring of dead cells there was a region with abnormal tracheids and parenchyma that was formed prior to the re-establishment of normal cambial activity.

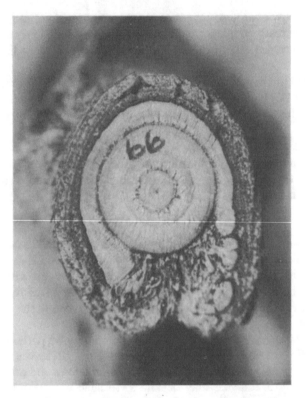

**Figure 1**   Frost ring formed in a young stem of *Thuja plicata* at the beginning of the 1967 growing season (D.H. Phillips)

Frost rings may be formed during both spring and autumn frosts and those produced at the end of one growing season or at the beginning of the next may be indistinguishable.

*Species susceptibility*
There is considerable variation within species and between species in susceptibility to frost damage. In Norway spruce (*P. abies*), for example, there are early and late flushing strains and the latter are less affected by frost. This

may be as a result of the late flushing strain escaping frost at a susceptible stage of growth rather than any inherent resistance to low temperature.

Peace (1962) has categorised the susceptibility to spring frost of the commoner British trees; his assessment is given in table 2.1.

**Table 2.1** Susceptibility to frost of common trees in Britain (after Peace, 1962)

|  | *Broadleaved species* | *Conifers* |
|---|---|---|
| Very susceptible | *Juglans* spp. | *Abies grandis* |
|  | *Fraxinus excelsior* | *Picea sitchensis* |
|  | *Castanea sativa* | *Picea abies* |
|  | *Quercus* spp. | *Larix* spp. |
|  | *Fagus sylvatica* | *Pseudotsuga menziesii* |
|  |  | *Tsuga heterophylla* |
|  |  | *Thuja plicata* |
| Moderately susceptible | *Acer pseudoplatanus* | *Pinus contorta* |
|  | *Aesculus hippocastanum* | *Pinus nigra* var. *maritima* |
|  | Some *Populus* spp. | *Picea omorika* |
| Hardy | *Betula* spp. | *Pinus sylvestris* |
|  | *Corylus avellana* | *Cupressus macrocarpa* |
|  | *Carpinus betulus* |  |
|  | *Tilia* spp. |  |
|  | *Ulmus* spp. |  |
|  | Many *Populus* spp. |  |

There is less evidence available on the susceptibility of different species to autumn frosts. Japanese larch (*L. kaempferi*) frequently suffers damage perhaps because it makes active growth in the autumn. Sitka spruce (*P. sitchensis*) is sometimes damaged in the nursery, and southern provenances (for example, Washington) may be more susceptible than northerly ones (such as Queen Charlotte Island).

*Amelioration of spring and autumn frost damage*
A number of measures can be taken to ameliorate the effects of this type of frost damage. A knowledge of the climatic conditions prevailing in an area and also of the local topography is important in order to select a suitably hardy species. Low and Greig (1973) made a detailed study of frost damage in Thetford Chase in eastern England and showed that such damage could be minimised by a variety of practices, including complete cultivation, deep ploughing, underplanting and strip felling. Measurements taken in different situations indicated that temperatures were higher over bare soil and within plough furrows than over a grass cover. Similar differences were found between sites under a tree cover or within felled strips and those in the open. Marked increases in the height growth and survival of Corsican pine (*P. nigra* var. *maritima*) were recorded when such ameliorative steps were taken.

In nurseries it is clearly important to initially select a site which is free from frosts but if this is not possible artificial overhead cover may give some protection.

### Winter cold

In Britain prolonged freezing conditions in winter usually occur when dry, cold, northerly or easterly air streams move over the country. Peace (1962) differentiates between damage due to low temperature alone and that due to cold, dry winds. In the former case less hardy species may be injured but in the latter the foliage of some of the more evergreen species is affected.

Forest species marginal to Britain such as *Pinus radiata*, some *Nothofagus* spp. and *Cupressus macrocarpa* are damaged in severe winters. Extensive death of foliage and dieback may occur. *Nothofagus dombeyi* and *N. procera* are particularly susceptible and, in addition to extensive dieback, character-istic cankers form on the main stem (figure 2). On the other hand, *N. obliqua* and *N. antarctica* seem relatively resistant to winter cold damage. Choice of suitable provenances of *Nothofagus* spp. is also likely to be an important consideration in this regard.

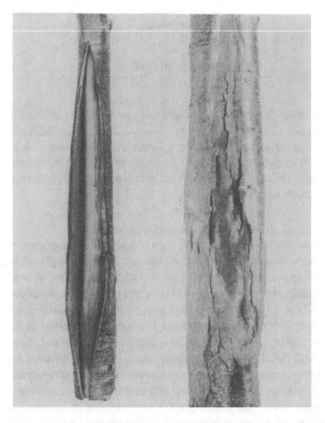

**Figure 2**   Cracking of the bark at the base of *Nothofagus* caused by cold in the winter of 1962–63 (Forestry Commission)

Browning of evergreen foliage has been observed on a number of tree species during or following cold winters. Such discoloration may be ephemeral, as in the case of bronzing of *Thuja plicata* in the nursery (plate 39), or more permanent and followed by some defoliation in a number of other coniferous species. *Sequoia sempervirens* is perhaps one of the most frequently affected species. Browning of foliage, particularly on the eastern side of trees, has been observed in spring on various species, including *Pseudotsuga menziesii*, *Abies grandis* and *Taxus baccata*, following cold winters. Peace (1962) suggests that such damage is not primarily due to winter cold but rather to excessive loss of moisture from the foliage under freezing conditions when it cannot be replaced from below. This hypothesis is supported by more recent observations that such symptoms are found when a particularly cold spell follows a very mild period (Redfern *et al.*, 1985).

Severe winters were recorded in various parts of Britain in 1981/82 and in 1986/87. Strouts and Patch (1983) report damage to a number of coniferous species following the severe winter of 1981/82. Strouts *et al.* (1987) and Redfern *et al.* (1987) both report winter desiccation injury to a wide range of coniferous species following the low temperatures and northerly to easterly winds which were prevalent from late January to early March 1986. Brown foliage was frequently found on the exposed sides of the trees.

Another form of winter cold injury is frost crack of stems (Kubler, 1988). This damage occurs more frequently on broadleaved species such as oak and poplar. The radial cracks, often on the south or southwest side of the trunk, are probably caused by differential contraction of tissues during freezing. Care must be taken to distinguish frost from drought cracks, which are found more frequently on conifers and in association with dry growing seasons.

## Mechanism of frost damage

This controversial topic has long been the subject of discussion between plant physiologists and even now there is no single straightforward explanation. There seems to be general agreement that cells are killed if ice crystals are formed within them. However, it seems that ice crystals form more frequently in inter- or extracellular spaces. Water is withdrawn from the cell during this process and the cell contents are thereby concentrated. It was thought at one time that the increased concentration of cell contents resulting from this and other cellular activities might depress their freezing point so as to avoid damage. It has now been demonstrated that this effect is small (Burke *et al.*, 1976). These authors suggest that in the more hardy species the plant tissue avoids freezing by deep supercooling to temperatures of $-20\,°C$ or less. Other workers, such as Levitt (1969), have placed emphasis on the structural nature of protein molecules when water is withdrawn from the cell during the formation of intercellular ice crystals. Levitt goes on to suggest that the accumulation of solutes within the cell at low temperatures may prevent the denaturing and aggregation of proteins, by osmotic retention of water. These considerations apply not only to the proteins within the cell but perhaps especially to those incorporated in the surrounding protoplasmic membrane.

If proteins associated with the membrane were irreversibly damaged there would be a loss of semipermeability and consequent death of the cell.

**Frost lift**
Frost lift or frost heave can sometimes occur in a nursery following a severe winter frost. Layers of ice form in the nursery soil and as a result the soil surface is raised. If the soil is moist water may be drawn by capillary action from below and the volume of water frozen in the surface layers can be considerable. Small plants are firmly held by the frozen soil and rise as it expands in volume. There may be some breakage of the roots of larger plants or they may remain unmoved. Once the soil thaws the small plants remain in their uplifted position and the soil returns to its original position. The upper roots of these plants are now exposed and if repeated frosts occur they may be completely uprooted.

Maximum frost lift occurs in moist soils but soil type may also be important. Clay soils give rise to greater frost lift than sandy soils. The presence of a mulch on the soil surface may help to insulate the soil below and reduce soil freezing. These considerations have a strong bearing on methods of minimising the risk of frost lift. Nurseries are not normally established on clay soils for a variety of reasons; one of these relates to frost lift. If frost lift is likely to occur then mulches of granulated peat or stone chippings can be applied to nursery beds to reduce the damage. Raised seedbeds may also help, perhaps, because they are drier than the surrounding pathways.

## EFFECTS OF HIGH TEMPERATURES

Some of the symptoms that appear during or after periods of high temperature are associated with lack of moisture in the tree rather than with excessive heat. Examples of this include drought crack on trunks of trees and leaf scorch on wilted trees. However, high temperatures can directly damage trees and the most commonly observed disorder is the killing of bark often referred to as sunscorch.

Direct sunlight damages the trunks of trees that have been growing under shaded conditions and are suddenly exposed to full insolation. This situation can arise, for example, when the side shelter is removed from a forested area or when gaps are made in woodland for building developments and road construction.

**Symptoms**
The bark and sometimes also the cambium are killed in strips or patches on the south or southwest sides of the trunk. Damage is particularly severe on thin-barked species such as beech but also occurs on sycamore and other maple species. Callus tissue normally develops at the margins of the areas of necrotic bark. The dead bark and the wood which is subsequently exposed may be invaded by *Nectria* spp. and decay fungi such as *Ganoderma applana-*

*tum*. At a much later stage the trunks may break, owing to the development of extensive decay within the woody tissues. Tattar (1978) differentiates between summer and winter sunscorch to the bark of trees. The symptoms are similar but Tattar suggests that winter sunscorch follows periods of rapid changes in bark temperature during cold and sunny winter days.

**Mitigation of damage**
Damage to bark from sunscorch often occurs following sudden exposure of trees to hot sun as, for example, when heavy thinning operations or total removal of side shelter are undertaken. However, if the operation can be staged over several years or thinning done less severely, less damage is likely to ensue. In the case of building developments where valuable trees are to be retained following the removal of immediate neighbours, it may be possible to protect stems by wrapping trees with insulating material or even by painting with white paint. However, the latter technique may not be aesthetically acceptable.

## DROUGHT

The term 'drought' is defined in Britain as a period of 14 or more consecutive days without rain. When used specifically in relation to the condition of trees, or of other plants, drought has a rather different meaning. A droughted tree is one suffering from a severe lack of moisture. This arises when the rate of transpiration exceeds the rate of absorption of water through the roots over an extended period.

Peace (1962) separates soil drought, the inadequate supply of water to the roots, from atmospheric drought, the excessive loss of water from the leaves. Jamison (1956), Kramer (1962), Salter and Goode (1967) and others point out that the internal water balance in trees or other plants depends on a complex of soil, climatic and plant factors.

The absolute water content of the soil does not give a reliable measure of the availability of water to the tree. As the soil dries out, the remaining water is held in the soil by increasingly greater forces until the plant roots are finally unable to extract further moisture. At this stage the plants wilt and the soil is said to be at permanent wilting point.

Different types of soil are able to withhold different quantities of water from the plant; clay soils are particularly retentive and sandy soils are not (Salter and Williams, 1969). However, the availability of soil moisture to the plants is complex and depends on a number of other factors, including the absolute water-holding capacity of the soil, the nature of the layers underlying the soil (a rocky substrate may prevent the replenishment of moisture in the topsoil) and the osmotic pressure of the soil water.

Climatic factors such as net radiation, temperature, humidity and especially wind can affect the quantity of water transpired by trees. Peace (1962) draws attention to two sets of conditions where wind can cause excessive transpiration. The first is where trees previously sheltered are suddenly exposed to greater moisture loss as a result of such operations as felling, thinning or other means of removal of side shelter. He further points out that the blast effect of wind in exposed situations is probably one of the most serious factors limiting tree growth in Britain.

Kozlowski (1964) emphasises that the ability of plants to survive moisture stress depends on their ability to absorb water, to reduce water loss and to endure dehydration. Thus the nature of a tree's root system can influence the uptake of water and deeper rooting species may have a clear advantage under drought conditions. Some tree species have special leaf characters such as thick layers of cutin or wax that limit transpiration from the leaves. Others may reduce moisture loss by abscission of leaves during dry spells.

**Symptoms**
No matter whether water deficits are caused by lack of soil moisture or by excessive transpiration from the tree or by some combination of the two, the symptoms on affected trees are essentially similar. Foliar symptoms include wilting followed by necrosis, usually starting at the leaf margins. Premature autumnal colorations may then appear, followed in severe cases by dieback of twigs and branches. Conifer needles turn brown, often starting at the tips, and young succulent shoots bend over and may die. In less severe cases the shoot may wholly recover its turgor or become bent near the base where some cells have been irreparably damaged. Under severe conditions symptoms may spread progressively down from the uppermost parts of a tree, dieback of shoots may develop and the tree may finally die. Where trees are subject to exposure to strong winds symptoms are more severe on the windward side and such trees may take on a permanently one-sided appearance. Small woods may also take on a characteristic wedge shape.

Young trees that have been transplanted from the nursery to their permanent site are susceptible to damage from water deficits, probably because their root systems have been severed, thus rendering them less able to take up water. Semimature trees that have been transplanted are particularly susceptible to such damage. On young conifers a reversible discoloration may occur and needles that turn red or purple under moisture stress can become green again once normal conditions return.

There have been a number of reports of damage to the bark of trees, both coniferous and broadleaved species, as a result of water deficits. True and Tryon (1956) reported that stem cankers were found on oak following a year of severe drought and that they were unable to isolate any pathogenic organisms. Zycha (1951a) suggested that drought (and winter cold) can cause death of bark and subsequent slime-fluxing on beech. He considered that these factors were causal agents of beech bark disease; however, it is now generally accepted that this disease is caused by the combined attack of *Cryptococcus fagisuga*, the beech coccus, and *Nectria coccinea* (Houston *et al.*, 1979). In this context it should be noted that death of bark in patches and subsequent slime-flux was observed in Britain on several species of broadleaved trees including birch, beech and sycamore following the severe drought in 1976. Strouts *et al.* (1985) report drought damage to Leyland cypress, western red cedar, giant *Sequoia* and a number of other species. The effect of drought on the bark of trees is still largely unexplained, though Bier (1959) has suggested that a lowering of the moisture content of the bark can allow weak pathogens, such as *Cryptodiaporthe salicina* on willow, to enter a weakened host.

The bark of several coniferous species cracks as a result of severe moisture stress. In Britain, this so-called drought crack has been observed during and following years of severe drought on *Abies procera, A. grandis* and *Picea sitchensis*; elsewhere in Europe it has been reported on *Picea abies*. A study of drought crack in *A. grandis* by Greig (reported by Phillips, 1965, 1966, 1967a), indicated that this damage could be closely correlated with drought years. The cracks were usually less than 1 m in length and limited to the bark, though the longer cracks, where present, extended into the wood (figures 3, 4).

**Figure 3**   Drought crack in Sitka spruce: tree showing external crack marks
(B.J.W. Greig)

Abnormal morphology in the xylem has been observed in some species following a severe drought. Peace (1962) describes for Japanese larch the formation of a ring of flattened cells with smaller lumens, the radial walls of which were sometimes bowed. These drought rings are distinct from the

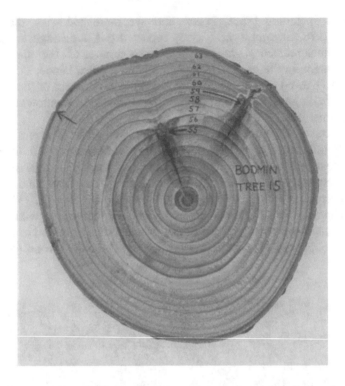

**Figure 4**  Drought crack in Sitka spruce : section of tree showing cracks within the wood (B.J.W. Greig)

reduction in overall ring width, also associated with drought, though the two may occur together.

### Species resistant to drought
There are many cases quoted in the literature where tree species have proved particularly resistant (or susceptible) to drought; a few examples are quoted below.

Parker (1968) suggested that deep roots of *Quercus* spp. and *Pseudotsuga menziesii* seedlings contribute to the successful establishment of these trees. Albertson and Weaver (1945) reported that the deeply penetrating roots of *Quercus*, *Juglans*, *Gleditsia* and *Maclura* helped these species survive extreme drought conditions in the American prairies. Pfeffer *et al.* (1948) found in Czechoslovakia that Douglas fir was resistant and Norway spruce was susceptible to drought, particularly in competition with other trees. In Britain, Peace (1962) noted that Japanese larch was particularly susceptible to soil drought on dry sites and places with locally shallow soils.

Cadman (1953) discusses the use of tree species in shelter-belts and considers the resistance of trees to excessive transpiration rather than to uptake of water from the roots. He lists the following species as particularly suitable for conditions of severe exposure:

*Abies procera*      *Acer pseudoplatanus*
*Pinus contorta*      *Fagus sylvatica*
*Pinus mugo*
*Pinus nigra*
*Pinus nigra* var. *maritima*
*Picea glauca*

## LIGHTNING DAMAGE

Most people recognise that trees are potential lightning conductors; few will shelter under trees during a thunderstorm because of the risk of a lightning strike. Damage to a tree when struck by lightning will vary according to the intensity of the discharge and a number of other factors such as the location, size and condition of the tree. Such damage may be limited to a single tree or small group of trees but can on occasion extend to an area covering a hectare or more. Damage often occurs on single, tall, isolated trees but it can also occur in post-thicket stage plantations and even in nurseries.

Diagnosis of lightning damage may not be entirely straightforward, particularly if the lightning has struck some months or years previously. At that stage a single dead tree or group of trees may provide little evidence of the cause. However, it is possible that the characteristic lightning scar may be evident on the trunk of one or more trees. Such lightning scars are more readily identifiable soon after the lightning strike and may take the form of superficial bark stripping or may penetrate deeper into the wood, forming a rough jagged furrow. In extreme cases the trunk or parts of it may be completely shattered. The lightning scars often follow a spiral course which in some cases follows the spiral grain of the tree. Damage can sometimes extend down into the roots and the superficial soil or litter may be thrown aside to leave the roots exposed.

When lightning strikes a group of trees, most die fairly rapidly but some may survive for a year or two longer. Where a tree remains alive there may be single dead branches in the live crown. In such cases the bark or cambium on the adjoining stem can be killed and the whole of the upper crown may subsequently die. Lightning may also be responsible for the presence of dead tops of tall *Sequoiadendron* trees so commonly seen in Britain (plate 2).

In some recently killed small groups lightning scars may be absent but careful examination of the still fresh bark may reveal the presence of long strips of dead bark running from the roots to the upper stem. Such hidden scars, which may be discontinuous and tapering in the higher parts of the stem, are useful diagnostic features when trees are examined fairly soon after the suspected lightning strike.

When trees have survived a lightning strike it is often possible to find evidence of abnormal tissue in the growth rings. These abnormalities may be indistinguishable from frost rings and care must be taken in the interpretation of the available evidence. A study of the metereological records for the area where the damage was observed should enable the investigator to determine whether the date of a thunderstorm coincided with the period of growth indicated by the abnormal tissue in the annual growth ring.

A detailed explanation of the electrical nature of lightning is beyond the scope of this book. It is often thought that a lightning discharge travels downwards from the atmosphere to the earth. However, one alternative theory which has been put forward suggests that there is a faintly luminous downward stroke, often with a number of branches; this is succeeded by a much brighter return stroke from earth to cloud which retraces the path of the leader stroke and its branches. The current in the return stroke may be up to a hundred times that in the initial leader. This explanation would fit the observation that the hidden lightning scars are wide at the base of the tree and taper further up the stem.

Peace (1962) reviews the susceptibility of different tree species to lightning damage. Damage is recorded more frequently among broadleaved species on trees such as oak and poplar and among conifers on pine. This is probably because these are among the commonest genera planted as individual isolated trees. However, there may be some specific differences in susceptibility to lightning damage and these are exemplified by beech, which is often grown as an isolated tree but does not appear to be damaged to the same extent as other species. Various explanations for this difference have been put forward, mainly relating to the smoothness of beech bark and the presence of a continuous film of water on the bark during thunderstorms. Further information is needed, however, before such explanations can be properly verified.

The overall damage to trees in Britain by lightning is probably small and measures are rarely taken to ameliorate the damage. Lightning conductors are very occasionally fixed to valuable amenity trees. A conductor can be seen, for example, on one of the oldest London planes in Britain, found in the grounds of the old Ranelagh Club at Barn Elms in London. More significant damage has been found in some European and North American forests, where large areas may be killed outright or where lightning damaged trees can be subsequently attacked by bark beetles and perhaps decay fungi. In North America and other parts of the world lightning strikes can occur without associated rainfall and there are many reports of forest fires that have been started by lightning. Such fires have only rarely been recorded in Britain.

## HAIL DAMAGE

Significant damage to trees by hailstorms occurs relatively infrequently in Britain. Symptoms of the physical damage to stems and foliage are clear enough at the time but recognition of the damage some time after the event may be a little more difficult.

Hailstones can perforate, tear or bruise the foliage of broadleaved species, especially in the early stages of growth. They can also cause small discrete lesions on young shoots of either conifers or broadleaved trees. Side shoots of conifers may be broken, though leaders are less susceptible to this form of damage, perhaps because of their upright position.

Damaged foliage may be retained throughout the growing season and provide evidence of hail damage earlier in the same year. The lesions on the stems may heal over and the callus tissue that develops may provide evidence of hail damage for some time after the event. In some cases the wounds may allow entry of pathogenic or decay-causing organisms. In either case, sectioning of the lesions (which are mainly on the upper side of the stem) will indicate that they all originated at one time in the annual ring. Although hailstorms are often local, meteorological records may well provide supporting evidence of their occurrence.

Dobbs and McMinn (1973) made observations on 2 to 4-year-old research plots of *Picea glauca* and *Pinus contorta* following a ten minute hailstorm in 1972 in British Columbia. The hailstones were 1–1½ inches (2–4 cm) in diameter, 2 per cent of the trees were killed and only 14 per cent escaped damage. The effects were more severe on *P. contorta* than on *P. glauca* and on older trees of both species. Competing vegetation provided some degree of protection against stem lesions and defoliation but not against damage to leaders or terminal buds.

A number of experiments have been undertaken in Europe to protect crops against the damaging effects of hailstorms. Guilivo (1973), in relation to apples, pears and vines, considers that passive control by the use of plastic nets, with mesh size 4 × 8 mm, is preferable to active methods using rockets and silver oxide which are intended to precipitate the clouds in a less damaging form. No such ameliorative measures, however, have proved justified in forestry.

## Soil factors

### NUTRIENT DEFICIENCIES

Nutrient deficiencies can cause symptoms in nursery or planted trees which are similar to those caused by some fungal or virus pathogens. It is clearly important that the tree pathologist should be aware of the symptoms of common nutrient deficiencies in order to distinguish them from pathogenic disorders. Information on deficiency symptoms in forest nurseries and plantations is well documented (e.g. Benzian, 1965; Aldhous, 1972; Binns *et al.*, 1980) and there is a large volume of literature on forest fertilisation (Baule and Fricker, 1970; Everard, 1974). However, much of this information relates to forest species (mainly conifers) and that on ornamental species is much more limited.

In this section the major concern is with methods available for the diagnosis of nutrient deficiencies. Although mention will be made of techniques for correcting deficiency symptoms, it is not appropriate here to enter into a full discussion of fertiliser treatments either for nursery plants or for established trees. Symptom expression depends on a number of factors, including the particular nutrient or nutrients involved, the tree species, the time of year and the local weather conditions. In general, symptoms include yellowing, purpling or withering of parts or the whole of leaves and needles. They often appear first at the tips of needles or the margin of leaves and may then progress to occupy a larger proportion of the foliage.

Symptom expression will vary on different species and may be more severe on some species than on others. Symptoms will also vary according to the time of year and may become more intense as the season progresses. Climatic factors such as drought or waterlogging may initiate or exacerbate deficiency symptoms. Surrounding vegetation may compete with the trees for nutrients and when competition is severe nutrient deficiency symptoms may develop.

**Diagnosis of deficiencies**
Nutrient deficiencies can be diagnosed in several ways, none of which alone is likely to give a conclusive result. It is important to weigh up all the evidence available before reaching any conclusion. The following diagnostic methods are available:

(1) Visual symptoms.
(2) Foliar analysis.
(3) Soil analysis.
(4) Experimental addition of nutrients.

Visual symptoms of nutrient deficiencies can be a very useful guide to determining the cause but they are rarely sufficient to provide a certain diagnosis. Severe deficiencies may cause poor overall growth, including reduced height increment, small leaves or needles and perhaps shoot dieback. Leaves or needles may be discoloured or necrotic and in addition plants may become more susceptible to damage by other agents such as pests, diseases or climatic factors.

A summary of the visual symptoms of specific nutrient deficiencies is given below.

*Copper*
Deficiency of copper is relatively uncommon but Benzian (1965) made some interesting observations at one forest nursery on a very light sandy soil. In most seasons, withered needle tips developed on Sitka spruce seedlings; symptoms varied in intensity from season to season and from one part of the nursery to another. The symptom was referred to as 'needle tip burn' and it normally affected only the more vigorous plants. The tips of the needles

turned a straw colour and were clearly demarcated from the healthy green portion of the lower part of the needle. The appearance of symptoms tended to coincide with warm sunny periods and if these conditions persisted the growing point would die and the needles turn brown or black. On the other hand, if the dry conditions were only short in duration the seedlings would grow and produce healthy needles, leaving a ring of affected needles lower down the stem. Benzian demonstrated experimentally that foliar applications of copper sulphate or Bordeaux mixture would remedy this deficiency. At the same nursery Benzian planted poplar cuttings and described symptoms of copper deficiency on this species. The first signs are a blackening of the leaf tips developing into a more general marginal scorch of the youngest leaves. The edges of the leaves become taut and brittle, forcing the whole leaf into a cup-like shape. Soon after this the whole leaf may develop an interveinal yellowing.

### Lime-induced chlorosis

Symptoms of lime-induced chlorosis are commonly found on soils formed from chalk or limestone. An excess of lime in the soil may severely limit the uptake or absorption of a number of other nutrients, especially iron and manganese.

Peace (1962) refers to a serious case of lime-induced chlorosis in Scots pine growing on oolite limestone. Symptoms developed first on trees when they were 12 years old. The needles turned chlorotic and then brown, growth of needles and shoots was reduced and by the time the trees were 20 years old many had died. The most serious damage occurred where the limestone was near the surface but the relationship between soil depth and symptom severity was not entirely clear. Corsican pine is less susceptible to lime-induced chlorosis and when planted with Scots pine it was unaffected.

Lonsdale *et al.* (1979) made observations on lime-induced chlorosis in beech plantations growing on chalk downs in southern England. Chlorosis was particularly marked where evidence of patterns of ancient land-use had persisted. The beeches were healthy on ancient field boundaries, where the soil had not been disturbed and symptoms of chlorosis were most severe immediately below field boundaries where the surface soil had been eroded away by ploughing and the remaining soil had been admixed with the underlying chalk.

### Magnesium

In many agricultural and horticultural crops magnesium deficiency is usually seen on older leaves and the same may also be true for broadleaved tree species. Symptoms of magnesium deficiency on *Thuja plicata* include yellowing of the older needles followed by browning and death of tissues. However, Benzian (1965) draws attention to a rather different situation in Sitka spruce seedlings. In this case a yellow discoloration first appears on the tips of the younger needles. The symptoms are sometimes referred to as 'hard yellows' and the yellow needles usually lie at a wider angle to the stem than healthy needles. On broadleaved species interveinal chlorosis and growth reductions are commonly associated with magnesium deficiency.

## Nitrogen

Nitrogen is one of the three major inorganic nutrients required for all plant growth. When nitrogen is deficient (plate 5), most plants are paler green in colour overall and show reduced growth as compared with healthy plants. In nursery beds, pine and spruce seedlings may grow better on the edges of the beds than in the middle. When nitrogen deficiency is more acute, the tips of the youngest needles of Sitka spruce seedlings may turn pinkish. The problem of nitrogen deficiency in Sitka spruce plantations has recently been reviewed by Taylor and Tabbush (1990). In broadleaved species suffering from nitrogen deficiency, reduced growth may be accompanied by premature autumn coloration.

## Phosphorus

Phosphorus is the second major inorganic nutrient. In upland Britain, phosphatic fertilisers are used on iron pan, peaty-gley and deep peat soils (Mayhead, 1976), and in the lowlands Everard (1974) has recommended its use on all heathland or former heathland soils and on Hastings beds in southeast England. However, visual symptoms of phosphate deficiency (plate 6) are difficult to distinguish and may only be evident by overall slow growth. In the nursery it is sometimes possible to recognise phosphate deficiency by a somewhat duller or 'lack-lustre' foliage; in young plantations poor height growth and much reduced needle length and weight are symptomatic of phosphate deficiency. In general, other techniques must be used to detect this deficiency and foliar analysis is probably the most valuable tool in this context (see Binns *et al.*, 1980).

## Potassium

Potassium is the third major inorganic element required for plant growth. One of the most detailed descriptions of potassium deficiency (plate 7) has been given by Benzian (1965) for Sitka spruce seedlings. A purplish colour develops in the needles from mid-summer onwards; the symptoms are most obvious on the youngest needles and may be limited to needle tips or extend the whole length of the needle. The purple discoloration may persist through the rest of the season or may change to reddish-yellow and eventually to yellow. At this stage the whole seedling becomes yellow and, therefore, can be distinguished from magnesium-deficient seedlings which are only partly yellow.

In young Sitka spruce trees on the most infertile peats there is poor development of the terminal bud, usually in the second or third year after planting (Binns *et al.*, 1980). Where the deficiency is moderate the needles at the apex of current shoots turn yellow from the tip downwards; later the tip may become purple or even brown. In more severe cases, all the needles on the current year's shoot may be affected with similar chlorosis developing from tip to base of individual needles. Potassium fertiliser is used in forestry only on deep peat or peaty-gley soils (Mayhead, 1976).

On broadleaved trees discoloration of the foliage is often followed by marginal necrosis and die-back.

## POOR SOIL AERATION

Adequate soil aeration is essential for the maintenance of root growth. The concentration of oxygen in the atmosphere of well-aerated soils is somewhat lower than that in the air above ground. However, if the concentration is below 10 per cent, tree roots may be damaged and if it drops below 3 per cent, root growth stops entirely (Tattar, 1978). A number of different factors can cause poor aeration of the soil, including an excess of water, compaction of the soil and sealing of the soil surface.

Few trees will survive if the ground is permanently waterlogged. However, damage to trees caused by excess water generally occurs when land becomes temporarily waterlogged, owing to flooding, blockage of drains or earth-moving operations which change the natural drainage flow. The effects on tree growth will largely depend on the species present, the time of year and the duration of the inundation. The best method to deal with problems associated with the presence of excess water is to ensure that an efficient drainage system is present.

Compaction of the soil also prevents the free interchange of oxygen between soil and air. Man, animals and machines can all be causes of serious soil compaction. The increased use of forests for recreation purposes has led to soil compaction on well-used paths. Regularly used animal tracks can also cause damage to neighbouring trees. Some of this damage may be difficult to avoid but the judicious use of fencing to control access may alleviate the situation. Damage caused by heavy machinery used for harvesting forest produce, or for restoring mineral workings prior to forest planting, can be relieved by deep cultivation using specially designed tines (Binns and Fourt, 1980).

The commonest cause of surface sealing of the soil is the introduction of tarmac or other soil coverings around established trees in car parks and on new roads and footpaths. The extent of damage to the tree will depend on the proportion of the ground that is covered close to the trunk. Surface covers such as tarmac or paving will not only prevent gaseous interchange but also prevent free entry of water to the soil and contribute to soil compaction, thus adding to the stresses imposed on the tree roots. Nevertheless, it should be pointed out that many trees do grow successfully in paved areas. In urban areas, for example, many such trees were planted in locations covered by paving slabs or tarmac and therefore have been able to adapt to this situation. Most damage occurs when the soil surface around established trees is covered and roots which previously had access to well-aerated soil are suddenly deprived of adequate oxygen.

**Symptoms**

The symptoms of poor soil aeration are similar, regardless of the reason for the lack of oxygen (Peace, 1962). In all cases these conditions can lead to death of roots. As a result, leaf margins and needle tips become necrotic and a progressive leaf fall and dieback may follow. The severity of the symptoms will depend upon the proportion of the root system affected. Where damage is slow to develop, growth of shoots may be stunted. In severe cases the entire tree may be killed.

# Chemical factors

## AIR POLLUTION

It has long been recognised that air pollution can have damaging effects on plants. In 1874, the American *Garden Chronicle* reported failure of crops and death of trees due to the adverse effects of air pollution (Cameron, 1874). During the 1930s a classic study of fume damage to vegetation, including trees, was undertaken at the Trail Smelter on the borders of the United States and Canada (Anon., 1939). In Europe until the 1960s, the sources of pollution tended to be local and problems such as acid smogs were especially associated with industrial conurbations. The enactment of clean air legislation led to dramatic improvements in air quality. At the same time it became clear that certain pollutants, especially sulphur dioxide, could be transported great distances in the atmosphere. This process, probably assisted by the construction of very tall chimneys at power stations and other major industrial installations, was the subject of considerable study in the 1970s. The Scandinavians were seriously concerned at the acidification of their rivers and lakes and the associated loss of fish. Sweden presented a case to the United Nations in 1971 on the long-range transport of sulphur dioxide. They showed that the sulphate concentration and acidity of rain in Sweden were higher than could be accounted for by local sources. This case was the subject of long and somewhat acrimonious debate but was finally accepted and has led, along with many subsequent events, to a concerted effort to reduce sulphur dioxide emissions in west European countries.

The term 'acid rain' was introduced to cover the wet deposition of sulphur dioxide at a distance from its source. It has now become common parlance for most types of air pollution and it no longer has a clear and specific definition.

Useful reviews of the air pollution literature may be found in, among others, Innes (1987), Kozlowski and Constantinidou (1986) and Longhurst (1989).

**Causal agents**

There are many toxic gases produced by a range of industrial and other processes. Some, termed primary pollutants, originate during such processes in a form which is toxic to plants. Perhaps the best-known examples are sulphur dioxide, nitrogen oxides (NOx) and fluorides. Others, known as secondary pollutants, have undergone chemical change after their initial production. These include the photochemical pollutants such as ozone and peroxyacetyl nitrate (PAN).

The major sources of sulphur dioxide are fossil fuel-burning power stations, industrial boilers and metal smelters. On the other hand, nitrogen oxides are produced mainly by internal combustion engines but also from fossil fuel-burning power stations.

The major pollutants mentioned above and a number of other important but less common pollutants are discussed below in more detail.

*Sulphur dioxide*
Sulphur dioxide is the best-known and most widespread of all air pollutants. Fowler and Cape (1982) gathered information on the distribution of sulphur dioxide in different parts of Europe and showed that the highest concentrations were to be found in the English midlands and in West Germany. In the USA highest concentrations were found in the industrial states of Indiana, Ohio and Pennsylvania. The OECD summary of environmental data (Anon., 1989) shows that the UK is the largest emitter of sulphur dioxide in Europe, followed by France, Italy and Germany.

Sulphur dioxide can be absorbed by plants either in gaseous form, termed dry deposition, or in solution either as sulphite or sulphate ions, termed wet deposition. Sulphate can also be deposited as ammonium sulphate.

Crane and Cocks (1989) report that emissions from power station chimneys are usually carried on the wind for hundreds of kilometres within a layer of air rarely more than 1 kilometre deep. These authors calculate that in dry weather with a wind speed of 7 metres per second, nearly half of the sulphur dioxide emitted will have travelled 600 kilometres in one day after emission.

However, the rates of removal of sulphur dioxide from the atmosphere depend on the occurrence of rainfall and are likely to be much greater during periods of precipitation. Measurements have shown that the sulphur content of rain was far above that which could be accounted for simply by the dissolution of sulphur dioxide in rain. Sulphur dioxide is oxidised to sulphate prior to deposition and by this means higher concentrations of sulphur could be explained. Oxidation by naturally occurring and photochemically generated oxidants is regarded as the most likely mechanism for the production of sulphate in the aqueous phase.

*Oxides of nitrogen*
The most important oxides of nitrogen found in the atmosphere are nitric oxide (NO) and nitrogen dioxide ($NO_2$). Because these two oxides interact with one another they are collectively referred to as NOx. In the UK roughly half man-made NOx is produced from internal combustion engines and a third from power stations, with other industries providing the remainder. According to the OECD summary of environmental data (Anon., 1989),

Germany is the largest emitter of nitrogen oxides in Europe, followed by the UK, France and Italy.

The concentrations of NO and $NO_2$ in the atmosphere vary diurnally and seasonally. During the day NO is converted to $NO_2$, some of which is subsequently consumed in the production of ozone, PAN and other constituents of smog. The principal mechanism for the removal of nitrogen dioxide from the atmosphere is direct deposition and oxidation to nitrate followed by deposition (Metcalfe *et al.*, 1989).

*Ozone*

Ozone has been recognised as a pollutant for at least 100 years. It can be found in the upper atmosphere, in the vicinity of electrical storms, and can also arise as the result of a photochemical reaction. During the 1950s it was discovered that ozone was present in the smog which developed over many American cities, especially during periods of very hot weather when stable air masses resided over polluted areas. It is now recognised as an important pollutant in the neighbourhood of many urban areas.

A series of chemical reactions lead to the production and destruction of ozone. Oxides of nitrogen are produced by the combination of nitrogen and oxygen at the high temperatures developing in vehicle engines and fossil fuel-burning power stations. In the presence of sunlight, these oxides react with the oxygen in the air and unsaturated hydrocarbons from vehicle engines to produce ozone.

*Peroxyacetyl nitrate*

Peroxyacetyl nitrate (PAN) is also an important photochemical pollutant found in urban smog. It is produced in a similar manner to ozone by the reaction of oxides of nitrogen and unsaturated hydrocarbons emitted by vehicle engines in the presence of sunlight.

*Fluorides*

Fluorine compounds such as hydrogen fluoride and silicon tetrafluoride are important as air pollutants. Kozlowski and Constantinidou (1986) rank fluorides as fifth in importance as agents for plant injury in the United States, following ozone, sulphur dioxide, PAN and nitrogen oxides. Fluorides are found in the vicinity of a number of manufacturing industries, including aluminium smelters, phosphatic fertiliser plants, brick and pottery kilns and steel manufacture, and in rocket fuel combustion. It is interesting to note that fluorides are by far the most toxic of common pollutants and may induce injury at concentrations up to 10 000 times lower than that of other pollutants (Weinstein and Alscher-Herman, 1982).

Fluorides appear to be concentrated close to point sources and the atmospheric concentrations decrease with distance from the source.

*Ammonia*

Ammonia has been implicated in die-back of trees both in the Netherlands (Buijsman *et al.*, 1985) and in Sweden (Bergstrom and Gustafson, 1985). The source of ammonia is often associated with agricultural holdings, particularly where very intensive systems of cattle or pig production are in place. Innes (1987) considers that it is the most important alkaline gas commonly found in the atmosphere and it may act to neutralise some of the more acidic gases.

*Particulates*

Finely divided solids, known as particulates, can also damage plants, although they have been the subject of only limited study. The most important sources of particulates are coal- or oil-burning stoves, cement factories, lime kilns, incinerators and straw burning.

**Effects of air pollutants**

This section is divided into two parts with a general description of the effects of acid rain first and then a more detailed account by individual pollutants of the specific symptoms related to that pollutant.

*Acid rain*

The possible deleterious effects of acid rain on biological ecosystems have been intensively studied over the past decade and more. Harmful effects of acid rain on freshwater ecosystems have been well established. For example, increasing acidity has a marked effect on fish populations in lakes and streams susceptible to pH change (i.e. those in areas with acidic geologies).

However, there has been considerable debate about the possible effects of acid rain on trees. Bormann (1985) considered that there was only meagre evidence to support the view that acid rain alone causes damage to forest trees. Innes and co-workers (Innes *et al.*, 1986; Innes and Boswell, 1987, 1989) have not been able to find a strong relationship between various measures of acid rain and crown density assessments of the major tree species in the UK. On the other hand, similar surveys in other European countries have been interpreted to demonstrate that acid rain does cause a reduction of crown density of trees. Crown density is a useful measure of the health of a tree which can be determined by trained surveyors. While it is recognised that crown density may be some measure of damage from air pollution, it is also a reflection of damage from other biotic and abiotic factors. The interpretation of survey results is therefore difficult when using this criterion. Innes (1989) has reviewed the evidence linking air pollution and forest health and concluded that the situation is complex and, in many cases, explanations of forest decline are related to the particular site. Fears in general that forest health is declining may be exaggerated but nevertheless in many areas there is cause for concern.

Symptoms associated with specific pollutants are described below and useful descriptions can be found in a number of references, including Heggestad (1968) and Kozlowski and Constantinidou (1986).

*Sulphur dioxide*
On broadleaved trees sulphur dioxide causes bleaching or necrosis of the interveinal regions of the foliage, the veins remaining green. The bleached or necrotic areas are often not clearly demarcated and they are frequently surrounded by more or less chlorotic areas that merge into healthy tissue.

In conifers, the needles have brown necrotic tips, often with a banded appearance which is probably related to successive toxic emissions. Mesophyll cells are most susceptible. In *Picea* and *Pinus* spp., injured cells were first observed near the stomata or the endodermis. The cells with most injury were near the intercellular spaces (Kozlowski and Constantinidou, 1986).

Smith (1981) quotes a range of threshold concentrations for suppression of photosynthesis by sulphur dioxide from 262 $\mu g/m^3$ (over 2 weeks) for conifers to 2620 $\mu g/m^3$ (over 2–4 hours) for broadleaves. Threshold doses for injury were quoted as 1820 $\mu g/m^3$ for 1 hour and 468 $\mu g/m^3$ for 8 hours.

*Oxides of nitrogen*
There is little information available on the effects of oxides of nitrogen on trees. However, Taylor and Eaton (1966) found a marked depression of growth and darker green foliage in tomato plants subjected to 0.5 ppm nitrogen dioxide for 10–12 days. At concentrations in excess of 3 ppm damage was similar to that caused by sulphur dioxide. Kozlowski and Constantinidou (1986) quote threshold doses to injury of 38 $\times$ 10$^3$ $\mu g/m^3$ for 1 hour exposure.

*Ozone*
Symptoms of ozone damage on broadleaved trees are frequently limited to the upper surface of the foliage, which is covered with small irregular flecks varying in colour from white through yellow to reddish brown. Costonis and Sinclair (1967) give a detailed description of the development of the symptoms on *Pinus strobus*. At first, small water-soaked spots appear near the base of the needle; after 6–8 hours affected areas become yellow-pink and then purple in colour. The affected areas then coalesce to form chlorotic or necrotic bands and on highly susceptible trees damage may extend until tip necroses have developed. Kozlowski and Constantinidou (1986) quote threshold pollutant doses of 300–600 $\mu g/m^3$ for 2–4 hours or more for forest tree seedlings of various pine species.

*Peroxyacetyl nitrate*
In contrast to ozone, PAN typically produces symptoms on the undersurface of the leaves of broadleaved trees. The lower surface appears glazed or bronzed, owing to the collapse of the mesophyll cells around the stomata.

Symptoms on coniferous needles are less characteristic and may include some chlorosis or bleaching (Brandt and Heck, 1967). Kozlowski and Constantinidou (1986) quote injury threshold doses of 1000–3000 $\mu g/m^3$ for 8 hours.

## Fluorides

In some cases fluorides act as cumulative poisons, although this may not always be the case. Jacobson *et al.* (1966) have demonstrated that once fluorides have entered the leaf tissue the pollutant is rapidly translocated to leaf tips and margins. As a result a tip and marginal necrosis of the foliage develops on broadleaved trees often with a distinct narrow reddish-brown line of dead tissue separating healthy from affected parts. In conifers, tip burn is a characteristic feature with a similar well-marked differentiation between live and dead areas (plate 3). The dosage of fluoride which gives rise to damage varies according to the host species but levels of 0.5 parts per billion fluoride for prolonged periods will cause damage to plants. Brandt (1962) considers that concentrations of 50–200 ppm fluoride in the foliage will cause necrosis on susceptible plants. Kozlowski and Constantinidou (1986) report that navel orange trees are sensitive to atmospheric fluoride at a concentration as low as 2–3 ppb.

## Ammonia

Ammonia has been implicated in the dieback of trees in the Netherlands (Buijsman *et al.*, 1985) and in Sweden (Bergstrom and Gustafson, 1985). Innes (1987) indicates that about 20 per cent of the ammonia that is emitted is deposited within 5 km of the source, the remainder being converted into ammonium compounds.

## Particulates

Deposits of genuine soot, which are distinct from sooty mould associated with honeydew secreted by aphids, are commonly found on the foliage of trees close to industrial factories burning coal or oil. It was thought at one time that sooty deposits might interfere with gaseous exchange through stomata. However, it is considered by some that the most serious effect is the reduction of light reaching the foliage (Jennings, 1934). Dust deposited around cement factories may cause chlorosis and death in both broadleaved and coniferous species (Darley and Middleton, 1966).

### Diagnosis of point source pollution damage

The difficulties associated with the diagnosis of damage caused by acid rain and other pollutants dispersed over long distances have already been discussed. The symptoms of pollution injury described above may be useful indicators of the nature of the injury; such damage is readily confused with that caused by other agents such as disease, insects and cultural conditions. It is also important to recognise that pollution damage may only be evident at certain seasons when the foliage is particularly susceptible. It may also be limited to susceptible species. Brandt (1962) discusses the wide range of factors which must be taken into account in assessing the possible role of air pollutants in causing damage to plants. He stresses that competent observers, preferably with some previous knowledge of the local vegetation, are needed in order to make a proper assessment and interpretation of the symptoms. It is clearly important to establish that there are potential sources of pollution in an area and where possible to monitor levels of pollution at the source and at varying distances from it. The distribution of air pollution damage in relation to the source and to wind direction at the time of emissions can provide important evidence in the diagnosis of pollution damage. Foliar analysis can in some cases provide useful information on the presence of pollutants in the leaf tissues.

### Relative resistance of trees to air pollution damage

While there will be continuing efforts to reduce the levels of pollution in the atmosphere, it is important that foresters and arboriculturists recognise polluted areas and where possible plant tree species which tolerate air pollution. Kozlowski and Constantinidou (1986) drew up a series of tables illustrating the relativity susceptiblity of a range of tree species to a number of different pollutants. Tables 2.2 and 2.3 are based on this information.

**Table 2.2**   Tolerance of trees to atmospheric pollution largely attributed to $SO_2$ (from Kozlowski and Constantinidou, 1986)

| Sensitive | Intermediate | Tolerant |
|---|---|---|
| | Conifers | |
| *Cedrus atlantica* | *Larix kaempferi* | *Chamaecyparis* |
| *Larix decidua* | *Picea sitchensis* | *lawsoniana* |
| *Picea abies* | *Pinus contorta* | *Ginkgo biloba* |
| *Pinus sylvestris* | *Pinus nigra* var. *maritima* | *Taxus baccata* |
| *Sequoia sempervirens* | *Pseudotsuga menziesii* | *Thuja occidentalis* |
| | Broadleaves | |
| *Populus nigra* var. *italica* | *Acer pseudoplatanus* | *Acer platanoides* |
| *Quercus ilex* | *Aesculus hippocastanum* | *Fagus sylvatica* |
| *Salix nigra* | *Castanea sativa* | *Platanus* × *hispanica* |
| *Ulmus parvifolia* | *Fraxinus excelsior* | *Quercus petraea* |
| | *Populus tremula* | *Quercus robur* |
| | *Prunus avium* | *Tilia cordata* |
| | *Robinia pseudoacacia* | |

## HERBICIDE DAMAGE

A wide variety of herbicides is now used routinely in agriculture and forestry. Many are selective, that is they kill unwanted weed species and allow the growth of crop plants or trees to continue unhindered. However, there are

**Table 2.3**  Relative tolerance of trees to fluoride

| Tolerant | Intermediate | Susceptible |
|---|---|---|
| | Conifers | |
| *Juniperus* spp. | *Thuja* spp. | *Larix* spp. |
| | | *Pinus contorta* |
| | | *Pinus sylvestris* |
| | | *Pseudotsuga* spp. |
| | Broadleaves | |
| *Alnus glutinosa* | *Acer campestre* | |
| *Platanus* spp. | *Betula pendula* | |
| *Salix* spp. | *Carpinus betulus* | |
| *Sorbus aucuparia* | *Castanea sativa* | |
| *Tilia cordata* | *Fagus sylvatica* | |
| | *Fraxinus excelsior* | |
| | *Juglans* spp. | |
| | *Populus* spp. | |
| | *Quercus* spp. | |
| | *Tilia* × *europaea* | |

occasions when herbicides are applied at the wrong time, concentration or dose, or are mistakenly applied to a susceptible species and damage to crop plants and trees can follow.

Herbicide damage to trees can often be recognised on the basis of its distribution. Herbicides are often applied to eradicate weeds surrounding trees and if the latter are damaged it will only be those within or close to the treated area that are affected. In a tree nursery the distribution of damage may be related to the uneven application of the herbicide due to poorly adjusted spray machinery. In the case of total weedkillers, often used, for example, on paths and verges, damage to neighbouring trees may result from spray drift or from secondary distribution of the herbicide in rainwater or by death of roots under the treated area.

Some herbicides can cause characteristic symptoms on susceptible plants. Such symptoms may be diagnostic on their own but, in many cases, further evidence of the presence of a herbicide and of the distribution of the damage should be sought before a firm diagnosis is made, as somewhat similar symptoms can be produced by other agents. Therefore, a knowledge of the symptoms is useful and a description of some of the symptoms on tree species that have been found in association with the application of a number of herbicides is given below. For convenience the herbicides have been divided into the following three main groups.

(1) Herbicides used in nurseries.
(2) Herbicides used in newly planted areas.
(3) Herbicides used for total eradication of vegetation.

**Herbicides in nurseries**

Herbicides are used as a pre-emergence treatment in nursery seedbeds; they are applied after sowing but before emergence of the seedlings in order to kill rapidly germinating weed seeds. Strouts *et al.* (1987) reported death and severe stunting of many rising 1-year-old European and Japanese larch seedlings following application of a pre-emergence herbicide, diphenamid, a practice not normally damaging.

At later stages in the growth of plants in the nursery different herbicides are used. For example, simazine is commonly used to control weeds in transplant lines. If this is applied at rates or concentrations greater than those recommended, damage can ensue. Shoots and needles of Japanese larch turn white, needles of Scots pine become brown, with a reddish transition zone between healthy and affected tissues. On broadleaved species, leaves develop a marginal or interveinal yellowing or browning. These symptoms appear 2–6 weeks after herbicide treatment. If the plants are not killed, normal healthy growth may appear later in the season.

**Herbicides for newly planted areas**

There are many different sites in which trees may be planted, ranging from forest locations with grass, herbaceous and woody weeds, to urban streets or parks with predominantly grass weeds. A range of herbicides is available to deal with the varied weed species present before and after planting.

Four commonly used herbicides for the treatment of grasses and grass–herb mixtures are propyzamide, atrazine, chlorthiamid and dichlobenil. The correct method of application is critical for these herbicides and incorrect application of the chemical can cause very characteristic symptoms. Where granules of chlorthiamid make contact with the stems of Scots pine, Corsican pine, Japanese larch and Douglas fir, a swelling develops at the base of the stem (plate 9). Radial growth at soil level is strongly inhibited, while growth immediately above continues, giving rise to the swelling. The stem may be girdled at soil level, owing to complete death of cambium, in which case the plant will die. Symptoms on Norway spruce are a little different. A narrow, dead strip of tissue develops on the stem extending upwards for several centimetres, sometimes as much as 30 cm. At the same time, the needles may become chlorotic and the lower branches may die back. This is usually attributable to the uptake of the breakdown product of dichlorobenzamide in excessive quantity. In extreme cases the stem may be girdled at the base.

Glyphosate is a herbicide used for killing grass, herbaceous and broad-leaved woody weeds, for example, in coniferous plantings. However, if applications are made during active growth of conifers or at too high a concentration many species may be damaged (plate 10). For example, larch and spruce plants turn white and then brown from the tips downwards. When damage is severe, resin-bleeding from shoots and stems may occur. Glyphosate remains within the plant and symptoms can persist throughout the season following application of the herbicide. Glyphosate is also used in agriculture, for example, for killing *Agropyron repens* in winter wheat crops.

Spray drift can occur in these or other circumstances and damage to neighbouring trees, either coniferous or broadleaved, has been reported. Shoot growth is severely reduced and leaves are small, often with entire margins, thus giving the whole shoot a bunched or tufted appearance. In severe cases, or on susceptible species such as beech, the tree may be killed or completely defoliated.

Hormone weedkillers such as 2,4-D and triclopyr are sometimes used in forestry to eradicate woody weeds and some herbaceous weeds. They are also widely used in agriculture for killing herbaceous weeds in grass or cereal crops. On occasion, applications to agricultural crops can drift on to neighbouring trees and damage sometimes occurs. However, careful observation of the distribution and nature of the damage is needed before the proper reason for the damage can be established. Injury caused by hormone weedkillers is usually manifested by the cupping or crinkling of leaves, curling of needles and sinuous distortion of petioles and shoots in active growth; in more extreme cases die-back may occur. Broadleaved trees are generally more susceptible than conifers. On poplars, for example, growing shoots become twisted and leaf margins curl inwards. On pine the distortion of growing shoots is sometimes confused with symptoms caused by *Melampsora pinitorqua*, the pine twisting rust. However, the presence of fructifications of the fungus and the more sporadic distribution of the symptoms should enable the cause of injury to be identified.

### Herbicides for total eradication
Total weedkillers such as sodium chlorate, bromacil, imazapyr or a mixture of paraquat and simazine are often used to eradicate weeds on paths or fallow ground. Damage to neighbouring trees may occur if their roots extend under the treated areas or if the herbicide seeps into untreated areas. Dead roots or withered foliage may result, particularly on that side of the tree closest to the treated area.

## SALT DAMAGE

Sodium chlorate, as rock salt, is commonly used on roads and footpaths to melt ice and snow throughout Europe and North America. Increasing amounts of salt have been applied in Britain over the past three decades, especially in hard winters. For example, applications in the winter of 1979/80 were more than twice those in the appreciably more severe winter of 1962/63 (Dobson, 1990). During hard winters application of salt in Britain may exceed 3 kg/m$^2$ of road (Davison, 1971) and 5 kg/m$^2$ in some areas of Europe (Defraiteur and Schumacher, 1988). Hence, damage to trees and other vegetation may be greater in continental European countries than in Britain. Jordan (1971), cited by Fluckiger and Braun (1981), estimated that de-icing salt applications were directly responsible for the death of 700 000 trees annually in western Europe.

Roadside trees can be affected either by the presence of salt in the soil, usually introduced through run-off from treated highways, or by salt spray thrown up by passing vehicles. The most common cause of salt damage in Britain is run-off from a roadside salt dump which has inadvertently been placed close to trees.

**Mode of action**
The precise manner in which salt affects the growth of trees is not fully understood. However, it is generally accepted that high concentrations of salt in the soil water can prevent or reduce normal uptake of water and nutrients into the tree. Some plants are salt-tolerant and they may be able to absorb greater concentrations of salt than salt-sensitive species. Nevertheless, both sodium and chloride ions can be toxic to plant growth and the ability of some plants to absorb higher salt concentrations does not necessarily enable them to avoid salt damage.

In addition, the nutrient balance in the water taken into the plant can be affected. For example, the presence of high concentrations of sodium in the soil water can result in it replacing calcium, magnesium or potassium in the nutrient intake. Nutrient deficiencies may then occur in the plant.

The other method by which salt may damage a tree is through the deposition of salt on branches and foliage following drift of salt spray from neighbouring roads. These deposits may cause damage by the withdrawal of water from the leaf or needle tissues by osmosis or the salt may be directly toxic.

**Symptoms and diagnosis**
As we have seen above, salt can cause damage in several different ways and can reach a tree through soil water or through aerial spray drift. It is not surprising, therefore, that symptoms of salt damage vary widely according to circumstances. In general, trees injured by excess of salt in the soil or on the foliage show small leaves, early leaf fall and dieback of twigs and branches. In addition there may be marginal necrosis on leaves or tip dieback on coniferous needles. However, the distribution of the damage is probably one of the most important factors leading to a correct diagnosis. Damage is likely to be most severe close to the roads or drain outlets and will diminish rapidly further from the highway. On individual trees affected by spray drift symptoms are likely to be more severe on the side facing the road. In some situations where snow has covered the lower branches of roadside trees, such branches may have escaped damage from spray drift. In Britain special attention should be paid to the presence of roadside deposits of salt or salted grit. Run-off from such areas may cause serious damage to roadside trees.

Confirmation of salt injury is best achieved by the chemical analysis of soil and plant tissues. It is important to analyse both soil and foliage, as salt may not in some circumstances be absorbed into the plant. Much research has been undertaken on the level of salt needed in the plant to cause damage and the results of the research vary according to the species studied. In practice samples should be taken from affected trees of the same species growing in the vicinity. William and Moser (1975) suggested that when chloride concentrations reach about 2.7 per cent in plant tissues, visual injury was evident. Dobson (1990) on the other hand reports that London plane foliage is damaged when foliar concentrations exceed 1 per cent of the leaf dry weight.

**Prevention**
Salt damage can best be avoided by ensuring that salt run-off from roads does not occur close to trees. In order to avoid damage from salt spray it is recommended in the United States that trees should be planted at least 9 metres from the highway. However, this is not always practicable and Dobson (1990) has pointed out that there is considerable scope for reducing current rates of salt application. At present there are two rates in Britain, 10 $g/m^2$ for precautionary salting and up to 40 $g/m^2$ before or during severe cold weather. In practice there is no doubt that these recommended rates are regularly exceeded. Regular calibration of salt-spreading equipment could make significant savings in salt applications, at the same time giving a satisfactory treatment against ice and without damaging roadside trees.

There are no cheap and effective alternatives to rock salt. Where corrosion by salt is a particular concern, as on some bridges or on elevated motorway sections, urea is an alternative but it costs ten times as much as salt. Likewise, on airport runways and stands, glycols are used because of concern relating to aircraft corrosion. Another promising alternative is calcium magnesium acetate, but this is still very costly.

**Species tolerance**
In areas where it is likely that salt will be regularly applied, species tolerant to salt injury should be planted. Various lists of species indicating relative tolerance to salt injury have been published. Braun *et al.* (1978) give tolerances of a number of European tree species and Dirr (1976) gives those of some American tree species. Dobson (1990) has also summarised data on species tolerance and table 2.4 is based on these data.

## DAMAGE BY NATURAL GAS

Leaks of natural gas from underground pipes can be injurious to vegetation, including trees. However, unlike coal gas, which is toxic to plants, natural gas (methane) is not directly poisonous but acts indirectly on the soil environment so that root activity is impaired. Much information on leakage of natural gas has been obtained during periods when supplies were converted from coal gas to natural gas. Holland was one of the first countries where conversion was undertaken and serious losses of street trees occurred following conversion in some areas. Investigations showed that gas leaks were likely to be responsible and it was discovered that natural gas, with a lower moisture content than town gas, had caused the drying out of certain types of joints on old cast-iron

**Table 2.4**  Salt tolerance of some common trees

| Tolerant | Intermediate | Susceptible |
|---|---|---|
| *Pinus nigra* | *Acer pseudoplatanus* | *Carpinus betulus* |
| *Populus alba* | *Acer saccharinum* | *Crataegus crus-galli* |
| *Quercus petraea* | *Betula pendula* | *Fagus sylvatica* |
| *Robinia pseudoacacia* | *Gleditsia triacanthos* | *Rosa canina* |
| *Salix alba* | *Pinus sylvestris* | *Sorbus aucuparia* |
| *Salix fragilis* | *Platanus* × *hispanica* | |
| *Sophora japonica* | *Quercus robur* | |
| | *Tilia platyphyllos* | |

mains. Other countries which converted later were able to take note of these Dutch investigations and introduce preventative measures. Damage to trees in Britain by natural gas has therefore not been severe but nevertheless reports of such damage are received on occasion.

**Mode of action**

The exact nature of the effects of natural gas leaks on soil and vegetation is reviewed by Pankhurst (1980) and Tattar (1978). Natural gas acts on the soil in a number of different ways. For example, it replaces other gases in the soil atmosphere, including oxygen. It can also cause changes in the soil microflora and may enhance the activity of methane-oxidising bacteria. Such activity may contribute to a lack of oxygen and relative excess of carbon dioxide in the soil atmosphere. The roots of trees and other plants become deprived of oxygen and are unable to carry on normal respiration and water uptake.

**Symptoms**

Symptoms of damage to trees caused by leakage of natural gas are difficult to distinguish from symptoms of other conditions such as waterlogging and associated lack of soil oxygen. They are also similar to symptoms of drought. Leaves may show a marginal scorch or be subject to premature fall, a progressive dieback may develop, or buds may fail to flush if a gas leak has occurred during the dormant season.

It is probably important in cases of suspected gas leakage to establish that underground gas pipes are present, that the characteristic smell of natural gas has been detected or that the most severe damage occurs close to the probable or known line of the gas pipe. The extent of the damage will depend on the size of the leak or leaks and to a lesser extent on the type of soil. In sandy soils, which are more porous, gas tends to move fairly rapidly upwards in the shape of an inverted cone. On the other hand, in clay soils, which are more resistant to diffusing gases, gas will move more slowly and in the shape of a truncated cone.

Confirmation of gas leakage is best achieved by detecting the presence of methane in the soil using a gas sampling probe. These are available either commercially or through the national gas authority. However, it should be noted that there may be a few occasions when methane gas may be present in soil for other reasons, e.g. in refuse dumps incorporating decomposing vegetable material.

**Treatment**
Repair of gas leaks clearly takes priority for reasons of safety. Although the sealing of leaks will prevent further escape of gas, the surrounding soil may retain excessive methane for some time afterwards. The disturbance to the soil during the repair operations will go some way to releasing any gas held in the soil but it is possible in addition to aerate the soil by introducing compressed air through soil probes.

# 3 Diseases caused by living agents on a wide range of hosts

## Diseases of seeds; seed-borne diseases

Detailed information on diseases of seeds can be found in the standard work on seed pathology, by Neergaard (1979). A world check-list of micro-organisms associated with seeds has been prepared by Anderson (1986), and a list of cone and seed diseases of North American conifers by Sutherland *et al.* (1987). Rees and Phillips (1986) have prepared a summary note on the detection, presence and control of seed-borne pests and diseases.

Many organisms may be harboured on the surfaces of seed. Others, potentially the most damaging, because they are those that most often attack the embryo and young plants in the seedbed, occur within the seed tissues (Peno, 1983). They include many moulds and bacteria, but some virus and virus-like diseases are also carried by seeds (and also sometimes by pollen). Some of these organisms (especially moulds) interfere with routine seed testing, and are to be regarded as laboratory weeds. Others destroy the seeds, thereby reducing the value of seed samples. Others are pathogens, which may persist on or in the seeds, and be transmitted and give rise to disease in crops raised from the seed samples. Neergaard (1979) regards all organisms on seeds or within their tissues as *seed-borne*, and describes those which persist to carry infection into the subsequent crop as *seed-transmitted*.

A number of important tree disease organisms have been isolated from seed, though it has not always been demonstrated that they are seed-transmitted. Among those which cause damping-off, root rot and other diseases in nursery plants are *Pythium* spp. (Buxton *et al.*, 1962), *Fusarium* spp. (Buxton *et al.*, 1962; Vaartaja, 1964), *Rhizoctonia solani* (Kamara *et al.*, 1981) and *Pestalotiopsis funerea* (Motta and Saponaro, 1983). A fungus of particular interest is *Caloscypha fulgens* (Pers.) Boud. This was first noticed in Europe by Salt (1967), when he was investigating the effect of sowing seed in cold soils. He found that when he kept Sitka spruce seed imported from North America at 10 °C before germinating it, germination was reduced, and the reduction was associated with the presence of seed-borne *Rhizoctonia solani*, *Cylindrocarpon* sp., *Gliocladium roseum* and an unidentified endophytic fungus which, following Epners (1964), he called 'the psychrophilic seed fungus', because it caused damage only in cold conditions.

He described the fungus, naming it *Geniculodendron pyriforme* (Salt, 1974) (figure 5). Later, Paden *et al.* (1978), in Canada, found it to be the conidial stage of the orange discomycete *Caloscypha fulgens* (which so far does not appear to have been recorded in Britain). The fungus kills the seeds in the nursery beds before germination, especially when the soil temperature is at about 10 °C, which is too cool for rapid germination of the seed but warm enough for the fungus to grow. *C. pyriforme* may spread and do severe damage to infected seed given a moist stratification treatment to break dormancy, but losses can be reduced by dressing the seed with thiram or captan (Epners, 1964; Gordon *et al.*, 1976).

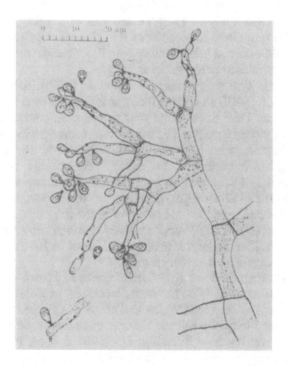

**Figure 5**   Conidiophores and spores of *Geniculodendron pyriforme* (G.A. Salt)

Among seed-borne organisms which affect more mature trees after the nursery stage are *Marssonina brunnea* (the anamorph of *Drepanopeziza punctiformis*) (Spiers and Wenham, 1983a,b), the cause of a damaging leaf spot of poplars, *Seiridium cardinale* (Motta and Saponaro, 1983), the cause of resin-bleeding and canker in *Cupressus* spp. and other members of the Cupressaceae, and *Sirococcus strobilinus* (Mitchell and Sutherland, 1986), the cause of a shoot blight of spruce and pine. Mitchell (1988) has published details of a very sensitive dot immunoassay for the detection of *S. strobilinus* on spruce seed.

Examination of seed samples generally shows that most of the associated organisms are common saprophytes, though in damp and otherwise adverse conditions many of these may become opportunist parasites which attack and damage seeds (Urosevic, 1979; Peno, 1983). When Batko (1954, 1955, 1956, 1957a, 1957b) examined samples of seed of a fairly wide range of conifers and some broadleaved trees, he found nearly fifty species of fungi. Though some were species of *Fusarium* which damage the roots of nursery plants, most were common saprophytes, including *Penicillium* spp., *Mucor* spp., *Rhizopus nigricans*, *Oedocephalum glomerulosum* and *Trichoderma viride*, many of which interfere with germination tests in the seed-testing laboratory. Similar fungi were found by Buxton *et al.* (1962).

From a detailed study of fourteen samples of known percentage germination, Batko (unpublished) concluded that there was an inverse relationship between mouldiness and germinability, and that moulds were a likely cause of the reduced germination of poor seed samples.

These findings agree with those of Garbowski (1936), Huss (1952), Gibson (1957), Schubert (1960) and Shea (1960), working elsewhere in Europe and in North America. Thus Shea found that fungal contamination of seed took place in cones stored in sacks. The longer the storage period, the greater was the contamination, and the more mouldy the seeds, the lower was their germinability. In experiments, he reduced the moulds and improved the keeping qualities of conifer seeds by fungicidal treatment of the cones. Work by Huss (1952) and Gibson (1957) pointed to the importance of mechanical damage, which rendered the seed more liable to attack by normally saprophytic moulds. Schubert (1960) reduced losses through moulding by collecting only mature seed, drying to a moisture content below 10 per cent, and storing at the lowest recommended temperature.

Whittle (1977) studied the mycoflora of cones and seeds of *Pinus sylvestris* from several sites in Britain. He made isolations at different stages in the development of the seed in the cone and at various stages as material passed through the seed extraction plant. He also examined seed stored at various temperatures and levels of humidity. He found that seed in the cones was contaminated at a very early stage, and for a long time the only contaminant was *Sclerophoma pythiophila* (the anamorph of *Sydowia polyspora*). During extraction, the seeds became contaminated also by various common 'storage moulds', such as *Penicillium* spp., *Trichoderma viride* and *Botrytis cinerea*. Over a storage period of 22 months, the incidence of *S. pythiophila* gradually fell, but that of the other moulds, especially the *Penicillium* spp., increased. Whittle's storage studies indicated, however, that the dry storage of seed of *P. sylvestris* at 2 °C, as normally practised, kept fungal contamination low enough to prevent significant damage.

Most of the work referred to above is concerned with seeds of conifers. Seed-borne organisms may also be found on seeds of broadleaved trees.

Thus *Ciboria batschiana* (*Stromatinia pseudotuberosa*), which is known to be one of the most important of the fungi that damage acorns in eastern Europe (Urosevic, 1957, 1958, 1961), is known to occur in Britain (Ramsbottom and Balfour-Brown, 1951; Rushton, 1977). In Italy, the same fungus causes black mould of sweet chestnut (Voglino, 1931). *Discula quercina*

(*Gloeosporium quercinum*) (anamorph of *Apiognomonia quercina*), again a cause of damage to acorns in Eastern Europe (Potlaychuk, 1953; Urosevic, 1957), has also been recorded on oak in Britain (Grove, 1937; Moore, 1959), but only on the leaves and shoots.

*Cryptodiaporthe castanea* (anamorph *Discella castanea*) was isolated by Wright (1960) in North America from fruits of Chinese chestnut (*Castanea mollissima*), and also occurs in Britain (Grove, 1935; Moore, 1959), but again has not been found there on fruits, but only on twigs and branches. *Monilinia fructigena* (anamorph *Monilia fructigena*) causes nut drop of hazel (Wormald, 1955).

It is also possible to isolate many bacteria from seeds of both conifers and broadleaved trees. Most of those found have been saprophytes or weak parasites, like *Bacillus subtilis*, *B. mycoides*, *Pseudomonas herbicola* and *P. fluorescens*, found by Peno (1983) on the testas of seeds of Norway spruce. The parasitic *Xanthomonas campestris* pv. *juglandis* may occur on walnut fruits (Wormald, 1930).

Finally, some virus and virus-like diseases may be seed-transmitted (and sometimes carried by pollen). Thus the elm mottle virus may be carried by seed (Cooper, 1979).

Control of seed-borne organisms must begin in the forest to ensure that contamination does not increase between seed collection and delivery of the seed to the seed processing plant. Shea (1960), who found harmful moulds spread from Douglas fir cones to the seeds, improved the keeping qualities of the seed by dipping the cones in a fungicide, and by paying careful attention to moisture conditions and temperature in storage. Huss (1952) and Gibson (1957) both refer to attack by moulds that follow mechanical damage to seed by rough handling, dewinging, etc., and this suggests that if damage of this kind is unavoidable, subsequent mould infection might be prevented by fungicidal treatment. As already noted, conifer seed infected by *Caloscypha fulgens* can be successfully treated with a fungicide. According to Uresevic (1958), losses of acorns through attack by *Ciboria batschiana* (*Stromatinia pseudotuberosa*) may be avoided by early harvesting, reduction of the moisture content of the seed to 45 per cent, and storage at a low temperature, but protecting them from frost.

Delatour and Morelet (1979) suggest the use of thiram or benomyl to protect acorns from the same fungus. Hangyal-Balul (1986) found that benomyl protected stored acorns from damage by various fungi, but the effect was not carried over into the nursery, where it was also necessary to spray seedlings with dithane or benomyl. Wells and Payne (1980) reduced fungal contamination and damage in sweet chestnut fruits by means of a hot water treatment. Spiers and Wenham (1983b) controlled infections of poplar seed by *Marssonina brunnea* and prevented subsequent seedling infections by treating the seed with benomyl, captafol, chlorothalonil, dodine, thiophanate or thiram. Beradze and Dzimistarishvili (1985) successfully controlled the rotting of hazel nuts by *Botrytis cinerea* by spraying the trees with Bordeaux mixture, zineb or benomyl. Motta (1984) greatly reduced contamination of seed of *Cupressus* and *Thuja* spp. by *Seiridium cardinale* by treatment with benomyl or thiophanate methyl.

## Nursery diseases

### Damping off and nursery root rot

Once the seed has been sown, it and the resulting seedlings and transplants may be attacked not only by some of the fungi on the seed itself, but also by others present in the soil. Three forms of damage result from these attacks (Hartley, 1921). In the first, known as *pre-emergence damping off*, the seeds and seedlings die within the soil. In the second, known as *post-emergence damping off*, attack is seen in the first few weeks after the plants appear above ground, and occurs on the succulent roots and hypocotyls of the growing plants. The seedlings, mainly those in the cotyledon stage, tend to wilt, and collapse at ground level, where the tissues are brown and shrivelled, and often show extensive rotting of the roots. This is typical damping off as commonly understood. As a result of pre- and post-emergence damping off, the crop may be sparse and poor, with scattered dead plants, or increasingly large bare patches may appear in the seedbed. At a later stage, a third form of damage may occur, when older, more woody seedlings and transplants do not collapse and show symptoms of damping off, but as a result of *root rot* may be stunted and sometimes killed (Warcup, 1952; Buxton *et al.*, 1962). Hence, damping off and succeeding root rot may reduce the numbers and size of the plants.

Perrin (1986), in a succinct review of some aspects of damping off in forest nurseries, points out that the commonest organisms found to cause damping off and nursery root rot are species of *Pythium*, *Fusarium* (especially *F. oxysporum*) and *Rhizoctonia solani*. The *Pythium* spp. tend to be most important in northern Europe, the *Fusarium* spp. in the south. Other fungi may also be involved, however, especially *Cylindrocarpon destructans* (the anamorph of *Nectria radicicola*) and species of *Alternaria* (Petaisto, 1982; Schönhar, 1984; Karoles, 1985; Lilja, 1986). The relative importance of these fungi varies from season to season and from nursery to nursery, depending on many factors which are still by no means fully understood.

Tree seedlings of all kinds may be affected, but conifers are more often and more severely damaged than broadleaved trees, and among the conifers Griffin (1965) found damping off in Scots, Corsican and lodgepole pines, Sitka spruce, western hemlock, Douglas fir, Japanese larch and beech, but members of the Cupressaceae, including western red cedar and Lawson cypress, appeared to be resistant. These findings agree with those of Hartley (1921), and accord with general European experience that pines, spruces and larches are especially susceptible and that members of the Cupressaceae tend to be resistant (Perrin, 1986). Nevertheless, substantial damage to seedlings of broadleaved trees may sometimes be experienced. Thus Hangyal-Balul

(1983) reported an attack on *Robinia* seedlings in Hungary, while Kessler (1988) described root rot of oak and beech plants caused by *Cylindrocarpon destructans* in German forest nurseries. Damage occurred when winter temperatures were exceptionally low.

Much of the work done on damping off in Britain was reviewed by Griffin (1965), who found that in Sitka spruce (on which most of the studies described have been carried out) in unfavourable circumstances damping off may cause losses of up to 50 per cent of the plants, though they are more often of the order of 5 per cent or less. In France, very high losses have also sometimes been experienced (Perrin, 1986).

Species of *Pythium* appear to be the main cause of damping off in British nurseries (Warcup, 1952; Griffin, 1955; Buxton *et al.*, 1962; Salt, 1964, 1965; Redfern and Low, 1972). Buxton *et al.* (1962) isolated *Pythium* spp. mainly from damped-off plants, and found them common on roots in the early part of the season. They began to disappear in July, and had gone by October. Salt (1964, 1965) found *Pythium* spp. early in the season, and in some experiments they killed all seedlings before emergence. The most important species are *P. ultimum* and *P. debaryanum* (Warcup, 1952), but *P. irregulare* and *P. mammilatum* may also cause damage on acid soils (Ram Reddy *et al.*, 1964). These fungi are commonly present in nursery soils, but have rarely if ever been isolated from seed, though Buxton *et al.* (1962) obtained indirect evidence that they might be seed-borne, but were not revealed by normal isolation methods.

The causes of the later root damage and stunting are not fully understood, and appear to be complex. Griffin (1965) found three types of root damage. In one he found extensive root injury on immature tissues from which he was able to isolate pathogenic *Pythium* spp., and above which new healthy lateral roots were sometimes produced. In another only the root tips were affected, and no pathogens could be isolated. These types occurred on heavy soils. On light soils in one nursery, the plants showed an extensive browning of the root cortex, only a small white root tip being left unaffected. Here the damage appeared to be caused by ectoparasitic eelworms. He considered that much of the damage on the heavy soils was due to adverse physical conditions. Experiments made by Vamos (1954) in Hungary support this view. Buxton *et al.* (1962) also studied plants showing leaf browning and root rot, and others showing only stunting. They found that the browning, seen mainly in July and August, was caused mostly by *Fusarium* spp., especially *F. oxysporum*. The leaf tips first became yellow, and eventually the whole plant turned brown, while the root cortex became brown and eventually black; the cortex finally sloughed off, leaving the vascular cylinder, which then also turned brown, sometimes with bright red patches. In those cases in which only stunting occurred, the damage became progressively worse as the season advanced, though the affected plants remained green and their roots usually showed no lesions. It appeared that some isolates of *F. oxysporum* and *F. roseum* could cause stunting without other symptoms, but the cause of much of this stunting was obscure, and seemed to be unrelated to earlier damping off or browning.

Similarly, Ram Reddy *et al.* (1964) found evidence that the cause of the stunting might be different in different nurseries. Salt (1965) carried out pathogenicity tests with isolates of *F. oxysporum, F. solani, F. sambucinum* and *Cylindrocarpon radicicola*, using seedlings in pots, and found these fungi capable of killing many of the plants after emergence.

Observations suggest that damping off is more common in Britain in heavy than in lighter soils. It is also much more frequent in old nurseries than in more newly established ones (Griffin, 1956, 1957), and is more commonly found in relatively alkaline nurseries than in those with a more acid soil reaction. Thus Warcup (1952, 1953) found that *Pythium ultimum*, one of the most important damping-off fungi, was abundant in soil from three alkaline nurseries whose soil reaction ranged from 6.8 to 7.7, but rare in or absent from soil with a pH of 5.3 to 5.5 from two acid ones. Similarly, Griffin (1955, 1956, 1957) found damping off in the more alkaline of the nurseries he studied, but not in two with acid soils of pH 4. In experiments with *Pythium ultimum*, he found acid heathland soils did not favour the development of damping off. Ram Reddy *et al.* (1964), however, found damping off associated with *P. irregulare* and *P. mammilatum* in one acid nursery. More widely in Europe, Perrin and Sampangi (1986) found that *Pythium* caused damage mainly on neutral or alkaline soils, *Fusarium* on acid ones, while attack by *Rhizoctonia solani* was independent of soil type.

From the above, it will be seen that damping off is one of a complex of disorders, sometimes grouped together as 'soil sickness', and grading also into seedling and transplant root rot. Its control is dependent on individual circumstances, but its cause is not always easy to diagnose in practice. The main organisms that cause the disease are almost always present in soils, and specific fungicides effective against one fungus have relatively little effect on others. In conditions favourable to infection the disease progresses so rapidly that by the time the fungus involved has been identified it is too late to make a useful application of such specifics. Perrin (1986) also cites a case in which the application of prothiocarb, which is active against *Pythium* spp., increased damping off because it altered the microbial balance of the soil in favour of *Fusarium*.

In general, in Britain, nurseries should not be sited on heavy and alkaline soils, and alkaline grits should not be used as a covering for seedbeds. In nurseries in which damping off by fungi makes its appearance, steps may be taken against soil-borne organisms.

As an immediate step in affected beds, an emergency drench with captan may be used (Carter and Gibbs, 1989), though the makers' instructions must be carefully followed, as at higher dosages the plants may suffer chemical damage (Kozlowski, 1986).

Some other chemicals have also given good experimental results. Perrin (1986) reported very good control using a mixture of fungicides, including benomyl, iprodione and prothiocarb (individually active respectively against *Fusarium* spp., *Rhizoctonia solani* and *Pythium* spp.). Duda and Sierota (1987) obtained satisfactory biological control by applying spores of the antagonistic fungus *Trichoderma viride* in sawdust 20 days before sowing Scots pine seed. Perrin (1986) has advocated the further development of biological control measures, pointing to work on antagonistic soils and on mycorrhiza. Some forest soils contain antagonistic fungi which suppress *Pythium* spp. (Bouhot and Perrin, 1980), while others similarly suppress *Fusarium* spp. Samples of such soils have been used to inoculate trial nursery beds with promising results. Certain mycorrhizal fungi have also been found to render the rhizosphere unfavourable to *Pythium* or *Fusarium* spp., and use might be made of this effect (Perrin, 1986).

Before resowing or replanting infested beds, further steps may be needed to protect the ensuing crop by means of seed dressings, or by soil fungicides or fumigants. Warcup (1951, 1952) and Faulkner and Holmes (1954) obtained excellent results with soil steaming, but this is expensive and difficult to use in the field. Therefore, most work has been concentrated on chemicals used on seed, or worked or watered into the soil.

Of the chemicals for use in the soil, those with a strong fumigant action have been found to be the most effective (Salt, 1964). Formaldehyde and chloropicrin have given good results, and so have methyl bromide (Perrin, 1986) and methyl isothiocyanate (Lopez, 1986). Carter and Gibbs (1989) recommend the application of dazomet, which is worked into the soil, and is easier and safer to use than many of the other materials. In Hungary, Hangyalne and Toth (1987) found that they could obtain excellent control of damping off in seedlings of *Robinia pseudoacacia* by dressing seeds with propamocarb before storage over the winter in plastic bags. The soil in which the seeds were then sown in the following spring was disinfected every second or third year. They also recommended that *Robinia* cuttings should be treated with a mixture of benomyl and mancozeb.

### Seedling blight and minor cankers and diebacks associated with *Diaporthe eres* Nitschke

*Diaporthe eres* is a large complex of host-specific species, the most important members of which are listed by Cannon *et al.* (1985). These fungi are commonly saprophytic, but as weak though sometimes very damaging parasites their anamorphs (which are species of *Phomopsis*) may affect a wide range of mainly woody hosts, including orchard, forest and ornamental trees, various shrubs, and some herbaceous plants. The perithecia of *D. eres* are embedded in the host tissues, their black necks protruding through the blackened host surface. The colourless, bicellular ascospores measure 10–14 × 2.4–5 μm (Ellis and Ellis, 1985). Species within this complex particularly associated with forest and ornamental trees include *D. conorum* (Desm.) Niessl, which causes a seedling blight of spruce (and sometimes of other conifers), and which is more fully described under its anamorph, *Phomopsis occulta* (Sacc.) Traverso, in chapter 4, and *D. scobina* Nitschke and *D. controversa* (Desm.) Nitschke ex Fuckel, both of which are associated with usually minor cankers and diebacks of ash, and which are discussed further under their anamorphs, *P. scobina* Höhnel and *P. controversa* Traverso, in chapter 12. The form on orchard and ornamental apples, pears and cherries is *D. perniciosa* Marchal (anamorph *P. mali* Roberts).

**Grey mould: *Botrytis cinerea* Pers. (anamorph of *Botryotinia fuckeliana* (de Bary) Whetzel, syn. *Sclerotinia fuckeliana* (de Bary) Fuckel)**
The grey mould *Botrytis cinerea* has a very wide host range, occurring as both saprophyte and parasite on almost every plant of temperate regions, and causing severe damage to many horticultural crops. It may be found on trees in both the forest and the nursery, and in the cold store. A general account of the disease on forest trees is given by Pawsey (1964a), and much of the literature on *B. cinerea* and some other species of *Botrytis* which may perhaps sometimes affect forest trees has been reviewed by Coley-Smith *et al.* (1980) and Mittal *et al.* (1987).

*The causal fungus*
On woody hosts the fungus produces a fine web of greyish or brownish mould (plate 4) with dense tufts of erect, branched conidiophores. The branches of the latter have inflated ends with minute sterigmata, each bearing subglobose or ovoid spores about 8–14 × 6–9 μm (Ellis and Ellis, 1985). The fungus also produces round or irregular black sclerotia about 1–2 mm across. These are usually partly embedded in the host tissues. In Britain these sclerotia usually germinate to produce tufts of conidiophores, but elsewhere some strains have given rise to the stalked apothecia of *Botryotinia fuckeliana* (Cannon *et al.*, 1985). In Britain these apothecia have been found on cupules of sweet chestnut fruits and dead leaves of oak and willow, and their colourless ascospores measure 8–10 × 4–5 μm (Ellis and Ellis, 1985). *B. cinerea* is a facultative parasite that readily colonises dead plant remains, and thence spreads to living plants. On these it often grows first as a saprophyte on dead leaves or lesions on stems caused, for example, by frost, by other fungi or similar agencies, and then in favourable conditions spreads to the living parts as an active parasite.

    *B. cinerea* is able to grow over a very wide range of temperatures. In studies on a strain isolated from snow-smothered nursery stock, Hartley *et al.* (1919) found the fungus grew at 0 °C. A strain from *Cryptomeria japonica*, whose seedlings in Japanese nurseries are liable to mould damage under snow, grew at 1–31 °C, but not at 37 °C, and its optimum temperature was between 20 and 25 °C (Ito and Hosaka, 1951). Sato *et al.* (1959) found that the conidia germinated between 0 and 25 °C, with an optimum between 15 and 20 °C. The mycelium survived several days exposed to −7 °C, and conidia were produced under snow. It is this ability to grow at low temperatures that enables the fungus to damage plants under snow cover and in cold stores (Sato *et al.*, 1959; Haas and Wennemuth, 1962).

*Losses caused*

Occasionally *B. cinerea* causes a dieback of trees in plantations, affecting the young developing twigs (Zycha, 1961/2).

In Britain, damage to trees is mainly confined to plants in nursery seedbeds, and even in these, attacks are sporadic. Hence, the fungus rather rarely causes heavy losses, though it does give trouble in some years in a few nurseries, and Peace (1962) mentions that on occasions it has destroyed half the seedlings of Sitka spruce in some beds, and almost all those of very susceptible species such as *Sequoia* and *Cupressus*. Similar losses may occur elsewhere in Europe, where, for example, Neergaard (1957) reported severe damage in forest nurseries in Denmark, and Lilya (1986) records considerable losses in dense pine seedbeds in Finland.

*Important tree hosts*

All the common conifers are known to be susceptible, though some are more often affected than others. There is no doubt that seedlings of *Sequoia* are very liable to damage, and in Britain more cases are recorded on this genus than on any other, in spite of the fact that it is little grown. The closely related *Sequoiadendron* is also susceptible. Many attacks are also recorded on Sitka spruce, though this is probably rather because of the large nursery area under this species than because of its absolute susceptibility. Damage has also occurred in beds of Norway spruce, Japanese larch, western hemlock, Douglas fir, Scots, Corsican and lodgepole pines, grand fir, *Cupressus* spp. and *Cryptomeria japonica*. Seedlings of hardwoods may also be attacked, though less often than those of conifers.

These observations, chiefly from experience in Britain, agree broadly with the data, partly from Europe but also from North America and Asia, collected together by Mittal *et al.* (1987).

*Effect of age of host*

Most of the damage affects young nursery stock, particularly in the summer, autumn and winter of the first year, after which the plants become more woody and less susceptible. Transplants are seldom affected. Thus Jamalainen (1961), working in Finland, found it easier to infect 1-year-old than 2-year-old pine and spruce plants. Similarly, Sato *et al.* (1959) in Japan found 1-year-old plants of 'Sugi' (*Cryptomeria japonica*) more susceptible than older ones.

*Symptoms of disease*

The symptoms on conifer seedlings commonly show as a top dieback, with a yellowing and later browning of the needles, which become covered with mycelium and conidiophores. Some of the leaves may fall, but many remain hanging on the shoots, and the black sclerotia of the fungus may develop on them. The shoots become twisted, flaccid and downward bent. Because of the loss of their tops, affected plants become bushy through the growth of side shoots from below. In some species, especially in western hemlock, and sometimes also in Japanese larch and lodgepole pine (Peace, 1962), attack may be lower down, the fungus in the seedbed spreading from the dead leaves killed by mutual shading of the plants.

On seedlings of broadleaved trees, necrotic patches occur on the leaves, and the tops of the plants wither. The characteristic grey mould develops on the damaged areas. On larger plants, dieback of shoots may take place.

When the disease affects forest trees in plantations, it attacks the young shoot tips, or if it causes lesions lower down, it may girdle the young shoots, and cause secondary wilting and the death of their ends (Zycha, 1961/2). Such symptoms are not easy to separate from those caused by late frosts, particularly as shoot tips damaged by frost are often later colonised by *B. cinerea* in damp weather. Dieback of this kind is unusual in Britain. Affected trees usually soon recover, and in a few years show no further signs of damage.

### Factors affecting the disease

There is little firmly based information on conditions favouring infection and spread in forest crops, though factors usually considered important are frost damage, wet weather or generally high humidity, density of the stand, and what are vaguely called 'adverse conditions'. Baker (1946) established that *Botrytis* spores do not germinate at a relative humidity below 93 per cent, though in studies in Britain in a number of nurseries, Murray (1962a) was unable to find any correlation between outbreaks of grey mould in seedbeds and humidity figures in the nurseries concerned. Cases of grey mould in British nurseries reach their peak in warm, humid spells in July and August, however, and a second, smaller peak occurs in October, when damage by autumn frosts tends to affect unhardened plants.

Murray, in many of his experiments with fungicides, found it difficult to induce infection, and so did Halber (1963), who worked on Douglas fir in Oregon. Halber reached the opinion that attack took place only within a relatively narrow range of conditions, and that unseasonable freezing weather, high humidity and density of stand all contributed to infection.

Sato *et al.* (1959) found evidence that a deficiency of phosphorus and potash may render *Cryptomeria* seedlings under snow more susceptible to attack, and that this effect is enhanced if nitrogen is increased. They noted that autumn transplants were susceptible because they lost most of their root systems. On the other hand, in experiments they found that root pruning seemed to impart a resistance by controlling excessive late growth. Root damage as a possible predisposing factor was also suggested by Hartley *et al.* (1919).

## Control

Attempts have been made to control the disease by means of fungicides, and some of these have been found to give control if applied early enough. Jamalainen (1961), in Finland, found quintozene effective on pine and spruce seedlings, and zineb less so. In Britain, some trials have been carried out with both quintozene and tecnazene, but neither gave any control. Magnani (1963a), in experiments with several species of *Eucalyptus*, obtained control with captan, which was also found effective by James and Woo (1984) (on western larch) and Alonso *et al.* (1985). Murray (unpublished) tested various fungicides in tree nurseries in Britain over several years, and obtained his best results with thiram. James and Woo, in the trials cited above, also obtained good results with chlorothalonil and iprodione. In attempts to control grey mould on hazel nuts, Beradze and Dzimistarishvili (1985) found that Bordeaux mixture gave the best results, followed by zineb and benomyl. Alonso *et al.* (1985) found that dichlofluanid also gave good results in their tests. In their general table of fungicides effective against grey mould, Brooks *et al.* (1989) list benomyl, carbendazim, chlorothalonil, diclofluanid, iprodione, thiophanate methyl, thiram and vinclozalin. The last of these they recommend to be used only after a pilot trial; in their experiments with this material, James and Woo (1984) found it proved phytotoxic to containerised western larch seedlings (though not to 2+0 bare root seedlings). Carter and Gibbs (1989) recommend spraying forest nursery crops as soon as the disease is seen, using either thiram, captan or benomyl.

Turning to chemicals other than fungicides, Japanese workers (Sato *et al.*, 1955, 1959) controlled root growth if they treated *Cryptomeria* seedlings with the growth regulator maleic hydrazide, and found they thereby increased resistance to *B. cinerea*.

Apart from the application of chemicals (which may be necessary as a routine in late summer and autumn on very susceptible genera such as *Sequoia* and *Cupressus*), some other steps can be taken. On those nurseries in which the disease is known to be troublesome, sowing densities may be reduced to produce a more open stand, and special care should be taken to protect the plants from autumn frosts. Only the minimum shading necessary to do this should be applied, however, or the humidity of the beds may be raised, and conditions made very favourable for the fungus. If any signs of the disease are seen, the beds should be sprayed as suggested above.

## Diseases caused by *Pestalotiopsis funerea* (Desm.) Steyaert and *Pestalotia hartigii* Tubeuf (*Truncatella hartigii* (Speg.) Steyaert)

In Europe and elsewhere, *Pestalotiopsis funerea* and *Pestalotia hartigii* have both been associated with disease in trees.

*P. funerea* is common on moribund leaves and twigs of many trees. It occurs especially on conifers, on which it may cause a usually minor needle blight and canker, but in France it has been found to be one of the causes of a severe leaf browning of various members of the *Cupressaceae* (Morelet, 1982). Similarly, in Germany, Urbasch (1989) found the same fungus on brown dead tips of twigs, and on ripe cones and seeds of several species of *Thuja*. It also appeared on seedlings which he raised from infected seeds. In Britain it is sometimes associated with *Phomopsis juniperovora* as a cause of shoot blight and canker of juniper plants (Brooks *et al.*, 1989) (described in chapter 9). Outside Europe, in India it produces a leaf spot of *Eucalyptus globulus* (Upadhyang, 1986). Its spores are formed in small black acervuli. They are five-celled, the three inner cells brown, the end cells colourless and prolonged into appendages, of which the apical cell bears three or four (sometimes up to six), and the basal cell one (figure 6). These conidia measure 25–33 × 8–13 μm (Ellis and Ellis, 1985). To control attacks by *P. funerea* on juniper plants, Brooks *et al.* (1989) recommend the application of a prochloraz–manganese complex just before or after the cutting over of nursery plants. If damp or wet weather follows, further applications may be needed every few weeks.

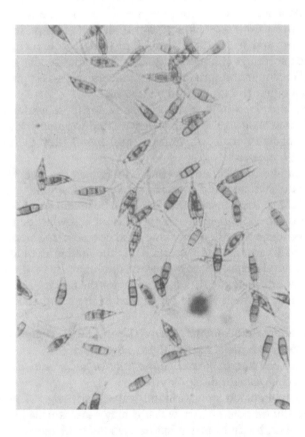

**Figure 6**  Spores of *Pestalotiopsis funerea* from a culture on malt agar (Forestry Commission)

*P. hartigii* appears to be a weak wound parasite. It has been found on conifer seedlings (chiefly on spruce, but sometimes on other conifers) with the so-called strangling disease. Its spores are four-celled, the two middle cells brown, the end cells colourless, the apical cell prolonged into one to four appendages or into one branching appendage (figure 7). They measure 18–20 × 6 μm (Guba, 1961). Affected seedlings are typically girdled at the base, with a swelling of the tissues just above the girdled zone (figure 8). Black pustules bearing the spores of *P. hartigii* are present on the diseased tissues. The affected plants yellow and die (Hartig, 1894). The cause of strangling disease is by no means certain, as many inoculation experiments have shown only that the fungus may sometimes in adverse conditions be induced to attack wounded plants. The disease is of only minor importance.

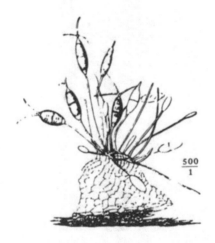

**Figure 7**   Spores of *Pestalotia hartigii* (K.F. Tubeuf)

**Smother fungus: *Thelephora terrestris* (Ehr.) Fr.**
The basidiomycete *Thelephora terrestris* is common on the soil surface in forests, and may often be found in nurseries on heathy land. It produces clusters of chocolate-brown, fan-shaped fruit bodies, the tops of which are covered by radiating fibrils (plate 13). Its dark brown, angular, warty spores measure 8–9 × 6–7.5 μm (Wakefield and Dennis, 1981).

The fungus is not parasitic, but in nurseries it may grow so vigorously around the young plants that they are completely enveloped, smothered and killed (Hartig, 1894; Cunningham, 1957). Older plants may be so closely fastened together by its growth that it is difficult to separate them without causing damage (Peace, 1962). It rarely causes significant losses, and no steps for its control have been elaborated. Soil disturbance and removal of fruit bodies when weeding are likely to discourage its growth.

**Figure 8** Strangling disease on Norway spruce: young spruce infected near ground level by *Pestalotia hartigii*, and showing the characteristic stem swelling above the girdled zone (K.F. Tubeuf)

## Post-nursery diseases

### DISEASES OF LEAVES AND SHOOTS

### The powdery mildew *Phyllactinia guttata* (Wallr.: Fr.) Lév. (*Phyllactinia corylea* (Pers.) Karst.)

The powdery mildew *Phyllactinia guttata* has a very wide host range, affecting many trees, including species of *Acer, Betula, Buxus, Carpinus, Castanea, Corylus, Fagus, Fraxinus, Quercus* and *Salix*. It is most important on *Corylus* (hazel), under which it is described in chapter 19.

## Leaf spots and shoot diebacks caused by *Apiognomonia errabunda* (Roberge) Höhnel (*Gnomonia errabunda* (Roberge) Auersw.)

A number of species of *Gnomonia* which cause leaf spots and sometimes shoot diebacks in various broadleaved trees were transferred to *Apiognomonia* by Barr (1978). They include *A. errabunda sensu stricto*, *A. quercina*, and others. Monod (1983) includes most of them under *A. errabunda* (but retains *A. veneta*, which causes anthracnose of plane trees, as a separate species). *A. errabunda sensu stricto* causes leaf spots of beech, under which it is described in chapter 11, and *A. quercina* causes leaf spots and dieback of oak, under which it is described in chapter 10.


## DISEASES OF SHOOTS AND STEMS

### Diseases caused by *Nectria cinnabarina* and *N. galligena*

Some species of *Nectria* have little or no pathogenic significance. Some, like *N. ditissima*, almost confined to *Fagus*, and *N. desmazierii* on *Buxus* attack only one or very few hosts, under which they are considered here. *N. coccinea* has a rather wide host range, but is most important on *Fagus* (under which it is therefore described) as the cause of beech bark disease. It also causes cankers on *Acer* and *Populus*, and has been found on *Alnus*, *Carpinus*, *Ilex*, *Morus*, *Quercus*, *Rhamnus*, *Sambucus*, *Taxus* and *Ulmus* (Booth, 1959). Two species of *Nectria*, however, are particularly well known as the causes of disease on a wide range of host trees. The first of these is *N. cinnabarina*, the coral spot fungus, which causes a dieback on many woody plants. The second is *N. galligena*, which causes canker of apples and pears, and a somewhat similar canker of many broadleaved forest trees. These and other species of *Nectria* have been discussed and described by Booth (1959), and Perrin (1976) has published a key to the European species of the genus.

### Coral spot caused by *Nectria cinnabarina* (Tode ex Fr.) Fr. (anamorph *Tubercularia vulgaris* Tode)

*Nectria cinnabarina* is a saprophytic fungus of world-wide distribution that sometimes invades living tissues as a facultative parasite. It then causes a dieback and bark necrosis in a large number of woody hosts.

It produces numerous pink, cushion-shaped conidial pustules, usually up to about 1 mm across, that burst through the host bark on dead and dying branches or main stems. The surface of each pustule becomes covered by a pink layer of elliptical conidia that each measure 5–7 × 2–3 μm. Later, usually in the spring, more or less globular red perithecia up to about 0.5 mm across arise either round the edges of the conidial pustules or on separate stromata (plate 14). The perithecia contain cylindrical to clavate eight-spored asci with thin apices. The two-celled, colourless ascospores measure 12–20 × 4.5–6.5 μm (Booth, 1959).

Optimum temperature for the growth of the fungus on agar is about 22.5 °C (van Vloten, 1943). The conidia need abundant moisture for germination, and the optimum temperature for germination is 18–20 °C (Mangin, 1894). Jorgensen (1952) and Booth (1959) found little evidence of host specialisation, though isolates vary in virulence (van Vloten, 1943; Uri, 1948; Schipper and Heybroek, 1957).

*N. cinnabarina* may attack by spreading from dead to living tissues, often passing in this way from small branches that have died back (Anon., 1969). It also commonly enters through wounds, including those caused by pruning, storm damage, frost and perhaps by insects (Jorgensen, 1952). Its spores may be carried on pruning tools (Uri, 1948).

It may cause dieback by growing down from wounds caused by pruning or hedge clipping, or upwards from wounds lower down. If growth is downwards from a dead stub into a larger branch or main stem, an eccentric dead area of bark develops round the stub base. Fruit bodies arise in large numbers on the dead bark. Inside affected plants, the wood may turn brown, or green (in *Acer* spp.), or violet (in *Fraxinus* spp.) (Jorgensen, 1952). *N. cinnabarina* may occur on almost any broadleaved woody host, and Jorgensen (1952) lists over 100 species on which it has been found in Denmark. On trees in Britain it is common especially on elms and *Acer* spp., and it is found fairly frequently also on beech hedges and on lime and horse chestnut trees. In North America it causes a canker of *Gleditsia triacanthos*, the honey locust (Bedker and Blanchette, 1984a,b). The same authors (1983) found that if the roots of the trees were pruned, the size of the cankers was increased, and they took longer to heal.

Within tree genera, species differ in their susceptibility. Thus Uri (1948) found sycamore very susceptible, field maple less so, and *Acer negundo* immune to attack. In Holland, among the elms, the cultivar Christine Buisman, bred for resistance to Dutch elm disease (*Ophiostoma ulmi*), was found to be severely damaged by coral spot (Schipper and Heybroek, 1957). In inoculation trials with *Gleditsia triacanthos* var. *inermis*, Bedker and Blanchette (1984b) found that the cultivar Thornless was more resistant than the cvs Sunburst, Skyline and Imperial.

*Control*

If control of coral spot in ornamental trees is necessary, the basic measures include the cutting out and destruction of diseased tissues, leaving no stubs open to infection. This treatment should be done only in dry weather, and Riecken (1985) suggests that it is best done in late summer rather than in winter or spring. Because of the effect of root pruning on canker development, Bedker and Blanchette (1984a) recommend that when *Gleditsia* trees are moved, as much as possible of the root ball should be dug up with the plant, to reduce transplanting check to the root system. They also indicate other cultural measures, such as attention to watering and the supply of fertilisers.

Pruning wounds may be treated with a dressing to prevent colonisation by *N. cinnabarina*. In the past such dressings have often given disappointing results, though Lonsdale (1986, 1987) reports that a sealant containing thiophanate methyl and the inoculation of wounds with spores of a species of the antagonistic fungus *Trichoderma* have shown promise.

In recent years a good many fungicides have been tested against *N. cinnabarina*, either as sprays for use on wounds or as general sprays on hedges and young trees. Bedker and Blanchette (1984b) found that benomyl could be used to protect *Gleditsia* trees from canker. Riecken (1986) in trials on dieback of sycamore, *Tilia tomentosa* and *Robinia pseudoacacia* also found that benomyl was the most effective of the chemicals tested, while other materials which reduced damage were captafol, fenarimol, nuarimol, pro-chloraz, propiconazole, thiabendazole and thiram.

## Cankers on broadleaved trees caused by *Nectria galligena* Bres.
## (anamorph *Cylindrocarpon heteronemum* (Berk. and Broome) Wollenw., syn. *C. mali* (Allescher) Wollenw.)

*Nectria galligena* is known in orchards in western Europe, parts of Asia, South Africa, North America, Chile, Argentina and Uruguay, New Zealand, and perhaps Tasmania, as a cause of cankers of apples and pears (CMI Map No. 38). Apple canker has been reviewed by Swinburne (1975). To the forester, particularly in western Europe and parts of eastern North America, it is also familiar as the cause of a canker of many broadleaved trees.

The bright red, ovate or globose perithecia of the fungus measure 250–350 $\mu$m across, and are scattered or in small or large groups. Their club-shaped, thin-walled asci measure 75–95 × 12–15 $\mu$m, and each contains 8 ascospores. These are colourless and two-celled, varying in shape from oval or ellipsoid to spindle-shaped (figure 9). They measure 14–22 × 6–9 $\mu$m. The conidia of the anamorphic *Cylindrocarpon* stage form on small white pustules, and are cylindrical, slightly curved and narrow at each end (figure 9). They measure 55–65 × 5–6 $\mu$m (Booth, 1959).

*N. galligena* is active at very low temperatures. Small numbers of asco-spores may be released and sometimes germinate at air temperatures of about 0 °C, though the optimum temperature for release is between 21 and 26.5 °C (Lortie and Kuntz, 1963; Lortie, 1964). Vegetative mycelium grows slowly at 2 °C (Munson, 1939). Different strains of the fungus vary in pathogenicity, but all have a wide host range, and most workers have found little evidence of host specificity (Lohman and Watson, 1943). In Northern Ireland, however, Flack and Swinburne (1977) found that the fungus from ash cankers had a somewhat different host range from that from typical cankers on apple and pear, though both forms had a wide host range with many hosts in common.

*Symptoms*
Characteristically, the fungus causes concentric, target cankers on stems and branches. The cankers often form round branch stubs, though they may also develop around other entry points open to the fungus. Their concentric form is due to their periodic growth; the fungus grows and kills the bark in the autumn and winter, but is overgrown by host callus in spring and summer.

The symptoms show first as dark reddish-brown or black water-soaked

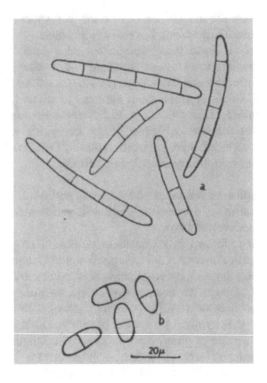

**Figure 9** (a) Conidia and (b) ascospores of *Nectria galligena* (H. Wormald)

spots around the point of entry. The canker develops to become a tumour-like swelling, and the surface bark then cracks and is then usually shed. According to Zeller and Owens (1921), however, cankers on apple and pear in Oregon and California in the USA almost always remain closed.

The mature target-like cankers are sunken and become oval or elongated up and down the stem for as much as several feet. Laterally their size is limited by the diameter of the host branch or stem, though the host tissues swell at the point of canker formation, probably because *N. galligena* produces growth-promoting substances (Lortie, 1964).

The fungus may also cause an eye rot of the fruit of apples and pears; it finally grows from the eye into the whole fruit, which eventually becomes mummified (Wormald, 1955).

Small conidial pustules, and, later, small groups of red perithecia, form on the edges of the cankers and on affected fruits.

*Spore production and host infection*
As noted above, *N. galligena* is active at low temperatures. Its ascospores may be dispersed all through the year, and various workers have found large differences in the times of maximum discharge. Thus Lortie and Kuntz (1962, 1963) found that on hardwoods in Quebec the conidia were produced in

greatest numbers in May and June, and production then quickly fell off to a minimum in August. They found that the conidial stromata later produced perithecia. These then shed ascospores at any time when the temperature rose to 0 °C or more, though the peak discharge was in August and September. The peak period in France and Germany is at about the same time, in summer and autumn (Swinburne, 1975). In Britain, Munson(1939) found that ascospore discharge reached a maximum in February, and fell until September, and it then rose rapidly in October and November, whereas Cayley (1921) found a maximum in spring. Swinburne (1975), in Northern Ireland, found a maximum in spring and early summer, and a second minor peak in the autumn. It is likely that the time of greatest discharge is much affected by local climate and weather.

The fungus may enter and establish itself through any opening that allows access to the living inner bark or cambium (Ashcroft, 1934; Brandt, 1964). Entry may be through lenticels and leaf scars, round branch stubs and dead buds, and through wounds caused by insects, weather damage by sun, frost, snow and hail, and by pruning (Marsh, 1939; Brandt, 1964). *N. galligena* may also enter through openings caused by other fungi (Zeller and Owens, 1921), notably the scab fungus, *Venturia inaequalis*. Wiltshire (1921), in Britain, found that the fungus entered mainly through small cracks that appeared in the leaf scars in autumn and again in the following spring.

It was at this time, too, in October and March, that Marsh (1939) found apples in Britain were most readily infected in his inoculation experiments. Winter inoculations were much less effective. Lortie (1969), who studied the canker on hardwoods in Quebec, similarly found that inoculations made from December to March were less successful than those at other times of the year. On apples in Northern Ireland, Swinburne *et al.* (1975) found that most infections took place between April and August.

### Host range and susceptibility
In apple and pear, cultivars vary in their susceptibility to canker. Thus among the well-known eating and cooking apples in Britain, Cox's Orange Pippin, James Grieve and Worcester Pearmain are very susceptible, as are Crispin and Spartan (Alford and Locke, 1989), but Bramley's Seedling and Lane's Prince Albert are more resistant. Cider apples also vary, and, among these, Ellis Bitter, Royal Wilding and Silver Cup are highly resistant. Among the pears, Fertility and Marie Louise are very susceptible (Wormald, 1955). Ornamental crab apples may also be attacked.

In France, Bulit (1957) found that several local and Canadian varieties were attacked, other varieties such as Golden Delicious were less affected, and a few others, including Boskoop and La Nationale, were resistant.

In Chile, Vergara Castillo (1953) found resistance in some apple varieties, including White Water Pearmain, and the wild apple Hoover, which, however, can be used only as stock.

Such information must be used with caution, as varieties apparently resistant in one area may be found susceptible in another (Swinburne, 1975).

Information on broadleaved forest trees is less detailed, comparisons being mainly between genera and species, especially from studies in the USA and Canada. Almost all broadleaved trees may be attacked. Those most affected in North America are maples, birches, American beech, poplars and oaks (Ashcroft, 1934; Brandt, 1964; Lortie, 1969). Willows may also be attacked (Zalesky, 1968), while hickory, ash, elm and black cherry are less affected (Grant, 1937; Brandt 1964). In Europe, *Nectria* cankers have been described on poplar, hawthorn, beech, ash and willow (Liese, 1931; Mooi, 1948; Booth, 1959; Flack and Swinburne, 1977).

*Factors affecting the disease*
Factors affecting the incidence of canker are many, and they are not fully understood.

In apple and pear, infection and development are encouraged by damp weather, impeded drainage, the presence of leaf scars, pruning wounds and wounds made by ice and snow, over-close planting, infestation by insects such as woolly aphids, and damage by other fungi, especially scab (*Venturia inaequalis*) (Vergara Castillo, 1953; Swinburne, 1975). Broadly similar factors affect the disease in broadleaved forest stands. In New England, Grant (1937) found cankers especially on exposed slopes at high elevations, while Brandt (1964) notes that at lower elevations they are commonest in cold pockets with poorly drained soil, and where trees have been damaged by wind, ice and snow. In forest trees, vigour may be important, the most vigorous trees recovering from canker damage (Nelson, 1940). On the other hand, in apple orchards, trees with soft, vigorous growth tend to be especially prone to cankers (Swinburne, 1975).

*Losses caused*
On apples and pears, badly affected orchards are unprofitable, and losses through eye rot may sometimes be severe (Wormald, 1955). In the case of broadleaved forest trees, the fungus causes loss mainly by stunting growth and reducing the number of saleable trees of good form. Cankers may also allow the entry of rot fungi, and by weakening the wood give rise to breakage in the crown or main stem (Grant, 1937; Brandt, 1964).

*Control*
Much work has been done on the control of *Nectria* canker in apple and pear orchards, in which the value of the crop justifies substantial expenditure. The danger of canker can be reduced by the growing of the less susceptible varieties, by the grassing down of orchards to reduce soft, over-vigorous growth, and by the drainage of wet sites (Anon., 1971). Where canker occurs, the most effective method of control is to remove inoculum by pruning and cutting out all cankered wood (Swinburne, 1975). Various paints have been formulated for use on individual cankers, to prevent spore production and spread, and reinfection of cut-out cankered areas. They contain mercuric oxide, octhilinone or phenyl-phenol, and should be used only over the

dormant season, and not on very young trees or on green wood or buds (Alford and Locke, 1989; Ivens, 1990).

In commercial orchards of any size, control is mainly by the use of fungicidal sprays to prevent the sporulation of the fungus on the cankers and so prevent its further spread. These fungicides should be chosen from among those, such as benomyl, carbendazim, copper hydroxide or thiophanate methyl, which also control scab caused by *Venturia inaequalis*, which itself causes cracks that assist the entry of *N. galligena*. They have generally been applied most effectively during the major infection periods. Hence, in southeast England, they have been used especially in autumn (to protect fresh leaf scars) and again in spring (to prevent infection through the cracks that then develop in leaf scars and elsewhere), though copper sprays should not be used at bud burst, when their use may lead to an increase in canker instead of a reduction (Alford and Locke, 1989).

Steps taken to eradicate cankers by the use of canker paints, sprays and other measures also help to prevent or reduce fruit infection. Fruit rotting may also be controlled by dipping freshly harvested fruit in benomyl, carbendazim or thiophanate methyl before storage (Alford and Locke, 1989).

In forest stands, chemical control of canker is impracticable. In badly affected stands in North America, it may be necessary to replace with conifers (Spaulding, 1938). Otherwise, canker-free trees should be selected to make the final crop, and the cankered trees should be removed and utilised when thinning (Roth and Hepting, 1954). As the stand increases in age, the number of new canker infections falls off sharply, and the trees in the final crop generally remain free from infection.

### Cankers associated with *Valsa kunzei* (Fr.) Fr. (anamorph *Cytospora kunzei* Sacc.)

In Europe the ascomycete *Valsa kunzei* is usually no more than a saprophyte, but it is sometimes associated with rather inconspicuous resinous cankers on various conifers. It is widespread in Britain and in mainland Europe, across Asia into China, and in North America.

Its anamorph, *Cytospora kunzei*, has sometimes been found in Britain on European, Japanese and hybrid larches, causing non-perennating cankers and girdling branches up to about 1 cm across (Young and Strouts, 1975). It is also known as a cause of cankering of several *Larix* spp. in China (Yuan, 1982). In Switzerland it has been found causing cankers of *Pinus griffithii* (Zinkernagel and Flückinger, 1987). It sometimes also occurs on species of *Abies* and *Tsuga* and on *Pseudotsuga menziesii*. It is best known, however, as the cause of a very damaging canker of *Picea pungens*, the Colorado blue spruce, in North America.

The conidia of *C. kunzei* emerge in yellow tendrils; they are one-celled, colourless and sausage-shaped, measuring $4-5 \times 1$ µm. The perithecia of the teleomorphic stage, *V. kunzei*, form in stromata with greyish-brown discs about 1 mm across, and the curved, colourless, one-celled ascospores measure $5-7 \times 1.5$ µm (Ellis and Ellis, 1985). The conidia may be shed all through the year in wet conditions, but the ascospores are released mainly in the spring (Kamiri and Laemmlen, 1981b).

*V. kunzei* appears to be a wound parasite which can attack only stressed trees, drought damage being the most important predisposing factor (Kamiri and Laemmlen, 1981a).

## Phomopsis disease of conifers caused by *Phacidiopycnis pseudotsugae* (Wilson) Hahn (anamorph of *Phacidium coniferarum* (Hahn) di Cosmo, syn. *Potebniamyces coniferarum* (Hahn) Smerlis)

The dieback and canker of conifers caused by the fungus first described as *Phomopsis pseudotsugae* (Wilson, 1920) is known in Britain, much of the rest of Europe, the Pacific northwest of the USA, and in New Zealand (CMI Map No. 320).

### The causal fungus

The fungus is usually found in its pycnidial state, though a perfect stage also occurs (Hahn, 1957; Gremmen, 1959; Smerlis, 1962). The imperfect stage is not a true *Phomopsis*, and has been renamed *Phacidiopycnis pseudotsugae* (Hahn, 1957). Its one- or many-chambered pycnidia grow in the bark, their apices protruding. They measure about 0.3–1 mm in diameter, and contain colourless, one-celled, fusiform spores measuring $4.5–8.5 \times 2–4$ μm (Wilson and Hahn, 1928). In damp weather the spores exude as white tendrils. The same organism is known as a blue-stain fungus of pine and spruce logs under the name of *Discula pinicola* (Naumov) Petr. (*Ligniella pinicola* Naumov) (Robak, 1952b).

The rarely found perfect stage forms black, disc-shaped ascocarps embedded in black stromata that develop under the bark but are later erumpent. The asci are cylindrical-club-shaped, $80–135 \times 8–12$ μm, and contain eight colourless, one-celled elliptical or fusiform ascospores measuring $10–19 \times 3–6$ μm. Filamentous paraphyses occur between the asci (Hahn, 1957). This teleomorphic stage is currently known as *Phacidium coniferarum* (Cannon *et al.*, 1985).

In culture the fungus grows at the rate of about 5 mm per day at 25 °C, and produces pycnospores in stromata within about 3 weeks to a month. With age the mycelium turns dark olive (Gross and Weidensaul, 1967).

### Mode of entry

*Phacidiopycnis pseudotsugae* is a rather weak wound parasite (van Vloten, 1952; Hahn, 1957; Gross and Weidersaul, 1967). It enters through wounds made when pruning and brashing, and occasionally those made in extraction. Its mode of attack on nursery and other small plants has not been studied, but it may perhaps enter these through leaf scars, and wounds made during cultivation and handling. Plasman (1953) suggested that in Belgium it attacked Douglas fir following frost damage; it has also been known to follow frost damage to Douglas fir in Britain (Redfern *et al.*, 1983) and to pines in New Zealand (Hood and Sandberg, 1985).

Abundant evidence from inoculation experiments has shown that except in rare instances it can invade wounded tissues only in the dormant season.

*Host range and susceptibility*

In Britain, Phomopsis disease has been found mainly on Douglas fir, but has sometimes been known to affect Japanese larch, and occasionally also other conifers, including European larch, *Abies grandis, A. alba (A. pectinata), Cedrus atlantica* and *C. libani,* and *Tsuga albertiana* and *T. sieboldii* (Anon., 1951b; Wilson, 1925; Wilson and Hahn, 1928). Elsewhere it occurs also on pines (Hahn, 1957; Hood and Sandberg, 1985), *Larix occidentalis* (Wicker, 1965) and *Tsuga canadensis* (Gross and Weidensaul, 1967).

Of the three forms (or subspecies) of Douglas fir, the green coastal form generally appears to be more susceptible than the blue and intermediate ones (Gerlings, 1939; Hahn, 1957).

*Symptoms*

The symptoms of Phomopsis disease vary somewhat with the host species. On Douglas fir in Britain the fungus causes dieback of shoots, girdles young stems, and produces stem cankers on older trees (Anon., 1951b). When leading and lateral shoots are attacked, they may die back for up to about 30 cm. Their outer tissues die, and are cut off by a cork layer from the healthy parts of the shoot below, so that the shoot suddenly narrows at the junction between the diseased tissues and the healthy, still growing part of the stem.

If the fungus succeeds in spreading down from a lateral shoot into the main stem, the latter may be girdled. The girdled zone, which may be up to about 15 cm long, then remains constricted at first, with wider healthy tissues both above and below it, until in time the stem above it also dies.

Stem cankers, found on older trees, may arise around the sites of wounds or after the passage of the fungus down a lateral shoot. Growth of the fungal mycelium is limited, and results in oval cankers up to about 15 cm long by 7.5 cm wide. The diseased bark is quickly killed, and is eventually forced off by callusing around its edges. Continued growth of the callus tissue results in the final healing of the canker.

The pycnidia of the fungus are commonly found in the bark of the dead tissues, but often they are accompanied by the fructifications of other fungi (Hahn, 1957).

On Japanese larch *P. pseudotsugae* causes cankers on stems and branches (Wilson, 1921; Anon., 1951b; Hahn, 1957). The cankers may be larger than those on Douglas fir, and are often resinous round their edges. Cracks appear in the affected bark, and healing is usually slow, the diseased tissues remaining on the tree. On western larch (*Larix occidentalis*) in the USA, Wicker (1965) described the disease as 'carrot top', because it turned the tops of the crowns of young trees bright yellow some 6 to 8 weeks before their normal autumn colour change.

Dieback of shoots occurs mainly in nurseries and young plantations not long planted out, but the stem canker form of the disease is found in older trees from about 5 to 25 years old, particularly on 10 to 11-year-old trees in the pole stage (Wilson, 1925; Anon., 1951b; Hahn, 1957).

*Losses caused*

It is rare for Phomopsis disease to cause serious damage. Occasional severe losses have been known to occur locally through dieback of leading shoots of nursery transplants and in young plantation trees up to 8 years old (Anon., 1951b). Young Douglas fir may die if extensive dieback of the leaders takes place, and those that survive often show a bushy growth (Wilson, 1925; Hahn, 1957). Wilson (1925) found some 6 to 10-year-old plantations in which about half the trees were lost. After beating up, 20 per cent of the plants were still affected the following year. Cankers on older trees are not of major import-ance, though they may somewhat reduce the quality of the timber.

*Control*

To avoid the disease, nurseries in which Douglas fir or larch are to be grown should not be placed close to plantations of these trees, which may otherwise be a source of the fungus spores. If Phomopsis disease appears in the nursery, the affected plants should not be used for planting, but should be collected together and burnt. Similarly, if young plants are affected after planting out, the dead and diseased plants should be burnt before beating up, to avoid infection of the replacements in their turn.

In older trees, care should be taken when pruning to leave no stubs, and to prune and brash (especially in the case of Japanese larch) only in the active growing season, when healing of wounds is rapid and infection rarely occurs. Extraction damage should also be guarded against.

## WILT DISEASES

### Verticillium wilt caused by *Verticillium dahliae* Kleb. and *V. albo-atrum* Reinke & Berthold

*Verticillium dahliae* and *V. albo-atrum*, the most important of a number of *Verticillium* species causing wilt diseases, attack over 350 hosts, including more than 70 tree species and varieties, and many shrubs (Engelhard, 1957). They are widespread in all continents, especially in the more temperate areas (CMI Maps Nos. 365, 366).

They are microfungi with erect conidiophores bearing whorls of branches that produce ovate-oblong, one-celled, colourless conidia at their ends (figure 10). The conidia accumulate in small droplets, and measure $3-7 \times 1. 5-3$ μm (Brooks, 1953). The two species can be distinguished from one another only in culture, when *V. dahliae* produces minute black groups of cells called *microsclerotia*, whereas *V. albo-atrum* produces dark resting mycelium (Berkeley *et al.*, 1931; Isaac, 1949). Indeed, some prefer to regard them both as forms of *V. albo-atrum* (Ende, 1958), and they have often not been dif-

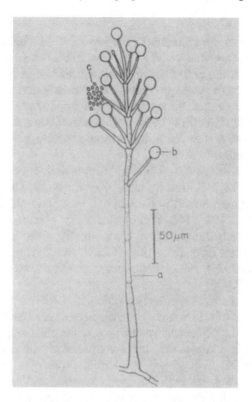

**Figure 10**  *Verticillium albo-atrum* : conidiophore (a) with conidia (c) in conidial droplet (b) (J.H. Western, after Sewell)

ferentiated from one another very carefully in records of wilt on trees. Their strains show little evidence of host specialisation, though they vary in pathogenicity (Donandt, 1932; Ende, 1958).

Most of the isolates from trees in Britain (as well as in North America) are of the microsclerotial type, and are here regarded as *V. dahliae*. In culture this fungus will grow at 30 °C, whereas *V. albo-atrum* has a lower temperature maximum for growth (Caroselli, 1955a; Dochinger, 1956a).

*V. dahliae* and *V. albo-atrum* are 'soil invading' fungi (Garrett, 1956a), and spread by the contact of plant roots with the remains of previously infested root material (Isaac, 1953; Ende, 1958). They may persist in soil infested in this way for at least 2 years (Caroselli, 1954). There is also evidence that common weeds such as groundsel (*Senecio vulgaris*) and black nightshade (*Solanum nigrum*) may act as symptomless carriers (Ende, 1958; Matta and Kerling, 1964). In addition, *V. dahliae* and *V. albo-atrum* may be wound parasites, transmitted from diseased to healthy trees by tools such as saws, axes and increment borers (Dochinger, 1956b).

## Symptoms

Within the plant, the fungus spreads in the sapwood, in which mycelium and tyloses can be found (Dufrénoy and Dufrénoy, 1927). The diseased sapwood becomes stained, and the colour of the stain varies somewhat with the tree species. Thus the stain in *Acer* spp. (plate 27) is usually dark green, though it may be brownish or eventually turn black (Meer, 1926; Goidanich, 1932; Strong, 1936), and that in elm·is brown. This staining is a helpful diagnostic feature, though it may be faint or absent in some species, and in large trees it may not occur throughout.

External symptoms are very variable, but they usually show as a wilting and yellowing of the leaves, followed by their premature fall. Death of branches may follow, and young trees often die within a year, and older ones become stag-headed (Gravatt, 1926; May and Gravatt, 1951) but may later recover, though sometimes only temporarily. Occasionally areas of bark may be killed, and a slime-flux ooze from the affected parts, from which the bark peels off (Gravatt, 1926). More commonly, however, no bark symptoms occur (Goidanich, 1934).

On maples and other *Acer* species (on which *V. dahliae* is a very important parasite), three forms of attack have been distinguished. The first is a *chronic wilt*. Progressive wilting of the foliage and crown dieback leads to death of the crown, usually within a few years, but sometimes within the season of the infection. The second is an *acute wilt*, in which the foliage of the whole crown suddenly wilts and dies, generally in late summer, and hangs, dried and withered, on the tree after normal leaf fall. These forms of wilt have been described by Caroselli (1957b). In the third form, described by Strong (1936), the trees are apparently healthy at the end of the season, but fail to flush in the following spring. All these forms of the disease have been observed in Britain by Piearce (1972), who called the third form of the disease *overwinter attack*.

## Losses caused and affected species

A list of the trees on which the disease has been found in Britain is given by Piearce and Gibbs (1981), who state that monocotyledenous plants and conifers are not affected, and that species of *Alnus*, *Betula*, *Carpinus*, *Fagus*, *Populus* and *Platanus* appear to be very resistant (Piearce and Gibbs, 1981). Verticillium wilt is of no importance on forest trees in Britain but it is important as an occasional source of severe losses in nurseries raising susceptible stock such as *Acer* spp., especially the Norway maple (*A. platanoides*) and various Japanese maples. It will indeed attack a wide range of *Acer* spp. and varieties. Among other important amenity trees it is most common on *Catalpa* (in which it affects large mature trees as well as nursery stock), but the limes (*Tilia* spp.) and the sumachs (species of *Rhus* and *Cotinus*) are also very susceptible, and so is the golden rain tree (*Koelreuteria paniculata*). In the USA, the Judas tree (*Cercis siliquastrum*) is often attacked, as well as the tulip tree or yellow poplar, *Liriodendron tulipifera* (Morehart *et al.*, 1980). In Britain Verticillium wilt has also been found as the cause of wilt and finally the death of a young tree of *Nothofagus obliqua*. It has sometimes been known to attack horse chestnut, sweet chestnut, *Robinia pseudoacacia*, *Sorbus torminalis*, and *Malus* and other fruit trees. In the case of *Catalpa bignonioides*, the Indian bean tree, it has attacked established trees up to 20 years of age (Piearce and Gibbs, 1981), and this tree, together with *C. bungei*, has been severely affected in Italy. Mature horse chestnuts have also been killed on rare occasions in Britain (Strouts *et al.*, 1982). In Poland, Verticillium wilt has caused heavy losses in oak seedlings (Lukomski, 1962).

Most of the information on relative susceptibility within genera and species applies to species and cultivars of *Acer*. Thus Hoitink *et al.*, 1979) screened seven species of *Acer* against attack by *Verticillium albo-atrum* and found evidence that *A. saccharinum* and all the tested varieties of *A. rubrum* were resistant. Valentine *et al.* (1981), with *V. dahliae*, found variations between plants raised from different seed samples of the Norway maple. Townsend *et al.* (1990), also using *V. dahliae*, examined various named cultivars of the Norway maple, and found marked variations between cultivars, with cvs Crimson King and Greenlace being the most susceptible and cvs Jade Glen and Parkway the most resistant, with others between also showing wide differences. According to Holmes (1967), in the case of elms, those most resistant to wilt are also those most resistant to Dutch elm disease, but this finding needs reappraisal since the emergence of new forms of the causal fungus of the latter disease.

*Factors affecting the disease*
Young plants may become diseased if raised in soil already infested with *Verticillium* spp., especially if the land has previously been used for the growing of susceptible crops such as potatoes, tomatoes, strawberries or hops (Piearce and Gibbs, 1981). The disease can also be brought into a nursery on stock raised on infested land elsewhere. Susceptible weeds may also harbour the *Verticillium* spp.

Caroselli (1955b, 1959) found that the severity of the disease was affected by soil moisture, and the amount of water in the sapwood, symptoms of wilt and the amount of sapwood staining both being increased in trees growing in dry soils, and in those with a low sapwood moisture level. Morehart and Melchior (1982) also found a correlation between stress and severity of the disease.

Nitrogen levels also affect the progress of wilt. When nitrogen supplies are low, the fungus may remain in the stem bases, but when ample nitrogen is available, the vessels may be obstructed, and wilting and yellowing of the foliage may occur (Donandt, 1932). Sometimes, however, with increasing nitrogen additional growth of the tree may mask the external symptoms, and the increased extent of the vascular staining is then found only by cutting into the tissues (Dochinger, 1956a). The provision of adequate potassium, on the other hand, may reduce the severity of the disease (Smith, 1983; Ashworth *et al.*, 1985).

*Control*
Severely affected trees can only be removed and burnt (McKenzie, 1954). Attempts are sometimes made to prune out the diseased parts of slightly

wilted plants, but as the fungus usually occurs widely throughout their tissues, they are often unsuccessful. Sometimes, however, if dead and wilted shoots are pruned out, the trees given fertiliser, and watered to keep the roots moist in dry weather, they may show some recovery, though the improvement is not always permanent (Strong, 1936). When pruning an infested stand, care should be taken to sterilise pruning tools when moving from tree to tree (Gravatt, 1926).

If a diseased tree has been removed and is to be replaced, one of a species or variety resistant to Verticillium wilt should be chosen.

Verticillium wilt is, however, most important as a disease of nursery stock, and a number of measures can be taken in nurseries to mitigate its effects.

In nursery beds in which Verticillium wilt has occurred, weeds should be kept down after the removal of the affected crop, and the beds either left fallow for several years, used for conifers or other resistant species (Lukomski, 1962), or treated with a soil fumigant such as chloropicrin (Wilhelm and Ferguson, 1953), methyl bromide, metham sodium or dazomet. So far, attempts to control the disease within the plant by means of systemic fungicides such as benomyl have been only partially successful (Piearce, 1972), though drenching young plants with benomyl, carbendazim, a prochloraz–manganese complex or thiophanate may sometimes reduce the symptoms of wilt (Brooks *et al.*, 1989). Such treatments cannot be recommended with confidence, however, as it rarely eliminates the disease, and may therefore lead to the sale of diseased but apparently healthy plants. Growing in containers, using sterilised soil, can be used to limit the disease in infested nurseries.

**Fireblight caused by *Erwinia amylovora* (Burrill) Winslow *et al.***

Fireblight, caused by *Erwinia amylovora* (CMI Descr. B 44), has long been known in North America and New Zealand, and is now also widespread in northern Europe. It was first recorded in Europe in Kent, in 1957, and became established in much of southern Britain in spite of vigorous efforts to eradicate it. It has since spread to much of France, the Netherlands, Belgium, Germany, Denmark and Poland, Cyprus and Egypt and across Asia. It is important to prevent its further spread into apple and pear-growing areas in the Rhone valley, northern Italy and Spain. A general account of the disease is given by Lelliott and Billing (1984).

*E. amylovora* is a Gram-negative rod, measuring $0.7–1.0 \times 0.9–1.5$ µm, with peritrichous flagella. It produces diagnostic bacterial ooze on slices of immature pear fruit. The causal bacterium and methods available for its isolation are described by Fahy and Persley (1983) and Lelliott and Stead (1987).

Fireblight is commercially important mainly as a disease of pome fruit trees, particularly of pears. It also affects many rosaceous ornamental trees and shrubs, and some native woodland and hedgerow trees of the subfamily Pomoideae of the same family, and these may become sources of infection to neighbouring fruit orchards.

*Infection and symptoms*

Primary infection takes place on the blossoms. On pears, these quickly wilt, and the bacteria spread to the shoots, which may also wilt, when the affected leaves turn dark brown or black. On very susceptible varieties (the worst affected of which is Laxton's Superb), infection quickly spreads in summer throughout the tree, which dies within 6 months of the initial attack.

After the killing of blossoms and leaves in summer, dark-green or brown cankers develop on the branches, and from these, glistening whitish bacterial slime oozes in wet weather. The bacterium overwinters in these cankers. A reddish-brown discoloration is present in the inner bark and outer wood of the cankered areas. The brown leaves remain on affected branches after general leaf fall in autumn. So do diseased fruits, which shrivel and become dark brown or black.

In spring and summer the bacterial slime from the cankers is carried to the blossoms by insects and rain, and on the wind as dried particles, or attached to pollen grains. In Britain most infection has taken place not in the spring, when inoculum production and flowering do not usually coincide, but from mid-June to September, when on some pear varieties summer blossoming occurs when abundant inoculum is also present. Hence the frequency of infection of the variety Laxton's Superb, which produces much summer blossom.

The disease symptoms on apple are broadly similar to those on pear, but are generally less severe, and the bacterial droplets are golden yellow.

On affected hawthorn (*Crataegus* spp.) and whitebeam (*Sorbus aria*) leaves on some of the branches turn yellow, then brown, and then shrivel and fall (plate 54). In the case of the whitebeam, which is very susceptible, further rapid invasion takes place from the point of infection into the main branches and to the trunk, and the tree is commonly killed, as in the case of pear, within 6 months of the first attack. There is evidence that *E. amylovora* can enter whitebeam trees through undamaged bark, probably through the lenticels (Strouts *et al.*, 1985).

Though *Sorbus aria* is so susceptible, *S. aucuparia* is more resistant, and *S. decora*, *S. intermedia* and *S. latifolia* seem to be immune under European conditions (Zeller, 1979). Some *Crataegus* spp. also show resistance, especially *C. coccinea* and *C. prunifolia* (Zeller, 1979).

*Control*

In Britain attempts at control of fireblight have concentrated mainly on the protection of fruit orchards by the destruction and replacement of the very susceptible pear variety Laxton's Superb, and the grubbing out or killing of adjacent infected hosts, especially hawthorn hedge plants. Infection of these hedge plants has also been much diminished by clipping, which reduces flowering.

Growers have also been recommended to break off secondary blossom, to remove beehives from orchards before the secondary blossom is formed (as bees may carry the causal bacteria) and to disinfect pruning tools with 3 per cent lysol between trees.

The disease is notifiable under the Fireblight Disease Order and outbreaks should be reported to the Plant Health Branch of the Ministry of Agriculture, Fisheries and Food. For commercial growers the Ministry also operates a fireblight warning scheme (Alford and Locke, 1989). The principles behind this scheme have been described by Billing (1984).

The most important measure when considering the growth of ornamental species of *Sorbus* or *Crataegus* in areas in which fireblight is prevalent is to select from among known resistant species.

## ROOT DISEASES

### *Phytophthora* root diseases
At least five different species of *Phytophthora* attack the roots of forest and ornamental trees in Europe. They are:

(1) *P. cambivora* (Petri) Buisman.
(2) *P. cinnamomi* Rands.
(3) *P. citricola* Sawada.
(4) *P. cryptogea* Pethybridge and Lafferty.
(5) *P. megasperma* Drechsler var. *megasperma* Waterhouse.

Of these, *P. cinnamomi* and *P. cambivora* are both associated with the very characteristic 'ink disease' of sweet chestnut (and sometimes of oak), which is described in chapter 12. Other *Phytophthora* spp. have been found on trees, though not necessarily on roots. Thus *P. cactorum* (Lebert and Cohn) Schroeter causes a seedling blight of beech (see chapter 11) and the bleeding canker disease of the horse chestnut and lime (see chapter 19), which may less often be caused by *P. citricola* and *P. syringae* Kleb.; the latter also causes leaf and shoot blight and wilt of lilac (see chapter 19), and a collar and fruit rot of apple, pear and other fruit trees.

*Phytophthora* root diseases, commonly and rather misleadingly called Phytophthora root rots and caused by the five species listed above, occur on a range of broadleaved and coniferous hosts in Europe. As a whole, in Britain they most often and most severely affect the trees listed in table 3.1, but from time to time they have caused serious damage to species of *Populus*, *Platanus*, *Sorbus* and *Ulmus*, Western red cedar (*Thuja plicata*) and young nursery plants of Douglas fir and silver fir (*Abies* spp.) (Strouts, 1981a).

The taxonomy of *Phytophthora* spp. is a complex subject, largely because the methods of classification are based on morphological characters that do not clearly discriminate between species (Ribeiro, 1978). Some of the problems concerned have been discussed by Brasier (1983). Only an outline of the principles of identification can be given here and the interested reader

**Table 3.1**   Some trees in Britain most often and most severely affected by Phytophthora root disease (after Strouts, 1981)

| | |
|---|---|
| *Acer* spp. | *Malus* spp. |
| *Aesculus hippocastanum* | *Nothofagus* spp. |
| *Castanea sativa* | *Prunus* spp. |
| *Chamaecyparis lawsoniana* | *Syringa* |
| *Eucalyptus* spp. | *Taxus baccata* |
| *Fagus sylvatica* | *Tilia* spp. |

is referred elsewhere for detailed information (Waterhouse, 1963, 1970; Newhook *et al.*, 1978; Hamm and Hansen, 1987).

Features frequently used in the identification of *Phytophthora* spp. include characters of the sporangia, such as their size and shape, their *l/b* ratio (the ratio of length to breadth), whether they are caducous (readily shed) or persistent, whether or not they have apical papillae, and whether they proliferate, with new sporangia forming within the remains of the older ones. In the case of the sexual stage, the size and wall characters of the oogonium and oospore are of importance, while the antheridium may be paragynous (growing at the side of the oogonium) or amphigynous (when the oogonium perforates the antheridium and grows through it, so that the antheridium finally becomes a collar round the base of the oogonium). The different species may be either heterothallic or homothallic. Other features of significance include those of the hyphae, the presence or absence of chlamydospores, growth in culture, etc. Some important characteristics of the species listed above are given in table 3.2.

*Short notes on individual species*
*P. cactorum* (CMI Descr. F. 111)   Sometimes involved in root diseases (for example, of sweet chestnut, Lawson cypress and ash), but more often known as the cause of seedling blight of beech (and sometimes of other broadleaved trees and of conifers) and of bleeding canker of horse chestnut and lime. It also causes collar rots of sweet chestnut, apple and other broadleaved trees, and cankers in horse chestnut, maple, etc. Its oogonia are mostly 25–35 μm across. Occurs throughout most of Europe, and most temperate areas of the world.

*P. cambivora* (CMI Descr. F. 112)   Occurs throughout most of Europe, chiefly as one of the fungi involved in 'ink disease' of sweet chestnut. In Britain it has also been found associated with damage to the roots of other trees, including *Acer, Aesculus, Chamaecyparis*, beech, *Sorbus* and *Ulmus*.

*P. cinnamomi* (CMI Descr. F. 113)   This has a wide host range and is almost world-wide, especially in the warmer parts of the world. It is important as one

## Diseases of forest and ornamental trees

**Table 3.2** Some characteristics of *Phytophthora* spp. attacking forest and ornamental trees in Europe

|  | (1) | (2) | Sporangium | | | | | | | Antheridium | |
|---|---|---|---|---|---|---|---|---|---|---|---|
|  | | | Apex | | | Persistence | | Proliferation | | | |
|  | (1) | (2) | (3) | (4) | (5) | (6) | (7) | (8) | (9) | (10) | (11) |
| *P. cactorum* | ★ | — | ★ | — | — | ★ | — | — | ★ | ★ | — |
| *P. cambivora* | — | ★ | — | ★ | — | — | ★ | ★ | — | — | ★ |
| *P. cinnamomi* | — | ★ | — | — | ★ | — | ★ | ★ | — | — | ★ |
| *P. citricola* | ★ | — | — | ★ | — | — | ★ | — | ★ | ★ | — |
| *P. cryptogea* | — | ★ | — | — | ★ | — | ★ | ★ | — | — | ★ |
| *P. megasperma* var. *megasperma* | ★ | — | — | — | ★ | — | ★ | ★ | — | ★ | — |
| *P. syringae* | ★ | — | — | ★ | — | — | ★ | — | ★ | ★ | — |

| | | |
|---|---|---|
| (1) Homothallic | (5) Non-papillate | (9) Not proliferating |
| (2) Heterothallic | (6) Caducous | (10) Mainly paragynous |
| (3) Markedly papillate | (7) Persistent | (11) Mainly amphigynous |
| (4) Not markedly papillate | (8) Proliferating | |

of the fungi which causes 'ink disease' of sweet chestnut (and in France also of red oak), but also causes root death and collar rot and consequent crown damage to many trees, including *Chamaecyparis, Cupressus, Juniperus, Larix, Picea, Pinus, Pseudotsuga, Taxus, Acer, Betula, Eucalyptus, Fagus, Juglans, Magnolia, Malus, Nothofagus, Platanus, Prunus, Pyrus, Robinia* and *Syringa*. In northern Europe it is especially important as a cause of severe damage to *Chamaecyparis* spp. There is a vast Australian literature dealing with severe attacks on *Eucalyptus* and other trees. In southwest France, Lung-Escarment *et al.* (1984) report that the growing of plantings of eucalypts, especially *E. gunnii*, is threatened because of serious damage by the same fungus. The oogonia of *P. cinnamomi* are about 40 μm across.

*P. citricola* (CMI Descr. F. 114). This has a wide host range. On trees in Europe as a whole, it is especially important in citrus crops. In Britain, however, it has been found from time to time as a cause of root disease on other trees, including horse chestnut and lime (on both of which it may also cause bleeding canker), *Nothofagus, Salix, Viburnum* and *Thuja plicata*. The oogonia are about 27–32 μm across.

*P. cryptogea* (CMI Descr. F. 592). This is another world-wide species with a wide host range. In northern Europe it causes much damage to ornamental conifers, but it may also attack other trees, including sweet chestnut, eucalypts and *Prunus, Malus, Populus* and *Viburnum*. Its oogonia measure 20–32 μm across.

*P. megasperma* var. *megasperma* (CMI Descr. F. 115)   This species is wide-spread, mainly in the warmer parts of the world. On trees it often attacks the root collar and spreads upwards, causing dieback in the crown. Its attacks are sporadic, but it has sometimes killed whole avenues of, for example, horse chestnut trees (Brasier, 1989). The *P. megasperma* complex has been studied by Hansen *et al.* (1986).

*P. syringae* (CMI Descr. F 32).   This fungus is also widespread in Europe, with a fairly large host range, but in connection with forest and ornamental trees it is mainly the cause of leaf and shoot blight and wilt of lilac (*Syringa*), though it also causes a fruit rot of apples, pears and other orchard trees.

*Symptoms*
The crown symptoms of Phytophthora root diseases are typical of root diseases in general (Strouts, 1981a). In broadleaved species, leaves may be abnormally small, yellow or sparse over parts of or all the crown. Some or all of the shoots growing in coppiced woodlands may exhibit similar symptoms, including sweet chestnut affected by ink disease. These symptoms may be followed by the ultimate death of the tree but in other cases trees may recover.

In conifers and especially in Lawson cypress, the foliage may lose its freshness or lustre and become drab, yellow and finally die.

Careful examination of the tree soon after the initial attack may reveal symptoms more specific to Phytophthora root diseases. Some or parts of the roots close to the main stem may be dead, whereas more distal parts may be live, perhaps with dead areas of bark on them. *Phytophthora* spp. do not decay the roots of woody plants but kill the host tissues in the cambial region.

The dead bark may extend up the main stem for some centimetres or even metres on big trees in a characteristic strip or tongue-shaped lesion (plate 15). On some cultivars of Lawson cypress the strip of infected bark may extend to the base of a lower branch and this may also be killed. Although these lesions are indicative of Phytophthora root diseases, they can also be caused by other factors such as lightning, bonfire or chemical injury, and care must be taken to achieve a proper diagnosis.

Confirmation of the presence of *Phytophthora* spp. can only be obtained by culturing from recently infected tissues or (less conclusively) from the soil around infected roots. Isolation of the fungus from diseased material is notoriously difficult and a variety of techniques have been developed for the purpose. One of the best-known techniques involves the use of apples or other fruits as a bait (see, for example, Newhook, 1959). Infected material, from either the host or the soil, is inserted into the apple and, if *Phytophthora* spp. are present, a firm rot develops in the tissues of the apple. The presence of *Phytophthora* spp. can be further substantiated by growing pieces of the infected apple tissue on a suitable medium such as potato dextrose agar. There has also been much research into the use of selective media for isolation of *Phytophthora* spp. and a number of antibiotics have proved useful (e.g. Tsao and Ocana, 1969).

*Biology*

Infection by Phytophthora root diseases is particularly favoured by wet and warm soil conditions. At these times large numbers of motile zoospores are released from the sporangia. The zoospores are chemically attracted to the elongation zone of young host rootlets, where most infection is thought to occur (Zentmeyer, 1961). The sporangia and zoospores are relatively short-lived and it is the chlamydospores and perhaps oospores which play an important role in longer-term persistence in the soil. When suitable conditions occur the chlamydospores germinate and give rise to sporangia and zoospores. Oospores may act similarly, though the proportion that germinate is often very low.

Long-distance transmission of Phytophthora root diseases can occur in drainage or irrigation water, through the transport of infected plants or in infested soil adhering to plants, vehicles, tools, footwear, etc.

These diseases are commonly found in water-retentive soils and in nurseries where very susceptible plants are grown year after year. They are also encouraged by the presence of farmyard manure and organic mulches. Low concentrations of oxygen and high concentrations of carbon dioxide produced under these conditions appear to encourage the production of zoospores.

*Control*

Phytophthora root diseases are difficult to control, but some measures can be taken to avoid or ameliorate them. Various methods of control in nurseries and ornamental plantings are available and these have been reviewed by Strouts (1981a).

The key to the control of Phytophthora root disease in the nursery is thorough sanitation. If an outbreak is identified in its early stages, all infected plants should be destroyed and the soil sterilised. If infection is widespread, it may be necessary to use only containerised material and to ensure that all planting stock and soil is clean. The containers should be isolated from infected soil by the use of raised gravel beds or standing areas over stout polythene sheeting. There are a few non-phytotoxic fungicides such as fosetyl-aluminium, etridiazole, furalaxyl and propamocarb hydrochloride (Ivens, 1990) which have been developed for the control of Phytophthora root diseases. In the main, the effect of these chemicals is to suppress the production of zoospores rather than to eradicate the fungus and therefore repeated treatments with the chemical are necessary and they are best used as protectants (Brooks *et al.*, 1989).

Control of Phytophthora root disease in plantings of ornamental trees is largely a matter of proper site selection and aftercare. Water should not be allowed to collect around the base of woody plants, and excessive irrigation or heavy applications of farmyard manure should be avoided. Livestock should not be permitted to congregate around susceptible species. Where sites are known to be infested by the disease, especially in hedge plantings, tolerant species such as Leyland cypress, western hemlock or *Chamaecyparis pisifera* cv. plumosa or cv. squarrosa can be used to replace the susceptible Lawson cypress.

## Group dying of conifers caused by *Rhizina undulata* Fr. (*R. inflata* (Schaeff.) Quél.)

*Rhizina undulata*, one of the larger ascomycetes, is sometimes known as the tea-break fungus because it grows on and around the edges of the sites of fires, many of which were at one time made by workers when brewing tea in their lunchtime breaks. It has long been known as the cause of group dying in conifers (Hartig, 1894), and has been recorded in Britain (Brooks, 1910; Murray and Young, 1961), Ireland (McKay and Clear, 1953), France (Lanier, 1962), Denmark (Yde-Andersen, 1963), Finland (Laine, 1968), the Netherlands (Gremmen, 1961a), Portugal (Neves *et al.*, 1986) and other parts of Europe, in the northwest and the east of the USA (Weir, 1915; Davidson, 1935a; Zeller, 1935) and Canada (Ginns, 1968), and in South Africa and Swaziland (Lundquist, 1984). A short account of group dying is that by Phillips and Young (1976).

### The causal fungus

The fruit bodies of *R. undulata* appear in June or later. Initially small, whitish and globular, they become disc-shaped when about 6 mm in diameter, and deep chestnut-brown on the upper hymenial surface, with a creamy margin. They then develop into irregularly cushion-shaped bodies with hollow undersides connected to the ground by many thick, hollow, whitish, root-like structures (figure 11). The creamy margin of the brown fruit body disappears when growth is completed, by which time single fructifications are up to about 5 cm in diameter, and clusters may have fused in crust-like groups up to 25 cm long.

The asci measure about $400 \times 20$ µm, and contain colourless, fusiform ascospores measuring $22\text{--}40 \times 8\text{--}11$ µm (figure 12). They are separated by unbranched, colourless paraphyses (Lange and Hora, 1963; Dennis, 1968). The fruit bodies are annual, but old fructifications may persist until the following season, and are black or dark greenish, with a slimy surface (Ginns, 1968). Fruit bodies are common in warm summers, but not in cool ones (Murray and Young, 1961). The mycelium of the fungus bears clamp connections similar to those of basidiomycetes.

*R. undulata* colonises the sites of fires made during thinning operations and those made to destroy lop and top when preparing for replanting after clear felling. Factors affecting the colonisation of fire sites were studied by Jalaluddin (1967a,b). From observations and the results of experiments Jalaluddin concluded that the fungus grew only on fire sites, and was able to colonise them only when they were made on acid soils, and when fresh conifer roots were present.

Colonisation followed the germination of ascospores, which could survive

**Figure 11**  Fruit bodies of *Rhizina undulata* growing on burnt ground under Scots pine trees. One fructification has been cut across to show the root-like structures connecting the underside to the ground (D.H. Phillips)

**Figure 12**   *Rhizina undulata* : asci and ascospores (Forestry Commission)

in the soil, small numbers persisting for as long as 2 years. The spores were stimulated to germinate by the heat of the fire, which, at its edges, heated the soil sufficiently in half an hour or more to cause germination but not to kill the spores. In Jalaluddin's laboratory experiments maximum germination occurred when the spores were heated to 37 °C for 3 days, followed by incubation at 22 °C. In agar culture, mycelial growth was most rapid at 25 °C, at a pH between 4.5 and 5.2, while growth at pH 7 was very slow. The rate of growth in the soil in East Anglia appeared unlikely to exceed 1 m/year.

## Progress and symptoms of the disease
Epidemiology of group dying in Britain was studied by Murray and Young (1961). Following colonisation of a fire site, the mycelium of the fungus ramifies outwards through the soil and attacks the roots of conifers with which it comes into contact. As a result, young newly planted trees may soon die. On pole-stage trees, the finer roots are quickly killed, and the larger roots become covered by a network of yellowish-white mycelium, which forms infection cushions over the lenticels (figure 13). As a reaction these fill with resin and appear white instead of their normal pink colour. In spite of this, the fungus usually succeeds in entering, and invades the host tissues. Its progress may still be halted by the formation of a cork barrier, and a discrete rounded area of dead tissue surrounded by a dark line is then visible if the bark scales of the root are removed (figure 14). Often the mycelium advances unchecked, however, and many coalescing lesions form over the root surface, and the root soon dies. Longitudinal cracks can then often be found in the bark of the roots, and extensive areas of dead cortex can be found if the bark scales are stripped off.

**Figure 13**  Mycelium of *Rhizina undulata* on root of Sitka spruce, with infection cushions over the lenticels (Forestry Commission)

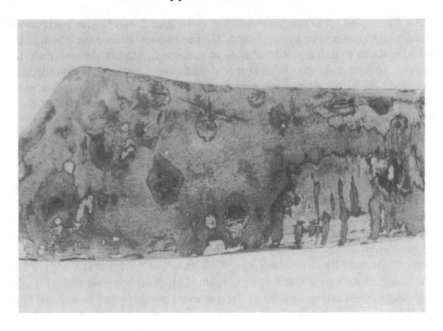

**Figure 14**  Lesions caused by *Rhizina undulata* on roots of Sitka spruce; the bark
scales have been removed (Forestry Commission)

The first symptoms seen on the tree, however, include prolific coning
(which may occur from 1 to 3 years before the tree dies), and the enlargement
of the lenticels on the trunk, with resin bleeding from the bark, and a thinning
of the foliage (McKay and Clear, 1953, 1955; Murray and Young, 1961). The
resin may exude as blobs on the bark, or sometimes a copious flow may run
down the trunk, from a height of as much as 3 to 4 m (10 to 12 feet) (McKay
and Clear, 1955). Many needles may fall while still green, and complete loss
of all the needles may occur suddenly in the middle of the growing season
(Murray, 1953; McKay and Clear, 1955).

By the time such symptoms have become visible, however, extensive root
death will have taken place. By the second year after the establishment of the
fungus, the symptoms usually show on many trees close to the original fire
site, and a few trees may already be dead. In favourable seasons, fruit bodies
appear, sometimes on the fire site (often around fresh stumps), but commonly
round its edges. The fungus advances through the soil in the manner of a fairy
ring, and so the fruit bodies appear in a wider and wider, complete or
discontinuous circle in successive years, the central area within the ring
becoming free of the fungus as it grows outwards into fresh soil. Extending
rings may be partly checked by drainage ditches and similar excavations.

With time, more and more trees are killed, and individual groups may
cover up to a quarter of an acre (0.1 ha). After about 6 years, further spread
normally ceases, but groups established by *R. undulata* may later be greatly
extended by windblow (Murray and Young, 1961).

*Host range and age of attacked trees*
R. *undulata* confines its attacks to conifers. In Britain it seems first to have
been recorded as a cause of damage by Brooks (1910), who found it in East
Anglia killing newly planted Scots and Corsican pine and European and
Japanese larch. Jalaluddin (1967a,b) noted similar losses among newly
planted trees in the same area, and found the killing of larger trees following thinning was unusual there. Deaths of young plants have also occurred
in the Netherlands (Gremmen, 1961a), Finland (Laine, 1968), Sweden
(Hagner, 1962), North America (Ginns, 1968), and South Africa and
Swaziland (Lundquist, 1984). Davidson (1935a) in Maryland, USA, found
the fungus killing red pines in nursery seedbeds.

In Britain, however, though, as already noted, R. *undulata* sometimes kills
young plants after planting out in East Anglia (and occasionally elsewhere in
the drier eastern parts of the country), this type of damage is rare, and the
fungus has been important mainly as a killer of trees in the pole stage (chiefly
between the ages of 22 and 32 years) in the wetter areas of the west (Murray
and Young, 1961).

Group dying occurs there especially in Sitka spruce, in which it tends to be
more severe than in other species, though it has also been found in Norway
spruce, in Scots, Corsican and lodgepole pines, and in European and
Japanese larch. So far it has not been found in Douglas fir, though it has been
known to kill seedlings (but not pole-stage trees) of this species in North
America (Ginns, 1968). Hartig (1894), in Germany, also recorded the fungus
in Douglas fir, and in *Abies alba*, *Tsuga mertensiana* and *Pinus strobus*, as
well as on some of the species mentioned above. In Portugal it has been found
on *Pinus pinaster* (Neves *et al.*, 1986), and in South Africa and Swaziland on
*P. canariensis*, *P. elliottii*, *P. kesiya*, *P. leiophylla*, *P. patula*, *P. pinea*, *P.
radiata*, *P. roxburghii* and *P. taeda* (Lundquist, 1984).

*Factors affecting group dying*
Many of the factors affecting the development of group dying, including its
dependence on fires and the presence of fresh conifer roots, and its association with acid soils, have already been noted. The disease appears to be
commoner on light, sandy soils and on peat rather than on the heavier
mineral soils (Murray and Young, 1961).

*Losses caused*
In Britain, losses of young plants soon after planting out are uncommon and
soon cease. Losses depend very much on the area of ground heated by the
fires. Hence, if lop and top is burnt on a few widely spaced 'spot' fires, losses
remain small, but if burning is done in shallow drifts and a large area is
heated, the risk of loss increases. This may be why Gremmen (1961a) found
that in young plantings in the Netherlands nearly all the young plants might
sometimes be attacked. In older crops in Britain, most groups in pole-stage
stands cover between about one-tenth and a quarter of an acre (0.04–0.1 ha),
and in groups about 30 yards (27 m) in diameter, up to 100 trees may be

killed, though losses of this order occur only in Sitka spruce (Murray and Young, 1961). On exposed sites, losses through attack by *R. undulata* may be followed by extensive windblow, which may eventually be more damaging than the fungus. Similarly, opening up of the crop, especially in the case of Norway spruce, may sometimes lead to excessive transpiration, needle fall and death of trees previously damaged but not killed by the fungus.

*Control*

As *R. undulata* commonly grows on and outwards from the sites of fires, its activities in the forest can be prevented if no fires are lit in the plantations. If workers are allowed to light fires on rides or waste areas, it is important to ensure that the fire sites are separated from the adjacent plantations by ditches or similar barriers, otherwise the fungus may grow out into the crops. It may perhaps also be possible to prevent the spread of newly established groups in the same way by trenching round them, but if ditches are not well-maintained the fungus may succeed in crossing them.

Because the lighting of fires in forests is now usually strictly controlled, this disease has generally become uncommon in Britain. In cases where discipline has been relaxed it has quickly reappeared.

**Violet root rot caused by *Helicobasidium brebissonii* (Desm.) Donk (*H. purpureum* Pat.) (anamorph *Rhizoctonia crocorum* (Pers.) DC, syn. *R. violacea* Tul.)**

Violet root rot, caused by *Helicobasidium brebissonii*, which is fairly common on sugar beet, asparagus, carrots and other agricultural crops (Moore, 1959), has occasionally been reported on forest nursery stock. It occurs throughout most of Europe and in many parts of the rest of the world (CMI Map No. 275). Its felt-like, purplish fructifications have a smooth hymenium of crozier-shaped basidia transversely septate to form up to four cells, each with a sterigma bearing an elliptical basidiospore measuring 10–12 (–15) × 6–7 μm (Wakefield and Dennis, 1981). They encrust the stems and lower parts of their host plants, spreading outwards over the soil, but usually occur only in damp weather in April. It was then that Watson (1928) found them in Scotland round the collar and lower branches of Sitka spruce seedlings, and growing over the surrounding soil. The plant roots were rotten and purple-brown, and covered with small black infection cushions. *H. brebissonii* has again been found occasionally in more recent years, in 1972, 1974 and 1975, causing severe damage to Sitka spruce transplants in Scottish nurseries established on old garden or agricultural land. McKay and Clear (1958) found the same fungus in Eire destroying the root cortex and causing yellowing and browning of the branch apices of three-year-old transplants of Douglas fir and lodgepole pine growing on land 4 years previously cropped with sugar beet. *H. brebissonii* has also been found attacking two-year-old Douglas fir plants in England. Affected plants should be destroyed.

## Root and butt rot caused by the *Armillaria* spp. (the honey, bootlace or shoestring fungi)

The honey fungus *Armillaria mellea* (Vahl:Fr.) Kummer has long been known to be very variable. Following work by Hintikka (1973) on the sexual cycle of the fungus, Korhonen (1978) collected and crossed European strains of *A. mellea*, and found that they fell into five intersterile groups, which he tentatively called A, B, C, D and E. Previous attempts had been made to divide *A. mellea* into separate species, based on the morphology of the fruit body (Romagnesi, 1970, 1973; Romagnesi and Marxmüller, 1983), and it has become clear that Korhonen's groups agree broadly with Romagnesi's species, and that *A. mellea* as it was once understood is in fact a group of species, as Singer (1956) had earlier suggested. The names now applied to Korhonen's groups A to E are as follows (Gregory and Watling, 1985; Guillaumin *et al.*, 1985; Roll-Hansen, 1985; Grieg *et al.*, 1991):

A. *A. borealis* Marxmüller and Korhonen.

B. *A. cepistipes* Velenovsky f. *pseudobulbosa* Romagnesi and Marxmüller.

C. *A. ostoyae* (Herink) Romagnesi (*A. obscura* (Pers.) Herink).

D. *A. mellea* (Vahl:Fr.) Kummer (*A. mellea sensu stricto*).

E. *A. lutea* Gillet (*A. bulbosa* (Barla) Kile and Watling, *A. gallica* Marxmüller and Romagnesi).

The fruit bodies of all these species are more or less honey-coloured toadstools with a ring on the stem and with white or off-white, adnexed or slightly decurrent gills and a white spore mass. A further species in the group, *Armillaria tabescens* (Scop.) Emel. (*Clitocybe tabescens* (Scop.) Bres.), lacks a ring. It has been found in southern Britain from time to time (Rishbeth, 1983), but in Europe is mainly a Mediterranean species attacking cork oak and *Eucalyptus* spp. (Guillaumin, 1988); it also kills or causes decay in trees in the southern USA. It may form yellowish mycelial strands under the host bark (Rishbeth, 1983).

Some of the above species and several others also occur in North America (Anderson *et al.*, 1980; Lin *et al.*, 1989), and yet others are known in Australasia (Kile and Watling, 1983), China and other areas.

The ringed European species listed above differ from one another in the morphology of their fruit bodies. These fungi also produce bootlace-like or ribbon-like, brown or black rhizomorphs in the soil and under the bark (under which some also develop mycelial sheets) and the rhizomorphs show characteristic differences in their form and abundance from species to species. The *Armillaria* species differ also in their host and geographical ranges and in their pathogenicity. In spite of these differences, it is by no means always easy to identify the individual species because they tend to intergrade, and mating tests with haploid tester strains may be needed to provide certain identification. Even mating tests are sometimes difficult to interpret (Holdenrieder, 1986; Gregory, 1989), especially in the case of *A. cepistipes* and *A. lutea*, which appear to be very closely related (Roll-Hansen, 1985). Guillaumin *et al.* (1985) and Rishbeth (1986), however, have shown that the various species differ in their growth characteristics and growth rates in culture. Of these five fungi, *A. ostoyae* and *A. mellea* are important pathogens, the former mainly on conifers, the latter especially but by no means exclusively on broadleaved trees. The other species are mainly saprophytic, but may be parasitic on trees weakened by other agencies.

*Armillaria mellea sensu stricto* is a highly pathogenic fungus with a wide host range. It is found from Denmark and Britain southwards to the Mediterranean (Guillaumin *et al.*, 1985; Roll-Hansen, 1985).

Its fruit bodies grow in tufts (plate 17). The smooth, honey-coloured cap has few scales, and the long stipe is smooth, without scales, and usually cylindrical or only slightly swollen at the base. The ring is well-marked. The rhizomorph system is much less extensive than that of *A. ostoyae* or *A. lutea*, and the rhizomorphs grown in sand from woody inocula are curved, thin, dichotomously branched, brittle, and yellowish-brown when young (Rishbeth, 1982). They grow rapidly on malt agar, on which they are flattened and ribbon-like (Guillaumin and Berthelay, 1981). Under the bark of affected trees, *A. mellea* forms mycelial sheets which are white when young, but often later become yellowish (Rishbeth, 1983).

*A. mellea* is particularly important as a parasite of broadleaved trees. In France, Guillaumin and Berthelay (1981) found no other *Armillaria* sp. on orchard trees or grape vines, on which *A. mellea* was a very serious damaging agent. In Denmark, *A. mellea* seems to occur only on broadleaved trees, especially beech (Marxmüller and Printz, 1982; Watling *et al.*, 1982), and it has also been found attacking broadleaved trees in France and Germany (Romagnesi, 1973; Jahn, 1979; Jahn and Jahn, 1980; Marxmüller, 1982). In England, Rishbeth (1983) also found that it killed many healthy broadleaved trees unaffected by stress, but it also often killed conifers such as *Sequoiadendron giganteum* (Wellingtonia) and members of the Cupressaceae, including *Thuja plicata* (western red cedar), *Chamaecyparis lawsoniana* (Lawson's cypress) and × *Cupressocyparis leylandii* (Leyland cypress). It was also among the species of *Armillaria* that killed suppressed broadleaved trees, and it sometimes colonised suppressed pines. Occasionally it attacked or colonised broadleaved trees killed by Dutch elm disease, beech bark disease or sooty bark disease of sycamore. It often attacked garden trees weakened by other agencies. It was also a common colonist of stumps of broadleaved trees.

When Rishbeth (1982) inoculated 5-year-old Corsican pine with *A. mellea*, infection was followed by extensive resin flow, which did not occur after inoculation with *A. ostoyae*.

*A. ostoyae* (*A. obscura*) is the second of the highly pathogenic European *Armillaria* species. It appears to occur in southern Norway and Sweden, Finland, Denmark, Poland, Germany, Britain, Spain and Portugal (Guillaumin *et al.*, 1985; Roll-Hansen, 1985), Ireland (Clancy and Lacey, 1986) and Italy (Intini, 1988). The caps of its fruit bodies are often pale at first, with a pinkish tinge but darken to reddish-brown, and are covered with rings of brown scales. The stipe is often more or less cylindrical, sometimes slightly bulbous at the base, and it is covered by many conspicuous brown-tipped scales. The ring is conspicuous and thick, often with a brown fringe at its edge. The fruit bodies usually form in clumps (Guillaumin *et al.*, 1985).

*A. ostoyae* produces mycelial sheets under the host bark. These sheets may be pure white when young, becoming yellowish with age, and they much resemble those of *A. mellea*. As the bark dries out and begins to lift, the sheets tend to be replaced by a network of flattened rhizomorphs (Rishbeth, 1983) (figure 15). Rhizomorphs developed in sand when Rishbeth (1982) was preparing woody inocula proved to be dichotomously branched, reddish-brown when young, and less curved and brittle than those of *A. mellea*.

**Figure 15**   Flattened rhizomes of *Armillaria ostoyae* on stem of Sitka spruce from which the bark has been removed (Forestry Commission)

The fungus may grow a long way up the stems of dead trees, and Rishbeth (1983) found its fruit bodies to a height of 6 m on 50-year-old Scots pine in Thetford Forest.

*A. ostoyae* is almost confined to conifers. It has killed young pines in Finland, and also attacked Scots pine in Poland and *Pinus pinaster* in Germany (Roll-Hansen, 1985). Guillaumin and Berthelay (1981) recorded it also on *Abies grandis*, *Picea abies*, *P. sitchensis* and *Pseudotsuga menziesii*, as well as on *Abies alba*, on which, however, they considered it to be saprophytic. It may sometimes become temporarily parasitic in stands of Scots pine and *Pinus uncinata* in southwest France, causing patches of dead trees which later regenerate (Durrieu *et al.*, 1985). Intini (1988) recorded it on dying trees of *Abies alba* in Italy, and in Mexico Shaw (1989b) found it on *Pseudotsuga menziesii* and *Abies durangensis*.

Culture experiments by Rishbeth (1986) indicated that it has a low tolerance to calcium carbonate, and in the field in Britain (1982) he found that it killed pines on former broadleaved woodland sites on acid soils but not in similar stands in alkaline areas. Gregory (1989) found it killing pines on sites which had not previously carried broadleaved trees, and Rishbeth (1988) recorded it killing first-rotation Sitka spruce. In inoculation trials Rishbeth (1983) found that it killed many 2-year-old Scots pines and 5-year-old Corsican pines. In the field it was by far the commonest *Armillaria* species killing both these pines in the absence of stress from any other cause, and in similar circumstances it was the only one killing Norway spruce, Douglas fir and European larch. It was also one of the species of *Armillaria* found on suppressed pines, and a cause of decay of Norway spruce. He also found it on one occasion killing ash, the only such example on a broadleaved tree, though it occasionally grew on stumps of such trees. It was almost invariably the *Armillaria* species found on pine stumps. Guillaumin and Lung (1985) studied *A. ostoyae* and *A. mellea* on a range of trees and other woody plants, and found that *A. ostoyae* was the more damaging of the two on Norway spruce and *Pinus pinaster* but less aggressive than *A. mellea* on *Quercus borealis*, peach and grape vines, while hybrid black poplars and *Eucalyptus dalrympleana* were resistant to both species.

The remaining *Armillaria* spp. are of much less pathological importance.

*Armillaria borealis* has a relatively northern distribution in Europe, apparently occurring in West Germany and northeastern France, northwards to Norway and Sweden and into the USSR (Guillaumin *et al.*, 1985; Roll-Hansen, 1985); in Britain it has been found in Scotland but so far not further south (Gregory and Watling, 1985).

It has been described in detail by Gregory and Watling (1985), and its main features have been indicated by Guillaumin *et al.* (1985). The fruit bodies grow in loose tufts. Their caps are pale honey-coloured with a pinkish tinge, with small brown or black fugaceous scales in concentric rings. The cylindrical stipe may be up to about 65 mm long, sometimes with a few yellow, brown-centred scales in ring-like zones, and it may be slightly thickened though scarcely bulbous at the base. The ring is distinct and double.

In culture, *A. borealis* may produce cylindrical rhizomorphs or undifferentiated mycelium, and fruit bodies may also be formed in culture (Guillaumin *et al.*, 1985).

Korhonen (1978) found that this fungus was common on birch, as well as on decayed wood of Norway spruce and other conifers, and it also caused a spruce butt rot. Marxmüller (1982) found it mainly on conifer stumps, but sometimes also on those of broadleaved trees and on dead standing birch trees. Gregory and Watling (1985) collected it from old birches and from stumps of Sitka spruce, and they also found it causing a butt rot of standing

Sitka spruce and the death of young trees of the same species. So far *A. borealis* seems to be mainly a saprophyte or a weak parasite affecting trees damaged by other agencies.

*A. cepistipes* was originally described by Velenovsky (1920). Korhonen (1978) found that his species B was very similar to Velenovsky's fungus, but not identical with it (Roll-Hansen, 1985). Romagnesi and Marxmüller (1983) therefore named Korhonen's fungus *A. cepistipes* f. *pseudobulbosa*, Velenovsky's fungus, which was found on beech in Czechoslovakia and in spruce forests in Germany, then becoming *A. cepistipes* f. *cepistipes*.

*A. cepistipes* f. *pseudobulbosa* has been found from Italy and eastern France northwards into Scandinavia and Finland (Guillaumin *et al.*, 1985; Roll-Hansen, 1985). Roll-Hansen (1985) points out that morphologically *A. cepistipes* and *A. lutea* (*A. bulbosa*) are closely similar. The same difficulty caused Guillaumin *et al.* (1985) in their table setting out the morphology of the fruit bodies of the European *Armillaria* species to combine the characters of *A. cepistipes* f. *pseudobulbosa* and those of *A. lutea* in a single column. The cap of these fungi is brown or olive-coloured, up to about 14 cm across, with widely scattered scales, and the stipe is smooth, with few or no scales, but bulbous and often yellowish at the base. The ring is thin and delicate. The rhizomorph system is early and profusely developed, the individual rhizomorphs being black, thick and tough, and monopodially branched (Guillaumin *et al.*, 1985). Fruit bodies often develop on these rhizomorphs on the ground some distance from the food base (Marxmüller, 1982; Rishbeth, 1982).

The rather little-known *A. cepistipes* f. *cepistipes* has a small, pale-coloured fruit body only about 1–5 cm across (Romagnesi and Marxmüller, 1983), and it appears to be saprophytic.

*A. cepistipes* f. *pseudobulbosa* seems to occur mainly on old stumps and dead trees (Korhonen, 1978; Romagnesi and Marxmüller, 1983). Rishbeth (1982, 1983) in inoculation trials was unable to infect two-year-old Scots pine with this fungus. Korhonen (1978) found that it was sometimes involved in butt rot of Norway spruce, and Gregory (1989) isolated it from basal decay of Norway spruce and western hemlock. Korhonen (1978) also recorded it on living *Acer* and lilac trees, while Guillaumin and Berthelay (1981) found it on living limes (*Tilia* spp.). At present the fungus appears to be mainly a saprophyte which sometimes becomes a weak parasite (Gregory, 1989).

In *Armillaria lutea* (*A. bulbosa*, *A. gallica*) the fruit bodies are similar to those of *A. cepistipes* f. *pseudobulbosa*, with a brown cap with discrete brown scales and a smooth stipe with a bulbous base and a rather delicate ring. Again like *A. cepistipes* f. *pseudobulbosa*, *A. lutea* has a richly developed system of tough, black, monopodially branched rhizomorphs (Guillaumin *et al.*, 1985; Rishbeth, 1988). In southern England, Rishbeth (1982) frequently found these in first-rotation conifer crops 20 m (or even 60 m in one case) from the edge of adjacent woodland containing broadleaved trees. Fruit bodies often develop from these rhizomorphs scattered over the ground around the food base.

Guillaumin and Berthelay (1981) found that *A. lutea* was the dominant *Armillaria* sp. in the French oak forests they examined. They considered that it was normally a saphrophyte which could become parasitic on trees first weakened by other agencies, including drought or defoliation by insect pests. Rishbeth (1982, 1983), who attempted to inoculate two-year-old Scots pine and five-year-old Corsican pine with little success, also concluded that it could attack trees weakened by other causes, and that it could cause extensive butt rot of broadleaved trees. Guillaumin *et al.* (1985) reported the fungus on apple trees and on pedunculate oaks, but again only when these were otherwise enfeebled, and considered it only a weak parasite. This view is supported by the observation of Rishbeth (1982) that freshly cut stumps of broadleaved trees colonised by *A. mellea*, which is strongly parasitic, rarely produced coppice regrowth, which was, however, profuse in such stumps colonised by *A. lutea* or *A. tabescens*.

*Spread into and through the stand*
It has proved difficult to demonstrate the mode of entry of the *Armillaria* species into uninfested stands. Leach (1937) failed to infect stumps or roots by inoculation with spores, and Gibson and Goodchild (1961) similarly failed to infect fresh pine billets. However, Rishbeth (1964) obtained a few successful inoculations of both pine and hardwood stumps, and concluded that although colonisation by spores might be relatively rare, it was probably important in giving rise to new infection centres. Similarly, Molin and Rennerfelt (1959) found that mycelium of *A. mellea* developed on the surfaces of exposed sections of spruce, pine and birch, and considered that it grew following the deposition of airborne spores. Further accumulated evidence from inoculation experiments and from a study of the distribution of clones of the fungi in a given area indicates strongly that the *Armillaria* spp. establish themselves by airborne spores which colonise freshly cut stumps and wounds such as those caused by extraction damage (Rishbeth, 1988).

Further spread within stands takes place partly by means of the rhizomorphs that grow through the soil from infested trees or stumps, and partly by contacts between infested and healthy roots (Clancy and Lacey, 1986; Rishbeth, 1985, 1988).

The rhizomorphs enter through roots (Gladman and Low, 1963) or at the root collar (Day, 1927a). Day found that rhizomorphs became attached to the host bark, and hyphae grew from their tips into the cork cells. They appeared to exert a toxic influence on the host cells before penetration. Small trees were often closely encircled by a web of rhizomorphs. The fungus could enter uninjured and apparently healthy trees, and often attacked some of the best and most vigorous ones, a fact confirmed by Boullard (1961). In support of this, Zeller (1926) and Thomas (1934) found that *A. mellea* could enter healthy, undamaged roots. Nevertheless, there has been much argument as to whether the fungus can in fact successfully invade healthy, unwounded trees, and Day (1929) later decided that it could enter and kill trees only if they were previously weakened by other causes, such as drought or defoliation. Results of experiments by Christensen and Hodson (1954) lend some support to this view, and Buckland (1953) in studies on Douglas fir in Canada found that when the fungus entered trees from the soil it did so only through mechanically or physiologically damaged roots. Pawsey and Gladman (1965) noted *A. mellea sensu lato* commonly associated with extraction wounds on conifer roots. Many of the above questions have been answered as a result of the realisation that the several *Armillaria* species as now understood differ in many respects, some being aggressive parasites and others only weak ones which can at times attack trees also damaged by other agencies. As far as the rhizomorphs are concerned, Rishbeth (1985) found that *A. lutea* (*A. bulbosa*), which has a richly developed rhizomorph system, spread more readily than *A. mellea*, the rhizomorphs of which form less extensively, while *A. ostoyae* was intermediate in these respects. It is interesting to notice that Marsh (1952), in England, found evidence that in the tree and bush-fruit crops he studied, in which it may be presumed that the pathogen was *A. mellea sensu stricto*, spread was by root contact. In Scandinavia, Molin and Rennerfelt (1959) reached the conclusion that infection of coniferous forest trees also took place mainly by root contacts. They considered that spread by means of rhizomorphs was much less important, and occurred over short distances of less than one metre. *A. mellea* seems to be absent from large parts of Scandinavia (Roll-Hansen, 1985). In Britain, and other temperate countries, however, infection of coniferous forest trees appears to be mainly by means of rhizomorphs (Day, 1927; Woeste, 1956; Patton and Riker, 1959; Redfern, 1978), though entry by root contacts also occurs in some stands (Prihoda, 1957).

Arthaud *et al.* (1980) studied the infection of isolated roots of *Pinus pinaster* by *Armillaria ostoyae*, and found that the mycelium penetrated the cortical parenchyma.

Entry into tree roots may be delayed by reactions of the host tissues (Buckland, 1953), which may check its growth by laying down callus and resin barriers (Tippett and Shigo, 1981). Various competing and antagonistic fungi may also hinder entry. Thus in Poland Orlos (1954, 1957) considered that *Polyporus borealis* and other fungi sometimes competed with *A. mellea* and prevented it from colonising stumps of Norway spruce. Plavsic (1979) also drew attention to a range of fungi in the root collars of usually dead trees, some of which hindered the growth of *A. mellea* but sometimes favoured it.

*Host range and relative resistance of various trees to killing and butt rot*
*Armillaria mellea sensu lato* has been recorded on over 650 hosts (Raabe, 1962), causing a root, collar and butt rot in a wide range of broadleaved and coniferous trees and shrubs, and sometimes attacking herbaceous hosts also (Moore, 1959). It causes much damage in woodlands and conifer forests and in parks and gardens. On hardwoods in Britain it was considered to be the cause of the death of many oaks in the early part of this century (Osmaston, 1927). At this time, *A. mellea*, on its own or in association with other causative agents, killed many oaks in continental Europe (Georgévitch,

1926). It has been recorded also on all other common hardwood trees, as well as many garden shrubs.

In Britain, however, though the fungus may sometimes cause damage in cricket bat willow plantings (Anon., 1968) and other hardwood crops, in forests the *Armillaria* species much more often cause losses in conifer crops (often those planted on hardwood sites) than on hardwood crops themselves (Day, 1929). The following comments supplement those given above under the individual species of *Armillaria*.

Among the conifers, most species may be affected, but spruces (including Sitka spruce) are especially liable to attack, while Douglas fir and grand fir are generally resistant (Redfern, 1978). However, even the more resistant species may be killed when young. The situation is complicated, because species often killed by *Armillaria* spp. are not always those most liable to butt rot. Some, such as Sitka and Norway spruce, western red cedar and western hemlock, are often killed, and also among the most susceptible to rot in older trees. On the other hand, Scots and Corsican pine are also frequently killed, particularly the former, yet pines are hardly ever affected by butt rot. Among the remaining species, the silver firs and Douglas fir are only occasionally killed, and are also fairly resistant to butt rot, and so are the most resistant to both aspects of the disease (Anon., 1967a). Baxter (1953), Peace (1962) and Greig and Strouts (1983) give more comprehensive lists of apparently resistant and susceptible species. Apart from the forest conifers mentioned above, Greig and Strouts (1983) quote Lawson cypress, Leyland cypress, monkey puzzle, the cedars, the Japanese cedar (*Cryptomeria japonica*) and the Wellingtonia (*Sequoiadendron giganteum*) as especially susceptible and often killed at any age. Of the broadleaved trees, birches are also very susceptible and so are the willows, apples and elms as well as the *Acer* spp. (except *A. negundo*), lilac and the *Prunus* spp. (except for cherry laurel and blackthorn).

Except once more for the forest conifers already mentioned, the firs (*Abies* spp.) and yew (*Taxus baccata*) (in Britain grown far more often as ornamentals than as forest trees) are resistant, especially the latter, which is scarcely ever affected. Among the broadleaved trees, Greig and Strouts list the box elder (*Acer negundo*) and the Californian black walnut (*Juglans hindsii*) as almost immune, and beech, oak, ash and lime as resistant; other broadleaved trees they have found to be resistant include the common box, the incense cedar (*Calocedrus decurrens*), the hawthorns, the sweet gum (*Liquidambar styraciflua*), the London plane, cherry laurel, blackthorn, the false acacia (*Robinia pseudoacacia*) and various garden shrubs and woody climbers.

*Symptoms and forms of attack: general observations*
Once one of the *Armillaria* spp. has succeeded in entering and girdling a tree root, it spreads distally from the point of entry along the cambial zone between bark and wood, and there produces its characteristic sheets of mycelium. Proximally, spread takes place against the host resistance in the bark scales, and the cambium is only entered and killed behind the advancing mycelial front (Woeste, 1956; Redfern, 1978). If the fungus reaches the

collar, it may girdle young trees and kill them. In conifers (particularly in spruces and pines), the first sign of attack is often a flow of resin from the base of the trunk (figure 16). The resin trickles down the bark, matting the litter on the ground at the foot of the tree into a hard mass. With time the whole root system may be colonised; the needles then yellow, and by the end of the summer they become brown and fall, and the tree dies. It is at this stage that the typical network of blackened rhizomorphs develops beneath the loosening bark.

**Figure 16**   Resin flow from trunk of *Picea omorika* attacked by *Armillaria mellea* (Forestry Commission)

On large trees, the fungus may progress only slowly, and for a long time may be confined to the roots. This is especially so in old hardwoods. Infections may be limited to small parts of the root systems for many years without obvious signs of infection above ground. With time and increasing root damage, however, particularly in drought conditions or following damage by defoliating insects, the trees tend to become stag-headed, and may eventually die. In many cases the fungus makes little headway until the trees are felled (Leach, 1937), but may form limited lesions on the roots, often delimited by the black zone lines described by Campbell (1934). Once the trees are felled, however, the fungus rapidly builds up in the stumps, and the latter may remain as sources of infection for at least 20 years. Though *A. mellea* may kill both coniferous and hardwood trees, and colonise their stumps in both cases, hardwood stumps generally appear to provide more suitable infection sources than those of conifers (Day, 1929). However, several cases have been recorded (Redfern, 1975) in which killing attacks were associated with conifer stumps alone.

Gladman and Greig (1965) have described the appearance of *Armillaria* decay, with particular reference to the butt-rot-susceptible species among the major conifers. When butt rot occurs, external symptoms, apart from occasional resin flow, are seldom seen, and infection is found only when the tree is felled, or following windblow resulting from root rot. In its earliest stages, infection shows in cross-sections of the stem base as a pale brown, grey or bluish, often water-soaked stain, that later becomes dark brown or blue-black. Later still, incipient decay becomes apparent, and water-soaked brown or orange areas appear. In longitudinal sections of the stem, scattered irregular white or cream-coloured pockets are visible in the wood. Black zone lines can often be seen bounding or within the decayed area. In the final stages of decay, the centre of the stem may be reduced to a soft, stringy wet mass, often containing secondary fungi and bacteria that follow the *Armillaria* sp. Surrounding this area, tissue as yet invaded by *Armillaria* alone usually appears as a soft, wet, orange-brown zone containing white pockets, and often also irregular black skins, the remains of the zone line tissue.

Butt rot caused by *Armillaria* spp. usually results in a conical zone of decayed tissue that does not ascend for more than about 60 cm from the base of the stem (figure 17).

*Factors affecting the progress of the disease*
The likelihood of initial infection by rhizomorphs is much affected by inoculum potential, which is governed by both the size of the stumps from which the rhizomorphs begin their growth and the distance they have to grow before encountering a suitable host (Garrett, 1956b). Stumps such as those of Scots pine, which soon decay, are less important as infection sources than those of more durable broadleaved trees (Rishbeth, 1983). If the nutrient status of the substrate is too low, rhizomorphs probably do not develop. As already noted, rhizomorph initiation may be hindered by the presence of various antagonistic soil fungi. In laboratory experiments, *Trichoderma viride* (Garrett, 1958) and *Pleurotus ostreatus* (Kunze, 1952–3) have been shown to exert this effect, whereas *Aureobasidium pullulans* may stimulate rhizomorph production (Pentland, 1965). In other conditions, the rhizomorphs are produced, but inoculum potential is relatively low, the fungus may make no progress against the resistance of the host, and it may be overtaken by other organisms (Garrett, 1956b).

The most important physical characters affecting the disease are generally considered to be soil temperature and soil moisture. In California, Bliss (1941, 1944, 1946) studied the fungus in agar culture and on citrus in soil

**Figure 17**  Longitudinal section through butt of Norway spruce showing conical zone of central decay and stain caused by an *Armillaria* sp. (Forestry Commission)

tanks. He obtained severe, moderate and slight infection of citrus at 10, 17 and 24 °C, respectively. At 31 °C the inoculum was killed. Soil temperatures are not likely to place any marked limitations on infection in Britain.

Drought conditions have often been considered to render trees more liable to infection. Gladman and Low (1963) studied a crop of Sitka spruce growing on a peaty flat containing a number of dry knolls, and found *Armillaria* butt rot in the trees on the knolls but not in the wet flat.

*Losses caused by the Armillaria species*

The *Armillaria* spp. are very important as causes of damage to many ornamental trees and shrubs in parks and gardens, and to fruit trees in orchards. It is difficult to assess the extent of losses caused by these fungi in the forest. In Britain, although they are absent from plantations established on land that was not formerly under forest, they soon become established. As already noted, they may kill trees, or rot the timber in their butt lengths, but they cause far less overall damage than *Heterobasidion annosum* (*Fomes annosus*). Thus Peace (1938) found 80 per cent of the rotten trees he investigated were infected by *H. annosum* and only 1 per cent by *A. mellea*. Nevertheless, the incidence in individual stands may sometimes be considerable.

Greig (1962) observed that *A. mellea sensu lato* frequently killed young conifers up to the first thinning when they were planted on ex-hardwood sites, but it usually caused little or no loss of saleable timber because the affected trees tended to be few and scattered. Locally the losses could be quite heavy, however, especially in the case of Sitka and Serbian spruce, western hemlock, Corsican pine and the larches, and if site conditions were unfavourable, groups of trees might be killed. On the other hand, Redfern (1978) has associated group killing with other factors, such as the history of harvesting operations on the site, and variations in the pathogenicity of different isolates (or species as now understood) of *Armillaria*.

Damage from butt rot is relatively low because the fungus rarely advances more than about 60 cm up the stem (though secondary fungi sometimes extend the rot further up the tree). In addition to this, rotting of roots may lead to windblow (Gladman and Low, 1963). Occasionally, large numbers of trees may show butt rot, and Greig (1962) noted one crop of Sitka spruce in which 90 per cent of the butts removed in one thinning had either butt rot or stain. As a rule, however, the fungus affects only a small proportion of the crop.

*Control measures*

It is very difficult to control attacks by the *Armillaria* spp. To prevent primary infections by basidiospores in newly established forests or gardens, it may be possible to kill fresh thinning stumps with an ammonium sulphamate solution, which also encourages harmless decay fungi which compete with *Armillaria* and break down the stumps (Rishbeth, 1983).

Once *Armillaria* spp. become established on a site, infested stumps become sources of infection to surrounding trees and new plantings. To reduce stump colonisation, biological control methods, intended primarily for use in plantation crops in tropical or subtropical areas, were worked out by Leach (1937) in tea plantations in Nyasaland (Malawi). Leach observed that after penetration the fungus tended to grow most rapidly on those parts of the tree richest in carbohydrates. He considered it possible that if he could reduce the carbohydrate content of the roots of trees before felling, their stumps would be less susceptible to invasion by *Armillaria* when the trees were felled, and, hence, they would be less likely to become sources of infection to a succeeding crop. He found experimentally that this was so, and developed a method for depleting the stump carbohydrate reserves by ring-barking them from 18 months to 2 years before felling. Other workers in Africa (Wiehe, 1952; Gibson and Goodchild, 1961) have confirmed Leach's results, and his methods have been widely used in tea and Tung plantations there. So far this method of control has been used only in areas being cleared for crops of high value, and not in general forestry. Studies by Redfern (1968) suggest that in Britain girdling hardwood trees before felling, or killing them by means of 2,4,5-trichlorophenoxyacetic acid does not hinder colonisation of the stumps by *Armillaria*, though it does appear to make the stumps much less effective as food bases for the fungus within five years of treatment. It is therefore possible that girdling or poisoning could be used to reduce infection in conifer plantings to be established on infested hardwood sites, provided a delay in replanting is acceptable.

Direct inoculation with fungi to compete with *Armillaria* has sometimes been proposed and attempted. Thus Rishbeth (1986) quotes a case in which elm stumps treated with ammonium sulphamate were also inoculated with *Coriolus versicolor* or *Phlebia merismoides*, which both became well established within three years. Federov and Bobko (1988b) found that *Phlebia gigantea* and *Pleurotus ostreatus* were both promising competing fungi.

A highly desirable measure where feasible is the physical removal of the stump mass with as much as possible of the infested root material by one of the methods described by Wilson (1981). This may be followed by some form of soil treatment.

Some workers have suggested treatments of roots or the root zone with various fungicides, including iron sulphate (Borg, 1935; Jenkins, 1952), iodine (Guyot, 1933), copper (Rayner, 1959) and a refined creosote preparation (Pawsey and Rahman, 1976). The latter concluded from their experiments that a number of materials applied to the soil may, depending on site and perhaps climatic conditions, cause death or suppression of rhizomorphs. However, unless the source of rhizomorphs is eradicated by removal of infested stumps and roots, rhizomorphs are likely to reinvade the treated area should treatment be allowed to lapse. Similarly, Federov and Bobko (1988a) tested various materials, including benomyl, carbendazim, copper oxychloride and thiophanate methyl, but found that none had more than a very temporary effect on the fungal mycelium or rhizomorphs. In laboratory trials on the control of *A. mellea*, *A. ostoyae* and *A. bulbosa* (*A. lutea*), Turner and Fox (1988) compared hexaconazole, flutriafol, fenpropidin and a guanide with a phenolic and with cresylic acids. All these materials, especially the first four, showed some promise under the conditions of their tests.

Soil fumigants are likely to be of greater value than more conventional fungicides. When dealing with valuable plantation crops in warm climates, the fungus may be eradicated by soil injection with carbon disulphide (Bliss, 1944, 1951; Darley and Wilbur, 1954). The chemical is poisonous and highly inflammable, however, and doubt has been cast on its effectiveness in Britain because good results are only obtained at fairly high soil temperatures (Anon., 1961). The soil fumigant methyl bromide was also found to be effective in citrus and fig plantations when tested by Larue *et al.* (1962). Escarment *et al.* (1985), in south-west France, obtained encouraging results from laboratory and soil treatment trials with methyl bromide used against *Armillaria ostoyae* (*A. obscura*), and Greig (1987b) has reported on trials in which metham sodium, dazomet and methyl bromide were applied to infested stumps through vertically drilled holes. These materials, especially methyl bromide, gave very promising results. Methyl bromide is highly toxic, and is controlled by poisons regulations. Formaldehyde drenches have also been recommended to fumigate infested areas (Brooks *et al.*, 1989). In gardens, it is often difficult to use such materials without causing damage to surrounding healthy plants.

In forests, some loss from *Armillaria* spp. is normally acceptable, but if it is necessary to replant areas of conifer in which losses through these fungi have been high, or to fill gaps caused by their activities, one of the relatively resistant species such as Douglas fir or grand fir should be chosen where silviculturally possible. Resistant trees and shrubs can also be used in parks and gardens, where individual plants are of special value.

### Root and butt rot caused by *Heterobasidion annosum* (Fr.) Bref.
### (*Fomes annosus* (Fr.) Cooke)

*Heterobasidion annosum* (*Fomes annosus*) is the most damaging of all the fungi that attack conifers in Britain and is equally important elsewhere in Europe. Peace (1938) found it to be the commonest cause of loss in a wide butt-rot survey, and Low and Gladman (1962a) found it was responsible for 90 per cent of the rot they had seen in conifers in Scotland. In some circumstances it may also kill mature trees, particularly pines, as well as young trees of this and other genera. It is widely distributed throughout the world (CMI Map No. 271) and has an extensive host range. *H. annosum* has been monographed by Negrutskii (1986).

### *The causal fungus*

The basidial fruit body is perennial, usually bracket-shaped (figure 18), but often irregular and becoming resupinate when, for example, growing on the underside of roots or lining the roofs of rabbit burrows made under infected stumps. It is usually about 5 to 10 cm across but may be up to 30 cm. The upper surface is covered by a bright reddish-brown crust that becomes dark brown or black with age, and is concentrically grooved. The thin, acute margin is white or cream when the fruit body is growing. The underside is white, becoming cream or biscuit-coloured, and pierced by fine pores, the openings of the tubes that bear the basidial hymenium (figure 19). The tubes are formed in distinct, normally annual, layers, each 1–10 mm deep. The flesh of the fructification is tough and leathery, and becomes hard and woody on drying. It is whitish, cream or pale wood-coloured. The basidiospores are white, globose or subglobose, $4$–$6 \times 3$–$5$ $\mu$m. As it grows the fruit body often encloses small objects such as small sticks and conifer needles, and holes may be left as these decay and become detached (Greig and Redfern, 1974; Wakefield and Dennis, 1981). Fructifications may be found at all times of the year if the humidity is high enough. They may be killed by extreme cold, but are resistant to moderate frost and dry weather, and may shed spores even after prolonged drought and at 0 °C (Rishbeth 1951a; Haraldstad, 1962).

They occur on stumps, sometimes at the bases of dying trees, and on infected roots exposed by windblow (plate 18). They generally form at or just above ground level, and are often hidden by brash and litter, or occur in deep sheltered cavities. Indeed, infected stumps may bear fruit bodies when covered with brash, or when well-sheltered, but show none when more exposed, so that absence of fructifications may not indicate the absence of the fungus (Rishbeth, 1951a). As already noted, resupinate fructifications may be present in rodent tunnels passing near diseased roots. They sometimes also appear in litter, and are then associated with fine infected roots, though they

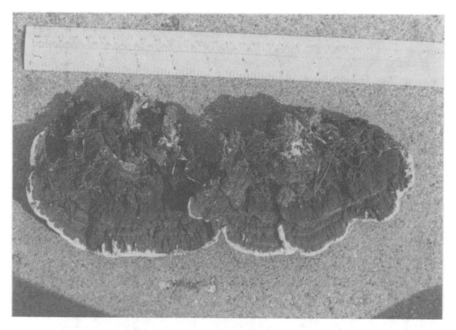

**Figure   18**   Typical   bracket-shaped   fruit   body   of   *Heterobasidion   annosum*
(D.H. Phillips)

may then occur several metres from the stem of the tree to which the roots
belong (Rishbeth 1951a). Small white pustules, sometimes with rudimentary
pore surfaces, are frequently found, especially on cut surfaces of stumps or
the base of dying trees and under roots (figure 20). Mycelium, when visible, is
found only as thin, white, tissue-like sheets under the bark of dead roots (Low
and Gladman, 1960).

Schönhar (1980) examined the survival of basidiospores in soil samples
collected from spruce plantations. He found that those held in soil at a
moisture content of 30 per cent were still alive after one year. Those kept at a
moisture content of 60 or 90 per cent showed a reduced capacity for germination after 3 months, and failed to germinate at all after one year.

Work in North America indicates that spore deposition varies from place to
place and may vary also with the time of year and the time of day. Thus James
and Cobb (1984) found that deposition was greater in a forest in northern
California than in one further south. In the northern forest, most spores were
shed in autumn and fewest in winter, while in the southerly one most were
shed in winter and fewest in summer. In the southern forest, where diurnal
spore deposition patterns were also examined, most spores were shed at
night. Further north, in western hemlock forests in Oregon and Washington,
spore deposition was usually greatest in autumn and spring, and at its lowest
level in winter and summer (Edmonds *et al.*, 1984b). Also in western hemlock
forests in Washington, Edmonds *et al.* (1984a) studied deposition in June and

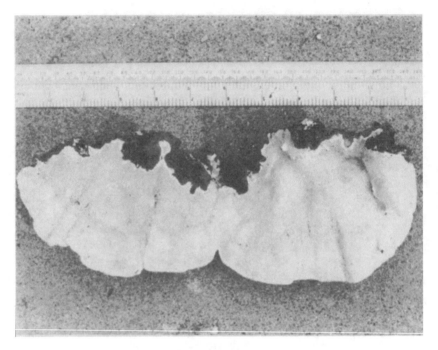

**Figure 19**  Lower surface of fruit body of *Heterobasidion annosum* seen in figure 18
(D.H. Phillips)

October. In June there was no particular deposition pattern, but in October most spores were shed at night and early in the morning.

An anamorphic stage, consisting of a fine turf of conidiophores with swollen heads, formed directly on the mycelium, may also occur, though it is rather rarely seen in nature and is easily overlooked. The conidia are sub-globose to ovoid and hyaline, 4.5–7.5 (–10.5) × 3.0–6.0 μm (Nobles, 1948), and borne on small sterigmata (figure 21). Rishbeth (1951a) occasionally found this stage in East Anglia round bark beetle holes at the sheltered bases of young pine trees, and under the bark of a dead pine just above ground level, and Bakshi (1950, 1952) discovered it in the tunnels of ambrosia beetles. It can often be found in damp weather when carefully sought for on the surfaces of infected stumps covered by litter and on the ends of stacked infected logs. Its colonies are recognisable on incubated infested wood after 4–5 days and so often form a useful means of identifying the fungus. It is readily produced in culture, and on infested wood inoculated in moist conditions, but it is usually thought to play little part in natural dispersal. Nevertheless, Hsiang *et al.* (1989) produced evidence that in western hemlock forests in the western USA conidia may form from one-third to one-half of the spores of *H. annosum* in the air.

The conidia are not produced in culture when the relative humidity is below 95 per cent, and they form at temperatures from 0 to 22.5 °C in light or darkness (Rishbeth, 1951a).

**Figure 20** *Heterobasidion annosum*:pustules (young fruit bodies) on roots of Douglas fir (Forestry Commission)

Rishbeth (1951a) found that growth of *H. annosum* in culture took place at temperatures from 0 to 28.5 °C, with an optimum at 22.5 °C, when he measured a growth rate of 7.8 mm/day on malt agar, and 7.3 mm/day in pine roots. Similarly at 10 °C the fungus grew 2.5 mm/day on agar and 2.8 mm/day in roots, while in pine stumps in the field it grew 2.6 mm/day (about 1 m/year) over a period when the mean soil temperature was also 10 °C. In culture, optimum growth takes place from about pH 4.0 to 5.5 (Etheridge, 1955). Much additional information on the physiology of the fungus is given in the review by Gundersen (1962).

In Europe and North America, *H. annosum* is now known to be hetero-thallic and to consist of at least two intersterility groups, while a third group is known in Australasia (Chase and Ullrich, 1983; Holt *et al.*, 1983; Siepmann, 1988; Stenlid and Swedjemark, 1988). So far in Europe, the P-isolates are known to attack Scots pine of all ages, as well as Norway spruce, Douglas fir and larch, while the S-isolates chiefly affect Norway spruce (in which they mainly cause butt rot), but they also attack young Scots pine (Siepmann, 1988; Stenlid and Swedjemark, 1988). A similar situation has been described in Italy (Moriondo *et al.*, 1988), where, however, isolates from *Abies alba* seem to differ from both the P- and S-groups. In this connection it is interesting that when Worrall *et al.* (1983) compared isolates from *Pinus ponderosa* and *Abies concolor*, the pine isolates proved to be more aggressive to pine then to *Abies*. Similarly the *Abies* isolates were more aggressive to

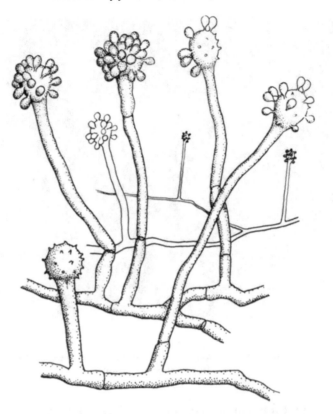

**Figure 21**  The microscopic anamorphic stage of *Heterobasidion annosum*, showing the spores borne on the swollen heads of the conidiophores (× 500 approx.) This stage has been called *Oedocephalum lineatum* (Forestry Commission)

*Abies* than to pine. Additional evidence of host specialisation towards conifers was produced by Hsiang and Edmondson (1989).

Stenlid and Wästerlund (1986) found that they could detect a fairly high proportion of the decayed trees in a Norway spruce stand by the use of an increment borer. Zollfrank *et al.* (1987) give details of an immunohistochemical assay technique for the detection of *H. annosum* in Norway spruce wood.

*Spread of the fungus in the forest*
Information on the rate of spread of the fungus in British forests relates only to pine, in which Rishbeth (1951a) found the growth was about 1.0 m/year. According to Fowler (1962), rate of spread through pine plantations in northeastern USA is about the same as in Britain, though it may be about twice as fast in some plantings of the very susceptible red pine (*Pinus resinosa*).

Rishbeth (1951b) and Gibbs (1967b) have shown that the roots of suppressed trees are relatively susceptible to invasion, yet they are rarely entered under natural conditions, perhaps because of competition by other fungi on the root surface, and also because the fungus is not commonly present in soils.

Braun (1958) found that soil supported the growth of *H. annosum* only after sterilisation, and concluded that the fungus could not grow freely in the soil. When Francke-Grosmann (1962) tried to grow the fungus in non-sterilised soil, it grew only when she provided it with a large food base, and even in these favourable circumstances it disappeared in a few weeks, apparently destroyed by other fungi, mites, springtails and other organisms.

Basidiospores of *H. annosum* have been shown to occur in soil (Harald-stad, 1962) and may sometimes be washed down to a depth of 50 to 90 cm (Rishbeth, 1951a). In laboratory experiments, however, Rishbeth found it was difficult to induce spores to produce growth on unsterilised root pieces. He therefore concluded that the fungus did not usually enter uninfested stands directly by the colonisation of roots by means of spores or free-living mycelium present in the soil, and sought other means by which it might enter the crop.

Hepting and Downs (1944), in a study of white pine plots in North Carolina, found root and butt rots (mainly caused by *H. annosum*) in 75 per cent of the trees in previously thinned plots, but in only 4 per cent where no thinning had been done. They suggested that butt rot was so extensive in the thinned areas because a build-up of *H. annosum* had taken place in the stumps of the thinned trees. Rishbeth (1951a,b) obtained experimental evidence that the fungus colonised freshly cut conifer stumps by means of airborne spores, and then grew down into the stump roots, and by root contact then invaded the healthy living roots of adjacent trees (figures 22, 23). As noted above, over a period when the mean soil temperature was 10 °C *H. annosum* grew in stumps at the rate of about 1 m/year, and Rishbeth estimated that when pines were planted about 1.4 m apart, the fungus could pass down into surface-colonised stumps and along their roots, and closely approach the roots of adjacent trees within a year of initial infection. Similarly, Meredith (1960), in inoculation experiments with pine stumps, found the fungus invading from the stump surface often reached the proximal ends of lateral roots in six months, and was present in their more distal ends 12 to 18 months after inoculation. Many other workers have also found an association between thinning stumps and attack by *H. annosum* (for example, Schönhar, 1981; Stenlid, 1987).

The surfaces of freshly cut stumps remain susceptible to invasion by *H. annosum* for only 3 to 4 weeks or even less after felling (Rishbeth, 1951a, 1959a; Cobb and Schmidt, 1964), though the period of susceptibility of the subsurface tissues may be reduced or prolonged by various chemical treatments applied to the stumps (Rishbeth, 1959a).

*Host range and susceptibility*
*H. annosum* attacks a very large number of woody hosts. Economically it causes loss only in conifer crops, but it may also attack hardwoods such as birch, beech, and rarely also oak and alder (Rishbeth, 1957), which are

**Figure 22** *Heterobasidion annosum*:transfer of disease from an old infected thinning stump to a nearby Japanese larch (Forestry Commission)

seldom severely affected, but (especially birch) may pass on the fungus to a succeeding conifer crop (Low and Gladman, 1960). Wagn (1987) reported on a Danish experiment which was continued over 23 years. Among the most susceptible trees were the conifers *Pinus contorta*, *Picea glauca*, and *Larix leptolepis* (though in this experiment *L.* × *eurolepis* proved resistant). Also very susceptible were the broadleaved trees *Alnus glutinosa*, *Betula pubescens*, *Corylus avellana*, *Pyrus communis*, *Quercus rubra* and *Sorbus intermedia*.

In conifers, symptoms of attack and the damage resulting vary with the tree species concerned (Peace, 1938; Low and Gladman, 1960; Hsiang and Edmonds, 1989). Pines are very rarely subject to butt rot, except in old age, but in dry areas with soils of high pH (often on limed, ex-agricultural land), many trees may be killed (figure 24). Pines are also killed on other sites with more acid soils but damage is generally less serious. However, their fresh stumps are readily colonised by *H. annosum*, and then transmit it by root contact to more susceptible species planted subsequently or in mixture. Some time before they die, the affected trees show a thinning of their foliage and a shortening of the needles (figure 25). Differences in susceptibilities between individual pine trees have been detected (Werner, 1987).

Trees of genera of conifers other than pines may also be killed, especially when young. As they become older, they are more usually affected by root and butt rot.

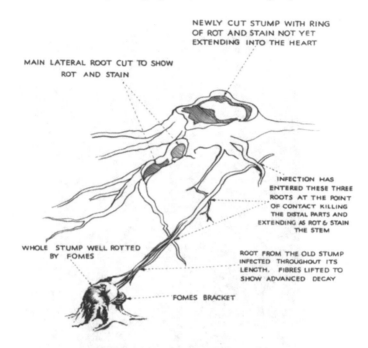

NEWLY CUT STUMP WITH RING
OF ROT AND STAIN NOT YET
EXTENDING INTO THE HEART

MAIN LATERAL ROOT CUT TO SHOW
ROT AND STAIN

INFECTION HAS
ENTERED THESE THREE
ROOTS AT THE POINT
OF CONTACT KILLING
THE DISTAL PARTS AND
EXTENDING AS ROT & STAIN
THE STEM

WHOLE STUMP WELL ROTTED
BY FOMES

ROOT FROM THE OLD STUMP
INFECTED THROUGHOUT ITS
LENGTH. FIBRES LIFTED TO
SHOW ADVANCED DECAY

FOMES BRACKET

**Figure 23** Diagram interpreting figure 22. *Heterobasidion annosum* has spread from the long-infected 10-year-old thinning stump (bottom left) to the 30-year-old healthy tree (top right) through the root contacts (Forestry Commission)

Some species of *Abies* (the silver firs) are resistant except when very young, when they may be killed by the fungus. Resistance varies very much with the species; for example, *A. alba* appears to be highly resistant on most sites, and *A. grandis* seems slightly less so (Low and Gladman, 1960), while *A. balsamea* is very susceptible (Roll-Hansen in Anon., 1962). Even *A. alba* has been severely affected on ex-agricultural sites in Italy (Bronchi, 1956; Capretti and Moriondo, 1983) and silver firs in Britain may also be attacked, particularly on alkaline soil, but the fungus is slow to advance up the stem. There is evidence from work by Coutts and Rishbeth (1977) and Worrall and Parmeter (1983) that *Abies grandis* and *A. concolor* may be resistant to attack by *H. annosum* because they react by the production of wetwood, which inhibits the fungus.

All the other conifers commonly planted in Europe may be severely affected by root and butt rot, though in Douglas fir (*Pseudotsuga menziesii*) decay rarely proceeds beyond the roots, which may be extensively rotted, rendering the tree liable to windblow (Low and Gladman, 1960, 1962b) (figure 26). Douglas fir has also been found to be resistant elsewhere. Thus, in Italy, Capretti and Moriondo (1983) reported that it remained free of infection even in mixed stands in which *Abies alba* was severely affected.

In other genera, the fungus spreads from the roots into the trunk, to cause an extensive butt rot (Gladman and Greig, 1965) (figures 27, 28, 29). At first,

**Figure 24** A group of second-rotation Scots pine killed by *Heterobasidion annosum* (Forestry Commission)

infection shows only as a stain, visible in the stem when the tree is felled. This stain is often excentric in position, and can be traced down into the roots from which it originated. Its colour varies with the tree species, though in all it becomes dark brown with time. Whereas in Lawson's cypress (*Chamaecyparis lawsoniana*), western hemlock (*Tsuga heterophylla*) and western red cedar (*Thuja plicata*), it is dark brown from the beginning, it is at first brownish grey in larch, and in the spruces, pink, lilac and bluish staining often mingles with the brown. Further growth of the fungus results in incipient and finally advanced decay. In incipient decay, the wood usually becomes pale in colour, and in longitudinal section begins to show black flecks, and later small, regular, narrow, rather lens-shaped pockets of white material. In the earlier stages the wood remains relatively firm and sound, and the decayed zones are surrounded by bands of stained tissue. Later, the white pockets tend to coalesce, and their white contents may disappear. The decay may have progressed about 4 m up the stem by the time the trees are 30 years old (Low and Gladman, 1960) and in older spruces may reach more than twice this height (Low, personal communication). Empty spaces occur in the decayed wood, which then shows extensive breakdown, and is light, dry and

**Figure 25**   Thinning of the foliage of Scots pine attacked by *Heterobasidion annosum* (B.J.W. Greig)

fibrous (figure 30). Affected trees may therefore become hollow and snap in high winds.

Timber showing the earliest stages of infection may be used, for example, for fencing material, but it must then be thoroughly impregnated with creosote before use, or it will act as a source of infection.

In one set of adjacent unreplicated 34-year-old plots, Gladman and Low (1963) found the percentage of infected trees of Scots pine, grand fir (*Abies grandis*), European larch, hybrid larch, western red cedar, Douglas fir and western hemlock to be respectively nil, 6, 25, 40, 50 and 96 per cent. They considered these figures gave a fairly good general impression of the relative susceptibility of those species, though on many sites they had found losses in western red cedar relatively higher and those in Douglas fir relatively lower than those quoted here. This was partly because the figures were based only on an examination of the stump surfaces of felled trees, and therefore gave an incomplete picture, as the fungus progresses at a different rate up the trunks of trees of different species and, while in Douglas fir it normally remains mainly in the roots, in western red cedar it moves fairly quickly up the stem.

Some indications of the relative susceptibility of a number of species to butt

**Figure 26**   Root system of windblown Douglas fir showing extensive decay caused by
*Heterobasidion annosum* (Forestry Commission)

rot caused by *Heterobasidion annosum* have also been given by Greig (1979).
In a sample of 6-year-old trees from a susceptibility trial, the incidence of
infection by *H. annosum* was 21 per cent in western hemlock, 19 per cent in
Leyland cypress, 11 per cent in Douglas fir, 7 per cent in western red cedar
and 1 per cent in grand fir and Norway spruce. The low incidence of infection
in Norway spruce was somewhat surprising and may have been related to the
poor growth of the tree on the experimental site rather than any inherent
resistance.

Heavy resin production by stumps hinders invasion by *H. annosum* (Meredith, 1959), and the ability to produce much resin as a reaction to invasion
may be a factor in resistance. The presence of resin also affects the passage of
the fungus from infested to healthy roots by root contact (Rishbeth, 1951b).
Johansson and Stenlid (1985) and Stenlid and Johansson (1987) found that
infection of wounds on the roots of Norway spruce caused an accumulation of
phenolic compounds, which inhibited the growth of the fungus. Tippet and

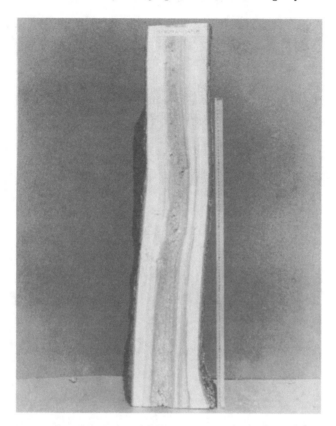

**Figure 27** Longitudinal section of European larch with extensive decay caused by *Heterobasidion annosum* (scale length = 1 m) (Forestry Commission)

Shigo (1980) reported that invasion of roots of *Pinus resinosa* by *H. annosum* led to compartmentalisation of the fungus and resin production, decay then being contained within barrier zones.

*Factors affecting development of the disease: competing fungi*
Competing fungi often prevent the colonisation of stump surfaces by *H. annosum*. Thus Moreau and Schaeffer (1962) state that spruce is seldom attacked by this fungus in the hills of the French Jura because of antagonism by *Polyporus borealis*. Most of the information on this subject, however, relates to species of pine, in which *Phlebia gigantea* (*Peniophora gigantea*) is much the most important competitor (figure 31) (Rishbeth, 1951a; Meredith, 1959; Gremmen, 1963), though other common fungi, such as *Trichoderma* spp., *Penicillium* spp. and *Stereum sanguinolentum* may also play a part (Meredith, 1960; Curl and Arnold, 1964; Gooding, 1964; Capretti and Mugnai, 1989).

Holdenrieder (1984) examined various fungi for antagonism but concluded that of them all only *P. gigantea* could be relied upon to prevent invasion by

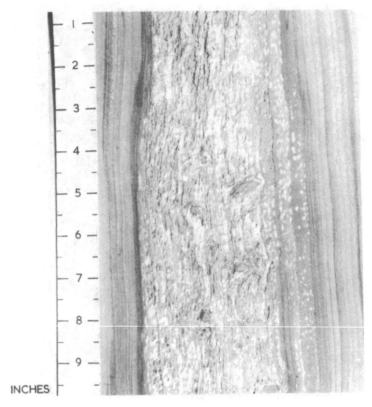

**Figure 28**   Detail of decay of European larch by *Heterobasidion annosum*, showing
          pockets of white mycelium (Forestry Commission)

*H. annosum.* Of the fungi they tested, Poleshchuk and Yakimov (1986) found
that the most promising antagonist, apart from *P. gigantea*, was the oyster
mushroom, *Pleurotus ostreatus*.

Spores of *P. gigantea* are often present in the atmosphere in sufficient
quantities to colonise fresh stumps and prevent the entry of *H. annosum*. At a
later stage, *P. gigantea* may displace *H. annosum* in stumps already invaded
by the latter (Rishbeth, 1951a; Meredith, 1960; Hodges, 1964; Towers, 1966).
Meredith (1960) found that when he inoculated fresh stump surfaces with
mixed suspensions of spores of *H. annosum* and *P. gigantea*, the latter
became the dominant colonist even when its viable spores were outnumbered
by those of *H. annosum* by as much as 10 or sometimes even 100 to 1.

Species of *Trichoderma* have been studied by various authors. Flemming *et
al.* (1982) examined *T. harzianum*, and Donnelly and Sheridan (1986) discovered that *T. polysporum* produced anthraquinones, which inhibit the
growth of *H. annosum*.

The spread of *H. annosum* in stumps on ex-hardwood sites may be restricted by *Armillaria mellea* invading either from infected roots or by means of
rhizomorphs at the root collar (Low and Gladman, 1962b).

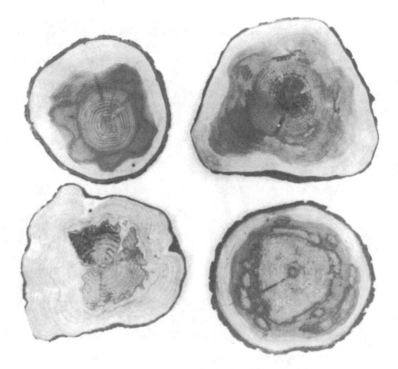

**Figure 29**  Cross-sections of trees showing decay by *Heterobasidion annosum* in Douglas fir (top left), Sitka spruce (top right), western hemlock (bottom left) and European larch (bottom right) (Forestry Commission)

Competing fungi also appear to play some part in limiting the passage of the fungus in the soil from root to root (Rishbeth, 1951b; Moreau and Schaeffer, 1962; Gibbs, 1967a).

*Other factors affecting the disease*
The likelihood of stump infection is much affected by the distance from spore sources. This is emphasised by the figures shown graphically in figure 32, which indicate a rapid fall-off in the number of spores in the air only a few hundred metres from the edge of a forest in which fruit bodies of *H. annosum* occur. Similarly, Gladman and Low (1963) sampled conifer needles from east Scotland (where conifer plantations were well established before 1900, and *H. annosum* fruit bodies were common), and found a mean count of 2652 viable spores of *H. annosum* per 10 g of needles, while on similar samples from west Scotland (where almost all conifer plantations were relatively young, with few fructifications), the counts averaged only 44 viable spores. Nevertheless, some viable spores of the fungus were found on almost every sample. Rishbeth (1959c), in spore trapping experiments with needle samples and with muslin square air samplers, obtained some evidence that spores of *H. annosum* could sometimes be found as far as 50 to 80 or more miles from the nearest sources.

**Figure 30**    Rot caused by *Heterobasidion annosum* in Scots pine: the final stage, when
the wood has become light, dry and fibrous (D.H. Phillips)

As noted earlier, in some places, spore levels in the air are low at certain
times of the year (Edmonds *et al.*, 1984b). Schönhar (1980) in Germany
considered that there was little stump infection in the area he examined
because the trees had been felled in winter, when there were few spores in the
atmosphere.

Shaw (1989a) inoculated young trees and stumps of western hemlock and
Sitka spruce in southeastern Alaska. Little infection took place, and he also
found few natural infections. The fungus also failed to survive for long in
naturally infected stumps. He concluded that this was due to the high water
content of the stumps in this area of high rainfall, and to the low temperatures
experienced there. Coutts and Rishbeth (1977) had earlier noted that in grand
fir (*Abies grandis*) wetwood may be formed in the sapwood in response to
attack by *H. annosum*. This may be at least in part the reason for the
resistance of *A. grandis* to attack by this fungus.

Once colonised by *H. annosum*, stumps may remain as sources of infection
for many years. Larch stumps have on one occasion been found to be infested
as long as 63 years after felling (Low and Gladman, 1960), though this is
unusual, and Curschmann (1960) found the fungus no longer present on land
cultivated and not replanted for about 50 years after forest clearance. Large
stumps, which are slow to rot and disappear, remain longer as sources of
infection than do small ones. Pine stumps, which are readily colonised by the

**Figure 31**   Fruit body of *Phlebia gigantea* (*Peniophora gigantea*) on pine stump. *P. gigantea* competes with *Heterobasidion annosum* in the colonisation of thinning stumps. Pine stumps may be protected from infection by *H. annosum* by artificial inoculation with oidia of *P. gigantea* (Forestry Commission)

fungus, rot more rapidly than those of other species, so *H. annosum* does not persist in them as long as it does in stumps of, for instance, larch and spruce.

Though fresh stumps provide the main point of entry into healthy crops, the fungus has sometimes gained entry also through root contacts with fence posts made from infested wood (Jorgensen, 1955). Occasionally infection of brashing, pruning and extraction wounds also occurs (Low and Gladman, 1960). In Europe this appears to be very uncommon (Braun, 1956, 1958; Pawsey and Gladman, 1965), though it occurs more often, particularly in *Tsuga*, in parts of North America (Englerth, 1942; Foster, 1962; Hunt and Krueger, 1962). Suppressed trees appear to be rarely attacked (Meredith, 1960; Anon., 1964), but windblown trees have sometimes become infested (Anon., 1962). Singling of double leaders has also provided a place of entry in some cases (Low and Gladman, 1960).

Various other factors affect the development of the disease, the most important being soil conditions, former land use, the age of the trees, and the amount of thinning or other cutting carried out in the crop. Much of the available information comes from studies in pine crops, and so does not necessarily always apply fully to crops of butt-rot-susceptible species. Of the

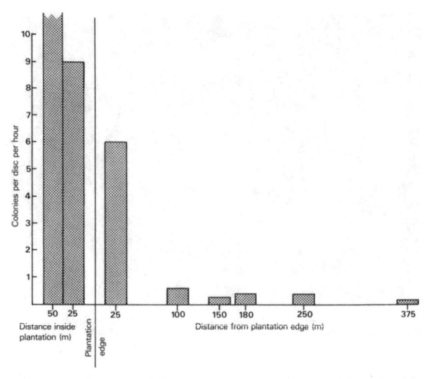

**Figure 32**  Spread of spores of *Heterobasidion annosum*. Spores of the fungus were trapped on Scots pine discs in a plantation of *Thuja plicata* and outside it at various distances from the plantation edge. The spore concentration is expressed as the mean number of colonies per exposed disc. Craig Phadrig Forest, Scotland, 1960 (data of J.D. Low and R.J. Gladman)

soil factors, pH, soil moisture conditions, soil texture, organic matter content and drainage appear to have some influence. Many workers have found a relationship between high soil pH and death of pine, and Wallis (1960) showed that in the East Anglian pine stands he studied, losses were high where the pH was above 6. On sites where the pH was lower than this, losses were negligible. Similarly, Low and Gladman (1960) found that pines on wet acid peats were not affected by *H. annosum*. In Culbin Forest (Morayshire) established on alkaline sand dunes, they found that pines of the first rotation were killed. After planting, the sandy soil fairly rapidly became more acid, and losses in the second-rotation crops appeared to be reduced. The situation in Culbin Forest is exceptional, and losses in pine crops on alkaline sites are usually long-continued. Soil moisture is also important, however, and *H. annosum* may kill trees even on acid soils in very dry areas. Thus death of both lodgepole pine (*Pinus contorta*) and Sitka spruce (*Picea sitchensis*) has been recorded even in acid areas in Tentsmuir Forest (Fife), where the rainfall is low and the soil light and well-drained, as well as in similar areas in

East Anglia (Rishbeth, 1957). Other workers have also noted more severe attack on light soils than on heavy ones (Jørgensen *et al.*, 1939; Anon., 1963; Pagony *et al.*, 1983). This may also be because the lighter soils tend to dry out, and the trees are then more liable to invasion. In this connection, Rishbeth (1951b) in pot experiments obtained some evidence that well-watered plants were less readily attacked than those with a poor water supply. Jørgensen (1962) got similar experimental results, and also concluded from field observations that lack of soil moisture in dry summers may predispose the trees to attack. Similarly, Rishbeth (1951b) also found more diseased trees on sites with soils low in organic matter, and pointed out that in the dry area in which he was working, the organic material may help to keep up the soil moisture level. The situation is complex, however. Maraite and Mayer (1966) found some evidence that in spruce crops in Belgium attack by *H. annosum* is more severe on well-drained sites than on moderately or poorly drained ones, while Peace (1938) in Britain found most rot in spruce on well-drained sites but most on larch where the drainage was poor.

Former land use is of major importance. In general, *H. annosum* appears first and develops most rapidly on sites previously occupied by conifer crops because on these the fungus is usually already established in the stumps of the previous crop by the time planting takes place. For the same reason the incidence of the disease and the height reached by the fungus in the affected trees tends to increase with the number of conifer rotations (Jørgensen *et al.*, 1939; Low and Gladman, 1960; Capretti and Moriondo, 1983). On the other hand, attack by *H. annosum* tends to be delayed on ex-hardwood sites, though it develops on these in time (Greig, 1962).

The effect of thinning is at least partly bound up with the number of susceptible stumps open to colonisation by *H. annosum* spores. Rykovski and Sierota (1984) compared unthinned plots of 35-year-old Scots pine with others subjected to various thinning regimes. The financial loss through decay was 11 per cent in the unthinned (control) plots, 33.5 per cent in those given selective thinning or line thinning to remove every sixth row and 44.5 per cent in those line thinned to remove every fourth row.

On sites other than those previously under conifers, *H. annosum* is most severe on those previously under agriculture (Anon., 1930, 1963; Bronchi, 1956; Cantiani, 1960; Driver and Dell, 1961; Capretti and Moriondo, 1983). Here there may sometimes be a strong connection with pH and former liming, as Wallis (1960) was able to show that in East Anglia many pines were killed on ex-agricultural land with a pH above 6, but where the soil reaction was lower, losses were no greater than on ex-Calluna heath. In other parts of Britain, however, pines may be killed where the pH is considerably lower than 6 (Low, unpublished). Soil pH appears to affect the killing of trees more than it does butt rot, which may develop extensively on relatively acid soils.

Literature on the effects of fertilisers on damage by *H. annosum* on Scots pine and Norway spruce has been reviewed by Jokinen (1983). Much of the available information is difficult to interpret, but in pot experiments with pines, Pobegailo *et al.* (1980) found that N alone reduced resistance to attack, but as part of NKP it increased it. Both P and K increased resistance, the effect of K lasting for up to four years.

Information on the effect of age of the individual trees is sparse. As the fungus takes time to spread, it is clear that there will be a tendency for older crops to be more affected than younger ones. Thus, in larch crops studied by Gothe (1957), infection was most prevalent in the oldest trees. On the other hand, on sites on which killing of trees takes place, death may be more rapid in the younger trees (Anon., 1930), no doubt partly because when small the stems are rapidly girdled. Rishbeth (1951b) in inoculation experiments on pines, found the fungus penetrated the roots of young trees more readily than those of old ones, causing more infections, and achieving a deeper penetration of the tissues in a given time.

Experiments and field observations by James *et al.* (1980a,b) indicate that pollution by photochemical oxidants may increase the susceptibility of trees and stumps of *Pinus ponderosa* and *P. jeffreyi* to colonisation by *H. annosum*. Such pollution appears to reduce the production of oleoresins and to hinder the growth of competing fungi such as *Trichoderma* and the blue stain fungi.

*Losses caused*

As already mentioned, Hepting and Downs (1944) noted an association between thinning and attack by *H. annosum*, and since then many other workers have confirmed this connection, which is associated with the method of entry of the fungus by the infection of fresh stumps by airborne spores. Thus, in pine in East Anglia, Rishbeth (1957) found a mean loss of 0.3 per cent in volume after the first thinning, and losses of 5.9 and 19.5 per cent after subsequent thinnings.

Turning to losses from butt rot, Dowden, in an unpublished study dealing partly with Norway but mainly with Sitka spruce aged 35 to 40 years, found loss in volume caused by *H. annosum* amounted to between 2 and 24 per cent. Further information on losses related to different forms of thinning are given above under 'other factors affecting the disease'.

Most, though not all, accounts of losses caused by *H. annosum* in Britain do not give figures for loss in volume or value, however, but only for numbers or percentages of affected trees, with sometimes additional information on the distance rot had progressed up the trunk of the trees examined. Thus Low and Gladman (1960) in three Sitka spruce crops found between 20 and 80 per cent of the stems affected, and in the worst-affected trees in one 28-year-old planting, the fungus had progressed nearly 5 metres up the stem. In two crops of western red cedar, 35 and 60 per cent of the trees were rotted, and they found 50 per cent affected in Douglas fir and western hemlock. In a number of plantations of larch, between 10 and 60 per cent of the trees showed decay, and in some trees rot had reached a height of nearly 4 metres. In another study, Gladman and Low (1963) found 96 per cent of the trees decayed in a plot of 34-year-old western hemlock.

Pratt (1979a,b,c) made a detailed study of butt rot in 484 trees in 27 stands of Sitka spruce in Britain. On average 9 per cent of the volumes of the trees were occupied by incipiently decayed or decayed wood. If the whole butt lengths of decayed trees were rejected after felling, losses in volumes were 33 per cent and in value 43 per cent. On the other hand, a computer-simulated sawmilling study indicated that losses in value would amount to about 10 per cent if such lengths were not rejected but were converted to pallet boards.

Similar losses have been recorded in many other countries, and in some cases evidence has been gained that loss has resulted not only through death of trees and through decay, but also because of loss of increment in diseased trees (Arvidson, 1954; Henriksen and Jorgensen, 1954). Further damage may result because gaps in the crop and rotten roots may lead to windblow. Loss may also be incurred because on some infested sites it may be impossible to replant with high-yielding but susceptible species, or it may be necessary to shorten the rotation of a crop infested by the fungus (Murray, 1962b). Thus Pratt and Greig (1988) examined a 35-year-old, first-rotation crop of Norway spruce on old agricultural land in Devonshire. This crop was badly affected eleven years after the first thinning, and had to be clear felled, leading to an estimated loss of 4.4–6.6 per cent of the potential volume. In another first-rotation crop of Norway spruce, in Aberdeenshire, eleven years after the removal of a Scots pine nurse crop, they found that 73 per cent of the trees showed butt rot, extending 3 metres up the stems. Again the crop had to be felled prematurely. Hence, in these cases there was a loss through actual decay, and a further loss of potential increment through premature felling. The sites now also contained infested stumps as a ready source of inoculum for any succeeding replacement crop.

*H. annosum* is therefore a great potential cause of loss, and one whose activities should be controlled as effectively as possible.

*Control*

A large proportion of the conifer forests in Britain are relatively young, and *H. annosum* has had little time to enter and spread through them. Much of the work to find control measures has therefore sought means to keep the fungus out of plantations that so far remain uninfested. It has been based on the finding of Rishbeth (1951a,b, 1952, 1959a) and others that *H. annosum* initially enters crops by means of airborne spores that colonise the surface of freshly cut stumps, and that much of this colonisation can be prevented by thoroughly painting the fresh stump surfaces immediately after felling with creosote, a mixture of this with tar, or with titanium white and zinc oxide paints. Following this work, creosote, which is cheap and easily procured, was widely used as a stump protectant.

Creosote, however, is unpleasant to handle and variable in constitution, and was found to give rather erratic results in the field (Rishbeth, 1957; Driver, 1963a,b; Berry, 1965). To be effective it must be applied very thoroughly, and immediately after felling. It acts as a general fungicide, forming a thin barrier that for a long time keeps all fungi from colonising the stump surfaces. Treated stumps are slow to die and decay, and the tissues just below the creosote layer remain susceptible to colonisation by *H. annosum* for an abnormally long period if the protective layer is broken (Rishbeth, 1959a). Hence, if treated stumps are later damaged, for example, by brash-chopping machines, the fungus may enter through the scars. Stumps treated with creosote are only very slowly colonised by *Phlebia gigantea* and other competing fungi, so if *H. annosum* is already established in the stump roots, it may grow freely through the woody tissues (Rishbeth, 1957, 1959a,b).

Therefore other chemicals have been tested to find materials without the disadvantages of creosote (Rishbeth, 1959b; Phillips and Greig, 1970). Several have been found to give better control than creosote, particularly those that do not exclude all fungi from the treated stumps, but allow the entry of some fungi antagonistic to *H. annosum* (Rishbeth, 1959b; Gundersen, 1963, 1967). The most generally effective of those so far most thoroughly tested are sodium nitrite, urea and disodium octaborate (Phillips and Greig, 1970). Borax has also shown promise (Witcher and Lane, 1980; Rosnev, 1983), and Mohr (1984) reported good results with hydrogen peroxide.

Sodium nitrite was used in Britain for a routine stump treatment in the 1960s but, owing to its toxicity to man and domestic animals, it was withdrawn in 1972 and replaced by urea. A 20 per cent solution of urea, together with a marker dye, is applied copiously to the freshly cut stump surface. A variety of applicators has been used; the most satisfactory is a plastic bottle with a brush attached to the screw-cap lid (figure 33). The brush is used to remove the sawdust and debris from the stump surface so that the entire surface is effectively treated. Urea has also been tested and found effective in Finland (Hallaksela and Nevalainen, 1981). A 10 per cent solution of disodium octaborate may be used as an alternative to urea.

As noted above, in natural conditions competing fungi may prevent the entry of *H. annosum* through the surface of fresh stumps. Their action is erratic, however, for when felling occurs, their spores may be too few in numbers compared with those of *H. annosum* for them to be effective. Following work by Rishbeth (1963), in the case of pine it is now possible to ensure the presence of a suitable competing fungus by inoculating fresh stumps with oidia of *Phlebia gigantea* (*Peniophora gigantea*). This fungus can be grown in culture and the oidia thereby produced can be made up in a sealed plastic sachet. These sachets, which are commercially available in Britain, can be kept in a cool store for several months without serious loss of viability. For use in the forest, the contents of a sachet are added to water according to the directions on the label and a marker dye is added. The spore suspension obtained in this way is applied to freshly cut pine stumps in the same manner as a solution of urea (Greig and Redfern, 1974). Good results with *Phlebia gigantea* have also been obtained in other parts of Europe (Fedorov and Poleshchuk, 1982; Rykowski and Sierota, 1983; Jokinen, 1984; Sierota, 1984).

The antagonistic fungus *Trichoderma viride* has been recommended for the control of some tree diseases (Corke, 1980). To control *H. annosum*, Seaby (1985) devised a technique for the application of spores of *T. viride* to fresh stump surfaces by adding them to the chain-saw lubricating oil.

**Figure 33** Treatment of conifer stumps to prevent the entry of *Heterobasidion annosum*, using a plastic bottle with brush attachment (Forestry Commission)

Once *H. annosum* has entered a crop, it begins to spread, and by the end of the rotation many stumps may be infested and so present sources of infection to a succeeding conifer crop. In Thetford Forest in eastern England, extensive death of pine has occurred in second-rotation pine crops where these have replaced crops seriously attacked by the disease. Greig and Burdekin (1970) report a series of experiments to test treatments which might overcome this problem. These treatments included removal of stumps of the first-rotation crop, delayed replanting after clear felling and treatment of stumps with *P. gigantea*. The most successful treatment was removal of stumps and this reduced infection in the second rotation after 11 years from 54 per cent in control to 20 per cent in stump removal plots.

Extensive surveys were undertaken in Thetford Forest (Greig, 1971) in order to identify those areas where serious losses from *H. annosum* were likely to occur in the second rotation. Approximately one-fifth of the area due to be clear felled in a five-year programme was severely affected and, following a thorough economic evaluation, stump removal was introduced as a routine treatment for these areas (figure 34). A management system based on these findings has been described by Greig (1984).

Stump removal may have other advantages, as once the stumps are removed soil cultivation is possible, and in frosty sites may give better establishment with less beating up of some species, and reduce later weeding costs. There is also some evidence that, in the case of pine, inoculation of the

**Figure 34**    Control of *Heterobasidion annosum* by removal of pine stumps prior to
replanting a second rotation (Forestry Commission)

stumps with *Phlebia gigantea* may be beneficial, as this fungus grows down
and occupies a large part of the stump body and limits the growth of *H.
annosum* present in the roots (Rishbeth, 1963).

Finally, the relative susceptibility of species may be taken into account
when replanting an infested site. Thus, except on alkaline or old arable sites,
Scots or other pines may safely be used to follow a species that has suffered
from *H. annosum* butt rot, though account has to be taken of possible
reductions in total volume production resulting from such a change of species.
For suitable sites, *Abies grandis* is a high-yielding alternative, though we still
need more information on the degree of resistance of this species in conditions of heavy infestation.

Work on breeding or selecting for resistance to *H. annosum* has not been
extensive, though screening of southern pines was started in North America
by Driver and Ginns (1966), and in Europe Werner (1987) has reported on
resistant progenies of Scots pine.


MISCELLANEOUS DISEASES

**Slime flux and bacterial wetwood**
Water may ooze, sometimes in large quantities, from tree trunks, for
example, from frost cracks or wounds made for injection of fungicides to
control Dutch elm disease (Murdoch and Campana, 1984). This water often

smells of fermentation, as it becomes colonised by bacteria and yeasts and other fungi, which convert it into a slimy outflow known as a *slime flux*. This flux is toxic, and kills considerable areas of bark as it runs down the stem (Hamilton, 1980) (plate 11). It is watery in consistency, and may leave a chalky deposit. It is never gummy (or finally crust-like), as is the case in the bleeding canker of, for example, horse chestnut trees attacked by *Phytophthora* spp. (see chapter 19).

Among the conifers, in *Abies* spp. the central heartwood commonly shows a conical zone of wet wood, which tapers upwards, is surrounded by a dry zone, and is acid in reaction. This cone of wet wood is more or less sterile, and its formation seems to be a normal part of the development of the heartwood (Etheridge and Morin, 1967; Coutts and Rishbeth, 1977). It may be beneficial in that the water-soaked tissues inhibit the growth of decay fungi (Worrall and Parmeter, 1982).

Similarly, Murdoch and Campana (1983) found wet wood of this kind in every American elm (*Ulmus americana*) that they examined, and concluded that the condition caused no damage to the trees unless they were wounded in some way. Such water-soaked elm wood is also resistant to attack by some decay fungi, such as *Coriolus versicolor* (Coleman *et al.*, 1985).

There is, however, a different condition, known as *bacterial wetwood*, which mainly affects broadleaved trees, especially elms, poplars, willows, maples and horse chestnuts (Ogilvie, 1924; Worrall and Parmeter, 1982; Murdoch and Campana, 1983). Much information on this disease was collected and reviewed by Hartley *et al.* (1961). In bacterial wetwood the tissues become water-soaked, and much more alkaline than those of healthy wood, which are slightly acid; they are also colonised by a large and mixed population of aerobic, anaerobic and facultative anaerobic bark-inhabiting or water- or soil-borne bacteria (Murdoch and Campana, 1983; Scott, 1984a).

Typically, wetwood forms in the heartwood as a water-soaked column which tapers upwards and may extend into the branches. This column is infected by many bacteria which cause fermentation, and this produces high pressures of methane and other gases. This fermentation also forms toxic products which may enter the transpiration stream and cause yellowing and death of leaves and dieback or even slow death of branches (and occasionally of whole trees). If the wetwood zone extends out to the cambium, the latter is killed. Gas pressure may cause radial cracking as far as the outer bark surface, when a toxic, evil-smelling slime exudes, like that described above, so that the flux-soaked bark becomes brown and dies. Sometimes cracks formed in this way heal over and show as well-marked ribs running up the trunk (Rishbeth, 1980; Murdoch and Campana, 1983). The affected wood becomes brown on exposure to air, perhaps owing to the oxidation of phenolic materials (Scott, 1984a).

Carter (1945) concluded that wetwood was a bacterial disease caused by *Erwinia nimipressuralis* (now considered to be synonymous with *Enterobacter cloacae*), but more recent work suggests that it is caused by the simultaneous activities of various bacteria, including species of *Clostridium*, *Enterobacter* and *Edwardsiella* (Schink *et al.*, 1981; Murdoch and Campana, 1983; Scott, 1984a). Many of these bacteria can degrade pectins (Schink *et al.*, 1981).

Scott (1984b) has published details of a method for detecting wetwood in live poplar trees by measuring the electrical resistance of the wood, using a Shigometer.

Little can be done to control wetwood and slime fluxes. In the case of trees in parks and gardens, the bark killed by fluxing is sometimes removed, and the wood dressed with a proprietary sealant to promote healing, and excess water is drawn off by tapping the trunk with a tube to prevent it from running down and killing more of the bark.

### Diseases associated with *Sydowia polyspora* (Bref. and Tav.) E. Müller (anamorph *Sclerophoma pythiophila* (Corda) Höhnel)

The ascomycete *Sydowia polyspora* (especially as its anamorph *Sclerophoma pythiophila*) occurs widely in Europe into the USSR and in North America, mainly as a saprophyte on needles, cones and seeds of many conifers, including *Abies*, *Larix*, *Picea*, *Pinus*, *Thuja* and *Tsuga*. The locules in its black conidial stromata contain colourless, pip-shaped conidia which measure 4–8 × 2–3 µm (Ellis and Ellis, 1985). The pseudothecia of the teleomorphic stage are black and globoid and contain a few large asci each of which contains 20–26 colourless ascospores (Miller, 1974). These measure about 9–28 × 3–8.5 µm (Holliday, 1989) and are mostly 3-septate, sometimes with one vertical septum (Miller, 1974). In culture the fungus often produces a stage resembling *Aureobasidium pullulans* (Batko *et al.*, 1958).

*Sclerophoma pythiophila* has sometimes been associated with a number of conifer diseases. In Britain, Batko *et al.* (1958) found it involved in the death of current year's needles of Scots pine. Some of the dead needles remained long caught up among other still healthy ones. The fungus appeared to be extending damage caused by the pine needle midge, *Contarinia baeri*. Jahnel and Junghans (1957) found it causing a similar disease of the current needles of Scots pine, again associated with *C. baeri*, and also found it causing the shedding of older needles. In Finland it has been associated with a dieback of *Pinus contorta*, and in Russia with both a dieback of *Pinus contorta* and the formation of witches' brooms, also in pines (Batko *et al.*, 1954). In North America Miller (1974) found it causing a tip dieback of the white fire, *Abies concolor*, grown in California as Christmas trees.

Damage by this fungus is usually only local and sporadic.

### Crown gall caused by the bacterium *Agrobacterium tumefasciens* (Smith and Townsend) Conn (*A. radiobacter* subsp. *tumefasciens* (Smith and Townsend) Kean et al.)

The species of *Agrobacterium* present many problems of taxonomy and nomenclature, which are discussed by Kersters and De Ley (1984). Some forms cause plant tumours, and these are placed in *A. tumefasciens* (Panagopoulos, 1988). All the *Agrobacterium* species are Gram-negative, non-sporing rods motile by one to six peritrichous flagella. Methods of identification are given by Fahy and Persley (1983) and Lelliott and Stead (1987). Within the species, forms able to induce tumour production possess Ti plasmids which can be passed from species to species, or lost, when the bacterium becomes non-pathogenic (Holliday, 1989).

The bacterium overwinters in the soil as well as in tumours, and enters plants by wounds. The affected plants produce rounded, irregular galls which are eventually woody and fissured. These tumours commonly develop just above the soil surface (hence the name 'crown gall') and are made up of a confused mass of parenchymatous and vascular tissues. They usually decay at the end of the growing season, to be replaced on woody hosts by new galls in the following year (Panagopoulos, 1988).

*A. tumefasciens* has a wide host range, but on forest and ornamental trees seems to be most important on poplars (Nesme *et al.*, 1987; Zhang *et al.*, 1988) and on *Sequoia* (Peace, 1982). Clapham and Ekberg (1986) induced tumour formation in *Abies nordmanniana* and Norway spruce (*Picea abies*). Clapham *et al.* (1990) discovered quite wide variations in resistance to tumour formation between open-pollinated families from a Swedish population of Norway spruce. In India galls have been found on *Eucalyptus tereticornis* (Jindal and Blardwag, 1986). Crown gall may be important in nurseries because it may stunt or kill affected plants, and galling of the better plants makes them unsaleable.

Kerr (1980) succeeded in controlling the disease on orchard trees and roses by dipping planting materials in a suspension of the cells of a non-pathogenic strain, strain 84, of an *Agrobacterium* which inhibited the growth of most but not all of the pathogenic forms (Kerr and Panagopoulos, 1977). Apart from this, control is by soil fumigation, the use of disease-free stock, avoidance of wounding of the plants, and sterilisation of tools and propagating equipment.

### Diseases caused by the bacterium *Pseudomonas syringae* pv. *syringae* van Hall

The species of *Pseudomonas* form straight or curved Gram-negative rods which are motile by polar flagella. The genus has been described by Palleroni (1984), and the methods for its separation from *Xanthomonas* and for the identification of its various subspecies and pathovars are given by Lelliott *et al.* (1966) and Lelliot and Stead (1987).

Of the forms affecting forest and ornamental trees, *P. syringae* subsp. *savastanoi* causes ash canker, and sometimes attacks oleanders and the Japanese privet; it also causes the disease of olives known as olive knot. Ash canker is described under ash in chapter 13.

*P. syringae* van Hall consists of many pathovars. Of these, the pv. *morsprunorum* causes a well-known bacterial canker of stone fruits, including almond and cherry (under which it is described in chapter 18, and on which it sometimes also causes a shoot wilt). *P. syringae* pv. *avellanae* Psallidas causes a dieback of the stems of hazelnut trees in Greece (Psallidas, 1987).

*P. syringae* pv. *syringae*, however, has a wider range, causing a number of symptoms on a variety of hosts. On stone fruits it gives rise to cankers which can be distinguished from those caused by pv. *morsprunorum* only by the isolation and identification of the causal bacterium. On apples it sometimes causes a blister spot, while on pears and lilacs it may produce a blight of the blossoms. It may also cause spotting of the leaves and dieback of the shoots of many trees, including species of *Acer*, *Betula*, *Crataegus*, *Juglans*, *Magnolia*, *Populus*, cherry laurels (*Prunus laurocerasus*) and other *Prunus* spp., *Quercus*, *Salix*, *Sorbus* and *Ulmus*.

It occurs not only in Europe, but also in North America and Australasia.

In North America, Malvick and Moore (1988) found that it could pass the winter on maple shoots in the nursery, the infected shoots becoming an inoculum source in the succeeding spring. In Canada, Boer (1980) described symptoms on cherry laurel leaves, the necrotic spots on which were surrounded by a reddish border which was itself surrounded by a yellow ring. Cankers also sometimes developed on the stems. Kam (1982c) examined bark cracking and other damage on poplars, occurring just above soil level and usually on the south-facing side. He considered that the damage was caused by a combination of frost damage and *P. syringae* pv. *syringae*, and that attack by the bacterium also made the leaves of the trees more sensitive to frost. In New Zealand, Haworth and Spiers (1988) found the same bacterium causing leaf spots, dieback and blackening of shoots, and cankers on the stems of poplars and willows, and decided that the severity of the attacks was connected with exceptionally heavy rains. In Britain, Strouts *et al.* (1983) attributed the deaths of many flowering cherries in 1982 to damage by *P. syringae* pv. *syringae* enhanced by the effects of a preceding cold winter.

In North America, Moorman *et al.* (1988) isolated the same bacterium from leaf spots on the English elm, *Ulmus procera*, and they produced similar spots on leaves of *U. americana* by inoculation.

Although *P. syringae* pv. *syringae* has been found mainly on deciduous trees, in New Zealand it has been known to give rise to dieback, wilting, stem cankering and death of nursery plants of *Pinus radiata* (Langridge and Dye, 1982; Dick, 1985). Dick (1985) found that frost accentuated the damage caused by the bacterium.

Where necessary, diseases caused by *P. syringae* pv. *syringae* can be controlled by pruning out diseased parts, and spraying with Bordeaux mixture or similar copper-based sprays.

## HIGHER PLANTS PARASITIC AND SEMIPARASITIC ON TREES

A number of phanerogamic plants in Europe are parasitic or semiparasitic on trees, though they are of relatively minor importance. The most widespread and best-known is the common mistletoe, *Viscum album* L. (a member of the Loranthaceae), which grows throughout Europe except in Iceland and

Finland (Polunin, 1969). In Britain it is common in the south of England, but it is rare north of the Humber, and absent from Scotland and Ireland (Clapham *et al.*, 1987).

The plant is rather woody, and evergreen, with dichotomously branching stems, and leathery, spatulate, pale yellowish-green leaves. It is dioecious, with small, inconspicuous greenish flowers, and the berry-like fruit is usually white.

In Britain it is considered to be most common on apple (especially in gardens) but it is frequent on lime and on poplars, and it has been found also on hawthorn, whitebeam, field maple and *Robinia pseudoacacia* (Lousley, 1976) and sometimes other broadleaved trees. In Britain it is at most very rare on conifers, but on the European mainland the subsp. *abietis* (Wiesb.) Abromeit grows on *Abies* and sometimes on other conifers, and the subsp. *austriacum* (Wiesb.) Vollman, with yellow berries, grows on *Pinus* and *Larix* (Blamey and Grey-Wilson, 1989), while another species, *V. cruciatum* Boiss., with red berries, occurs in southwestern Spain (Polunin, 1969). The common mistletoe is spread mainly by birds, which eat the fruit and get rid of the sticky seeds by rubbing them off their beaks on to tree branches, or swallow the whole fruit, eventually voiding the seeds in their droppings.

On germination the seed produces a primary sucker which penetrates the wood, and gives rise to further suckers parallel to the branch axis and between the wood and the phloem. As the tree grows, the suckers become more and more deeply embedded in the wood, from which the mistletoe extracts water and mineral salts. To the exterior, the rather shrubby mistletoe plant slowly develops, and usually persists for about 20 years (Anon., 1979). The wood is damaged by the growth of the suckers, but the damage is offset by the value of the mistletoe, which may be harvested and marketed for use as a Christmas decoration.

The forms of the common mistletoe attacking conifers on the European mainland may cause considerable damage, however, especially on Scots pine and *Abies alba* (Hartig, 1894), leading to breakage by wind, allowing the entry of decay fungi, and reducing increment. Delabraze and Lanier (1972) attempted with some success to control the plant on *Abies alba* and on *Populus* and *Salix* by spraying with 2,4,5-T, 2,4-D and related materials.

Another member of the Loranthaceae, *Loranthus europaeus* Jacq., grows in southeastern Europe, attacking oaks, and sometimes sweet chestnut and beech, on which it produces bushy growths up to 1 m across (Polunin, 1969). As in the case of *Viscum album*, its seeds are spread by birds, which eat its yellow berries. Its dull-green oval leaves are leathery and deciduous, and its small flowers are greenish-yellow. The root system of *L. europaeus* penetrates the host, which is stimulated to surround it by a gnarled swelling (Hartig, 1894).

Also usually placed in the Loranthaceae are the dwarf mistletoes. These, which have been monographed by Hawksworth and Wiens (1972), are highly important in North America, where they cause severe damage to many conifers. These American mistletoes are considered to be a threat to trees in

Europe, and are therefore covered by the plant health legislation of the European Community (Phillips, 1978). Only one of the dwarf mistletoes occurs in Europe, where it is of only minor importance. This is *Arceuthobium oxycedri* (DC) M. Bieb., which forms dense bushy growths on junipers in southern Europe and as far north as Austria (Hartig, 1894).

Two species of *Lathraea*, members of the Orobanchaceae, lack chlorophyll and attack trees in Europe. The first is the toothwort, *L. squamaria* L., which is found throughout the continent except in Portugal (Polunin, 1969), and is widespread in Britain. It attaches itself to the roots of various broadleaved trees, especially hazel, elm and alder, but also sometimes ash, poplar, beech and cherry laurel, but seems to have little effect on the hosts. In spring and early summer it produces one-sided, white or pinkish spikes with small white scale leaves and irregular white, pink-tinged flowers (figure 35).

**Figure 35**   Flowering spike of the toothwort, *Lathraea squamaria* (D.H. Phillips)

*L. clandestina* L. is native in southern Europe from Spain to Italy (Polunin, 1969), but has been naturalised in a few places in Britain. It is parasitic on the roots of various broadleaved trees, especially alder, poplar, willow and hazel (Grenfell, 1985; Blamey and Grey-Wilson, 1989), but again is of little importance. Its clusters of purplish flowers form in April and May on the soil surface below the host trees.

The large dodder, *Cuscuta europaea* L., one of the Convolvulaceae, mainly affects herbaceous plants, but may also attack willows, alders and elms (Polunin, 1969; Lousley, 1976). Its pale-green, thread-like stems attach themselves to the host plant by sucker-like bodies, and bear globular groups of pinkish-white flowers. The plant is widespread in Europe; in Britain it occurs mainly in the south, and it is absent from Ireland. The effect on the host seems slight.

# 4 Diseases of spruce (*Picea* spp.)

Two spruces, Norway spruce (*Picea abies*) and Sitka spruce (*P. sitchensis*), are important forest trees in Britain. Sitka spruce, a very fast-growing tree introduced from the west coast of North America, is now the most commonly planted of all our forest trees. Norway spruce, from continental Europe, is less important in the forest, but is produced in large numbers for the Christmas tree trade. On the continent it is an important forest tree, especially in mountainous areas from the Alps northwards to the north of Scandinavia. The Serbian spruce (*P. omorika*), a fast-growing species from Yugoslavia, has been grown experimentally by foresters, but has not become established in forest use. Many other species of *Picea*, particularly various cultivars of the blue spruce (*P. pungens*), are grown in parks and gardens.

## Nursery diseases

### GENERAL

Damping off caused by species of *Pythium* is common in beds of first-year seedlings of Sitka spruce in the more northerly parts of Britain. Especially in other parts of Europe, species of *Fusarium*, *Cylindrocarpon destructans* (the anamorph of *Nectria radicicola*) and *Rhizoctonia solani* (the anamorph of *Thanatephorus cucumeris*) are also involved. The grey mould *Botrytis cinerea* mainly affects nursery stock, and is commonly reported in beds of Sitka and less often on those of Norway spruce. On a few occasions young Sitka spruce plants in nurseries in Britain have been attacked by *Phytophthora citricola*. The fungus *Geniculodendron pyriforme* may infest spruce seeds. Seed diseases, damping off, grey mould and the species of *Phytophthora* are described in chapter 3.

### SEEDLING AND SHOOT BLIGHTS

**Seedling blight and canker associated with *Phomopsis occulta* (Sacc.) Traverso (anamorph of *Diaporthe conorum* (Desm.) Niessl)**
*Phomopsis occulta* is a common saprophyte on dead leaves and stems and fallen cones of many conifers. In Britain it has been found on Norway and Sitka spruce, Scots and Corsican pine, *Pinus radiata* and *P. patula*, larches,

Douglas fir and *Abies alba*, and elsewhere in Europe Lanier *et al.* (1978) record it also on *Cupressus* and *Thuja*.

On the two spruces it has sometimes been associated with a seedling blight which has occasionally caused severe damage in spruce seedbeds. The affected plants, which died back from the tops, were girdled at the base, which became covered by a white mycelium (Peace, 1962). Rather different symptoms have been observed on nursery plants of *Picea pungens*, the Colorado blue spruce, in North America (Sanderson and Worf, 1986). The plants showed a needle necrosis and tip blight, beginning with a chlorotic flecking of the lower halves of the needles near the shoot tips. Later, small cankers developed on the stems, often with some exudation of resin. Pathogenicity tests were made with four species of *Picea* and two of *Abies*. These showed that *P. pungens* was especially susceptible, but *P. abies*, *P. glauca* and *P. obovata* were also affected, while *Abies balsamea* and *A. concolor* remained healthy. The disease was most severe on plants kept at the high temperature of 25 °C and at a high RH between 79 and 95 per cent.

The black pycnidia of the fungus on the diseased plants contain colourless A- and B-spores that exude as a white or yellowish tendril. The A-spores are ellipsoid-oblong and curved, and measure 8–11 × 2–2.5 μm, and the B-spores, which are hair-like, straight or walking-stick-shaped, measure 20–27 × 1 μm (Grove, 1935; Ellis and Ellis, 1985). Grove (1935) also described intermediate C-spores.

*P. occulta* is the anamorph of *Diaporthe conorum* (Desm.) Niessl, one of the members of the complex of species centred on *D. eres* Nitschke (Cannon *et al.*, 1985), which is briefly discussed in chapter 3.

### Seedling blights associated with *Rosellinia minor* (Höhnel) Francis (*R. andurnensis* var. *minor* Höhnel) and *R. thelena* (Fr.) Rabenh.

A seedling blight of two-year-old seedlings of Norway spruce with symptoms much resembling those associated with *Phomopsis occulta* and found once in South Scotland was described by Wilson (1922) from notes by Borthwick. It was attributed to *Rosellinia aquila* (Fr.) de Not., which is well known as a common saprophyte on dead branches of many broadleaved trees and shrubs. The attack was severe, almost all the Norway spruce plants in the seedbed being affected. The upper parts of the root systems, the stem bases and the lower leaves were covered with grey mycelium bearing colourless oval conidia measuring 8 × 4 μm. On incubation, Borthwick obtained black, round perithecia up to 1 mm across, containing asci each with eight dark-brown, boat-shaped ascospores measuring 18–20 × 7–8 μm. He controlled the disease by destroying the affected plants and soaking the ground with copper sulphate solution.

Since then the disease has been found in southeast England on Norway spruce seedlings and once on *Picea omorika*, but it has generally been of minor significance. Similar attacks, mainly on seedling Norway spruce, have been recorded in France (Vegh, 1984), Belgium (Vanderwalle, 1970), Germany (Burmeister, 1966, on *Picea omorika*) and Romania (Georgescu and Gasmet, 1954). Heavy losses of plants have sometimes ensued, especially in dense and crowded seedbeds. The disease has usually been attributed to attack by *Rosellinia herpotrichioides* Hepting and Davidson (a fungus found in North America causing needle blight on branches of *Tsuga canadensis*) or *R. aquila*. In 1985, young plants of *Picea orientalis* cv. *aureospicata* affected by a *Rosellinia* needle blight were imported from the continent into Britain (Redfern *et al.*, 1986). From a study of these plants and of herbarium material and of the literature, Francis (1986) concluded that the European needle blight of conifers associated with a *Rosellinia* was in fact, at least in almost every case, caused by *R. minor* (Höhnel) Francis. The more or less spherical fruit bodies of this fungus form in the dense, whitish-brown mycelial web which covers the host shoots and needles. The greyish-brown ascospores are inequilateral and at first have a gelatinous sheath. They measure 20–25 × 5.5–7 μm. Its anamorph has colourless, ovoid or elliptical conidia measuring 5–7 x 3.6 μm (Francis, 1986).

The danger of attack by *R. minor* should be reduced by judicious sowing to avoid overcrowding in the seedbed. Any diseased plants should be culled and burnt.

*Rosellinia thelena* is another fungus often found growing saprophytically on fallen material of various broadleaved and coniferous trees. The asci in its black perithecia contain dark brown ascospores measuring 20–26 × 7.5–8.5 μm, and each spore has a colourless appendage at each end. The spores of its conidial stage measure 5–8 × 2.5–3 μm (Ellis and Ellis, 1985). It has been found fruiting abundantly on dying cuttings of Norway and Sitka spruce in Scotland, but inoculation experiments failed to prove its pathogenicity (Redfern *et al.*, 1975).

### Black snow mould caused by *Herpotrichia juniperi* (Duby) Petrak (*H. nigra* Hartig)

*Herpotrichia juniperi* occurs in Europe in Britain, France, Austria, Germany, Poland, Denmark, Norway, Sweden, Finland, Switzerland, Romania, Yugoslavia, and in Siberia in the USSR, as well as in North America. It causes the disease known as black snow mould, which is most important on Norway spruce, but may also affect *Picea engelmannii* and *P. omorika*, and sometimes also species of *Abies*, *Pinus*, *Juniperus* and *Tsuga* (Spaulding, 1961; CMI Descr. F. 328). The disease causes serious loss only in mountainous areas with long-lying winter snow, mainly in nurseries but sometimes in plantations of young trees. Venn (1983) found that it could also affect spruce plants in cold store.

The disease has been described by Hartig (1894), Gäumann *et al.* (1934), Sivanesan and Gibson (CMI Descr. F. 328, 1972), Hanso and Törva (1975) and Bazzigher (1956a). The fungus grows even at 0 °C or just below, on plants under a cover of snow, producing a greyish mycelium. It grows rapidly once the snow melts, matting the host needles together with a now dark-brown or blackish, thick-walled, drought-resistant mycelium. Dark-brown or black pseudothecia partly embedded in this mycelium contain asci each containing 8 eventually four-celled brownish ascospores which measure 25–33 × 7–11 μm (Ellis and Ellis, 1985) and which can germinate as soon as they are shed.

The disease is of relatively local importance, but where it does occur control is difficult. It is not easy to make effective use of fungicidal sprays in the conditions in which black snow mould is damaging, except perhaps in cold stores. In areas in which the disease is troublesome, nurseries should if possible be sited in more lowland areas not subject for long periods to a cover of snow.

### Shoot blight caused by *Sirococcus strobilinus* Preuss (*S. conigenus* (DC) P. Cannon and Minter, *Ascochyta piniperda* Lindau, *Septoria parasitica* Hartig)

*Sirococcus strobilinus* occurs widely in Europe and parts of Canada and the USA. In Europe it may sometimes cause defoliation and dieback, mainly of the current shoots of Norway and Sitka spruce in nurseries, and occasionally in young trees in plantations. It was first described on these trees in 1890 by Hartig (see Hartig, 1894). In Britain (where attacks of shoot blight have occurred mainly in Scotland) it has also been recorded on seedlings of lodgepole pine (figures 40, 41, chapter 5) (Redfern *et al.*, 1976) and as a saprophyte on its cones (Ellis and Ellis, 1985). In Germany it has been reported on *Picea pungens* (Schneider and Paetzholdt, 1964). It has also been recorded on *Larix* (Lanier *et al.*, 1978) and *Pseudotsuga* and *Abies* (Kujala, 1950a). In North America shoot blight affects many spruces, including Norway spruce, white spruce (*Picea glauca*) and red spruce (*P. rubens*) (Hepting, 1971), blue spruce (*P. pungens*) (Sanderson and Worf, 1986), black spruce (*P. mariana*) (Wall and Magasi, 1976) and Engelmann spruce (*P. engelmannii*) (Sutherland *et al.*, 1981). It also occurs there on lodgepole pine (Illingworth, 1973), and on red pine (*Pinus resinosa*) (O'Brien, 1973) and on western hemlock (*Tsuga heterophylla*) (Funk, 1972). In the case of *Pinus resinosa* large trees 70 feet (*c.* 21 m) high have been severely affected, the disease spreading upwards in the crown from the basal branches, while young trees sheltered below the larger ones were also infected and killed (O'Brien, 1973). Sutherland *et al.* (1981), who studied the disease on container-grown stock in British Columbian nurseries, showed that the fungus could be seedborne and the disease seed-transmitted in Sitka, white and Engelmann spruces and in the hybrid between the last two.

Because attacks have been sporadic and local, the disease has so far been of relatively minor importance. In Europe it seems to have been found most often in the north of Scotland and in Scandinavia.

Infection shows first of all in the early part of the summer at the base or the middle of the current shoots, where the needles turn brown and soon fall. The disease tends to spread upwards from the lower needles towards the ends of the shoots and some movement may also take place downwards into the previous year's growth, often leading to the death of some of the side shoots (Hartig, 1894). For a time the shoot tips remain green, but later wilt, droop and finally die, often curling over and assuming the form of shepherds' crooks (Funk, 1972). The pith of the diseased stems may become brown (Laing, 1929; Lagerberg, 1933). Hartig (1894) pointed out that the symptoms of *Sirococcus* shoot blight much resembled those caused by spring frost, but in the case of the latter many shoots were suddenly affected, and all the frost-damaged leaves died simultaneously. Later in the summer on plants affected by *S. strobilinus* the fungus produces dark-brown to black, usually multilocular pycnidia in the dead bark. Whitish tendrils of sticky, colourless, 2-celled, rather fusiform spores, each measuring 12–16 × 3 μm (Ellis and Ellis, 1985) emerge from the irregular openings of these fructifications.

The fungus does not germinate or grow below 10 °C (Smith and Graham, 1973), though Wall and Magasi (1976) found that it can survive freezing temperatures over the winter; they also concluded that it is encouraged by below-average temperatures and wet cloudy weather during the summer.

Peace (1962) suggested that the fungus on pine might be a different variety from that on spruce. This suggestion receives some support from an observation by O'Brien (1973) that spruce trees growing under infected red pines (*Pinus resinosa*) in the Lake States in North America seemed unaffected. However, Redfern (1973) found that an isolate of the fungus from lodgepole pine seedlings in Scotland would also attack wounded (but not unwounded) Norway and Sitka spruce plants. In the case of lodgepole pine, he found that plants of a provenance from southern interior British Columbia, Canada, were more susceptible than either of the spruces, but plants of two other provenances, one from central interior British Columbia and one from north central Oregon, USA, remained unaffected. Illingworth (1973), who studied the disease on lodgepole pine in a provenance trial in a nursery in British Columbia found similar differences in susceptibility.

The fungus can enter nursery crops from nearby older trees on which it is already established, and, as already noted, it may also be transmitted by infested seed. Hence, at least in areas in which shoot blight is a problem, nurseries should not be sited close to infested plantations; and *S. strobilinus* should be looked for in tests in seed laboratories. It is possible that the treatment of seed with thiram or captan (or some other fungicide), as suggested for *Geniculodendron pyriforme* (q.v., under 'seed diseases', in chapter 3) would eliminate the fungus in infested samples, and thus would be worth further investigation. Diseased nursery plants should be culled and burnt. Smith *et al.* (1972) tested a number of fungicides against the disease in nursery beds of Jeffrey pine. They obtained good control by spraying with chlorothalonil, captafol and maneb with zinc. More recent trials have been made by Kliejunas (1989), who tested a number of chemicals against the disease on *Pinus jeffreyi* in the USA. Results showed that propiconazole, penconazole and triadimefon applied at intervals of eight weeks gave good control, which was as effective as that given by chlorothalonil applied twice as often at intervals of only four weeks.

ROOT ROTS

**Violet root rot caused by *Helicobasidium brebissonii***
Violet root rot (described in chapter 3) has sometimes been found causing severe damage to Sitka spruce transplants in nurseries in Scotland.

# Post-nursery diseases

NEEDLE DISEASES

**Needle casts associated with *Lophodermium piceae* (Fuckel) Höhnel and *Lirula macrospora* (Hartig) Darker (syn. *Lophodermium macrosporum* (Hartig) Rehm)**
A fungus usually known as *Lophodermium piceae* is occasionally associated with a needle cast of Sitka and Norway spruce in Scotland and northern England, mainly on nursery plants. It occurs also in other parts of Europe and in North America (Tehon, 1935). In France it is known also on needles of *Abies alba* (Gourbière *et al.*, 1986) and in India it has been found on *A. pindrow* (Sharma and Sharma, 1981).

Diseased needles turn reddish brown, often showing dark transverse bands (diaphragms). The black, elongated, boat-shaped ascocarps of the fungus form in longitudinal rows on the needles, between the dark bands, and are up to about 2 mm long and ½ mm wide. The asci each contain eight hair-like, colourless, non-septate ascospores each measuring 85–115 × 1.5–2 μm, and embedded in a thick gelatinous sheath. A spermagonial stage, *Leptostroma abietis* Darker, with brown, elliptical pycnidia (spermagonia) about ⅓ mm across also occurs. The pycnidia contain oval spermatia measuring 3–4 × 0.8 μm. Both the ascocarps and the spermagonia open by a slit. The ascocarps ripen in summer, and the ascospores are then shed and infect the needles, which soon show pale-green bands. The spermagonia and ascocarps develop in the succeeding spring, occasionally on needles still attached to the tree, but more often on those fallen to the ground (Ferdinandsen and Jørgensen, 1938–9).

According to Barklund (1987), *L. piceae* is in fact confined to *Abies*. The fungus on spruce is unnamed, but may be the same as *L. nanakii* Conran and Minter. There is no doubt that the fungus commonly called *L. piceae* is only a weak parasite. Studies by Barklund (1987) in Sweden indicate that it is not pathogenic, and is more common in young, healthy stands than in diseased ones. In West Germany, Suske and Acker (1987) found it a common colonist of symptomless Norway spruce needles, and Butin and Wagner (1985) found evidence that it was no more than a saprophyte which followed needle damage caused by abiotic factors. Similarly, in France, Gourbière *et al.* (1986) concluded that the fungus on *Abies alba* was also only saprophytic.

According to Campbell and Vines (1938), *Lirula macrospora* (Hartig)

Darker (*Lophodermium macrosporum* (Hartig) Rehm), the ascospores of which measure 81–168 μm (Terrier, 1953), has also been known to attack Norway spruce in Scotland. In a more recent study in West Germany, Schütt (1985) found no evidence that this fungus had any significant pathogenic effect.

### Needle cast associated with *Rhïzosphaera kalkhoffii* Bubak and other *Rhizosphaera* spp.

*Rhizosphaera kalkhoffii* is a common and widespread needle fungus recorded in Britain (Moore, 1959) and throughout much of Europe (Bubak, 1914; Schönhar, 1959), in North America (Hawksworth and Staley, 1968), Japan (Kobayashi, 1967) and New Zealand (Hood and Sandberg, 1985). It is very common on the blue spruce (*Picea pungens*) and its varieties, and on Norway and Sitka spruces, and has also been recorded on *Abies procera*, *A. alba*, Douglas fir, Austrian pine and *Pinus mugo* and (in New Zealand) on *Pinus jeffreyi* and *Sequoia sempervirens*.

In parts of North America, in some circumstances the fungus appears to be an important cause of needle cast of blue spruce (Hawksworth and Staley, 1968; Waterman, 1947). Waterman reported successful control on ornamental trees by spraying with Bordeaux mixture. In Britain some sixty years ago, Wilson and Waldie (1926) also considered that *R. kalkhoffii* sometimes caused a spruce needle cast which they described. In May, affected needles became discoloured and soon turned a purple-brown, and contained large amounts of fungal mycelium. The fungus later produced rows of small, black, stalked pycnidia, the stalks of which emerged through the stomata on the needle surfaces.

As they grow, the pycnidia carry up with them from the stomatal opening a small, shining white plug of wax, which may later disappear. The pycnidia contain ovoid, colourless, one-celled spores, measuring 7–10 × 3–4 μm (Bubak, 1914).

In general, however, in both Britain and the rest of Europe, *R. kalkhoffii* is considered to be doubtfully pathogenic. As noted in connection with top dying of Norway spruce, neither Went (1948, 1949) nor Murray (1957) succeeded in inoculating Norway spruce with the fungus. More recently, in the German Federal Republic, Schütt (1985) found it as frequent in healthy as in declining Norway spruce, and concluded that it was not a cause of spruce decline, and studies by Butin and Wagner (1985) also led to the conclusion that the fungus was simply a colonist of needles killed by abiotic factors. Hood and Sandberg (1985) in New Zealand were of the same opinion.

The related fungus *R. pini* (Corda) Maublanc emend. Kobayashi has been found associated with a needle browning and loss of older foliage of Sitka spruce in northern England (Redfern *et al.*, 1987). This fungus, the colourless (sometimes later brownish) pycnospores of which measure 15–21.5 × 7–12 μm, has also been associated with a needle blight of *Abies* spp. in France and of *Tsuga diversifolia* in Japan, but it has generally been of only minor significance (Kobayashi, 1967).

**Needle rusts:** *Chrysomyxa abietis* **(Wallr). Unger and** *C. ledi*
**(Alb. & Schwein.) de Bary var.** *ledi* **and var.** *rhododendri* **(de Bary)**
**Savile (***C. rhododendri* **de Bary)**
The needle rusts *Chrysomyxa abietis* and *C. ledi* both occur sporadically on
the foliage of spruce. *C. abietis* is autoecious, and produces only pustules of
teliospores on infected needles. The diseased needles remain on the tree over
the winter. *C. ledi* is a heteroecious rust which exists in a number of varieties
with spermagonia and aecia on spruce needles. Two of these varieties attack
spruce in Europe. In the first, var. *ledi*, the uredinia and telia occur on *Ledum
palustre*; in the second, var. *rhododendri*, they are on species of *Rhododendron*.

   *C. abietis* is fairly common, and in some seasons causes epidemics in
Scotland and northern England; it has also been found in Northern Ireland,
south Wales and southwest England. It is also widespread in other parts of
Europe and eastwards throughout Asia as far as Japan. It occurs mainly on
Norway spruce, but has also been found on Sitka spruce, and on *Picea rubens*,
*P. pungens* and *P. engelmannii* (Murray, 1955). In Japan, where a separate
variety may be involved, Takahashi and Saho (1985) found that Norway and
Sitka spruce and the native *Picea jezoensis* were resistant, while *P. engelmannii* and *P. rubens* were severely affected and the native *P. glehnii* was slightly
susceptible.

   Basidiospores (sporidia) infect the young developing needles in the spring,
gaining entry through the stomata (Grill *et al.*, 1980). In June and July the
infected needles show conspicuous, lemon-coloured, transverse bands, which
become a deeper yellow as the season advances (plate 19). The mycelium of
the fungus is hyaline with orange or yellow droplets, and can be found in
sections of the discoloured zones (Delforge, 1930; Grill *et al.*, 1984). In late
summer, telia begin to appear on the undersides of the needles, but they do
not complete their development until the following spring. In May and June
they then finally erupt by a slit through the lower epidermis of the diseased
needles to form elongated yellow or orange pustules up to 10 mm long. The
pustules consist of chains of orange or red-brown teliospores that individually
measure 20–30 × 10–15 μm (Wilson and Henderson, 1966). The teliospores
germinate to produce the basidiospores that infect the new foliage. Soon after
the germination of the teliospores, the affected needles die and fall off, so
that the branches have their current needles, but those of the previous year
are partly or completely lost. A degree of natural control of *C. abietis* may
sometimes be exercised by the parasitic fungus *Darluca filum*, which attacks
the telia (Murray, 1953). In Austria, Grill *et al.* (1983) also found midge
larvae feeding on the teliospores.

   *C. ledi* is less common than *C. abietis*, and its varieties *ledi* and *rhododendri*
occur on spruce only in the vicinity of their alternate hosts, on which,
however, they can persist in the absence of spruce. Hence, as wild and
cultivated *Rhododendron* species occur throughout Europe, *C. ledi* var.
*rhododendri* is also widely found. *Ledum palustre*, however, is almost confined to parts of northern Europe from Scandinavia eastwards into the USSR,

and *C. ledi* is similarly limited in its distribution. These varieties on *Ledum* and *Rhododendron* are morphologically indistinguishable on spruce, and their certain identification depends on a knowledge of the presence of their alternate host, and should if possible be confirmed by cross-inoculation. The variety *ledi* occurs in North America on both spruce and species of *Ledum*, while the variety *rhododendri* is known in North America and New Zealand on rhododendron species only (Ziller, 1974).

In Britain, *C. ledi* var. *rhododendri* has appeared from time to time in Scotland and Northern Ireland on Norway spruce, and occasionally, as in 1986 (Redfern *et al.*, 1987), also on Sitka spruce. The stages on *Rhododendron* have also been found quite widely in southern England (Moore, 1959). Basidia formed on *Rhododendron* leaves infect the young, current spruce needles in early summer, and the infected needles soon show yellow bands similar to those caused by *C. abietis*. In the case of *C. ledi* var. *rhododendri*, however, infection is followed by the development of spermagonia, which appear as small, reddish-brown spots. Later in the summer, white, cylindrical aecia develop (plate 20), and burst open in August to shed yellow aeciospores. The infected leaves fall in the autumn.

Both the spermagonia and the aecia may be found on both leaf surfaces. The aeciospores measure 24–28 × 20–22 μm (Wilson and Henderson, 1966). They infect leaves of *Rhododendron*, and the resulting mycelium overwinters in the leaves. The following spring it gives rise at first to uredinia, which form as small orange spots on the undersides of the leaves. The urediniospores spread the disease further among the rhododendrons. The uredinia may be succeeded, also on the undersides of the leaves, by telia, which form as laccate, brownish-red areas under the epidermis. In the early part of the summer, the epidermis ruptures over the telia, and the teliospores give rise to the basidia that complete the cycle by reinfecting the young spruce needles. In many areas no telia form on the rhododendron plant, and no spread to spruce then takes place. On spruce in Europe as a whole, the fungus has caused most damage in areas in Switzerland, Italy and the USSR where telia are produced (Ziller, 1974). In many places, indeed, the rust is more damaging to *Rhododendron* than to spruce.

Though *Ledum palustre* is known as a rare plant in a few places in Britain, *C. ledi* var. *ledi* does not seem to have been recorded there. This fungus, as noted above, occurs in Europe in the area from Scandinavia eastwards into the USSR. Its life history on spruce and the alternate host and the measurements of the aeciospores in spruce needles are broadly similar to those of *C. ledi* var. *rhododendri*.   As *C. abietis* and *C. ledi* on spruce cause severe loss of one year's needles, they are potential causes of marked loss of increment, but their actual effect is greatly reduced because they give rise to epidemics only in occasional years, and then usually fairly locally. In the case of both these fungi this appears to be partly due to the fact that the spruce needles are susceptible in only the early stages of their growth (Hartig, 1894), and epidemics occur only when basidiospores are produced during the period when the needles are still liable to infection (Fischer, 1933). There is evidence that, at least in the case of *C. abietis*, cold spring weather delays flushing of spruce needles to the time when maximum production of basidiospores is taking place. In more normal years, flushing is earlier, and the basidiospores are shed when most of the needles are no longer susceptible (Murray, 1953). According to Delforge (1930), *C. abietis* is commonest and most severe in Belgium in warm, humid conditions. Hence, it tends to occur in dense stands where the humidity is high. In the case of *C. ledi*, infection of spruce is also dependent on the presence of the appropriate alternate host and the production of the telia on its leaves.

As these needle rusts are so sporadic in occurrence, and so cause relatively little damage, no steps to control them are usually taken. If severe attacks occur in dense stands, thinning out of badly affected trees may be recommended. In the case of *C. ledi*, it is likely that much could be done to prevent attacks by eradicating rhododendrons and *Ledum* spp. in and around spruce plantations.

## DIEBACKS AND SHOOT BLIGHTS

### Spring frost damage

The developing shoots of many spruces, especially Sitka spruce, are liable to damage by frosts in spring amd sometimes also in autumn. Young Sitka spruce planted in frosty sites may in their early years be trimmed annually by frost until they assume a bush-like form (plate 1). In such places their growth is slow until they reach a height above the general frost level, after which their leading shoots are able to grow without further damage. Norway spruce crops are generally less affected because their populations form a mixture of early- and late-flushing individuals, some of which escape the frosts. The grey mould *Botrytis cinerea* may sometimes colonise buds and shoot tips killed by spring frost, and extend the dieback. Frost damage is further considered in chapter 2.

### Bud blight caused by *Gemmamyces piceae* (Borthwick) Casagrande (*Cucurbitaria piceae* Borthwick)

This disease was first described and the causal fungus named from Scottish material of *Picea pungens* (Borthwick, 1909). It has since been found in England and Wales (where, as in Scotland, it appears to be commoner than was at one time supposed), and in Switzerland (Müller, 1950; Bazzigher, 1956), Denmark (Ferdinandsen and Jørgensen, 1938–9), Germany (Mehlisch, 1938), and Austria, Finland and Russia (Shoemaker, 1967).

*Gemmamyces piceae* attacks the buds, which usually then die soon after the start of elongation. The buds often become curved or twisted, claw-like (Ferdinandsen and Jørgensen, 1938–9) or 'snail-like' (Mehlisch, 1938) (plate 21). When the damaged buds are not killed, they may produce thin, deformed shoots bearing short needles that may soon die (Müller, 1950), though some often survive. If attacks are repeated and severe, affected parts of the crown may become sparsely branched and irregular in shape.

Some time after infection, the fungus forms a black granular stromatic crust on the surface of the buds. The granulations are due to the presence of densely clustered fruit bodies embedded in and emerging from the stroma. The perfect fructifications are black, spherical perithecia about 1 mm across (plate 22), with cylindrical asci measuring 180–250 × 25–30 μm, each containing eight broadly spindle-shaped, yellow or brownish muriform ascospores measuring 36–50 × 13–15 μm (Müller, 1950). Two pycnidial forms have also been described; both outwardly resemble the perithecia, but the first (*Megaloseptoria mirabilis* Naumov) contains long, colourless, more or less curved, thread-like, multiseptate conidia measuring about 250 × 8–10 μm, and the second (*Camarosporium strobilinum* Bomm. Rouss. and Sacc.) contains brown, oval to pear-shaped, muriform spores measuring 25–30 × 15–17 μm (Ferdinandsen and Jørgensen, 1938–9).

Little is known about the epidemiology of the disease. Ferdinandsen and Jørgensen (1938–9) in Denmark have suggested that infections by conidia and ascospores may take place in winter, and that the mycelium at first grows saprophytically on the bud scales before penetrating and killing the elongated buds. A long time seems to elapse between infection and the development of symptoms of disease (Müller, 1950), but the dead buds and the fungal stroma may persist for as long as 9 or 10 years (Young, unpublished).

Bud blight has been recorded on *Picea pungens* (Borthwick, 1909), *P. alba* (*P. glauca*) (Ferdinandsen and Jørgensen, 1938–9), Norway spruce and *Picea engelmannii*, and (very rarely) Sitka spruce.

In Britain, the disease has been found on pole-stage Norway spruce at a number of sites widely scattered over the country. In some crops it had been present for up to ten years. It was also found in a 2.5 ha stand of 13-year-old *Picea engelmannii* in North Wales. In this case about 90 per cent of the trees were infected, some were severely damaged, and about 10 per cent had been killed by a combination of the effects of *G. piceae* and the green spruce aphid, *Elatobium abietinum*. In Denmark it has been recorded on 30- to 60-year-old Norway spruce and on 20- to 25-year-old *Picea pungens* trees (Ferdinandsen and Jørgensen, 1938–9). In Germany it has damaged nursery stock (Mehlisch, 1938).

Bud blight has attracted little attention, and appears to cause only slight loss. In Britain, even heavily diseased trees appear to grow with little loss of vigour, although Ferdinandsen and Jørgensen (1938–9) found that in a few affected Norway spruce crops in Denmark bud blight had almost stopped growth in up to 25 per cent of the affected trees. The disease could be troublesome if it attacked Christmas tree crops, because of the deformation it causes, but so far it has not been recorded in these.

### Black snow mould caused by *Herpotrichia juniperi*

The black snow mould caused by *Herpotrichia juniperi* may affect trees in the forest, but is much more important in the nursery, and is therefore described above under 'nursery diseases'.

### Shoot blight caused by *Sirococcus strobilinus* (*Ascochyta piniperda*)
Shoot blight caused by *Sirococcus strobilinus* may occur in young and even mature plantations, but as it is of most importance in nurseries, it is treated above under 'nursery diseases'.

### Shoot blight caused by *Pucciniastrum areolatum* and bud blight caused by *Chrysomyxa woroninii*
The rust *Pucciniastrum areolatum* sometimes causes a shoot blight. It is, however, much more important as a cause of damage to cones. The rust *Chrysomyxa woroninii* sometimes causes a stunting of spruce needle buds. Both these fungi are described further below, under 'cone rusts'.


## BRANCH, BARK AND TRUNK DISEASES AND DISORDERS

### Drought crack
Under forest conditions, drought crack is rather common in Sitka spruce (see chapter 2).

### Bark necrosis associated with *Nectria fuckeliana* Booth
Occasional cases of bark necrosis of spruce have been reported widely over Britain as well as in many other European countries. In Britain the disease has been found mainly on Sitka spruce, but in various parts of Europe it has been found also on Norway spruce, and on *Larix* and *Abies*. Various fungi have been isolated from the damaged tissues, and of these *Nectria cucurbitula* has most often been considered to be the cause. Zycha (1955), in inoculation experiments in Germany, failed to prove the pathogenicity of this organism, but more recently Lang (1981) showed that the fungus could cause the death of plants of *Abies concolor* and its hybrids with *A. procera* and *A. nordmanniana*. Day (1950) considered that bark necrosis symptoms on Norway spruce in Britain were caused by frost. Several cases of this disease on Sitka spruce in Britain have been traced back to the winters of 1946–7, 1954–5 and 1961–2, all of which were generally cold or contained exceptionally bitter spells.

There has been some confusion as to the identity of the *Nectria* sp. that has so often been found on the lesions. The name *N. cucurbitula* is a *nomen confusum*, and the fungus on spruce appears to be that renamed *N. fuckeliana* by Booth (1959). This fungus has a *Cephalosporium* stage with colourless, oval microconidia measuring 6–9 × 3–9 μm, a second imperfect stage, *Cylindrocarpon cylindroides* Wollenw. var. *tenue* Wollenw., with curved, colourless, three-septate, *Fusarium*-like conidia measuring 33–40 × 4–5 μm, and red perithecia with cylindrical asci each containing 8 two-celled, colourless or pale-brown ascospores measuring 13–16 × 5–6 μm (Booth, 1959).

In Norway, Roll-Hansen and Roll-Hansen (1980) found that this fungus frequently invaded stem wounds in Norway spruce.

The symptoms of bark necrosis as he found it in Sitka spruce in Scotland were described by Laing (1948). The trees he studied were about 33 years old, and damage in them had first been noted about 9 years previously, after the trees had been brashed and pruned. In the early stages, sunken patches of dead bark (resembling those occurring in Norway spruce following invasion by *Stereum sanguinolentum*) appeared round the brashing wounds or round the bases of dead branches. Death of the bark often continued until long, narrow, vertical strips were produced. Mycelium, conidial pustules and the perfect fruit bodies of the *Nectria* were present on and in the dead tissues. Growth of the stem continued between strips, so that in severe cases the stems became conspicuously fluted. A few of the affected trees had died or were dying, others had become ill-shaped, while resin was exuding from the bark of most of the remainder.

Crops affected to this degree have been rarely seen, and overall losses through this disease have been small. The diseased stems are best removed when thinning.

### Branch and trunk cankers associated with *Valsa kunzei*

*Picea* spp. are among the trees sometimes affected by cankers associated with the weak parasite *Valsa kunzei*. This fungus also affects larches, pines (and sometimes other conifers) and is therefore further described in chapter 3.

## ROOT DISEASES

### Group dying caused by *Rhizina undulata*

Group dying caused by the ascomycete *Rhizina undulata*, which affects a range of conifers, is commoner and more severe on Sitka spruce than on any other tree. Norway spruce is sometimes also attacked. The disease is described in chapter 3.

## DECAY FUNGI

Most of the information on the susceptibility of the spruces to attack by the common decay fungi concerns Sitka spruce, Norway spruce and *Picea omorika*. All of these are very susceptible to damage by *Armillaria* spp. Sitka spruce is also especially liable to attack by *Heterobasidion annosum* (*Fomes annosus*), *Phaeolus schweinitzii*, *Sparassis crispa* and *Stereum sanguinolentum*. Norway spruce is also very susceptible to damage by *H. annosum* and *S. sanguinolentum*, and also by *Tyromyces stipticus*. *P. omorika* is more resistant to most of these fungi, but is very susceptible to attack by *S. sanguinolentum*. These fungi are described in chapter 3 (*Armillaria* spp. and *H. annosum*) and chapter 20 (*P. schweinitzii*, *S. crispa*, *S. sanguinolentum* and *T. stipticus*).

## CONE RUSTS

### *Pucciniastrum areolatum* (Fr.) Otth (*Thekopsora areolata* (Fr.) P. Magn)

*Pucciniastrum areolatum* occurs scattered across northern Europe and throughout Asia as far as Japan. In Britain it has been known to attack the cone scales of Norway spruce in Yorkshire and a few places in Scotland. Elsewhere it has been found also on other *Picea* spp., including *P. engelmannii*.

In spring its basidiospores infect the female spruce flowers, and flat crust-like spermagonia exuding a strong-smelling, sugary secretion soon form on the undersides of the cone scales. Later, from August to November, many small, brownish, hemispherical aecia form, usually on the upper sides of the scales. They are about 1 mm across and 1 mm high, and split open at the top to release clouds of aeciospores that measure 20–30 × 16–22 μm (Wilson and Henderson, 1966). In wet weather, scales of healthy cones close up, but those of affected ones remain open (Murray, 1955).

The aeciospores infect the alternate hosts, the bird cherry (*Prunus padus*) and the cultivated *Prunus virginiana* and *P. serotina*. On reddish spots on the undersides of the leaves of these, yellow or brown uredinia are formed, followed by red to dark-brown telia which occur on red spots mainly on the upper sides of the leaves. The teliospores germinate in the following spring, producing the basidiospores that reinfect the spruce flowers.

Roll-Hansen (1965) in Norway has described terminal and sometimes lateral shoot infections of Norway spruce and *P. engelmannii* by the same fungus. The affected shoots were usually bent towards the attacked side, and died back to the point of entry of the fungus.

Occasional heavy losses of spruce seed may occur on sites in which susceptible *Prunus* spp. are grown in the vicinity of spruce plantations.

### Spruce cone rust, *Chrysomyxa pirolata* Winter (*C. pyrolae* (DC) Rostr.)

The heteroecious rust *Chrysomyxa pirolata* sometimes attacks spruce cones and destroys the seed in an area across Europe and Asia from Scandinavia eastwards into Russia and beyond as far as Japan, as well as in North America. In Europe it occurs mainly on Norway spruce, but in North America it is found on *Picea engelmannii*, *P. glauca*, *P. mariana*, *P. pungens* and *P. sitchensis*. In Britain (where it is rare) and in Iceland, Japan and many parts of western Canada, it is known only on its alternate hosts, which are species of *Pyrola* (Wintergreen) and the related *Moneses uniflora*, on which it produces its uredinia and telia. Its large, whitish aecia occur on the outside of the spruce cone scales. The warty aeciospores measure 17–35 × 22–37 μm (Ziller, 1974). The uredinia occur on all the overground parts of the *Pyrola* and *Moneses* plants, and the warty urediniospores measure 29–35 × 14–24 μm (Ellis and Ellis, 1985). The telia are mainly on the underside of the leaves, and the teliospores measure 7–9 × 12–19 μm (Ziller, 1974).

The rust is found on spruce only when the spruce and the alternate hosts grow closely together. In *Pyrola* and *Moneses* plants the fungus becomes systemic and perennial. Uredinia and telia form on the diseased plants in the spring and early summer of the year following that of infection (Nelson and Krebill, 1982; Sutherland *et al.*, 1984). The germinating teliospores produce basidiospores which infect the spruce cones, and Sutherland *et al.* (1984) noted that in western Canada cones of *Picea glauca* showed spermagonia about 6 weeks after telia were found on adjacent *Pyrola* plants, and aeciospores appeared on the spruce cones after a further 4 weeks. The aeciospores complete the life cycle by infecting the plants of *Pyrola* and *Moneses*.

For successful infection of the spruce cones, basidiospore production must coincide with the opening and pollination of the female spruce cones, and infection is also assisted by intermittent rain (Nelson and Krebill, 1982; Summers *et al.*, 1986). Suitable conditions for infection do not often occur, and epidemics caused by *C. pirolata* are relatively uncommon; in Europe this rust is not considered to be of major importance. Nevertheless it may cause heavy losses of seed from time to time, especially in North America, where 20–90 per cent of the seeds of infected crops may sometimes be destroyed. Further, resinosis of infected cones, distortion of the cone scales and the presence of masses of aeciospores between them may hinder the dispersal of the remaining viable seeds (Nelson and Krebill, 1982; Ziller, 1974).

Control measures are likely to be required only in seed orchards, and there only occasionally. Experiments in Canada showed that spraying with ferbam over a period during which the spruce cones are open and susceptible gives control of infection (Summers *et al.*, 1986). If possible, the alternate hosts should be eradicated from such areas and in their vicinity.

### Spruce bud and cone rust, *Chrysomyxa woroninii* Tranzschel

Like *Chrysomyxa ledi* var *ledi*, *C. woroninii* alternates between spruce and *Ledum* spp., and it occurs only where the two hosts are found adjacent to one another (Ziller, 1974). It occurs in Europe in the USSR and across Asia to Japan, as well as in Alaska and the northern parts of Canada (Ziller, 1974; McBeath, 1984). In Europe it attacks *Picea abies*, and in North America it affects *P. glauca* and *P. mariana*.

On spruce, infection takes place in late spring when humidity remains high, and the rust affects the leaf buds and the female cones. When infected needle buds open, they produce stunted shoots which have small, bright-yellow needles and look rather like male cones. Spermagonia form near the tips of the diseased needles and on the edges of the cone scales, and are followed in mid- to late summer by white, blister-like aecia which emerge from the stomatal bands along the remaining parts of the needles and on the cone scales. The warty aeciospores are orange-yellow in the mass, and are very variable in size, measuring 19–34 × 33–62 μm (Ziller, 1974). Infected cones produce no viable seed, and after the shedding of the aeciospores the stunted shoots blacken and die. The aeciospores infect the leaves of *Ledum* spp., on which the fungus overwinters, forming a perennial mycelium. In the following spring, witches' brooms arise on the infected plants, and, on these, brown crusts of telia are produced on the undersides of some of the leaves. In late spring the teliospores germinate and form yellow basidiospores which reinfect the spruce buds and young cones and so complete the life cycle (Ziller, 1974).

## VIRUS AND VIRUS-LIKE DISEASES

### Spruce virosis

What appears to be a virosis, which was described on Norway spruce in Czechoslovakia (Cech *et al.*, 1961), has been found on Sitka spruce in Britain (Biddle and Tinsley, 1968). The affected trees show shortening of the needles, which at first become yellowish green, and later fall.

Cech *et al.* (1961) inoculated seedlings by aphid transmission by *Adelges abietis* and by grafting, and observed hexagonal, rod-shaped particles 22–49 nm wide and often *c*. 625 nm long. Biddle and Tinsley (1968) found similar particles in exudates from twigs of affected Sitka spruce in Britain. Available information on this disorder has been summarised by Cooper (1979).

Ebrahim-Nesbat and Heitefuss (1985) have described a Rickettsia-like organism (RLO) present in the fine roots of young Norway spruce trees in Germany. The organism may have been associated with an observed decline of the infected trees. The same authors (1989) isolated flexuous filamentous particles, apparently of a virus of the potyvirus group, from acutely yellowed needles of a Norway spruce tree in Bavaria.

## MISCELLANEOUS DISEASES

### Top dying of Norway spruce

'Top dying', which occurs almost exclusively on Norway spruce (though it has also been recorded on *Picea glauca* and *P. omorika*) (Murray, 1957; Redfern *et al.*, 1985), appears to be a complex physiological disorder related in part to the susceptibility of Norway spruce to water stress and exposure. Typically it affects trees in the pole stage and beyond, though occasionally it has been found in young plantations in the thicket stage. Since 1950 it has been reported in all parts of Britain, on a wide variety of sites, as well as in other areas along the coasts of northern Europe in Denmark, Germany, Ireland, the Netherlands, Norway and Sweden (Diamandis, 1978a).

The first symptoms have been recorded in both the late summer and late winter. The initial damage appears as a spectacular browning or reddening of the foliage in parts or over the whole of the crowns of the trees (plate 16). On individual current and second-year shoots this browning progresses down-wards from the shoot tip. Affected needles fall, and thereafter the crowns decline or recover at varying rates according to the severity of the initial damage. Trees in decline bear sparse, sickly foliage, and they usually die within two or three years from the tops downwards, the roots being the last parts to succumb. Sometimes only edge trees are affected, and trees in the interior of the stand are always less affected than those at the edges. The dead and dying needles are often invaded by *Rhizosphaera kalkhoffii*, but the evidence does not suggest that this fungus is the cause of top dying. Went (1948, 1949) and Murray (1957) both failed to inoculate spruce trees with *R. kalkhoffii*, which at most seems to be only a weak parasite that causes no more than a minor leaf cast, and never leads to the progressive decline and death of trees under British conditions. Similarly Diamandis (1978b) was unable to show that this fungus (or any other he isolated) was the cause of the disorder.

Careful study of affected crops has shown, however, that the needle browning is preceded by a disorganisation of the growth pattern of the tree. The effects on growth are associated with mild winters, which tend to be followed by a reduction in growth in the following season. The growth reduction may be accompanied or followed within the next few years by needle browning and needle loss in a proportion of the trees.

Young (reported by Phillips, 1965, 1967b) suggested that two factors or groups of factors might be involved. The first, which could be called a *conditioning factor*, operated during relatively mild winters, and caused the growth reduction in the following season. The intensity of the conditioning factor, and, hence, the extent of the reduction in growth, varied from site to site. If the effect of the conditioning factor was sufficiently marked, part of the crop became susceptible to one or more *precipitating factors*, and if one of these operated near the time of the fall in growth, top dying might follow. There was evidence that top dying could be precipitated in this way by various factors that gave rise to increased air movement through the crop, and so to an increase in transpiration and water loss. Among such factors were brashing, thinning and removal of side shelter. Top dying is also liable to occur where there is usually considerable air movement, as in exposed situations or where gaps and uneven growth in crops make for turbulence. Summer drought also seems to be involved in some cases (Young, reported by Phillips, 1963a).

In the last of a series of six papers, Diamandis (1979) summarised the results of studies of the disorder in a number of forests in northeast Scotland, in which he attempted to elucidate the factors responsible. He confirmed earlier suggestions that top dying was associated with mild winters and exposure, and sometimes with drought in the early part of the growing season. From a study of weather data in the area of his investigation, he concluded that when mild winters were accompanied by long windy periods between December and March, the trees became water-deficient. The water deficit caused a growth reduction in the following season and, if severe, needle browning and loss also occurred. Drought in the period from May to August intensified the effect. Presumably operations such as brashing and thinning were similar intensifying factors. The disorder therefore resulted from the combined effects of atmospheric and soil drought. Liss *et al.* (1984) studied the root systems of eight Norway spruce trees with top dying from two sites in Bavaria. They found an association between crown damage and a decline in the fine root systems and their mycorrhiza. They suggested that the decline could have resulted from the weakening of the trees by air pollution.

In some crops of Norway spruce, top dying has caused severe losses, but in most cases the proportion of affected trees has been relatively small. Nevertheless, partly because the onset of the disorder has been so unpredictable, forest managers in Britain and Denmark have tended to reduce the planting of Norway spruce (Diamandis, 1978a).

So far, little can be done to prevent top dying, though it may be suggested that sudden changes in the environment of the crops should be avoided. Thus, particularly following mild winters, heavy thinnings and sudden removal of side shelter should not be carried out.

# 5 Diseases of pine (*Pinus* spp.)

The European *Pinus sylvestris* (Scots pine) and *P. nigra* var. *maritima* (Corsican pine) and the North American *P. contorta* (lodgepole pine) are the major species of pine grown in British forests. Many other introduced pines are grown in parks and gardens. *P. sylvestris* grows almost throughout Europe and into the western parts of Asia and as far south as Greece. Various forms of *P. nigra* are found from Austria into the Balkans and into the Mediterranean. *P. cembra* grows in the Alps and into the USSR. A number of other pines are found in Europe, especially in the Mediterranean area.

## General

Pines are subject to a number of disorders caused by non-living agencies such as frost and lime-induced chlorosis, and these are discussed in detail in chapter 2. Some of the pathogens that infect pines, including *Heterobasidion annosum* (*Fomes annosus*), *Armillaria* spp. and *Rhizina undulata*, are described in chapter 3. A number of important diseases, including rusts, root diseases, cankers and needle casts, attack pines in Britain. Some, such as Brunchorstia or Scleroderris dieback and Lophodermium needle cast, are equally or even more important in mainland Europe and North America. Others are seldom found in Britain, but may be common elsewhere; an example is Dothistroma needle blight, which has rarely been found in Britain, but has caused much damage in plantations of *Pinus radiata* in many parts of the world.

## Nursery diseases

NEEDLE DISEASES

**Needle cast caused by *Lophodermium seditiosum* Minter, Staley and Millar**
Prior to 1978, most published references to pathogenic *Lophodermium* spp. refer to *L. pinastri* (figure 36) as the species causing damage to Scots pine. However, as will be seen below, a new species, *L. seditiosum*, has been shown to be the main pathogen (Diwani and Millar, 1987).

Peace (1962) indicates that, in Britain, Lophodermium needle cast can cause serious defoliation of Scots pine in the nursery and young plantations. Older stands have also been attacked. Melchior (1975) reports that no new plantations of Scots pine have been established in northern Schleswig-

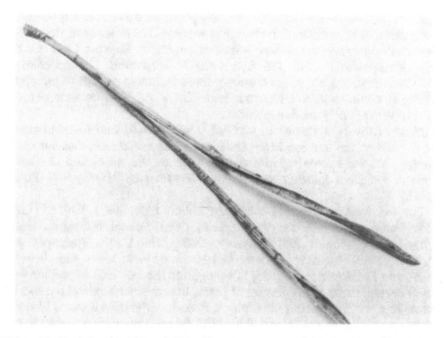

**Figure 36** Fruit bodies of *Lophodermium pinastri* on Scots pine needles. Note the distinct black lines across the needles (see table 5.1) (Forestry Commission)

Holstein, Germany, because of the severe losses which have occurred there. The first serious outbreak in North America, attributed to *L. pinastri*, was reported in 1966 and severe damage has subsequently been reported in nurseries of both Scots and red pine and in Christmas tree plantations of Scots pine (Skilling and Nicholls, 1975).

Since 1978, Minter and others (Minter *et al.*, 1978; Minter and Millar, 1980) have undertaken detailed studies into the taxonomy, ecology and biology of *Lophodermium* spp. on Scots pine. Minter *et al.* (1978) identified four species of *Lophodermium* on Scots pine needles: *L. pinastri* (Schrad. ex Hook) Chev.; *L. conigenum* Hilitzer; *L. pini-excelsae* Ahmad; *L. seditiosum* Minter, Staley and Millar.

The four species can be distinguished by the manner in which the ascocarps are embedded in the host tissue and in part on their ability to produce stromatic lines of different types across the needle. Their main characteristics are briefly summarised in table 5.1.

*L. pini-excelsae* is mainly a saprophyte which is sometimes found on Scots pine but occurs mainly on five-needled pines.

Minter and Millar (1980) studied the ecology and biology of *L. pinastri*, *L. conigenum* and *L. seditiosum* on Scots pine. They found that each of the three species occupied a distinct habitat: *L. pinastri* occurred on senescent needles in litter; *L. conigenum* on needles that had died in consequence of live branches cut off or broken, or of death of roots; and *L. seditiosum* as a parasite on second-year needles of nursery plants, in which role it appeared to be causing death of plants. It therefore seemed that in many of the earlier reports of outbreaks the disease had been incorrectly diagnosed and that the fungus responsible could well have been *L. seditiosum*. Other studies of *Lophodermium* on pines, for example, those including records of ascospore dispersal, could relate to different species. Therefore, interpretation of much of this research may need reappraisal.

Minter (private communication; CMI Descr. F. 568) later found instances of needles on lop and top being colonised by *L. seditiosum*, and not by *L. conigenum*. Such colonisation could account for the association of disease with debris from thinning and pruning recorded by Murray and Young (1956).

Further work in Britain (Diwani and Millar, 1986, 1987), Finland (Lilja, 1986), Germany (Wuhlisch and Stephan, 1986), Poland (Kowalski, 1982), Yugoslavia (Uscuplic, 1981; Lazarev, 1983a), the USSR (Raspopov and Petrova, 1989), the USA (Adams and Roberts, 1988; Ostry and Nicholls, 1989) and China (He *et al.*, 1985) has supported the view that *L. seditiosum* is an important cause of needle cast of pines, the other *Lophodermium* species occupying a minor, mainly saprophytic role. *L. seditiosum* appears to occur widely throughout Europe, and in the USA (especially where pines are grown for the Christmas tree trade) (CMI Descr. F. 568), and probably across Asia to China.

*Lophodermium seditiosum* (CMI Descr. F. 568)   The main differences between this fungus and other *Lophodermium* spp. on pines are summarised in table 5.1. The boat-shaped ascocarps of *L. seditiosum* are totally subepidermal and measure 800–1500 μm long. A few diffuse brown stromatic lines are sometimes present across the needle.

The asci of *L. seditiosum* are cylindrical, eight-spored, 140–170 μm long and 11–13 μm wide, and filiform paraphyses are present. The ascospores are filiform and measure 90–120 μm long.

The pycnidia of the anamorphic *Leptostroma* stage are subepidermal, often coalescing, 300–500 μm long and the bacillar conidia are 6–8 μm long.

A full description of the fruiting structures of *L. seditiosum* and other *Lophodermium* spp. on Scots pine needles is given by Minter (1981).

The first symptoms of the disease appear in summer as pale spots on the needles and these gradually enlarge and merge so that the whole needle turns brown. Needles are often infected in groups, and when infection is severe, the whole of the previous year's needles turn brown and fall in winter or early spring.

Infection occurs mainly through the cuticle and appears to be solely from ascospores. Millar (1975) reviewed the periods of ascospore release reported by various authors. Peak periods of release varied greatly and it is now apparent that records were probably taken from a range of *Lophodermium* spp. *In vitro* studies of sporulation periods of *L. seditiosum* by Minter and Millar (1980) indicated that in Britain ascospore release started in July and reached a peak between September and December. This corresponded well

**Table 5.1** *Lophodermium* species on pines

| Species | Ascocarps | Ascospores | Habitat |
|---------|-----------|------------|---------|
| *L. pinastri* (on many 2- and 3-needled pines; occasional on 5-needled pines) | Partly subepidermal 700–1200 μm long; lips usually red or orange; many thin black lines across needle | 70–110 × 2 μm | On senescent leaves in litter, and on cones |
| *L.pini-excelsae* (mainly on 5-needled pines) | Resemble those of *L.pinastri* but lips grey; few black lines across needles | 50–70 × 2 μm | Usually on needles of 5-needled pines, rarely on *P.sylvestris*; mainly saprophytic |
| *L.seditiosum* (on *P.sylvestris*, *P.nigra*, *P.halepensis*, *P.virginiana*, etc.) | Subepidermal 800–1500 μm long; lips colourless green or blue; sometimes diffuse brown lines across needle | 90–120 μm long, slightly wider than *L.pinastri* | Parasite on second-year needles, especially of nursery plants of *P.sylvestris*; sometimes on lop and top and on cones |
| *L.conigenum* (mainly on 2- and 3-needled pines, including *P.sylvestris*, *P. nigra*, *P.pinea*, *P.contorta*, *P.halepensis*) | Partly subepidermal; 900–2000 μm long; lips olive green; few brown lines across needle | 90–130 × 2 μm | On dead needles on branches and lop and on cones |

with the appearance of opened ascocarps on infected needles. In Yugoslavia, Lazarev (1983a) found that ascospores were shed from July to the April of the following year but the critical period for infection was from the middle of August to October.

Many outbreaks of needle cast have occurred in nurseries which are sited within or very close to pine plantations. Such plantations may not themselves be seriously affected by the disease but they may harbour sufficient inoculum to act as sources of infection for nearby seedlings or transplants. In North America, outbreaks of the disease have occurred in Christmas tree plantations of Scots pine following the introduction of plants from an infected nursery (Skilling and Nicholls, 1975). Such situations should be avoided wherever possible, but when infection does occur, approved fungicidal sprays can be applied.

A number of chemicals have given good control in the nursery, including dithiocarbamates in Finland (Lilja, 1986), BMC and other chemicals in

Russia (Vedernikov and Fedorova, 1986) and chlorothalonil in China (He *et al.*, 1985). In North America, Nicholls (1973) and Skilling (1974) have shown that properly timed sprays of maneb or chlorothalonil gave good control on red pine in nurseries and on Scots pine in Christmas tree plantations. Boudier (1987b) has reported on trials in Normandy, France, which showed that the best control of the disease on Scots pine and the Austrian, Corsican and Calabrian forms of *Pinus nigra* was given by the application of carbendazim and triadimefon given four times at fortnightly intervals from early June to the end of August. Good results were also given by flutriafol, which also enhanced growth. Minter and Millar (1980) have reviewed the recommended spray times given by various authors. The great majority suggest that sprays should be applied in the period July–September. This coincides with the periods of ascospore release of *L. seditiosum*.

Diwani and Millar (1986) describe methods of plant sanitation, cultural and chemical control used in Scots pine nurseries in Scotland. It is clearly important to site pine nurseries away from established pine plantings which may be sources of inoculum. There is also much potential for selection and breeding of pines resistant to needle cast. There are many reports of differences in susceptibility between provenances of Scots pine (e.g. Stephan, 1975) and in the progeny of plus trees in Poland (Korczyk, 1984), Germany (Wuhlisch and Stephan, 1986) and in the USSR (Prokazin and Kurakin, 1983). However, other factors, including environmental conditions, can also influence susceptibility and further research may demonstrate their significance.

**Needle rust: *Coleosporium tussilaginis***
The needle rust *Coleosporium tussilaginis* may occur in the nursery, but it is more commonly found in young plantations, and it is therefore described under 'post-nursery diseases'.

**Needle blight caused by *Dothistroma pini* (*Dothistroma septospora*)**
In Britain *Dothistroma pini* was found for a few years causing quite severe damage to pine transplants in one nursery, but has since been of no importance. In other countries, however, this fungus has caused heavy losses in pine plantations. Therefore, it is described below under 'post-nursery diseases'.

**The snow mould *Phacidium infestans* Karst.**
*Phacidium infestans* is a pathogen which mainly affects Scots pine, and is often reported from the Scandinavian countries, into Finland and the USSR, and also further south in the more mountainous regions (CMI Descr. F. 652). Closely related species attack a wide range of conifers, especially spruces in North America. The literature on *P. infestans* has been reviewed by Roll-Hansen (1989).

*P. infestans* is serious only at high elevations where prolonged snow cover is common. Small plants, particularly those growing weakly, may be subject to attack in the nursery, in plantations and in areas of natural regeneration (Roll-Hansen, 1975). The mycelium of the fungus spreads at low temperatures under the snow cover, and as the snow disappears infected plants appear with withered needles matted together by a mass of white mycelium.

The apothecia of the fungus are embedded in the needles, and develop over the summer. The host epidermis splits open into teeth and folds back to reveal the grey apothecial disc. The asci ripen in autumn, and the hyaline ascospores measure 15–28 × 5–8 μm (Ellis and Ellis, 1985).

Snow mould is most commonly found in dense seedbeds or in dense natural regeneration, and these situations should be avoided where attack is likely. The North American *Pinus contorta* (especially its northern provenances) is more resistant than Scots pine, but is by no means immune, especially if plants suffer damage in severe weather (Karlman, 1986).

On forest sites in which the disease is known to be damaging to young trees, more resistant species such as *Pinus contorta* or Norway spruce may be substituted for Scots pine. In the nursery, copper sprays have been found moderately successful, and quintozene, chlorothalonil and thiophanate-methyl have been used in Finland (Lilja, 1986). Carbendazim, benomyl and zineb are among materials which have also shown promise in the USSR (Vedernikov and Fedorova, 1986). Attempts to predict optimum times for spraying through the forecasting of outbreaks of snow mould have been made in the USSR (Vedernikov, 1985, 1986).

### Other snow moulds: the black snow mould *Herpotrichia juniperi* and the pine brown felt fungus *H. coulteri*

*Herpotrichia juniperi* (CMI Descr. F. 328) occasionally produces a dark brown mycelial felt on conifer shoots under snow in northern and mountainous parts of Europe, into Asia, and in North America. It may then affect and kill pines in nurseries and young plantations and damage plants in cold stores, but it is of relatively small significance. It attacks a number of conifers, but is of most importance on Norway spruce, and is therefore described under 'spruce' in chapter 4.

Another of these European snow moulds is *H. coulteri* (Peck) Bose (CMI Descr. F. 327). This is confined to pines, chiefly in southern Europe, again attacking plants in nurseries and young plantations under snow, covering them with a brown mycelial felt, and killing affected tissues. Its brown, two-celled ascospores measure 20–28 × 7–10 μm. It is of very little pathological importance.

## Post-nursery diseases

### NEEDLE DISEASES

#### Needle cast caused by *Lophodermium seditiosum*

*Lophodermium seditiosum* is clearly an important pathogen on Scots pine in nurseries and in young plantations up to 10 or so years of age, including Christmas trees. This observation is confirmed by Raspopov and Petrova

(1985) in Russia and by Diwani and Millar (1987), who both point out the importance of *L. seditiosum* as a pathogenic needle cast fungus and the largely saprophytic role of *L. pinastri*. *L. seditiosum* is described above under 'nursery diseases'.

### Minor needle casts

*Lophodermium pinastri*, *L. conigenum* and *L. pini-excelsae* have been described above in relation to needle cast in nurseries. They also occur in plantations, where they are largely saprophytic and of little significance as pathogens.

A number of minor (but sometimes locally serious) needle casts occur on pines, mainly in the post-nursery stage. They are caused by the ascomycetes *Lophodermella sulcigena* (Rostr.) Höhnel, *L. conjuncta* Darker and *Cyclaneusma minus* (Butin) Di Cosmo *et al.*, and the deuteromycetes *Hendersonia acicola* Munch & Tub. and *Sclerophoma pythiophila* (Corda) Höhnel (the anamorph of *Sydowia polyspora* (Bref. and Tav.) E. Müller). The Hypodermataceae occurring on conifers have been well described by Darker (1932, 1967).

*Lophodermella sulcigena* (*Hypodermella sulcigena*) (CMI Descr. F. 562) infects first-year needles of both *P. sylvestris* and *P. nigra* var. *maritima*, and sometimes *P. contorta* and *P. mugo*, causing a light pinkish-brown discoloration of the needles followed by defoliation and subsequent reduction of growth (Watson and Millar, 1971). Its elongated ascocarps are embedded in the needle tissues and open by a slit. Their asci, which are interspersed by paraphyses, contain colourless, narrowly clavate ascospores which measure $30-35 \times 4-5$ μm, and have gelatinous sheaths (Ellis and Ellis, 1985). The fungus occurs throughout most of Europe into the USSR. Both Mitchell *et al.* (1976a) in Scotland and Lazarev (1983) in Yugoslavia found that the critical period for infection was in July and August. The disease may sometimes cause severe damage in Scandinavia (hence one of its common names, 'Swedish needle cast'), and in Finland Jalkanen (1985) found that it was especially important in plantations ranging from 10 to 20 years of age on abandoned agricultural land. At one site in Scotland, Corsican pine trees attacked by *L. sulcigena* suffered a 59 per cent reduction in stem volume growth over 12 years compared with healthy ones (Mitchell *et al.*, 1976a). Needles damaged by *L. sulcigena* may be secondarily infected by *Hendersonia acicola* (which is described below). If this infection takes place soon enough, *L. sulcigena* fails to fruit, its further development is then prevented and affected trees may recover (Mitchell *et al.*, 1976b; Jalkanen, 1985; Jalkanen and Laakso, 1986).

*Lophodermella conjuncta* (CMI Descr. F. 658; Mitchell *et al.*, 1978) infects first-year and older needles of two-needled pines. Bright yellow bands appear on the needles and, later, the whole infected area of the needle turns brown.

The long grey apothecia open by a slit, and their colourless, narrowly clavate ascospores have gelatinous sheaths and measure 75–100 μm (Ellis and Ellis, 1985). Damage caused by *L. conjuncta* seems to be slight.

*Cyclaneusma minus* (Butin) Di Cosmo *et al.* (*Naemacyclus minor* Butin) (CMI Descr. F. 659) may be found, usually as a saprophyte, on many pines. In Europe it mainly affects Scots pine, and as a cause of needle cast there it has been troublesome chiefly in Poland and Yugoslavia, where it has resulted in the loss especially of two-year-old needles of trees up to 20 years old (Karadzic and Zoric, 1981; Kowalski, 1988). In North America, the same fungus is important as a cause of needle cast of Scots pine grown for the Christmas tree trade, but it also affects many other pines (Sinclair *et al.*, 1987). In New Zealand it has been damaging on *Pinus radiata* (Gadgil, 1984; Pas *et al.*, 1984a,b).

Pine needles are susceptible from an early stage, and turn yellow and fall in the year after infection. Inconspicuous pycnidia embedded in the needles contain rod-shaped spores measuring 6.0–9.5 × 1 μm; they form on the needles on the tree and are followed, most often after needle fall, by elongated jelly-like, whitish to yellowish apothecia which open by two epidermal flaps and contain colourless, worm-like, often curved ascospores, sometimes with one or two septa, and measuring 65–98 × 2.5–3 μm (Butin, 1973; Ellis and Ellis, 1985).

The closely related *C. niveum* (Pers. ex Fr.) Di Cosmo *et al.* (CMI Descr. F. 660) also occurs on pine needles, and has often been confused with *C. minor* (Butin, 1973) but it is at most only weakly pathogenic.

Work in New Zealand has shown that *C. minus* may cause a loss of volume (Pas *et al.*, 1984a,b), but may be controlled by spraying with dodine, anilazine or benomyl (Hood and Vanner, 1984).

*Hendersonia acicola*, the pycnidia of which contain brown, mostly two-septate conidia measuring 11–15 × 4–5 μm (Ellis and Ellis, 1985) infects pine needles (especially those of Scots and Corsican pine) previously attacked by *Lophodermella sulcigena* (interactions with which have been mentioned above). Needles infected by *H. acicola* turn a pale whitish grey, and become brittle and fall, but as a pathogen *H. acicola* is of small account.

*Sclerophoma pythiophila* (the anamorph of *Sydowia polyspora* (Bref. and Tav.) E. Müller) is a common saprophyte on dead needles of pines (and many other conifers). It may, however, be associated with shoot dieback and needle cast. Jahnel and Junghans (1957) observed an association between infection by *S. pythiophila* and damage by the pine gall midge, *Contarinia baeri*. Batko *et al.* (1958) confirmed this observation, finding evidence that *S. pythiophila* colonised wounds made by the insect, and probably extended the damage it produced.

*S. polyspora* is found on many conifers, and is therefore briefly described in chapter 3.

### Needle rust: *Coleosporium tussilaginis* (Pers.) Lév.

Several rusts of the genus *Coleosporium* occurring in central and northern Europe have been treated as separate species, largely based on the generic identity of the alternate host (e.g. *C. tussilaginis*, *C. senecionis*, *C. campanulae*). However, it is not possible to distinguish these species on the basis of morphological characters and it is now considered best, as far as Europe is concerned, to include them as races or race groups of one species (Wilson and Henderson, 1966). The spermagonia and aecia occur on two-needled pines and the uredinia and telia chiefly on members of the Compositae, Campanulaceae and Scrophulariacae.

The sporidia, which are sometimes called basidiospores, infect the current year's pine needles in the autumn and spermagonia and aecia appear during the following summer. Spermagonia develop as small yellowish spots and spermatia are exuded in a sticky fluid. Soon afterwards aecia appear, initially as white columnar blisters 1–5 mm high, whose covering membrane breaks to release orange-coloured aeciospores (figure 37). These can infect the leaves and stems of a number of herbaceous species, the commonest of which are probably *Senecio* spp. and *Tussilago* spp.; other hosts include species of *Campanula*, *Melampyrum*, *Petasites* and *Sonchus*. On groundsel, uredia can be found throughout the year as orange-brown spots on stems or the undersides of leaves. Several generations of the orange uredospores may be produced during the winter, so that the fungus can maintain itself on a herbaceous host without necessarily returning to a pine host. During the autumn telia may appear on the undersides of the groundsel leaves or on the stem; these develop as dark red-brown waxy crusts.

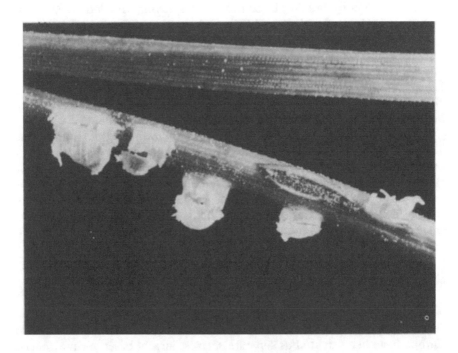

**Figure 37** Aecia of *Coleosporium tussilaginis* on needle of Scots pine (Forestry Commission) (× *c.* 3)

The aeciospores vary in shape from globoid to obovoid, 20–40 × 16–27 μm (Wilson and Henderson, 1966), and have a colourless wall 2–3 μm thick which is densely verrucose. Urediniospores are similar in shape and size to the aeciospores and measure 20–40 × 16–25 μm (Wilson and Henderson, 1966), but they have a thinner verrucose wall, 1–1.5 μm thick. The teliospores are clavoid to cylindric, smooth-walled and measure 60–105 × 15–24 μm (Wilson and Henderson, 1966). They are at first unicellular but later become four-celled and greatly thickened and gelatinous at the apex. This rust is quite commonly found on young pine plants either in the nursery or in the early years of establishment in the forest. In Britain it occurs on both *P. sylvestris* and *P. nigra* var. *maritima*, though it has not so far been recorded on *P. contorta*. The conspicuous white aecia sometimes give rise to concern but damage is rarely serious and no control measures are normally required. Good weed control in the nursery and its environs should eliminate the alternate hosts.

### Needle blight caused by *Dothistroma pini* Hulbary (*D. septospora* (Dorog.) Morelet) (anamorph of *Scirrhia pini* Funk and Parker)

Dothistroma blight is largely a disease of pines, although *Pseudotsuga menziesii* and *Larix decidua* are also slightly susceptible to attack. It has also been found on Norway spruce in southern Germany (Lang, 1987). It has caused serious defoliation in plantations of *Pinus radiata* in several parts of the world and this is by far the most important susceptible species from an economic point of view. The disease has been found on many pine species and is widely distributed throughout the world, having been recorded in Europe, South and East Africa, North and South America, Australia and New Zealand. In Britain, the disease caused severe damage to pine transplants in nurseries at Wareham Forest, Dorset, during the 1950s. Extensive beds of *Pinus nigra* var. *maritima* and small beds of *P. ponderosa* and *P. bungeana* were notably affected, while *P. contorta* was moderately damaged and *P. sylvestris* appeared immune. The disease was not found in plantations of pine at Wareham, and indeed it has been found only twice in Britain on such trees. On both occasions, in 1958 and 1989, it occurred in South Wales on Corsican pine, which showed severe browning (Strouts *et al.*, 1989). Butin and Richter (1983) identified both *D. pini* and its teleomorph *S. pini* on *P. nigra* stands in southern Germany and Lang and Karadzic (1987) found the pathogen on several pine species, including *P. nigra*, *P. cembra*, *P. aristata*, *P. koraiensis* and *P. tabuliformis*. *P. sylvestris* appeared resistant.

Hulbary (1941) was the first to name *D. pini* and he described it on *P. nigra* in Illinois. In a comprehensive review of Dothistroma blight of *P. radiata*, Gibson (1972) discusses the somewhat confused early history of the nomenclature of the pathogen. Several authors, including Siggers (1944) and Murray and Batko (1962), re-examined a range of material that had previously been disposed to fungi such as *Actinothyrium marginatum* and concluded that they corresponded most closely to *D. pini*.

The lesions on the pine needles are band-like and bear 1–12 stromata which are black, hypodermal in origin and erumpent when mature. The stromata vary in size from 300 to 750 × 150–400 μm, usually with one locule. The substromatal hyphae are septate, many-branched and are found only in the mesophyll. Conidia are hyaline, filiform, one- to five- , mainly three-septate when mature, often non-septate when immature. The size of conidia varies and three varieties have been distinguished. Thyr and Shaw (1964) described *D. pini* var. *pini* with conidia 22.4 (15.4–28.0) × 3.2 (2.6–4.0) μm and var. *linearis* 31.9 (23.0–42.0) × 2.4 (1.8–2.9) μm. Ivory (1967) added a third variety, *keniensis*, which measured 28.7 (15.0–47.5) × 2.6 (1.5–3.5) μm. This variety has been found only in Africa, whereas the other varieties have been found in North America only (var. *linearis*) or in all continents except Africa (var. *pini*). However, there is some doubt concerning the validity of these varieties; Gadgil (1967), for example, considers that the characters used for differentiation are insufficiently reliable.

The perfect stage of *D. pini* var. *pini* has been identified as *Scirrhia pini* on a range of pine hosts including *P.radiata* on Vancouver Island. This stage has also been found in France and in Germany (Butin and Richter, 1983). The ascospores formed in its black multilocular fruit bodies are colourless and two-celled and measure 11–16 × 3–4 μm (Ellis and Ellis, 1985).

Infection (the mechanism of which has been studied by Karadzic, 1989a) appears as a chlorosis and necrosis of needles on the main stem and base of lower branches. Under favourable conditions and on a susceptible host defoliation may be so severe that only needles at the extremities of the branches remain (Gibson *et al.*, 1964). Needle infections are first evident as yellow flecks which extend to become bands around the needle. As necrosis develops, these bands take on a characteristic red tinge and this has led to common names such as 'red band' being applied to the disease in some parts of the world. The black stromata develop within these bands and necrosis often extends throughout the needle, followed by needle cast. In Yugoslavia, Karadzic (1989b) found that conidia were shed from the beginning of April until the end of October, and ascospores were airborne from the second part of June to the end of September. The critical infection period was usually from early May to late June or sometimes the end of July. Needle infection was first seen in October, and was clearly visible by November.

In his review of the factors affecting dispersal and infection Gibson (1972) indicates that research in Chile, East Africa, New Zealand and the USA has shown the need for light rain or heavy mist in the effective dispersal of conidia. The conidial germ tube penetrates the needles through a stoma, a process taking 3 days or longer, depending on temperature and humidity. Macroscopic symptoms do not appear on the needles for 5–10 weeks. This relatively slow rate of growth is also a feature of the fungus in culture.

The first trials to test fungicides for the field control of the disease were undertaken in East Africa (Gibson *et al.*, 1966) and they demonstrated that fungicides based on copper compounds were particularly effective. These results have found practical application in New Zealand, where adequate protection can be achieved by three or four aerial applications of copper-based fungicides at 2.24 kg per hectare equivalent spread over the first fifteen years' life of a *P. radiata* plantation (Bassett and Zondag, 1968). After this period the crop develops sufficient resistance for further spraying to be unnecessary. Kershaw *et al.* (1982) wrote a handbook covering ground and aerial spray techniques for *D. pini* in New Zealand. Pas *et al.* (1984c) questioned whether the copper sprays at $22.50 per ha per spray were warranted with timber volume increases of 3–3.5 $m^3$ per ha per spray. Successful control by the use of copper sprays has also been reported by Petersen (1967) on *P. ponderosa* and *P. nigra* in the USA. In trials in Yugoslavia with various fungicides to control the disease on *P. nigra*, Karadzic (1987) found that applications of captan, zineb and benomyl were all of value, but copper oxychloride was still the most effective fungicide of those tested.

The planting of *P. radiata* in East Africa ceased soon after the disease was first recognised and its place was taken by *Cupressus lusitanica* and *P. patula*. The latter species had proved immune to attack by *D. pini*. Ivory (1968) has compiled a comprehensive list which shows the relative susceptibility to *D. pini* of pines growing in East Africa (Ivory and Patterson, 1969). Gibson (1972) points out that field resistance to *D. pini* appears to take two different forms; some species such as *P. radiata* become increasingly resistant with age, while others like *P. ponderosa* remain equally susceptible at all ages. Power and Dodd (1984) found that seedlings of Guadaloupe Island provenance of *P. radiata* showed less infection than seedlings of two other provenances. Burdon and Bannister (1985) describe a progeny trial of four provenances and compare a number of characters, including disease resistance, in seedlings and cuttings 8–9 years after planting.

### Needle blight caused by *Lecanosticta acicola* (Thüm.) H. Sydow (anamorph of *Scirrhia acicola* (Dearn.) Siggers)

The needle blight caused by *Lecanosticta acicola* is the cause of a damaging needle cast of many pines in North America (Sinclair *et al.*, 1987). The conidia of the fungus are olive-brown, two- to four-celled, and measure 15–35 × 3–4 µm; the two-celled, colourless ascospores of the teleomorph *Scirrhia acicola* (sometimes also called *Mycosphaerella dearnessii* Barr) measure 9–16 × 2.5–4 µm (CMI Descr. F. 367; Holliday, 1989). The teleomorphic stage is found only in dead tissues. In Europe the fungus has been found in Yugoslavia, where it caused severe damage (but was almost eliminated by a control campaign) and in Austria (where it proved of little importance) (EPPO, 1979). The disease has also been found in China (Zhu *et al.*, 1984). Kais *et al.* (1986) controlled it on young longleaf pines (*Pinus palustris*) in the southern United States by dipping transplants in benomyl.

### The snow moulds *Phacidium infestans*, *Herpotrichia juniperi* and *H. coulteri*

These moulds may attack trees in young plantations, but are of most importance in the nursery. They are therefore described above under 'nursery diseases'.

## SHOOT BLIGHTS AND DIEBACKS

**Brunchorstia dieback caused by *Ascocalyx abietina* (Lagerb.) Müller**
**(*Gremmeniella abietina* (Lagerb.) Morelet, *Scleroderris lagerbergii* Gr.;**
**anamorph *Brunchorstia pinea* Karst.)**

In Britain, Brunchorstia dieback occurs mainly on *Pinus nigra* var. *maritima*
and occasionally on *P.sylvestris*. It was reported to have devastated Austrian
pine (*P.nigra* var. *austriaca*) in Scandinavia in about 1880 and to have
severely attacked *P.cembra* in Switzerland and young *P.sylvestris* in Sweden
and Finland in more recent times. It has been found on a variety of pine
species in France, where damage was most severe in Verdun (Morelet, 1980)
and also in Poland (Kowalski, 1983) and Austria (Tomiczek, 1985). It was
first identified in North America in 1962 and has subsequently caused serious
damage to red pine (*P.resinosa*) and to *P.sylvestris* in the United States
(Skilling *et al.*, 1979). It has also been found in Japan (Yokota, 1983) on
*P.strobus* and *Abies* spp.

Gremmen (1968b) established that initial infection by *A. abietina* occurs on
developing shoots in the spring. However, the first symptoms, at least in
Britain and Holland, do not appear until the following winter, when resin
bleeding can be observed on the buds. Brown necrotic areas develop at the
base of the buds and in the cortex of the current year's shoot. In the spring
many infected buds fail to flush and the one-year-old needles turn brown at
the base and eventually die. A characteristic yellow coloration of the xylem
tissues can be seen (Read, 1967). When the bud is only partially infected, a
poor distorted shoot may be produced. Infected shoots may also survive, in
which case areas of depressed necrotic tissue may be found. At this stage
affected trees have many dead shoots in the crown, and if the attack is severe,
some trees will die. However, trees often survive and adventitious buds
develop below the point of dieback to provide new growth.

Somewhat different symptoms of the disease have been described in
Scandinavian countries and in the United States. In Norway, Roll-Hansen
(1964) reported the presence of girdling cankers on small Scots pine 0.5–2 m
in height. Ohman (1966) described the formation of elongated cankers at the
base of stems of *P.banksiana* in the United States which girdled and killed the
trees. On *P.resinosa*, also in the United States, Benzie (1958) mentions dead
branches, cankers and dead tops as the commonest symptoms. In a more
general description of the symptoms in the United States, Skilling *et al.* (1979)
mention death and dieback of shoots (plate 37), rapid girdling and death of
young trees and deforming cankers on the branches of older trees. The
European strain of the fungus (see below) causes more severe symptoms than
the so-called 'Lake States' strain.

Black, spherical pycnidia, up to 1 mm diameter, emerge through the
surface of infected stems, needle bases and buds. The conidia are hyaline,
sickle-shaped, mostly three-septate and measure 25–40 × 3–3.5 μm (CMI
Descr. F. 369). Black apothecia occur in groups on stems and in the axils of
the needles. They are about 1 mm in diameter and have short stipes. The asci

are inoperculate, eight-spored and measure 100–120 × 8–10 μm; ascospores are hyaline, ellipsoidal, three-septate and measure 15–22 × 3–5 μm. Apothecia have rarely been seen in Britain but are frequent in some parts of Europe and in North America.

At least three strains of the fungus, from Europe, North America and Asia, have been identified using immunological and other methods (Dorworth and Krywienczyk, 1975). In 1977 the European strain was reported for the first time in New York State (Skilling, 1977) and is more virulent and has a wider host range than the North American Lake States strain (Skilling *et al.*, 1979). In addition, it appears that the European strain produces very few apothecia or ascospores in the field. In spore-trapping studies in the United States there was a marked absence of ascospores in areas where the European strain was prevalent. This contrasted strongly with areas occupied by the Lake States strain, where ascospores can be trapped in large numbers during May and June. Long-distance dispersal of the fungus is thought to occur largely via the windborne ascospores and their absence in the European strain clearly has implications for disease spread. Transport of infected nursery stock or movement of infected Scots pine Christmas trees may provide alternative means of long-distance dispersal.

Serious outbreaks of Brunchorstia dieback occur from time to time in European countries and often appear to be related to climatic conditions. Butin and Hackelberg (1978) considered that a series of wet spring and summer months, over a period of six years, favoured the development of the epidemic. Disease centres were found in southern Finland associated with water drainage channels and increased relative humidity (Aalto-Kallonen and Kurkela, 1985). The same authors reported losses in radial increment of between 7.4 and 54 per cent. In the United States, on the other hand, greater emphasis is placed on the identity of the fungal strain in relation to the severity of outbreaks.

Experiments in the United States (Skilling and Waddell, 1974) have shown that the disease may be controlled in the nursery using the fungicide chlorothalonil applied about seven times from late May to mid-August. The use of chemicals is not practicable in plantation crops, where careful selection of planting sites at some distance from affected plantations is an important consideration. In Britain the planting of Corsican pine is not recommended in upland regions in the north and west of the country, where serious outbreaks have occurred in the past. Manka (1986) has suggested that thinning of Scots pine stands can reduce the incidence of this disease. Blenis *et al.* (1984) suggested that alternate freezing and thawing may be more favourable for disease development than temperatures slightly below freezing.

### Shoot dieback associated with *Cenangium ferruginosum* Fr.

*Cenangium ferruginosum* is usually only a saprophyte, most often found on fallen branches, but it occasionally causes dieback of pine shoots, especially on trees weakened by some other agency. It was found in England in 1979 following cold injury, on Scots and Corsican pines. It was associated with a dieback resembling that caused by *Ascocalyx abietina* (Gibbs *et al.*, 1980). The fungus forms dark-brown apothecia up to about 3 mm across, with yellowish discs, and the colourless ascospores measure 11–14 × 5–6 μm (Ellis and Ellis, 1985).

**The pine twisting rust *Melampsora pinitorqua* Rostr. (*M. populnea* Rostr.)**
The pine twisting rust *Melampsora pinitorqua* produces its aecial stage on two-needled pines and uredinial and telial stages on aspen (*Populus tremula*) and less abundantly on white and grey poplars (*P.alba* and *P.canescens*). Wilson and Henderson (1966) consider that this rust is indistinguishable morphologically from several other rusts which occur on poplars but which have different alternate hosts such as *Larix* spp. (see chapter 6) and *Mercurialis* spp.

Longo *et al.* (1985) consider that *M.pinitorqua* and *M.larici-tremulae* are two separate species, as they differ in pathogenicity, host range and a few morphological characters. These rusts are all often considered to be different forms of *M.populnea* but the name *M.pinitorqua* will be retained here for the form on pine.

Pale-yellow patches appear in June on the developing shoots of the current year. Spermagonia appear as small yellow flecks on these patches and are followed by yellow aecia which soon have no covering membrane (figure 38). Shoots girdled by the fungus hang down and wither, but if the lesion is confined to one side, normal growth on the opposite side results in a bending over of the shoot (figure 39). Subsequently the tip turns upward again, thus causing a characteristic twist in the stem. These symptoms might superficially be confused with damage by the pine shoot moth (*Rhyacionia buoliana*), where the shoot is hollowed out by the insect or with damage by hormone weedkillers, which can also deform the shoot. However, the presence of aecia on the shoot is a useful diagnostic character of infection by the pine twisting rust.

The spermagonia are subcuticular, yellow and up to 130 μm wide and 15 μm high. The aecia are erumpent through the cortex of young shoots, golden yellow in colour and reach 20 × 3 mm in size. Aeciospores are globoid or ovoid, pale reddish yellow, 14–20 × 13–17 μm and with a finely verrucose wall about 2 μm thick (Wilson and Henderson, 1966). Details of the uredinial and telial stages on poplar are described in chapter 16.

In Britain *M.pinitorqua* has been recorded on a number of pine species, commonly on *P.sylvestris*. *P. pinaster* is also very susceptible, even more so than Scots pine. Corsican and lodgepole pine are practically immune in Britain and can safely be planted with aspen. However, Longo *et al.* (1979) indicate that *P.nigra* was more susceptible in Sweden and Finland than in Italy. The phenology of *P.nigra* in relation to infection may be important.

The damage done by the fungus can be quite serious, as it causes deformations of the main stem and multiplication of leaders. Outbreaks, however, are sporadic and occur only where pine and aspen are growing in close proximity. This may happen when pines are planted in an area previously

**Figure 38**  Aecium of *Melampsora pinitorqua* (the pine form of *M. populnea*) on stem of Scots pine (Forestry Commission)

occupied by broadleaved species including aspen and the latter produces suckers following clear felling (Murray, 1955). Martinsson (1985) observed that Scots pine provenances severely affected in their early years were 37 per cent shorter than those which were unaffected.

It is clearly prudent to avoid planting susceptible pines on old aspen sites. Control involves the eradication or constant weeding out of aspen and would rarely be justified, since appreciable damage ceases once pine reaches the late thicket stage, and as a rule any economic loss is confined to some reduction of revenue from first thinnings.

### Shoot dieback caused by *Ramichloridium pini* de Hoog & Rahman

Shoot dieback of lodgepole pine has long been known in Britain but it was not until Rahman (1982) investigated the disease that a previously undescribed fungus, *Ramichloridium pini*, was shown to be the cause. In 1989, this dieback was also found on *Pinus peuce* in South Wales.

The disease takes the form of progressive killing of one-year-old shoots in thicket- and pole-stage crops. The leading shoots and topmost whorl usually

**Figure 39**   Twisting of stem of Scots pine caused by *Melampsora pinitorqua* (Forestry Commission)

remain unaffected but in severe cases the whole tree may be killed. The first sign of the disease is the yellowing of the needles at the tip of the newly developed shoot. Necrotic tissue can be found at the base of the yellow needles and in the pith. Necrosis develops during the winter months and by March buds and much of the shoot may be dead. The dead needles are retained and they turn orange or brown, making the disease very conspicuous.

The development of infection is similar to that of Brunchorstia dieback of Corsican pine. Rahman (1982) proved the pathogenicity of *R. pini* by inoculating lodgepole pine shoots in May and June.

*R.pini* is a hyphomycete with erect conidiophores producing one-celled conidia in sympodial order. Conidia in culture are pale, olivaceous, thin and mostly smooth-walled, obovate or obconical, 3–8 × 2–3 μm (de Hoog *et al.*, 1983).

Rahman (1982) suggested that the occurrence of the disease may be correlated with climatic factors such as air frost and high winds. There may also be a relationship with provenance, but no studies of control measures have been undertaken.

## Shoot blight caused by *Sirococcus strobilinus* (*Ascochyta piniperda*)

The shoot blight caused by *Sirococcus strobilinus* is most important on spruce, and is therefore described under *Picea* in chapter 4, though in Britain (especially in Scotland) it seems to be more common on pines (particularly *Pinus contorta*) (figures 40 and 41) than on spruce.

**Figure 40** Drooping of the needles of seedling one-year-old lodgepole pine, caused by *Sirococcus strobilinus* (Forestry Commission)

## Shoot blight caused by *Sphaeropsis sapinea* (Fr.) Dyko and Sutton (*Diplodia pinea* (Desm.) Kickx.)

*Sphaeropsis sapinea* (CMI Descr. F. 273), a common fungus found almost throughout the world, in Europe usually acts only as a saprophyte, but it may occasionally be associated with a locally damaging dieback of pines, and less often of other conifers. It is found chiefly on plantation trees, but sometimes

**Figure 41**  Pycnidia of *Sirococcus strobilinus* on needles of lodgepole pine. These pycnidia also occur on the bark of affected shoots (Forestry Commission)

appears also in the nursery (van Dam and de Kam, 1984). Elsewhere, especially in South Africa (where it follows hail damage), it is much more important (Swart *et al.*, 1985). Besides various forms of shoot blight, the fungus may also cause a collar rot of seedlings (Palmer and Nicholls, 1985). Its dark-brown pycnidia contain dark-brown, rough-walled conidia, some of which are finally unequally two-celled. They measure 30–45 × 10–16 μm (Holliday, 1989). If chemical control is necessary against *S. sapinea*, captan and benomyl have been shown to be the most effective fungicides (Palmer *et al.*, 1986; Marinkovic and Karadzic, 1987).

BRANCH AND STEM RUSTS

**Resin top caused by *Peridermium pini* (Pers.) Lev. (*Endocronartium pini* (Pers.) Y. Hiratsuka and the blister rust *Cronartium flaccidum* (Alb. and Schw.) Winter (*C. asclepiadeum* (Willd.) Fr.)**
The blister rust *Cronartium flaccidum* occurs throughout Europe and into the USSR and across Asia as far as China (Ju *et al.*, 1984) and Japan (Kikishuma *et al.*, 1984). *C.flaccidum* attacks a wide range of two-needled pines, in which its mycelium becomes perennial and on which it produces its spermagonia and blister-like aecia.

The rust in its complete form is heteroecious. Its most important alternate hosts are *Vincetoxicum hirundinaria* (*V. officinale*) and *Paeonia* spp., but there are many others, including species of *Pedicularis*, *Tropaeolum*, *Gentiana*, *Nemesia* and *Impatiens*. The fungus has been divided into several form species which differ in the ranges of their alternate hosts (Gäumann, 1959). On these alternate hosts the fungus produces its uredinia and telia on the undersides of the leaves. The teliospores germinate to produce basidio-spores, which infect pines.

There is, however, especially in the more northerly European parts of the range of *C.flaccidum*, a second fungus, *Peridermium pini* (*Endocronartium pini*), which also attacks two-needled pines. The aeciospores of *P.pini* germi-nate to infect pines directly, without the intervention of an alternate host, and this fungus and the symptoms it produces on pines are indistinguishable morphologicaly from those of *C.flaccidum*. *P.pini* may therefore be regarded as a non-alternating form of *C.flaccidum*. Work by Gibbs *et al.* (1988), however, indicates that this may be an oversimplification. The uredinial and telial stages of *C. flaccidum* have been found only a few times in Britain, on *Paeonia* and *Tropaeolum* (Moore, 1959; Reid, 1969). *P.pini* has been known as the cause of resin top of Scots pine in East Scotland for over a century. About 40 years ago it was also found in Thetford Forest, East Anglia, where over the past 20 years it has become increasingly important (Greig, 1987; Gibbs *et al.*, 1987). A few smaller outbreaks of resin top have been found in other parts of England and Wales. Gibbs *et al.* (1988) have found evidence that *P.pini* in Britain exists in two forms. Their studies showed that a relatively high percentage of the aeciospores from Thetford Forest had only one nucleus; these spores germinated to produce a short, usually septate germ tube, and they could successfully infect both Scots pine and a species of *Paeonia*. A relatively high percentage of the aeciospores from Scotland and from parts of England and Wales other than Thetford Forest, on the other hand, had two nuclei; these spores germinated to produce much longer germ tubes, the main parts of which remained non-septate, though they often developed a subterminal vesicle cut off by septa; and they could successfully infect only Scots pine. In northern Europe, this Scottish form of the fungus appears to occur also in Sweden, while the Thetford form is known also in the Netherlands; in Norway, two types of the fungus seem to be present, one resembling the Scottish form, and one like that found at Thetford.

Infection of pines by aeciospores of all these forms occurs mainly through the needles (van der Kamp, 1970; Moriondo, 1980). Infected needles turn brown as the fungus invades the tissues and spreads down into the branch, which becomes swollen. Browning of foliage on individual branches in other-wise healthy crowns is frequently the first sign of the disease. A black lesion develops at the base of the dieback, branches below that point remaining healthy. At a later stage the disease progresses into the main stem, which is then girdled, with consequent dieback of the crown. The lower crown remains healthy for a time, but eventually the whole tree may die. Girdling lesions on the main stem appear as blackened cankers 1 foot (30 cm) or more in length.

The black appearance of the cankers is due to copious production of resin from infected parts, and this symptom gives rise to the name 'resin top'. During May and June (or from March to June in the case of *C.flaccidum* in Italy, according to Moriondo, 1980) creamy-white aecia appear as blisters at the edge of the cankers. Cankers are often found at the base of branch whorls on the main stem but are sometimes intercalary. The development of large cankers on the main stem usually results in the death of the whole tree.

The spermagonia are flat and irregular in shape, and form blisters up to 3 mm in diameter. They are followed by the aecia, which erupt through the bark to form groups of creamy-white bladders each 2–7 mm long, about 2–3 mm across and 2–3 mm high (Gremmen, 1959) (plate 12). The pale-orange aeciospores which they contain measure 24–31 × 16–23 μm (Wilson and Henderson, 1966). The aeciospore walls are hyaline, verrucose and 3–4 μm in thickness. They germinate directly to infect pines (in the case of *P.pini*) or form basidiospores which infect the alternate hosts (in the case of *C.flaccidum*).

The uredinia of *C.flaccidum* form as small pustules on the undersides of the leaves of *Paeonia*, *Tropaeolum* and other plants, the lesions showing on the upper leaf surface as yellowish or purplish spots or blotches (Gäumann, 1959; Reid, 1969). The ovoid or ellipsoid, sparsely echinulate urediniospores measure 18–30 × 14–20 μm with a hyaline wall 1.5–2.5 μm thick. The telia appear on the site of the uredinia, emerging as straight or curved, bristle-like waxy masses which may eventually cover the entire underside of the leaf. They are made up of ellipsoid or cylindrical-oblong teliospores measuring 20–60 × 10–16 μm with a smooth wall up to 1 μm thick (Wilson and Henderson, 1966). In Britain records are few, but the uredinia seem to occur from July to October, followed in the late summer and autumn by the telia, while in Italy the uredinia are found from May onwards and the telia from July to September (Moriondo, 1980).

The pines affected in Europe include *P.sylvestris* (on which the disease occurs in Britain), *P.pinaster*, *P. pinea* and varieties of *P.nigra* (Moriondo, 1980), while many other two-needled pines are attacked in the Far East (Liu and Teng, 1986). The various pines differ in their susceptibility to resin top. Thus *P.sylvestris* is susceptible in Britain and other parts of northern Europe, and further to the southeast it has been much affected in Greece (Diamandis and de Kam, 1986). In Italy, however, it is resistant, the most susceptible species there being *P.nigra* var. *calabrica* (Ragazzi *et al.*, 1986), *P.nigra* var. *laricio* (*P.nigra* var. *maritima*) and *P.pinea*, while *P.pinaster* is more resistant than these, though less so than *P.sylvestris* (Moriondo, 1980; Ragazzi *et al.*, 1987).

Observations in Europe have also indicated that susceptibility to infection by *P.pini* and *C.flaccidum* is closely linked to genetic factors (van der Kamp, 1968; Raddi and Fagnani, 1981). The progeny from infected trees is more susceptible to the disease than that from healthy trees, and marked differences in susceptibility occur between crops of different provenances.

In a survey of Scots pine plantations in northeast Scotland, the disease was found in about half the stands over 30 years of age (Murray *et al.*, 1969). Disease incidence was low in crops less than 30 years old, but nearly 2 per cent of 40 to 50-year-old trees had stem cankers. The proportion of infected trees declined in crops older than this, although approximately 4 per cent of trees over 100 years old, including remnants of old Caledonian pinewoods, were diseased. Greig (1987a) reported that the incidence of disease in Thetford Forest, East Anglia, increased from 1 per cent in 1964 to 10 per cent in 1979, but that no appreciable increase had occurred in northeast Scotland over the same period. Gibbs *et al.* (1987) made a detailed study of infected trees, also in Thetford Forest, and concluded *inter alia* that 9 per cent of trees then infected would die within five years. In Greece, where the causal fungus was identified as *C.flaccidum*, Diamandis and de Kam (1986) found that young Scots pine trees of 10–20 years of age might be affected, though most losses were among trees 30–40 years old. In a natural pine forest of 1000 ha a volume loss of more than 5000 m$^3$ was experienced. The importance of *C.flaccidum* on various pines in Italy is indicated by the substantial Italian literature on the subject.

The hyperparasite *Tuberculina maxima* Rostr. has been found in locations in Britain and van der Kamp (1970) reports that it caused a one-third reduction of aeciospore production in one stand. The same fungus has long been known also in other parts of Europe (Tubeuf, 1895).

*Control*
No control measures are at present recommended against resin top caused in northern Europe by *Peridermium pini*. Volume loss may, however, be reduced in severely affected stands by more frequent thinning or by the selective removal of diseased trees during routine thinning operations. Timber in those trees killed by the disease rapidly deteriorates as a result of beetle infestations and the activity of blue stain and other fungi (Pawsey, 1964b). On a small scale it may be possible to prune out diseased branches.

Salvage fellings are also important in the case of *C. flaccidum*. In areas where this causes serious loss the planting of pines in areas where the ·alternate hosts are common may be avoided, or these host plants may be destroyed by herbicides.

The use of resistant provenances, and selection and breeding from resistant pines could be worth while in the case, for example, of Scots pine and *Pinus pinaster* (Raddi and Fagnani, 1978).

In China, some control of the disease on affected pines has been obtained by scarifying lesions and applying creosote and machine oil, creosote and corn oil, or pine tar oil in the spring (Ju *et al.*, 1984; Zhang and Fang, 1984), but these treatments are likely to be possible only on a small scale.

### The white pine blister rust *Cronartium ribicola* J.C. Fisher
White pine blister rust is one of the best-known diseases caused by rust fungi and it has caused extensive damage to five-needled pines in Europe, Asia and North America. It is a heteroecious rust whose alternate hosts are *Ribes* spp.,

on which it produces urediniospores and teliospores. Spermagonia and aecia occur on infected pine stems, which are often girdled, resulting in the death of a branch or whole tree (figure 42).

*Cronartium ribicola* is considered to be an Asiatic fungal species (Wilson and Henderson, 1966). The fungus was first found in Europe on *Ribes* in 1854 and on *P.strobus* in 1887, though the alternate host relationship was not established until a year later. The first record in Britain was made at King's Lynn on *Ribes* in 1892. It is generally agreed that the disease was introduced into the United States about 1900 on young plants of *P.strobus* from Germany and that a single shipment of the same species from France was responsible for its introduction into eastern Canada in 1910.

*C.ribicola* occurs on a number of five-needled pines in Britain and the rest of northern Europe, including *P.strobus* and *P.monticola*. It alternates on a wide range of wild and cultivated *Ribes* spp., including blackcurrant, which may be severely defoliated, and redcurrant and gooseberry, which are more resistant to the disease.

The first symptoms appear on pine needles as small discoloured spots, some three months after initial infection. A year or more later, the bark at the base of the needles turns a yellow-orange colour. This coloration spreads and the bark becomes swollen. Two to four years after initial infection, the white aecial blisters (figure 43) erupt through the surface, releasing their orange aeciospores into the atmosphere. The fungus is perennial on pine, quickly girdling and killing small shoots; the distal needles turn reddish brown and are readily visible on dying shoots. At a later stage, the main stem is invaded and girdled, leading to the death of the whole tree.

The sporidia infect pine needles through their stomata and the mycelium grows down into the shoot, where small, yellow spermagonia, 2–3 mm diameter, are produced beneath the bark. Aecia, 2–4 × 1–3 mm across and up to 3 mm high, are formed and exposed following rupture of the bark. The aeciospores are globoid, ellipsoid and polyhedroid, orange in colour and measure 22–29 × 18–20 μm (Wilson and Henderson, 1966). These spores are airborne and can travel for great distances (over 100 km) before germinating on the undersurface of *Ribes* leaves. Uredinia then develop within a few weeks as small yellow pustules. Several generations of urediniospores can be produced before teliospores appear on the lower leaf surfaces of the *Ribes* spp. Unlike the aeciospores, the urediniospores can travel only short distances of up to 1 km. The telia (figure 44) are brownish, bristle-like structures which bear the teliospores, in their turn giving rise to the sporidia (basidiospores), which can infect only the pine host.

The effects of climatic conditions on the epidemiology of white pine blister rust and a number of other North American pine rusts have been reviewed by Petersen and Jewell (1968). Temperature requirements for the formation and germination of basidiospores on pine hosts appear to be particularly important. Limited formation of basidiospores occurs above 13 °C and Van Arsdel *et al.* (1956) have found that at 16 °C basidiospores take 36 hours to germinate and 48 hours to infect the pine host. Such information can be useful in the

**Figure 42** Canker on main stem of *Pinus strobus* attacked by *Cronartium ribicola* which has killed the upper crown (Forestry Commission)

prediction of rust distribution and intensity. Under field conditions an average July temperature of 70 °C seems optimal for the disease. Variations in the local topography and small openings in the forest canopy can have important effects on temperature and moisture and therefore on the development of the disease.

One of the reasons that white pines have not been planted on any scale in Britain is the serious risk of rust infection. *Ribes* spp., both native and cultivated, are very widely distributed and it is difficult to find sites where they are not growing close by. In the USA, where there are large natural forests of susceptible pines, *Ribes* eradication campaigns have been undertaken in an attempt to control the disease. This control method has been reviewed by Petersen and Jewell (1968) and they concluded that, while this technique can effectively control the disease, it is expensive and complete eradication is essential.

Other methods of control, including chemical treatment by aerial spraying, and pruning of infected branches, have also been attempted with only limited success. Research into the possible use of the hyperparasite *Tuberculina* to control the pathogen has not so far met with success (Wicker, 1981). Perhaps

**Figure 43**   Tree of *Pinus strobus* attacked by *Cronartium ribicola*. Aecial blisters of
the fungus have erupted from the trunk of the tree, which also shows resin bleeding
(Forestry Commission)

the most promising approach to control is through the selection and breeding
of resistant pines. Hoff *et al.* (1980) discuss six mechanisms of resistance
which need to be taken into account as selection criteria. The possible
development of new races of the pathogen, able to attack previously resistant
selections of pine, must always be borne in mind in such programmes.
Stephan (1985) presents the results of inoculation experiments on 17 five-
needled pine species and shows percentage attacks ranging from zero in
*P.cembra* to 100 in *P.strobus*.

## STEM AND BRANCH CANKERS

**Cankers caused by *Crumenulopsis sororia* (Karst.) Groves (*Crumenula
sororia* Karst.) (anamorph *Digitosporium piniphilum* Gremmen)**
*Crumenulopsis sororia* was first recorded by Karsten (1871) on *Pinus sylvestris* in Finland, where it has also more recently been found on *P.contorta* by

**Figure 44** Teliospore horns of *Cronartium ribicola* on leaf of *Ribes* (Forestry Commission)

Kujala (1950). The first record in Britain was on *P.nigra* var. *maritima* in Wales, and it has subsequently been found on *P.sylvestris* and *P.contorta* at various locations (Batko and Pawsey, 1964; Hayes, 1973). Stephan and Butin (1980) reported its presence on *P.contorta* in Germany. It has also been found in France, the Netherlands and the USSR. Hayes (1975) indicates that provenances of *P.contorta* differ in their susceptibility to artificial inoculation and to natural infection by *C.sororia*. In addition to these species, the fungus has been recorded as a pathogen on *P.cembra*.

The disease occurs mainly on young, slow-growing trees on sites low in nutrients, with shallow soils with poor structure, and affected by exposure (Batko and Pawsey, 1964).

Grove (1969) transferred the genus to *Crumenulopsis* from *Crumenula*; Van Vloten and Gremmen (1953) had previously reviewed the literature on the nomenclature of *C.sororia* and concluded that it was distinct from *C. pinicola*, a fungus with which it had previously been confused.

The first symptom of the disease is the production of small resin pustules in the bark of stems and branches. Other agents, including insects, can cause similar symptoms but in the case of infection by *C.sororia* resin bleeding may occur, fungal mycelium penetrates the cambium and wood, and canker formation follows. On poorly grown *P.nigra* var. *maritima* the cankers are irregular in appearance and the area of bark in which the fungus is active is rather indeterminate (Batko and Pawsey, 1964). Well-established cankers usually cause a marked flattening on one side of the stem and on young trees this may lead to a complete girdling of the stem and subsequent death. On more vigorous *P.nigra* var. *maritima* copious resin flow can be observed below the initial area of infection, which may be either nodal or internodal (Manap and Hayes, 1971). Longitudinal splits appear in the bark, resin production continues and canker development soon becomes evident. The subsequent flattening of infected stems usually takes four to five years to become obvious. In both *P.sylvestris* and *P.nigra* var. *maritima*, with cankers of all ages and sizes, sections through the stem show a distinctive blackening of the bark and wood. In *P.contorta* canker development is somewhat similar to that on *P.nigra* var. *maritima*, but the black discoloration of the bark and wood is absent (Hayes, 1973). On recently infected *P.contorta*, cankers are found only at nodes on the main stem but at a later stage internodal cankers may appear. The black apothecia of *C.sororia* are irregularly round or oval in outline, 1–1.5 mm across and are found on the surface of the canker. When fresh there is a black raised margin at the periphery of the grey hymenium but the apothecia tend to shrivel and distort on drying out. They are often difficult to recognise against a background of blackened resin on broken cankered surfaces. The mature asci are 90–140 × 11–13 μm and the ascospores, which are commonly three-septate, measure 17–23 × 5–6 μm (Batko and Pawsey, 1964). The pycnidia of the anamorph, *Digitosporium piniphilum*, are black and measure 0.4–0.7 mm across, and the forked or fingerlike, multicelled conidia show a considerable size range (Gremmen, 1968a).

No measures have been developed so far for the control of this disease. As it tends to affect trees on poor sites, its effects are best reduced by site amelioration techniques, particularly by drainage and the use of fertilisers.

## ROOT ROTS

The root rot fungi *Heterobasidion annosum* and *Armillaria* spp. affect pines as well as many other trees. They are considered in chapter 3.

## HEART ROTS

The most important heart rot fungus affecting pines is *Phellinus pini*, but *Phaeolus schweinitzii* and *Lentinus lepideus* also cause heart rot of pine; *Sparassis crispa* causes a heart and butt rot. These decay fungi are described in chapter 20.

MISCELLANEOUS DISEASES

**Blue stain caused by *Ceratocystis* spp. and other fungi**
The sapwood of pine is subject to a discoloration known as blue stain (figure 45) which can be caused by a number of different fungi. It can be differentiated from other stains by its bluish-grey colour and by the presence of fungi which discolour the wood without decomposing it. It occurs in logs, sawn timber and sometimes in standing trees. Although the blue stain fungi do not decay wood, they can reduce its value, particularly when the appearance of the product is important. Thus in sawn timber used for packing cases, or in logs to be converted into chips, the presence of blue stain is often unacceptable. On the other hand, in sawn timber which is to be painted or used for carcassing in buildings, blue stain may not be important.

**Figure 45** Blue stain on the cut ends of billets of Corsican pine (Forestry Commission)

Blue stain fungi, especially *Ceratocystis* and *Leptographium* spp., can cause the death of pine seedlings (Rane and Tattar, 1987; Owen *et al.*, 1987). These authors report that the same fungi are also carried by bark beetles which attack pine species. The possible role of blue stain fungi in causing the death of pines in bark beetle attacks in the forest is discussed by Gibbs (1981).

Several species of *Ceratocystis* and the related imperfect genera

*Leptographium* and *Graphium* are commonly found associated with blue stain. The following species were found in pine logs by Dowding (1970) in Britain:

> *Ceratocystis pilifera*
> *Ceratocystis piceae*
> *Ceratocystis coerulescens*
> *Ceratocystis minor*
> *Graphium aureum*
> *Leptographium lundbergii*

Many of these fungi have been found in association with blue stain in pine in Europe and North America (Lagerburg *et al.*, 1927; Davidson, 1935b). Rumbold (1936) draws attention to another species, *Ceratocystis ips*, which is important in the USA.

There has been discussion in the literature for many years about the nomenclature of *Ceratocystis* and related genera. Hunt (1956) made a thorough study of the genus and he provided a detailed key for the identification of species.

In addition to *Ceratocystis* and related genera, there are a number of deuteromycetes which can cause blue stain, usually limited to a surface discoloration. In this context, Butin (1965) mentions *Cladosporium* spp., *Aureobasidium pullulans*, *Alternaria* spp., *Epicoccum nigrum* and *Sclerophoma pythiophila*. Superficial staining is in general due to the growth of fungi over the surface of the exposed wood, the colour coming mainly from the dark mycelium or spores.

Those fungi, particularly *Ceratocystis* spp., which penetrate into the rays and tracheids, have dark hyphae that sometimes produce pigments that diffuse into the xylem but more frequently produce a visual appearance of stain by diffraction of light from the pigmented hyphae.

In a review of the methods by which blue stain fungi enter the host tissues, Dowding (1970) indicated that many research workers in both Europe and America have proposed wind and insects as the main agents for distribution. Airborne infections occurred mainly at the cut ends of logs, on parts of logs from which the bark had been removed, or on sawn timber. Other infections, particularly those by *Ceratocystis* spp., were closely associated with the breeding galleries of bark beetles. Dowding (1973) made a special study of these associations and found that although *Tomicus piniperda* carried spores of *C.piceae*, colonisation of the tunnels was limited by prior establishment of basidiomycetes. Following secondary invasion by dipterous larvae, other *Ceratocystis* spp. were introduced and they became established in the well-developed larval galleries where basidiomycetes were absent. Cartwright and Findlay (1958) emphasise the role of bark beetles in the spread of blue stain fungi to logs in the forest. Scheffer and Lindgren (1940) attribute an important role in the dispersal of staining fungi to mites and small beetles on sawn timber. However, airborne dust from the sawing of logs and fragments of fungal mycelium may be another source of infection in timber mills. In the USA, Rumbold (1936) described *Ceratocystis* spp. disseminated by *Dendroctonus* spp. and correlated the incidence of *C.ips* with infestations by *Ips* spp.

Both moisture content of the wood and temperature can influence the growth of blue stain fungi. Findlay (1959a) indicates that the moisture content of the sapwood in a live vigorous tree is too high to permit the growth of these fungi. However, if the tree should lose vigour as a result of attack by insects or other agents, the moisture content of the wood may be sufficiently reduced for infection to occur. The minimum moisture content for the growth of most staining fungi is 27–28 per cent and this is clearly an important consideration in relation to the development of blue stain in felled logs or timber.

Blue stain fungi grow very much more rapidly under warm conditions than cool (Findlay, 1959a; Miller and Goodall, 1981). Lindgren (1942) reports that the optimum temperature for growth of most *Ceratocystis* spp. in Canada is about 25 °C.

Protection against infection by blue stain fungi, where this is required, can be achieved in a number of ways. Sawlogs should be extracted from the forest and transported to the sawmill as rapidly as possible, thereby reducing the opportunity for the airborne and insect-borne infection and subsequent development of blue stain. Where bark beetles are involved, a knowledge of their life cycle will indicate the period when eggs are laid in the bark of sawn logs and the time of adult emergence. Felling and subsequent forest operations may be organised so as to avoid the presence of sawlogs in the forest at egg-laying times or 'hot logging' may be practised and this will additionally ensure the removal of produce from the forest before adults emerge. The total bark beetle population may thereby be kept under control. An alternative practice is to store logs under water or beneath a water sprinkler so as to maintain a high moisture content which blue stain fungi cannot withstand (Findlay, 1959b).

Savory *et al.* (1970), in a study of chemical control of blue stain in Scots pine logs in the forest, reported that appreciable blue stain occurred only beween March and August. The most effective control was achieved by a single application, between April and August, of a mixture of 0.75 per cent lindane and 5 per cent tribromophenol in a light mineral oil. Findlay (1959b) reports similar experiments from North America and other parts of Europe.

In the case of sawn timber, the moisture content can be reduced to 20 per cent by kiln drying immediately after sawing, thereby preventing the growth of many blue stain fungi. Thorough air drying of sawn timber can achieve similar results, though severe staining may develop during a favourable summer. Where appropriate, timber can be chemically treated immediately after sawing, with aqueous solutions of sodium pentachlorphenate and borax. The concentrations required depend on a number of factors, including species and geographic areas (Findlay, 1959b; Reyna, 1983).

# 6 Diseases of larch (*Larix* spp.)

Two larches, the European larch (*Larix decidua*) and the Japanese larch (*L.kaempferi*, syn. *L.leptolepis*), together with the hybrid between them (*L. × eurolepis*), are important forest trees. As European larch is very susceptible to canker (caused by *Lachnellula willkommii*) and dieback, it is now little planted. It has therefore been overtaken in importance by the Japanese and hybrid larches. Hybrid larch would be more often planted if sufficient seed could be obtained. More European larch would also be planted in some areas if seed of the Sudeten provenances resistant to canker was more readily available.

## General

### FROST AND DROUGHT DAMAGE

Japanese larch is often damaged by autumn frosts, but observations in Britain indicate that larches are less liable to spring frost damage than many other conifers.

Japanese larch is particularly susceptible to drought damage, and young trees on thin soils or planted over a thick grass mat may be killed or badly affected in hot dry summers.

Frost and drought damage are further considered in chapter 2.

### OTHER MISCELLANEOUS DISEASES

*Ascocalyx abietina* (*Gremmeniella abietina*) and *Scirrhia pini*, which are most important on pines (and are therefore described in chapter 5), may sometimes cause minor damage to larches.

*Phomopsis juniperovora*, which causes a shoot dieback, mainly of nursery and sometimes of rather older plants of *Juniperus* and other members of the Cupressaceae (and which is therefore described in chapter 9), has also been known to affect larches.

Larch needles are also hosts to the common, mainly saprophytic fungus *Sydowia polyspora* (anamorph *Sclerophoma pythiophila*), which because of its wide host range is briefly described in chapter 3.

# Nursery diseases

## NEEDLE DISEASES

### Needle cast caused by *Meria laricis* Vuill. (*Allescheria laricis* Hartig, *Hartigiella laricis* (Hartig) Sydow)

Needle cast caused by *Meria laricis* is widely distributed throughout Europe, including Great Britain and the Republic of Ireland, and into the USSR (penetrating into Asia), and occurs also in the Pacific northwest of the USA and Canada and in New Zealand (CMI Map No. 379; Peace and Holmes, 1933; Batko, 1956; Maruyama, 1984; Myren, 1984; Nifren, 1984). In Britain the disease is important mainly in north and west Scotland and sometimes in west Wales, though it sometimes appears elsewhere.

### *The causal fungus*

*M.laricis* was considered by Vuillemin (1893) to be a simple member of the Basidiomycotina because its spore-bearing hyphae resemble those of the smuts and rusts (and of some other Heterobasidiomycetes). Most later authors have preferred to place it in the Deuteromycotina (Fungi Imperfecti) (Batko, 1956), and in North America another *Meria* sp. has been shown to be the anamorph of a species of *Rhabdocline*, a member of the Ascomycotina (Sherwood-Pike *et al.*, 1986).

*M.laricis* enters the host needles through the stomata (Biggs, 1957). Within the plant it at first produces branched septate hyphae, some of which have mucilaginous walls (Hiley, 1921). Soon, however, knots of hyphae form in the substomatal spaces, and from these, usually only on the lower sides of the leaves, bundles of conidiophores emerge through the stomata. Under a low magnification these bundles of conidiophores look like small, white, pustular spots, not unlike the white waxy plugs on the stomata, but they stain characteristically with cotton blue. Under a higher magnification they can be seen at first to be made up of groups of cylindrical or club-shaped hyphae, which later become transversely one- to three-septate. A sterigma develops on each of the segments so formed, and spores measuring $8–10 \times 2–3$ μm (Batko, 1956) are cut off from the sterigmata. The conidiophores then resemble the basidia of some of the Heterobasidiomycetes.

The spores themselves are colourless and slightly constricted at the centre. Just before germination, a transverse septum is laid down across the middle of the spore, and one end of the spore then puts out a germ tube which may then cut off many small secondary conidia (Peace and Holmes, 1933).

*M.laricis* exists in at least four strains, but these all appear to be equally pathogenic, and are unstable within the host. The fungus grows over a temperature range of $0–25$ °C, with an optimum between 17 and 20 °C. Spore germination also takes place from 0 to 25 °C, with the optimum at 17 °C. Mycelial growth occurs within a pH range from 3.1 to 10, with an optimum between 5 and 7 (Biggs, 1957). The spores cannot survive drying (Peace and

Holmes, 1933) and they germinate only when the relative humidity is higher than 90 per cent (Biggs, 1959).

### Infection of the crop

The fungus survives on infested needles that remain attached to nursery plants over the winter. Spores are produced in spring on these needles, and quickly infect the new developing leaves. Fructification also takes place on fallen needles, but on most of these the fungus dies out before the spring.

Infection appears to enter nurseries almost entirely on infested transplants, and spread by airborne spores does not seem to take place for more than a few hundred metres (Peace and Holmes, 1933).

### Host range

The host range of *M.laricis* is narrow. As a cause of significant damage the fungus is important only on European larch, though it has sometimes also severely affected hybrid larch. On rare occasions it has attacked and caused minor damage to Japanese larch, and it has also been recorded in Britain and the USA on Western larch (*Larix occidentalis*) (Batko, 1956; Cooley, 1984), and in Norway and the USSR on *L.sibirica* (Robak, 1946; Shafranskaya, 1960).

Schöber and Fröhlich (1967) studied a number of provenances of European larch, and found that all those susceptible to *Meria laricis* (as well as to larch canker and to frost) were those originating in high altitudes, at 1400 metres or above.

### Symptoms

When larch is infected by *M. laricis*, the leaves become discoloured and finally brown. On current year's needles on already established nursery plants, the symptoms may first be seen in May, but on first-year seedlings, which are infected from the plants around, signs of disease are not evident until June or later. Infection may take place at any point on a needle, which then dies from the tip down, as far as the infected zone. In cold weather, growth of the fungus is slowed down, however, and then for a time only isolated brown spots may be visible on the diseased needles (Peace and Holmes, 1933).

Damage by *M.laricis* has often been confused with that caused by frost, but the presence of the minute fructifications already described confirms the diagnosis of needle cast. Certain other differences between needle cast and frost damage may also be detected. Frost usually kills all the needles on the frosted shoot, the needles being much shrivelled, with those at the shoot tip particularly badly affected, whereas *M.laricis* first attacks only a proportion of the needles, usually leaving those at the tip undamaged. Needles affected by *M.laricis* are browned rather than shrivelled, and at first they are attacked towards the middle or base, and only later die back from the tips. Needles affected by frost usually remain long on the plant, but needles on shoots with needle cast mostly fall early, leaving relatively few on the plant as further sources of infection.

*Factors affecting the disease*
*Meria* needle cast is of importance mainly in the wetter highland zone of Britain, and its progress is favoured by warm, damp weather (Peace and Holmes, 1933; Biggs, 1959). In dry seasons it is of little significance. The age of the plants is also important, for although the fungus may occur on quite large trees, it has not been known to cause serious damage on trees more than eight years old (Peace and Holmes, 1933), and is usually of concern only in the nursery. There it does most damage to second-year plants, which are early infected by spores from diseased needles persisting on them (or on nearby plantation trees) from the previous season. First-year plants are very susceptible, but suffer badly only in wet seasons, if infected early.

*Losses caused ˙*
As a rule *M.laricis* does not kill plants, but weakens and stunts them so that a high proportion of the stock may be classed as culls, and so rejected when lifting for planting. Cooley (1984) in Washington, USA, reported a case in which about one-third of a crop of 140 000 two-year-old seedlings had to be destroyed.

*Control of the disease*
Needle cast may be largely avoided if larch is raised only from seed in nurseries sited away from larch plantations from which initial infection may come. If the disease appears in a nursery, it may be controlled by spraying, and its carry-over into the following year is known to be reduced by transplanting the affected stock.

Various spray materials may be used. Those most commonly used in Britain are colloidal and wettable sulphur preparations (Phillips, 1963b), several of which were tested by Peace (1936). Elsewhere, good results have been obtained with captan, zineb, phygon and copper oxychloride (Schönhar, 1958; Shafranskaya, 1960), though the last of these caused some needle scorch. More recently Khabibullina (1984) in the USSR has tested various formulations and mixtures of benomyl, carbendazim, chlorothalonil, thiophanate, colloidal sulphur and zineb. Three spray applications over the season at 30-day intervals gave satisfactory control, and the best results were obtained with mixtures of the fungicides and alternating, e.g., benomyl and zineb.

# Post-nursery diseases

## NEEDLE DISEASES

### Needle rusts: *Melampsoridium betulinum* and *Melampsora* spp.
A number of needle rusts in Europe have their spermagonial and aecial stages on larch, but they are of minimal significance on it. All are heteroecious, and

are better known and more important on their alternate hosts. They are listed in Table 6.1.

*Melampsoridium betulinum* (Fr.) Kleb. produces spermagonia and aecia from April to June on the needles of European and Japanese larch. The aecia, which are borne on the undersides of the needles, are fairly prominent, with a white peridium up to 1.5 mm wide by 1 mm long and 0.5 mm high. The spores measure 16–24 × 12–18 μm (Wilson and Henderson, 1966). The stage on larch appears to be rare, but the uredinia and telia are common on birch, and sometimes occur also on alders (Roll-Hansen and Roll-Hansen, 1981).

The other needle rusts are species of *Melampsora*, and on larch they produce yellow spermagonia and yellow or orange aecia. The latter are small and inconspicuous, and of the pustular type lacking a well-marked peridium, and known as caeomata. They are described by Wilson and Henderson (1966), Lanier *et al.* (1978) and Ellis and Ellis (1985). They are difficult to differentiate one from another, their aeciospore measurements all falling broadly within the range 15–25 × 10–20 μm. Their distribution on larch is therefore imperfectly known. Three of them, *M. populnea* R. Hartig f. sp. *laricis* R. Hartig, *M. larici-populina* Kleb. and *M. medusae* Thüm., have their uredinia and telia on poplars, while in a further three, *M. capraearum* Thüm., *M. epitea* var. *epitea* Thüm. and *M. larici-pentandrae* Kleb., these stages develop on willows.

*M. populnea* (syn. *M. tremulae* Tul.) exists in a number of races or form species, one of which, f. sp. *laricis* Hartig (also variously called *M. larici-tremulae* Kleb., *M. laricis* Hartig and *M.tremulae* f. *laricis* Hartig), has been recorded on European larch in Scotland. It also occurs sparingly on European larch in other parts of Europe and across Asia as far as China, where it has been shown that other larch species, including *Larix gmelinii* and its variety *olgensis*, *L. kaempferi* (Japanese larch), *L. principis-rupprechtii* and *L. russica* (Siberian larch) are also susceptible (Shang and Pei, 1984). It has pale-yellow spermagonia and caeomata.

The race of this rust that occurs on pine (the pine twisting rust, *M. pinitorqua* Rostr.) is more important than that on larch.

*M. larici-populina* has minute spermagonia and small orange aecia on larches. It is said to occur on larch in Britain (Ellis and Ellis, 1985); it is also known (though not always on larch) in other parts of Europe and Asia eastwards to China, in parts of Africa, South America and Australia and New Zealand. It is certainly present on larch in Poland (Krzan, 1981a), China (Pei and Shang, 1984) and New Zealand (Spiers and Hopcroft, 1985). It attacks European and Japanese larch, and can also affect *L. gmelinii* and its var. *olgensis*, *L. principis-rupprechtii* and Siberian larch (Pei and Shang, 1984).

*M. medusae* Thüm. is a North American rust which has spread to South America, Japan, Australia, South Africa and Europe. In North America, where the fungus causes considerable damage to conifers and to poplars, the spermagonia and aecia occur on various species of larch as well as on Douglas fir and species of *Abies*, *Picea* and *Tsuga* (Ziller, 1974). In Europe the fungus

**Table 6.1** Needle rusts with aecia on *Larix* in Europe

| Needle rust | Hosts with uredinia and telia |
|---|---|
| *Melampsoridium betulinum** | *Betula, Alnus* |
| *Melampsora capraearum** | *Salix* |
| *M.epitea* var. *epitea* | *Salix* |
| *M.larici-pentandrae* | *Salix* |
| *M.larici-populina** | *Populus* |
| *M.populnea* f.sp. *laricis** | *Populus* |
| *M.medusae* | *Populus* (in Europe only in SW France) |

* Recorded on *Larix* in Britain (Ellis and Ellis, 1985).

is of no importance on larch (or any other conifer) and is known on poplars so far only in France.

The remaining three needle rusts, with uredinia and telia on *Salix*, are little known on larch, and only two, *M. capraearum* (*M. larici-capraearum* Kleb.), with pale orange aecia, and the larch strain of *M. epitea* var. *epitea* (often called *M. larici-epitea* Kleb.), with bright orange aecia, appear to have been recorded on larch in Britain (Ellis and Ellis, 1985). The third, *M. larici-pentandrae*, sometimes occurs on larch elsewhere in Europe.

All these needle rusts are better known and more important on poplar and willow, and are further dealt with in chapters 16 and 17.

**Needle cast caused by *Meria laricis***
This disease may sometimes attack young trees in the forest but causes little damage to those more than eight years old. It is much more important in nurseries, and is treated above under 'nursery diseases'.

## DIEBACKS

**Phomopsis disease caused by *Phacidiopycnis pseudotsugae***
Phomopsis disease, caused by *Phacidiopycnis pseudotsugae* (which is more common on Douglas fir) sometimes attacks Japanese and occasionally European larch. This disease is described in chapter 3.

## STEM AND BRANCH DISEASES

**Larch canker caused by *Lachnellula willkommii* (Hartig) Dennis (*Trichoscyphella willkommii* (Hartig) Nannf., *Dasyscypha willkommii* (Hartig) Rehm, *Lachnella willkommii* Hartig)**
Apart from butt rot caused by *Heterobasidion annosum*, the most damaging disease of larch is the canker caused by *Lachnellula willkommii*. This discomycete is known in Great Britain and Ireland, and throughout the rest of

Europe eastward into Russia, as well as in China (Pan and Liu, 1985) and Japan (Ito *et al.*, 1963) and in eastern Canada and the USA (Magasi and Pond, 1982; Ostaff, 1985; Miller-Weeks and Stark, 1983).

The perfect fructifications of the fungus, which occur only on cankers, are small, stalked, white, hairy apothecia, with an orange or buff hymenial disc usually up to 4 mm across (plate 23). The hymenium contains club-shaped asci mostly measuring 135–165 × 9.5–11 μm, and filiform paraphyses scarcely swollen at the tip and extending beyond the asci. Each ascus contains eight 1-celled, colourless, elliptical ascospores mostly measuring 15–23 × 5–7 μm. The apothecia may be found more or less throughout the year, and differ only slightly from those of the related saprophyte *Lachnellula hahniana* (Seaver) Dennis. They are preceded, usually in the first half of the year, by an imperfect stage in which colourless spermatia measuring 2–8 × 1–2 μm are produced in a white, waxy stroma (Hahn and Ayers, 1934; Manners, 1953; Ito *et al.*, 1963). *L. willkommii* on cankers may be followed by *L.hahniana*, which also grows on fallen larch material, and on dead branches affected by dieback. *L.willkommii* itself, however, will grow only on cankers on living trees (Hahn and Ayers, 1943).

The two fungi have been much confused, and accounts of the growth and physiology of *L.willkommii* given in the earlier literature are therefore difficult to interpret. The fungus grows slowly in agar culture (Manners, 1953). Its ascospores germinate between 0 and 30 °C with an optimum at 20 °C, and will not germinate when the relative humidity falls to 92 per cent or below (Ito *et al.*, 1963). The fungus is resistant to freezing, and the ascospores are resistant to drying (Gaisberg, 1928; Plassman, 1927). Ascospore discharge is closely related to rainfall, and may take place at any time of year, ceasing only when the temperature drops below 0 °C (Pan and Liu, 1984; Sylvestre-Guinot, 1981).

The avenue by which *L.willkommii* enters the tree has been much discussed. Cankers often begin on young stems and branches from 2 to 5 years old (Hopp, 1957; Zycha, 1959), but many are initiated after seven or more years of stem growth (Pawsey and Young, 1969; Buczacki, 1973a). Most of the cankers are centred round short shoots, while some arise round the base of long shoots and around dormant buds (Zycha, 1959; Ito *et al.*, 1963). Zycha (1959) has suggested that the fungus may enter through leaf scars, while Buczacki (1973c), on the other hand, studied the morphology of short shoots and suggested that, when moribund, these could be infection courts for *L.willkommii*. Sylvestre-Guinot (1986) studied the infection of European larch seedlings up to the age of 10 years, and found that by the end of the experiment 59 per cent of the cankers formed had followed gnawing by weevils such as *Hylobius abietis*, 10 per cent had resulted from the infection of short shoots, and 11 per cent had followed damage by the moth *Cydia* (*Laspeyresia*) *zebeana*.

The cankers first appear as small round or oval depressions, mostly around a branch or a dead twig or short shoot, and they usually begin in the winter. Once within the bark, the fungus kills the cortex and the cambium. Hence, in the following season the affected cambium forms no new wood or bark, but the surrounding tissues grow with increased vigour, so that in transverse section subsequent annual rings appear eccentric, and some inward growth may take place over the edges of the diseased area (Pawsey and Young, 1969) (figure 46).

**Figure 46**   Cross-section of canker on European larch, caused by *Lachnellula will-kommii* (Forestry Commission)

The fungus is contained by corky tissues during the growing season, but in the following winter it usually penetrates or bypasses the cork layer, and kills a further zone of cambium round the edges of the original diseased area. This process continues, and usually results in a perennial oval, stepped canker with a depressed centre, surrounded by deformed, swollen tissues. Resin often exudes from the cankers, and runs down the stem and branches of diseased trees. The surfaces of the cankers in time become blackened, and fruit bodies of *L. willkommii* grow around the edges (figure 47). Development of cankers is sometimes checked, and the diseased areas become covered by callus tissue. Such old healed cankers remain as pockets of damaged tissue within the tree.

Cankers may appear on the main stem of the tree (when they cause the greatest damage) or on the branches, when they sometimes cause girdling and subsequent dieback, though dieback of larch is caused not only by *L. willkommii* but also, for example, by winter cold and windchill and apparently also by aphids (Pawsey and Young, 1969).

**Figure 47**  Canker on European larch, caused by *Lachnellula willkommii*. The disc-like apothecia of the fungus can be seen on the cankered tissues (and see plate 23)
(Forestry Commission)

The range of hosts affected by larch canker is small. In Britain, as in most other parts of the world, it is usually severe enough to cause serious damage only on European larch, but it has occasionally caused some loss in hybrid larch, and occurs also on Japanese larch, on which its effects are normally slight. In fairly recent years, however, it has caused severe losses in stands of Japanese larch in Japan (Ito *et al.*, 1963). Margus (1959) in Estonia found *Larix sukaczevii* susceptible and *L.kurilensis* resistant, while according to Plassmann (1927) *L.sibirica* (*L.russica*) is immune. The recent cases in Canada and the USA have been on the tamarack, *Larix laricina* (Miller-Weeks and Stark, 1983; Ostaff, 1985). Within European larch, different provenances vary greatly in their susceptibility to the disease. It has long been known that most alpine larch provenances from high elevations are very susceptible, and the Carpathian larches are more resistant. Of the Carpathian larches, certain Sudeten provenances show a good combination of canker resistance and good growth (Plassmann, 1927; Edwards, 1952; Schöber and Fröhlich, 1967; Pawsey and Young, 1969).

Factors affecting the severity of the disease on susceptible larches have not been clearly defined. Berkeley (quoted by Hiley, 1919) as long ago as 1859 suggested that canker occurred particularly in damp places. This suggestion receives some support from Hiley's observation that cankers are more frequent inside plantations than on their edges. The possible importance of frost has been stressed by Day (1931), though it has been shown (Hahn and Ayers, 1943; Manners, 1957; Zycha, 1959; Ito *et al.*, 1963) that *L.willkommii* may attack without prior damage by frost, and symptoms ascribed to frost by Day (1931) and Latour (1950) occur also in trees not affected by frost, but attacked only by *L.willkommii*, or even wounded mechanically (Manners, 1957). The theory that spring and perhaps autumn frosts were responsible for the annual extension of canker lesions (Day, 1931) has now been discounted. Buczacki (1973b) found no evidence to support this theory but suggested that alternating cold and warm temperature during the winter encouraged canker extension. Ostaff (1985) in New Brunswick, Canada, found most cankers on tamarack where the rainfall was high and the snowfall low, in areas with a mean maximum temperature above 0 °C and with the greatest number of degree days from November to March more than 4.4 °C.

Canker epidemics tend to occur in certain years (Ito *et al.*, 1963; Leibundgut *et al.*, 1964). The factor or factors common to these 'canker years' is not known, though Leibundgut *et al.* considered that high March temperatures were involved. Some evidence in Britain suggests that these 'canker years' are those in which heavy adelgid infestations occur.

The amount of damage caused by the disease varies greatly, depending partly on the susceptibility of the species or provenance, partly on the as yet obscure environmental factors mentioned above, so that the effects even on the same provenance may vary greatly from site to site. It depends also whether most of the cankers are on the branches, in which case the losses caused may be small, or whether many cankers occur on the main stem, in which case the stand may be almost worthless.

On some sites in Europe many provenances of European larch may be severely damaged by larch canker. If this species is to be grown in such places, canker is best avoided by growing a silviculturally suitable Sudeten provenance of known high resistance to the disease. Otherwise, Japanese larch or the hybrid *Larix* × *eurolepis* may be grown, and heavy loss is then unlikely.

## Cankers associated with *Valsa kunzei*

The weak parasite *Valsa kunzei* may be associated with branch and stem cankers in larch. It is also found on other conifers, and is considered further in chapter 3.

## ROOT DISEASES

### Group dying caused by *Rhizina undulata*

Larches are among the conifers affected by group dying caused by *Rhizina undulata*, which is described in chapter 3.

## ROOT AND BUTT ROTS

The larches are rather susceptible to the major butt-rot fungus *Heterobasidion annosum*. *Armillaria* spp. may kill young trees, but rarely cause butt rot in larch, which, however, may be rotted by *Phaeolus schweinitzii* and *Sparassis crispa*. It is also sometimes attacked by *Tyromyces stipticus*. The wound parasite *Stereum sanguinolentum* may also cause a slow decay. *H.annosum* and the *Armillaria* spp. are discussed in chapter 3, and the other decay fungi are described in chapter 20.

## VIRUS AND VIRUS-LIKE DISEASES

### Larch witches' broom and stunt disease
An apparently soil-borne disease of larch trees caused by a Rickettsia-like organism (RLO) has been described in West Germany (Nienhaus, 1985). The affected plants are stunted, with reduced root systems, and the side shoots grow out to form witches' brooms.

### Larch decline
Nienhaus (1985) also found a mycoplasma-like organism (MLO) in declining nursery larch trees in north Germany.

# 7 Diseases of Douglas fir (*Pseudotsuga* spp.)

Species of *Pseudotsuga* occur in North America south to Mexico, and in China and Japan. The only species extensively grown in Britain (and in the rest of northwest Europe) is the Douglas fir, *P.menziesii*, a native of western North America. *P.menziesii* exists in three main forms or varieties, sometimes separated (though intermediates may be found), as var. *menziesii* (the coastal or green Douglas fir), var. *caesia* (the Fraser River, intermediate or grey Douglas fir) and var. *glauca* (the Colorado or blue Douglas fir). The green Douglas fir is a fairly important forest species, which grows well on suitable sites. The blue Douglas fir is grown in some parks and gardens, but is not extensively planted because of its slower growth and susceptibility to the needle-cast fungus *Rhabdocline pseudotsugae*.

## General

Forest trees of Douglas fir are liable to wind damage, and also to damage by frost, both in the nursery and in the earlier stages of establishment in the plantation (Macdonald *et al.*, 1957), when needles may be killed and lesions form at the stem base. With regard to frost, provenance differences have been noted, inland provenances being generally more susceptible to damage than coastal ones, which are later to flush (Day, 1930a; Soutrenon, 1986). Damage by wind and frost are discussed further in chapter 2.

## Nursery diseases

### GENERAL

Nursery plants of Douglas fir are among those sometimes attacked by the grey mould, *Botrytis cinerea*. Young plants have also been found with basal swellings ('carrot root'), resembling those said to be associated with *Pestalotia hartigii* in some other conifers. This disorder is known in spruces as 'strangling disease'. Douglas fir plants in the nursery are also very susceptible to Phytophthora root rot caused by *Phytophthora cinnamomi*, and have also been attacked by *P.citricola*.

Douglas fir is also one of the trees affected by Phomopsis disease, caused by *Phacidiopycnis pseudotsugae*. This disease, which occurs both in the nursery and on older trees, also attacks Japanese larch and many other trees. All the above diseases are discussed in chapter 3.

## Post-nursery diseases

NEEDLE DISEASES

Douglas fir is affected by two well-known needle cast diseases, the first caused by *Rhabdocline pseudotsugae*, and the second by *Phaeocryptopus gaumannii*. Neither of these diseases is of great importance in Britain.

### Needle cast caused by *Rhabdocline pseudotsugae* Syd.
The needle cast caused by *Rhabdocline pseudotsugae* occurs throughout Britain, and is widespread on the continent of Europe, as well as in Canada and the USA (in western parts of which it originated) (CMI Map No. 52), where other *Rhabdocline* spp. also attack Douglas fir.

*R.pseudotsugae* has caused serious damage in many areas, but it has been important mainly because it has greatly restricted the growing of the blue and grey varieties of Douglas fir, which are very susceptible to its attack.

It rarely occurs in nurseries, but causes damage mainly on relatively young stands between the ages of 10 and 30 years (Rohde, 1932; Jancarik, 1964). In North America, however, it is a problem to growers of Douglas fir for Christmas trees (Brandt, 1960; Collis, 1971).

*The fungus*
The fruit bodies of the fungus are elongated ascocarps up to about $3 \times 0.3$ mm, embedded below the epidermis on each side of the mid-rib, usually only on the undersides of the needles (plate 24). In North America (Brandt, 1960) and in Europe (Peace, 1962), a form with larger ascocarps, up to 10–15 mm long, has also been found. When the ascoscarps ripen, the overlying epidermis splits open and folds back as a lid to reveal the orange-brown (later brown) interior which contains the club-shaped asci. These, which are interspersed by slender, often forked paraphyses, measure up to $130 \times 20$ μm, and each contains eight elliptic-cylindrical, colourless, one-celled ascospores, which may be slightly swollen at each end, and measure $15$–$20 \times 6$–$9$ μm. Each spore is surrounded by a thick gelatinous sheath (Parker and Reid, 1969). After discharge the spores become two-celled, one cell becoming dark brown with a thickened wall (Dennis, 1968). An anamorphic stage, *Rhabdogloeum hypophyllum* D.E. Ellis and Gill, has been described (Cannon *et al.*, 1985).

*Symptoms and disease development*
The ascocarps ripen and shed their ascospores in May and June (and in some cases into July), when the sticky spores adhere to the needles of the newly developing shoots. Spore discharge and infection take place only when the

humidity is high (with an optimum of 100 per cent) and when the temperature is between 1 and 15 °C (with the optimum of 10 °C) (Parker, 1970; Collis, 1971). Only the young, developing leaves are subject to infection.

After infection, the fungus grows within the needles, which, however, show no obvious external symptoms until the autumn. Then pale yellow blotches 1–2 mm across become visible, usually on the undersides. These blotches darken, and by the following spring become red-brown or purple-brown, and on them the new fructifications arise again in May and June to complete the cycle. The more severely affected needles quickly fall, but others remain on the tree and are slowly shed throughout the year.

*Host range and varietal susceptibility*
*R.pseudotsugae* affects only species of *Pseudotsuga*. It is of importance only on *P.menziesii*, but in Canada it has also been recorded on *P.macrocarpa* (Collis, 1971).

In the case of *P.menziesii*, the green variety *menziesii* is much less susceptible than the grey and blue varieties *caesia* and *glauca* (van Vloten, 1930; Liese, 1932; Collis, 1971) (though it is the most susceptible of the three varieties to attack by *Phacidiopycnis pseudotsugae* (Hahn, 1957)). Indeed, the effects of *R.pseudotsugae* on the grey and blue forms may be so severe that in many parts of Europe these varieties cannot be extensively grown (Fischer, 1938), and only the green variety is grown in British forests.

Nevertheless, resistant individuals and provenances can be found even in populations of the susceptible varieties (Liese, 1932; Anon., 1956a; Collis, 1971). In a trial of 10 provenances of the green and 66 of the blue variety of Douglas fir, Soutrenon (1986) found that 9 of the provenances of the var. *menziesii* were resistant to *R.pseudotsugae*. Of the provenances of the var. *glauca*, those from the coastal states of British Columbia, Canada, and Washington and Oregon, USA, were all resistant. Among the provenances from inland states, those from Idaho and Montana were resistant, but those from Colorado, Utah, Arizona and New Mexico were all highly susceptible.

*Factors affecting the disease*
Experiments and observations in Canada suggest that spread within susceptible stands is greatest during periods of high humidity and at relatively low temperatures at or below 13 °C (though the humidity is more important than the temperature) (Parker, 1970; Collis, 1971).

*Damage caused*
In North America, *R.pseudotsugae* is very important as a cause of damage to Christmas trees (Brandt, 1960; Collis, 1971). At best, affected trees grow slowly and take longer to reach a marketable size. In more severe cases, trees may lose all needles except those of the current year and they then become useless for the Christmas tree trade. Large forest trees, however, are usually little affected.

In Europe, older trees from about 10 to 30 years of age may be affected, especially in dense stands (Anon., 1956a). Trees heavily infested for several successive years may lose almost all their needles, cease to grow, and eventually die. Because of the disease, the growing of the susceptible grey and blue forms of Douglas fir on any scale is prevented. In Britain, although the disease may be found all over the country, it is of little importance because only the green Douglas fir is grown to any extent, except in parks and gardens. Kurkela (1981) in Finland found that the fungus reduced height growth and radial increment to a degree dependent on the severity of infection. Badly affected codominant and suppressed trees were killed, but dominant trees eventually recovered from the disease.

*Control*
At present the only practicable way of controlling the disease on a forest scale is to grow the more resistant green variety of *P.menziesii*, though it may be possible to make use of resistant provenances or clones of other varieties. Fungicides could be applied only to small trees of high value, and so are likely to be of use only in nurseries (where, however, the disease hardly ever occurs) and in North American Christmas tree plantations. Among the fungicides tested, lime sulphur and Bordeaux mixture have given promising results (Fischer, 1938; Collis, 1971), and good control has been obtained with benomyl (Morton and Miller, 1982).

**Needle cast caused by *Phaeocryptopus gäumannii* (Rohde) Petr.**
**(*Adelopus gäumannii* Rohde)**
Needle cast (sometimes called Swiss needle cast) caused by *Phaeocryptopus gäumannii* occurs in Britain (mainly in the wetter, more westerly parts), and throughout the rest of Europe, in western Canada, in both western and eastern parts of the USA, in New Zealand and in Tasmania (CMI Map No. 42).

In Britain, the overall effect of the disease is slight, in spite of the fact that the fungus can be found in a large number of stands of Douglas fir in the west of the country, where on rare occasions it may cause severe defoliation (Strouts *et al.*, 1984).

The disease generally affects trees between the ages of 10 and 40 years (Gaisberg, 1937; Buchwald, 1939). It may be found on younger and older trees, but on these it seems to cause little damage (Liese, 1939; Krampe and Rehm, 1952). In North America it causes serious depreciation of Douglas fir crops grown as Christmas trees (Michaels and Chastagner, 1984a). The needles of the affected trees are dehydrated and many are lost (Chastagner et al., 1984).

*The fungus*
The minute black, smooth, rounded pseudothecia of the fungus emerge from stromata on the undersides of the needles, and when numerous give the needle surface a sooty appearance. They measure up to about 80 μm across, and contain club-shaped asci measuring up to 40 × 15 μm, each containing

eight spores. The ascospores are colourless, two-celled, rather club-shaped, and measure 10–15 × 3.5–5 μm (Dennis, 1968).

### Symptoms and disease development
In Europe the asci ripen and shed their spores in May and June and into July, when the young shoots are developing (Gaisberg, 1937; Anon., 1956a).

In western North America Michaels and Chastagner (1984b) found that the ascospores were released between April and September, with a maximum in June and July. Ten times as many spores were formed on 1-year-old needles as on 2-year-old ones. Ascospore release occurs when the pseudothecia are damp, between temperatures of 5 and 30 °C, with a maximum at 20 °C. Once spore release has begun, 75 per cent of the spores are discharged within the first 20 minutes, and discharge is completed within 4 hours. After infection of the young needles, which takes place just after bud break (Chastagner and Byther, 1983b), the fungus grows within the needle tissues, and towards the end of the summer the affected needles become yellowish green. Over the winter, dark plugs of fungal mycelium form below the guard cells of the stomata. From these plugs the perithecia develop outwards and eventually form on the outside of the leaf above the stomatal pores, until in May they produce ripe asci and shed their spores to complete the life cycle (Krampe and Rehm, 1952; Anon., 1956a).

Infected needles may bear many fruit bodies and yet remain green and apparently little harmed, while in other cases, on more susceptible trees, they may become more and more yellow with age and finally die, turn brown and fall.

### Host range and varietal susceptibility
*P.gäumannii* affects only species of *Pseudotsuga*. It attacks all the main forms of *P.menziesii*, though different provenances as well as individual trees within the species show differences in resistance (Gaisberg, 1937; Liese, 1939). In Germany, Krampe and Rehm (1952) found apparently resistant trees in stands of the green Douglas fir (var. *menziesii*) which retained their needles for several years, while adjacent susceptible trees were almost bare. Some of these resistant individuals were tolerant, and retained their needles in spite of a dense covering of fructifications, while others appeared to be resistant to the fungus and showed only slight infection. Nelson *et al.* (1989) in the USA crossed trees with varying densities of pseudothecia on their needles and differing in the lengths of needle retention. They concluded that it should be possible to breed resistant trees to improve the quality and saleability of Douglas firs grown as Christmas trees.

### Factors affecting the disease
Infection by *P.gäumannii* seems to be encouraged by high summer rainfall and high humidity (Gaisberg, 1937; Merkle, 1951).

*Damage caused*
In Britain *P.gäumannii* can easily be found in a high proportion of the stands of Douglas fir in the western parts of the country, but it rarely causes any significant damage. Elsewhere it appears to be of greater importance. In Germany it may sometimes cause almost complete defoliation (Gaisberg, 1937; Krampe and Rehm, 1952). Liese (1939) and Merkle (1951), however, both indicate that the fungus rarely kills trees, but may render them liable to attack by *Armillaria mellea* (*sensu lato*) and other secondary invaders. It has recently been suggested that attack by *P.gäumannii* has been one factor in a decline of Douglas fir in New Zealand (Anon., 1975). As noted above, it causes severe damage to Christmas tree crops in North America.

*Control*
In Britain this disease is not generally sufficiently severe to merit any special control measures. Where control is needed, the best ultimate method would appear to be by selection and breeding from resistant trees. In New Zealand promising experimental results were obtained by spraying with copper sprays and benomyl, but lasting control was not achieved (Anon., 1975). Good control of the disease in North American Christmas tree crops has been obtained with chlorothalonil, preferably applied when the shoots are elongating in the spring (Chastagner and Byther, 1983a,b). The possibility of breeding resistant trees for use in Christmas tree plantings in the USA has been noted above.

**Needle cast associated with *Rhizosphaera kalkhoffii***
Douglas fir is one of the hosts of *Rhizosphaera kalkhoffii*, a fungus usually no more than a saprophyte, but sometimes associated with needle cast. *R.kalkhoffii* is more closely associated with species of *Picea*, under which it is therefore described (chapter 4).


DISEASES OF SHOOTS

**Phomopsis disease caused by *Phacidiopycnis pseudotsugae***
As noted earlier, Douglas fir is one of the trees susceptible to Phomopsis disease (caused by *Phacidiopycnis pseudotsugae*), which because of its wide host range is dealt with in chapter 3.


DISEASES OF THE TRUNK

**Resin bleeding**
From time to time Douglas fir trees in Britain may be affected by a profuse resin bleeding, the cause of which is still unknown. Although numerous fungi have been found on the lesions, which occur on the trunk and at branch bases, they all appear to be saprophytes. It has sometimes been considered that the

disorder is caused by drought or some other climatic factor, but insect damage has also sometimes been suspected as either a main cause or a contributory factor.

## ROOT DISEASES AND BUTT ROTS

Douglas fir is rather resistant to attack by *Armillaria* spp. and *Rhizina undulata*. *Heterobasidion annosum* rarely penetrates the butt, though it may kill young trees, and kill and rot the roots of older trees to such an extent that they readily become victims to windblow. Douglas fir is one of the conifers that may be severely affected by *Phaeolus schweinitzii*, and *Sparassis crispa* may also rot the roots and the butt. These fungi are discussed in chapters 3 and 20.

# 8 Diseases of minor forest conifers (*Abies, Cupressus* and *Chamaecyparis, Thuja* and *Tsuga* spp.)

## Silver fir: *Abies* spp.

There are about 50 species of *Abies*, but in Britain only one, *A.grandis*, has shown marked promise as a forest tree. It is a very high-yielding species, and shows considerable resistance to attack by *Heterobasidion annosum*. The quality of the timber is relatively low, however, and so far *A.grandis* has remained only a minor species in British forests. On the European mainland, the native silver fir, *A.alba* (*A.pectinata*), is a forest species. In Britain, a good many *Abies* species are grown in parks and gardens, the commonest being grand fir, noble fir (*A.procera*) and the Caucasian fir (*A. nordmanniana*).

## General

Frost and winter cold may damage plantation trees of *Abies* spp., causing browning and death of needles and sometimes of buds and shoots. Frost may also sometimes cause lesions on the lower stems of young trees. Nursery beds of *Abies* plants are also liable to damage by spring frosts. Frost and winter cold damage are discussed in chapter 2.

The black snow mould *Herpotrichia juniperi*, the canker fungus *Nectria fuckeliana* and the weakly pathogenic needle fungus *Rhizosphaera kalkhoffii* sometimes affect *Abies* spp., but are more important in spruces, and are therefore described in chapter 4. The canker and dieback fungus *Ascocalyx abietina* (*Gremmeniella abietina*) and the snow mould *Phacidium infestans* may also attack *Abies*, but are more important on pines, and so are described in chapter 5. *Sydowia polyspora* (anamorph *Sclerophoma pythiophila*), usually only saprophytic, is associated with a number of conifer diseases, and has been known to attack *Abies* spp. It is described in chapter 3.

# Nursery diseases

## DAMPING OFF AND 'CARROT ROOT'

*Abies* plants in nurseries are liable to damping off, and young plants have also been found with basal swellings like those associated with *Pestalotia hartigii* in spruce, Douglas fir and some other conifers. This disorder is sometimes described as 'strangling disease' or 'carrot root'. These diseases are described in chapter 3.

# Post-nursery diseases

## NEEDLE DISEASES

### Needle blight and dieback caused by *Herpotrichia parasitica* (Hartig) Rostrup (*Acanthostigma parasiticum* (Hartig) Sacc., *Trichosphaeria parasitica* Hartig ; anamorph *Pyrenochaeta parasitica* Freyer and v.d.Aa)

*Herpotrichia parasitica* has been found in west Scotland and east England causing a needle blight in *Abies alba* (Watson, 1933; Batko, 1974). It also occurs, mainly on *A.alba*, in alpine parts of Europe and in the Black Forest and Bavaria (Viennot-Bourgin, 1949). Its anamorph is *Pyrenochaeta parasitica* (Cannon *et al.*, 1985).

The fungus produces minute, dark-brown pseudothecia, 100–200 μm across, on small mycelial cushions on the undersides of affected needles. The pseudothecia bear brown, bristle-like septate hairs on the upper half, and contain hairlike paraphyses, and asci each containing eight ascospores. These are colourless or later greyish or brownish, ovoid, usually four-celled, and measure 15–20 × 4–5 μm (Hartig, 1894; Viennot-Bourgin, 1949; Batko, 1974).

The ascospores are shed, and germinate and cause infection in the autumn (Watson, 1933). The mycelium is white at first, but later brownish, and grows in and over the tissues of the undersides of the branches and needles. The fungus growing within the needles kills many of them, and they become brown, but remain matted together by the mycelium, and so do not at first fall from the tree.

The mycelium, which becomes perennial, remains dormant over the winter, but grows and spreads in the following spring to attack further shoots and leaves. It forms minute cushions on the undersides of infected leaves, and it is on these cushions that the superficial pseudothecia arise (Hartig, 1894).

The fungus may sometimes cause severe defoliation and dieback, and may destroy all the needles except those of the leading shoot (Watson, 1933). It is fortunate, therefore, that the disease is uncommon.

It is generally considered to be encouraged by moist conditions and dense planting. Hence, if control measures are needed, it is usually suggested that the effects of the disease can be reduced by thinning to open up the stand to admit more air and light and render the atmosphere drier.

### Needle blight caused by *Delphinella abietis* (Rostrup) E. Müller (*Rehmiellopsis abietis* (Rostrup) O. Rostrup, *R. bohemica* Bubak and Kabat; anamorph *Dothiorella* sp.)

Needle blight caused by *Delphinella abietis* has been found in Britain, Denmark (Wilson and Macdonald, 1924; Boyce, 1927), Norway (Nedkvitne, 1966) and British Columbia (Waterman, 1945).

The fungus produces small, black, globose pseudothecia mainly on the upper surfaces of the needles. The pseudothecia are at first immersed but later emerge to appear on the leaf surface, and measure up to 200 μm across. They contain a few club-shaped, thick-walled asci measuring up to 90 × 22 μm, each containing about 16–24 ascospores. These are colourless, two-celled, rather club-shaped, and measure 11–21 × 4–6.5 μm (Dennis, 1968). A pycnidial stage, a species of *Dothiorella*, also occurs.

Infection takes place soon after expansion of the young needles, which then shrivel and turn red, and later become dark brown or grey-black, and finally deep black. In the case of *Abies procera* the diseased shoots become twisted (Wilson and Macdonald, 1924).

In Scotland, Wilson and Macdonald (1924) recorded *D. abietis* on *Abies alba*, *A. procera* (*A. nobilis*), *A. pinsapo*, *A. pindrow* and *A. cephalonica*. The fungus has also been found on *A. alba* in Norway (Nedkvitne, 1966) and Denmark (Boyce, 1927), and on *A. lasiocarpa* in British Columbia (Waterman, 1945).

The disease is most severe on young trees, and on the lower branches of older ones (Wilson and Macdonald, 1924; Nedkvitne, 1966). Wilson and Macdonald found that 50 to 80-year-old trees of *A. alba* appeared to remain healthy even when exposed to infection from nearby diseased plantations.

In Norway, Nedkvitne (1966) found the disease was most severe on sites exposed to low late spring temperatures. However, infection could occur in the absence of injury by frost or other factors. In the 1920s, *D. abietis* appeared to be spreading rapidly in Scotland, where *A. alba* was severely attacked (Wilson and Macdonald, 1924; Boyce, 1927). More recently in England it has damaged noble fir trees grown as Christmas trees (Strouts *et al.*, 1986). Though it may sometimes kill trees (Nedkvitne, 1966) and the browning it causes may render Christmas trees unsaleable, it is usually of only minor significance.

If an attack occurs, severely affected young trees should be removed and burnt. Diseased lower branches of older trees should be treated in the same way if this can reasonably be done. Waterman (1945) found that attack by the related *Rehmiellopsis balsamea* in North Ameria could be controlled by three applications of Bordeaux mixture at 12-day intervals, beginning as the new growth first emerged.

**Needle blight associated with *Rhizosphaera oudemansii* Maubl.**
The fungus *Rhizosphaera oudemansii*, one of the Deuteromycotina, has been found in Spain in association with a needle cast of *Abies pinsapo*, the Spanish fir (Martinez and Ramirez, 1983). The ovoid, one-celled, colourless conidia measure 9–16 × 5.5–9 μm, and the fungus is known also in France, Holland, Germany and Alaska, USA (Sutton, 1980).

**Needle cast caused by *Lirula nervisequia* (DC:Fr.) Darker (*Hysterium nervisequium* DC:Fr., *Lophodermium nervisequium* (DC:Fr.) Cher.)**
*Lirula nervisequia*, which is described by Minter and Millar (CMI Descr. F. 783), occurs throughout most of mainland Europe, where it has a wide geographical and climatic range from France eastward through Germany to Denmark and Sweden, and south through Czechoslovakia, Austria, Hungary, Romania and Switzerland to Italy and Greece. It has been found on the common silver fir, *Abies alba*, and the Grecian fir, *A.cephalonica*. As a cause of needle cast it is of only local importance, and then only on *A.alba*. Towards the end of the nineteenth century, Hartig (1894) considered that the fungus was present throughout the range of silver fir but he had found it giving rise to a serious needle blight only in the Erzgebirge. It first became noticeable there from May to July, when 2-year-old needles entering their third year began to turn brown. A few months later conidial fruit bodies formed in two long, corrugated brown or blackish ridges on the upper sides of the needles, to be followed by elongated ascocarps on the undersides of the needles. The ascocarps ripened in the following April, by which time the shoots were three years of age, and many of the affected needles had already fallen. The colourless, one-celled, ellipsoid conidia measure 2–3 × 1 μm, and the ascocarps contain asci each with eight colourless, aseptate, filiform ascospores which measure 75–90 × 3–4 μm and which are covered by a thick mucous sheath. The fungus exists in two forms: the conidial fructifications of the var. *nervisequia* are inconspicuous, but those of the var. *conspicua* Darker fuse to form large multilocular bodies which often run throughout the length of the needle (CMI Descr. F. 783).

NEEDLE RUSTS

A number of needle rusts in Britain and elsewhere in Europe have their spermagonia and aecia on *Abies* spp., on which, however, they are pathologically insignificant. They include *Milesina blechni*, *M. kriegeriana*, *Puccini-astrum epilobii*, *P. goeppertianum* and *Hyalopsora aspidiotus*. *Melampsorella caryophyllacearum*, a much more damaging fungus, also forms spermagonia and aecia on needles of *Abies*, but it is important because it causes witches' brooms and stem cankers; it is considered below under 'diseases and disorders of branches and trunks'.

**Milesina blechni** Syd.

Though *M. blechni*, which is known throughout Europe, produces its spermagonia and aecia on *Abies* spp., on these it is rare and of little pathological importance. Its uredinia and telia are produced on the common fern *Blechnum spicant*, on which it appears to be frequent, though commonly overlooked.

The flask-shaped spermagonia are embedded in the surface of the current needles, usually on the underside. The white, cylindrical aecia, which are also on the current needles, in two rows, one on each side of the midrib, are about 0.3–0.4 mm across, and rupture at the top to shed the white, ellipsoid, ovoid or globose aeciospores. These measure 27–36 × 21–27 μm, and are covered in coarse warts over most of their surface, but on one side the warts are minute (Wilson and Henderson, 1966).

**Milesina kriegeriana** (Magn.) Magn.

The spermagonia and aecia of *M.kriegeriana*, which also occurs throughout Europe, have been found abundantly on *Abies* spp. in Britain on a few occasions, notably on *A.alba* in Scotland in 1932 and on *A.grandis* in Northern Ireland in 1952. They have also been found in Britain on *A.cephalonica* and *A.nordmanniana* (Wilson and Henderson, 1966). The rust is of little pathological importance. The uredinia and telia occur on ferns of the genus *Dryopteris*.

The spermagonia are found beneath the cuticle, mainly on the upper sides of the current needles. The aecia, also on the current needles, appear mainly on the undersides, on yellowish areas in two irregular rows. They are whitish and cylindrical, measuring 0.3–0.8 mm across and 0.5–1.3 mm high. They rupture at the apex to release the white aeciospores. These measure 22–48 × 20–36 μm, and are covered with fine warts.

**Pucciniastrum epilobii** Otth. (*P. pustulatum* (Pers.) Diet.)

*Pucciniastrum epilobii* is known in Europe, Asia, North America and New Zealand (Kupevich and Transhel, 1957). Its spermagonia and aecia occur on the undersides of the needles of *Abies* spp. in June and July. The uredinia and telia are frequent on species of *Chamaenerion* and *Epilobium* (the willow herbs).

The spermagonia form under the cuticle. The aecia, which occur in midsummer, in June and July, are whitish, cylindrical, about 1 mm high and 0.25 mm wide, opening at the top or sides and soon disappearing. The aeciospores are more or less rounded, ovoid or ellipsoid, covered in small spines but with a smooth spot on one side, and measure 14–21 × 10–14 μm (Wilson and Henderson, 1966).

In most areas *P. epilobii* is of little pathological importance, though locally it sometimes causes a severe needle cast of *Abies* stands infested with the alternate hosts. In Germany it has been reported as an occasional cause of needle cast and sometimes of dieback of trees up to 2 metres in height. Schönhar (1965) found that it was most severe following warm, moist spring weather. It does not spread to *Abies* more than about 50 metres from its alternate hosts.

### *Pucciniastrum goeppertianum* (Kuhn.) Kleb. (*Calyptospora goeppertianum* Kuhn.)

*Pucciniastrum goeppertianum* occurs in Europe (but apparently not in Britain) and across Asia to Japan, as well as in North America, and may occasionally cause needle cast of *Abies* spp. Its spermagonia may arise on both sides of the *Abies* needles, on the undersides of which its white, cylindrical aecia are formed in two rows. They may be found from June to September and the aeciospores, which have no smooth spot, measure 18–30 × 10–18 μm (Ziller, 1974). The aeciospores infect species of *Vaccinium*, the stems of which become reddish (later brown), erect and swollen (figure 48). The fungus is perennial within these plants, and on the swollen stems reddish-brown, crust-like telia form within the epidermis. On germination in late spring or early summer, the teliospores produce basidiospores which reinfect the young needles of adjacent *Abies* plants. If necessary, the disease can be prevented by the destruction of the conspicuous infected *Vaccinium* plants in *Abies* plantations (Hartig, 1894).

### *Hyalopsora aspidiotus* (Magn.) Magn.

*Hyalopsora aspidiotus* is found in Eurasia, North Africa and North America, and produces spermagonia and aecia on needles of *Abies* spp. (though these stages are not known in Britain, and the stages on *Abies* are generally rare). The spermagonia are large and yellow-brown, and form on 1-year-old needles, while the short cylindrical aecia arise on 2-year-old needles. The aeciospores measure 16–19 × 21–25 μm (Ziller, 1974). The diseased needles are eventually lost, but the trees are little affected. The uredinia and telia arise on the Oak Fern, *Gymnocarpium dryopteris* (*Thelypteris dryopteris*), on which the fungus can survive in the absence of *Abies* spp.

## DISEASES AND DISORDERS OF BRANCHES AND TRUNKS

### Bark necrosis associated with *Nectria fuckeliana*

A bark necrosis associated with *Nectria fuckeliana* is described under 'spruce' in chapter 4. The same disease also occurs in Europe on *Abies* species. In the western parts of the USA the same fungus has been found on *A.concolor*, causing cankering and sometimes death, especially of suppressed trees (Schultz and Parmeter, 1990).

### Witches' brooms and cankers caused by *Melampsorella caryophyllacearum* Schroet. (*M.cerastii* (Pers.) Schroet.)

*Melampsorella caryophyllacearum* occurs almost wherever *Abies* is found across Eurasia and in North America, and elsewhere also on the alternate hosts, in eastern Canada, Alaska, Siberia, China and Japan. It is also said to

**Figure 48** Plant of *Vaccinium vitis-idaea* (cowberry) affected by *Pucciniastrum goeppertianum*. Parts of the plant above the original infection at (a) show abnormal swelling (b), and the shoot at (d) is now dead (R. Hartig)

have been introduced into South America (Ziller, 1974). In Britain it has been recorded mainly in Scotland and Ireland, especially in the west (Wilson and Henderson, 1966).

The spermagonia and aecia occur on *Abies* spp. and the uredinia and telia on chickweeds and related species of *Cerastium* and *Stellaria*.

The minute, honey-coloured spermagonia are embedded under the cuticle, mainly on the upper sides of the needles, while the hemispherical or shortly cylindrical aecia occur on the undersides in two rows, one on either side of the midrib. The aeciospores form a reddish-yellow mass and are subgloboid, ellipsoid or polygonoid, measuring 16–30 × 14–20 μm. Their walls are densely covered with minute points (Wilson and Henderson, 1966).

Once established, the fungus becomes systemic in both *Abies* and the alternate hosts, and on the former gives rise to witches' brooms and cankers on branches and main stems.

Basidiospores from *Cerastium* and *Stellaria* infect *Abies* in May and June. The rust mycelium grows in the *Abies* shoots and by the autumn these show elongated swellings. Early in the following summer, buds on these swollen areas give rise to proliferations of erect shoots (witches' brooms) bearing spirals of pale soft needles. These needles carry the spermagonia and aecia (the spores from which infect *Cerastium* and *Stellaria*), and then by the end of the summer become yellow and fall.

The perennial mycelium continues to grow and enlarge the swollen tumour, the buds on which produce further pale-green erect shoots to enlarge the broom. So each year the broom and tumour tend to increase in size, and the needles on the broom produce further crops of spermagonia and aecia. As the tumour grows it tends to crack and become canker-like, and may be invaded by decay fungi. The brooms are susceptible to frost damage, and may be killed and fall, but while live buds remain on the tumours, fresh brooms arise, and some may persist for 15–20 years or more (Viennot-Bourgin, 1949; Peterson, 1964; Wilson and Henderson, 1966; Pupavkin, 1982).

Peterson (1965) in western parts of the USA found that though some infection could take place in almost any year, there were certain peak years when many infections occurred.

## Host range

*M. caryophyllacearum* will attack many species of *Abies*. In Britain it has been found on *A. alba*, *A. cephalonica*, *A. concolor* var. *lowiana* and *A. pinsapo* (Wilson and Henderson, 1966). Elsewhere, in Europe, the USSR and North America, it has been found also on *A.concolor*, *A. magnifica*, *A. lasiocarpa*, *A. religiosa*, *A. grandis*, *A.sibirica* (*A. pichta*), *A. sachalinensis*, *A. balsamea* and *A. procera* (*A. nobilis*) (Mielke, 1957; Viennot-Bourgin, 1949; Peterson, 1964; Kupevich and Transhel', 1957).

## Factors affecting the disease

According to Viennot-Bourgin (1949) the rust is encouraged by moist atmospheric and soil conditions, and as conditions tend to become less humid with distance from the soil surface, it occurs mainly at the base of the tree and decreases with height. Wet spring weather has been said to encourage the fungus (Pupavkin, 1982). Suitable moisture (and temperature) conditions, however, are not the only factors involved. Peterson (1964) found that abundant infection took place only when the basidiospores were released in numbers at the time when young fir shoots had developed to a suitable stage. Hence, in some parts of the USA many basidiospores might be present in the atmosphere but no infection took place because the fir shoots had not reached a receptive stage.

## Damage caused

In Britain, *M. caryophyllacearum* is of only minor significance. Elsewhere its importance varies, but in some areas (as in parts of central France) it may cause serious damage by reducing growth and volume, and sometimes, when many brooms form in the crown, by killing trees. The wood becomes brittle around burls and cankers, and trees may break at these points as a result of wind, or the weight of snow and ice. Cankered areas may also be invaded by secondary rot fungi.

According to Peterson (1964), broom rust commonly infects between 2 and 15 per cent of the trunks of silver firs in Central Europe. In the USSR Pupavkin (1982) found that in the Krasnoyav region up to 30 per cent of the Siberian fir trees (*Abies sibirica*) could be infected, with most disease on mature and overmature trees. In some parts of Europe cankered sections must be cut from the logs when marketing, and a loss of 5 per cent of the wood may result. Nevertheless, overall losses in European forests are usually fairly small (Viennot-Bourgin, 1949).

In North America the rust is of little or no importance in the eastern states, but in the west, in the Rocky Mountains and the Great Basin in the USA, almost every tree in *Abies* stands may be infected and may bear dozens of brooms. Disfigurement and loss of growth is of particular concern to growers of *A.concolor* for Christmas trees (Peterson, 1964).

*Control*
Among obvious methods of control are the destruction of the alternate hosts and the removal of broom-infected branches. Large-scale control of the alternate hosts is exceedingly difficult because they are so common and widespread. Their growth is in any case restricted as the trees close canopy, but some infection can still take place from plants on adjacent open land. The pruning out of diseased branches is not always economically possible on a forest scale and may not be necessary (Heck, 1927) except in small stands grown for Christmas trees (Peterson, 1964). It may still be be worth while in areas where the rust causes serious loss (Pupavkin, 1982). In most forests the only practicable control measure is the removal of diseased trees when thinning takes place. Special attention should be paid to cankered trees and those with brooms on or near the trunk (Peterson, 1964).

**Drought crack**
Trunks of *Abies* may develop large cracks which result from drought in hot, dry summers. A survey has indicated, however, that overall losses from this cause are small. Drought crack is further discussed in chapter 2.

**The mistletoe *Viscum album* L. ssp. *abietis* (Wiesb.) Abromeit)**
In France and Germany the subspecies *abietis* of the common mistletoe *Viscum album* may cause substantial damage to conifers, especially species of *Abies*. Affected trees, if heavily infested, are stunted in growth, and show crown dieback and wood defects. Decay fungi may enter through cracks in the bark made by the growth of the mistletoe plants. Delabraze and Lanier (1972) describe attempts to control mistletoe on *Abies alba* using 2,4-D and 2,4,5-T. Mistletoe is discussed further in chapter 3.

## ROOT DISEASES AND DECAY FUNGI

*Rhizina undulata* may attack the roots of *Abies* spp. as well as many other conifers. It is described in chapter 3.

*Abies* is one of the few conifers showing resistance to attack by *Heterobasidion annosum*, though the degree of resistance appears to vary with the species. *H. annosum* is considered in chapter 3. Among other decay fungi which attack *Abies* are *Fomitopsis pinicola* and the form of *Phellinus robustus* known as *P.hartigii*. These two fungi are described in chapter 20.

# Cypress and 'false' cypress: *Cupressus* spp. and *Chamaecyparis* spp.

Of the many species of *Cupressus* and *Chamaecyparis*, only Lawson cypress, *Chamaecyparis lawsoniana*, has been grown in Britain as a forest tree, and then only as a minor species. Many cultivars of *C.lawsoniana* with a great range of form and colour are grown in parks and gardens, where *C. nootka-tensis* and cultivars of *C. pisifera* are also common. The hybrid Leyland cypress, × *Cupressocyparis leylandii*, is also much planted in parks, gardens and shelter belts. Of the true cypresses, *Cupressus macrocarpa* is common in lowland parks and gardens, especially in coastal areas of the west. The Mediterranean Italian cypress, *Cupressus sempervirens*, is not generally considered hardy in northern Europe.

## General

Species of *Cupressus* and *Chamaecyparis* (especially *Cupressus macrocarpa*) tend to be susceptible to frost and damage by winter cold, though × *Cupressocyparis leylandii* is resistant to (but not altogether immune to) damage by both cold and exposure. *Cupressus macrocarpa* is resistant to sea winds, and this is almost the only reason it is grown so much in coastal districts. Some clones of Leyland cypress, especially 'Stapehill 20', are liable to defoliation in times of drought. *Cupressus* spp. are occasionally attacked by the fungus *Macrophomina phaseolicola* (Tassi) Goidanich, which produces brown or black lesions on their roots. This fungus affects many plants in the tropics, and reaches north into South and East Europe. *Pestalotiopsis funerea*, a weak wound parasite of various conifers, on which it causes needle blight and other symptoms, has been particularly damaging to members of the Cupressaceae in France (Morelet, 1982). It is described in chapter 3. *Phomopsis juniperovora*, best known on species of juniper and so described in chapter 9, has also been found causing shoot dieback (especially on nursery plants of species of *Cupressus*, *Chamaecyparis*, × *Cupressocyparis leylandii*, as well as on *Thuja*, *Thujopsis* and *Larix*).

The needle blight fungus *Didymascella thujina*, well known on *Thuja*, under which it is described below, has also been recorded on *Chamaecyparis lawsoniana*, on which, however, it has caused little damage.

## Nursery diseases

### Grey mould (*Botrytis cinerea*), and Phytophthora root rots
The cypresses are among the conifers most often damaged in nursery beds by the common grey mould, *Botrytis cinerea. Phytophthora* spp. (especially *P.cinnamomi*, but also *P.cactorum and P.cambivora* and sometimes other species) are also very damaging to nursery beds of cypress trees, particularly to the various forms of Lawson cypress. Phytophthora root rots may cause such heavy losses in beds of Lawson cypress that the grower is forced to change to container growing. These *Phytophthora* spp. may also attack and kill trees after planting out. Both *Botrytis cinerea* and the Phytophthora root rots are further considered in chapter 3.

### Canker and dieback caused by *Kabatina* spp.
In 1971 young nursery plants of Leyland cypress in southern England were found with short girdling cankers on the shoots. A species of *Kabatina* was isolated from the cankers and from similar lesions on old juniper plants from Gloucestershire (Young and Strouts, 1971). The same fungus was isolated the same year from cankers associated with a dieback of Leyland cypress and *Chamaecyparis nootkatensis* in Scotland. In one research plot of Leyland cypress, nearly all the trees, which were up to 1 metre tall, were affected, and some were killed. The cankers were often at branchlet bases, and dieback occurred most often on the windward side of the trees, which suggested that the fungus may be a wound parasite. Inoculation tests showed it to be pathogenic (Redfern, 1971). The fungus has also sometimes been found in southern Britain on other species of cypress, including *C.macrocarpa* 'lutea'. Only one-year-old shoots have been affected. No subsequent spread has taken place, and although the attack has made the plants unsightly, recovery has generally been good (Young and Strouts, 1971).

In 1975, *Kabatina thujae* (CMI Descr. F. 489) was found causing dieback of Lawson cypress in an Aberdeen nursery. The fungus, which was first described on *Thuja occidentalis* in Germany, forms black acervuli appearing as scattered spots at the bases of the dead shoots. The spots measure about 50–170 μm across, and bear colourless, one-celled, egg-shaped conidia measuring 4.8–8.0 × 2.3–3.5 μm (Schneider and von Arx, 1966). The same fungus has sometimes caused a severe leaf browning of species of *Chamaecyparis, Cupressus* and *Thuja* in France, where Morelet (1982) found that it could be controlled by spraying with mancozeb.

# Post-nursery diseases

## DISEASES OF BRANCHES AND TRUNKS

### Resin bleeding and canker caused by *Seiridium cardinale* (Wagener) Sutton and Gibson (*Coryneum cardinale* Wagener)

*Seiridium cardinale* (*Coryneum cardinale*), first described by Wagener (1928, 1939) as the cause of a highly destructive resin bleeding and canker of *Cupressus macrocarpa* in California, USA, is now known also on many other *Cupressus* spp. and on various other members of the Cupressaceae. It is still important in California, and is now present also in Argentina, Australia, England, Chile, France, Georgia (USSR), Greece, Israel, Italy, New Zealand, Northern Ireland, Portugal, South Africa, Spain, the USSR and Yugoslavia (Strouts, 1973; Caetano, 1980; Boesewinkel, 1983; Solel *et al.*, 1983; Graniti, 1986; Valdivieso and Luisi, 1987). Because of the earlier name of the causal fungus, the disease is often called Coryneum canker.

In Britain it has been found mainly on *Cupressus macrocarpa* (on which in some places it has caused severe damage), in the southern half of the country (where most of the trees of *C.macrocarpa* occur), but it is now known there also on Leyland cypress (Strouts *et al.*, 1984).

### *The causal fungus*

*Seiridium cardinale* (CMI Descr. F. 326) produces irregularly rounded to lens-shaped, erumpent black acervuli about 0.3–1.5 mm across on the dead host bark. The acervuli contain rather spindle-shaped conidia each divided by five transverse septa into six cells. These conidia usually measure 21–26 × 8–10 μm. The four inner cells are olive-brown, and the two end cells are colourless, with rounded ends (Wagener, 1939). The production of evanescent filiform, colourless, slightly curved spermatia within the acervuli has been described by Motta (1979). The fungus grows at temperatures from just below 5 °C to about 34 °C, with an optimum at 26 °C. In dry, protected conditions, the spores remain alive for many months, and even on exposed leaves they have been found to show 29 per cent germinability after 48 days (Wagener, 1948).

In culture, the reverse of colonies grown on 2 per cent Bacto malt extract + 2 per cent Bacto agar is salmon or saffron (or occasionally buff) coloured, with a pale olivaceous zone near the margin (Strouts, 1973). Culture filtrates of the fungus contain various toxic metabolites which may play some part in the production of the symptoms of the disease (Mutto and Panconesi, 1987).

In California, a perfect stage of the fungus, a species of *Leptosphaeria*, has been found (Hansen, 1956), but so far this does not appear to have been described, and it has not been recorded elsewhere.

### *Symptoms*

*S.cardinale* causes strongly resinous, girdling cankers on branches and sometimes on trunks, when they may be up to 20–30 cm long (Barthelet and Vinot,

1944; Strouts, 1970). The acervuli of the fungus appear scattered on the dead bark over the surface of the cankers.

The most conspicuous symptom is a browning and death of the foliage in parts of the crown beyond the girdled areas (Wagener, 1939) (plate 38). In Italy the fungus has been found on cypress cones (Grasso, 1969) and on the seeds of *Cupressus sempervirens*, *Thuja orientalis* and *Chamaecyparis lawsoniana* (Motta and Saponaro, 1983).

Within the host, the fungus grows vigorously in the outer tissues as far as the cambium and less abundantly also in the wood (Moriondo, 1972). A detailed study of the growth of the fungus within the plant has been made by Motta and Panconesi (1987).

### Spread of the disease

*Seiridium cardinale* is probably spread mainly by wind and rain, as well as by birds and bark beetles and other insects. It may also possibly be spread on nursery stock and on pruning tools (Wagener, 1939). Frankie and Parmeter (1972) found evidence that in the USA the moth *Laspeyresia cupressana* may play some part in the spread of the fungus. As noted above, the fungus has been found on the seeds of various members of the Cupressaceae.

### Host range and susceptibility

On trees in the field, *S.cardinale* is found most widely and causes the greatest damage on *Cupressus macrocarpa* and some of its cultivars. It is also damaging in Greece, Israel, Italy, Portugal and California on *C.sempervirens* (Xenopoulos and Diamandis, 1985; Mendel *et al.*, 1983; Grasso *et al.*, 1979; Caetano, 1980). In Italy, too, Grasso (1952) found that *C.arizonica* was very susceptible.

Field observations and inoculation experiments have indicated that other species of *Cupressus* are to some degree susceptible, among them *C. lusitanica* (Caetano, 1980). Viennot-Bourgin (1981) found evidence that *C. glabra*, *C. bakeri*, *C. dupreziana* and *C. torulosa* were tolerant to the disease. Even within *C.sempervirens*, resistant clones have been found (Raddi and Panconesi, 1984). In his inoculation trials in Britain, Strouts (1973) failed to induce canker formation in *C. lawsoniana*, and most of the evidence available (which he reviews) indicates that this species is immune to the disease. He did produce severe and perennating cankers on × *Cupressocyparis leylandii* and *Chamaecyparis nootkatensis* (though Coryneum canker has not been found on the latter in the field in Britain). The disease has since been found on × *C. leylandii* in Britain (Strouts *et al.*, 1983), Italy (Parrini and Panconesi, 1981) and New Zealand (Boesewinkel, 1983). Ponchet and Andréoli (1989), who studied the histopathology of the disease on clones of six *Cupressus* spp., associated resistance with thickness of the phelloderm. Phelloderm thickness in the more resistant clones exceeded 100 μm, while that of the more susceptible ones was less than 60 μm.

*S. cardinale* has also been known to attack some species of *Thuja* (including *T. plicata* and *T. orientalis*), and species of *Juniperus* and *Libocedrus*, though they are not often much affected in the field (Edson, 1938; Smith, 1938; Wagener, 1948; Strouts, 1973; Benetti and Motta, 1979).

*Damage caused*

A tree affected by *S. cardinale* often shows so many brown, dead branches that it quickly becomes too unsightly to have any ornamental value. Though spread throughout the trees is rather slow, the fungus eventually kills even large specimens of susceptible species such as *Cupressus macrocarpa*. In parts of Italy and Greece in which cypress trees form an important element in the landscape, this disease has had an effect comparable to that of Dutch elm disease in more northern countries, the appearance of whole regions being changed by the death of so many trees (Graniti, 1988).

*Control of the disease*

If seed is infested by *S.cardinale* it can be successfully treated with slurries or aqueous suspensions of benomyl or thiophanate-methyl (Motta, 1984).

Many workers, especially in Italy, France and the USA, have attempted to control the disease with various fungicides. Thus Wagener (1948) found that Bordeaux mixture had some controlling effect, and copper oxychloride has also given good results (Govi and Deserti, 1980; Marchetti and D'Aulerio, 1982). Some control has also been obtained with benomyl (McCain, 1984; Panconesi and Raddi, 1986; and others), carbendazim (Mathon, 1982), chlorothalonil (McCain, 1984), thiophanate-methyl (Viennot-Bourgin, 1981) and various dithiocarbamates (Caetano, 1980). Generally, however, these chemicals can be used effectively only in the nursery and on small and especially valuable ornamental trees, and they tend to give control only as protectants or in the early stages of infection; the treatments are expensive, and several applications are needed over the season (Govi and Deserti, 1980; Viennot-Bourgin, 1981; Marchetti and D'Aulerio, 1982; Panconesi and Raddi, 1986; Ponchet, 1986). Magro *et al.* (1984) and Marchetti *et al.* (1986) have suggested that isolates of the antagonistic fungus *Trichoderma viride* might be used to control *S.cardinale*.

Usually, in populations of larger trees, affected trees can only be removed and replaced by other, more resistant ones. Of these, the most obvious replacements for *Cupressus macrocarpa* in Britain are *Chamaecyparis lawsoniana* and *Thuja plicata*; × *Cupressocyparis leylandii* is also a possible choice, as it is rarely affected in the field. Elsewhere in other parts of Europe it may be possible to use resistant clones of *Cupressus sempervirens*, or some of the tolerant *Cupressus* species listed by Viennot-Bourgin (1981).

On individual trees it may be possible to remove infected branches and canker tissue and spray around the treated parts with benomyl, or treat the wounds with either Santar SM or Vinavil with benomyl (Nembi and Panconesi, 1982) or with Santar SM followed by Lac Balsam (Marchetti and D'Aulerio, 1983b).

**Resin bleeding and canker caused by *Lepteutypa cupressi* (Nattrass *et al.*)**
**Swart (*Rhynchosphaeria cupressi* Nattrass, Booth and Sutton)**
**(anamorph *Seiridium cupressi* (Guba) Boesewinkel)**

*Lepteutypa cupressi*, which causes a disease similar to that produced by *Seiridium cardinale* (to which it is closely related) has long been known in East and South Africa, Japan, Australia and New Zealand, where it has affected species of *Cupressus*, *Chamaecyparis*, × *Cupressocyparis* and *Juniperus*. It has recently been found on cypress trees in Greece on the island of Kos. It produces stromata immersed in the dead bark round the edges of the cankers to which it gives rise. Its oblong or ellipsoid ascospores are brown at maturity, and 4-celled, measuring 14–19 × 6–7.5 μm. The acervuli of its anamorph, *Seiridium cupressi*, produce more or less fusoid, 6-celled spores measuring 22–32 × 6–9.5 μm. The cells at either end are colourless, with long, unbranched appendages, while the four median cells are brown in colour (Graniti, 1986).

## ROOT DISEASES AND DECAY FUNGI

**Phytophthora root rot**

As noted above, Phytophthora root rot may affect cypresses beyond the nursery stage. Indeed attack by *Phytophthora* spp. is one of the commonest causes of the death of Lawson cypress trees of all ages, and sometimes also of that of other cypresses.

Members of the Cupressaceae may be attacked by the root- and butt-rot fungi *Heterobasidion annosum* and *Armillaria* spp., which are described in chapter 3. Young and Strouts (1977) in Britain found *Leptoporus ellipsosporus* on *Cupressus macrocarpa*, apparently in association with a severe tubular stem decay. *Phaeolus schweinitzii* has been found on *Chamaecyparis*.

# Red cedar or arbor-vitae: *Thuja* spp.

The introduced western red cedar (*Thuja plicata*) is one of the minor species in European forests. It is commonly grown as a specimen tree in parks and gardens, and as a hedge plant. Other *Thuja* spp., especially cultivars of the white cedar, *T. occidentalis*, are also grown as garden trees.

## General

*Kabatina thujae*, the cause of dieback, leaf browning and canker of various members of the Cupressaceae, and described above under '*Cupressus*' and '*Chamaecyparis*', has been found damaging *Thuja* in France. *Thuja* spp. also act as hosts to *Pestalotiopsis funerea*, which attacks many conifers (and is described in chapter 3). They may also be attacked by *Phomopsis junipero-vora*, described under '*Juniperus*' in chapter 9, and by *Sydowia polyspora* (anamorph *Sclerophoma pythiophila*), which has a wide host range, and is described in chapter 3.


## Nursery diseases

### Winter bronzing and winter cold damage

In winter, western red cedar, particularly in nursery beds, may turn a striking bronze colour, but in the spring, with warmer weather, the plants turn green again (plate 39). In very severe winters, however, the foliage may be killed, and it then turns a deep, glossy black (Day and Peace, 1934).


### Needle blight caused by *Didymascella thujina* (Dur.) Maire
### (*Keithia thujina* Dur.)

Needle blight caused by *Didymascella thujina* (*Keithia thujina*) affects *Thuja* trees of all ages, but causes serious damage only on nursery plants, on which its effects are so severe that in Britain it has for long been a limiting factor in the production of *Thuja* nursery stock. The disease occurs in Great Britain, Northern Ireland and the Irish Republic, and in France (especially in Normandy and Brittany), Belgium, Holland, Norway, Denmark, Italy, Canada and the USA (CMI Map No. 149), and in Germany (Burmeister, 1966).

Needle blight as it affects *Thuja plicata* in Britain and in France, has been described by Alcock (1928), Pawsey (1960) and Boudier (1983a). The fungus occurs almost entirely on the foliage, producing its small ascocarps embedded in the leaf tissue below the epidermis, which eventually splits, and round the edges of the apothecium curls back to form a scale-like flap (plate 40). The apothecia are red-brown (dull brown when dry), cushion-like, oval or irregular in shape, 0.5 to 1.25 mm across, and one, or some-times two or three, form on each diseased scale leaf. The hymenium contains asci and unbranched paraphyses (Pawsey, 1960; Soegaard, 1969). The asci each contain two oval, pitted ascospores that become dark brown when mature, though they may be almost colourless when shed. The asco-spores measure 15–24 × 15–16 μm, and when shed are surrounded by a sticky layer of mucilage 2 or 3 μm thick. They are two-celled, but one cell is minute and represented by little more than a mark at one end of the spore. No imperfect stage has been discovered, and the fungus has not been grown successfully in culture. When ascospore discharge has been completed, the ascocarps dry up, and appear as oval or rounded depressions and later as holes in the leaves.

It has been suggested that the fungus may be carried with seed in small pieces of leaf debris (Alcock, 1928), though this should be easily prevented by proper cleaning of the seed. The disease normally enters the crop by means of airborne ascospores from previously infested trees or nursery beds. These spores may be found in the air in Britain from May to October, with peaks usually in June and July and in September (Phillips, 1967b; Burdekin, 1968). In Denmark Soegaard (1969) found similar peaks, in June and July, August and September, and sometimes also in mid-October.

Needle blight is not often seen in first-year seedlings, but when found on these it usually occurs on the juvenile leaves low on the stem (Anon., 1967). Older nursery plants, in both seed beds and transplant lines, may be severely affected, and at first show browning of scattered leaflets. The fungus usually grows only in the mesophyll and the epidermis, and it cannot as a rule spread from one leaflet to another (Pawsey, 1960), but through multiple infections large areas of the fronds, particularly at the bases of the plants, may become brown and die. In a few cases severe damage has also been observed in which death of tissues through girdling of stems and shoots was closely associated with prior infection of a leaflet on the girdled zone, on which fruit bodies were present. Whether the fungus directly caused the girdling was not determined, but it was evident that in some circumstances severe damage could result when the number of leaflets infected was relatively small (Burdekin, 1970). In the brown leaflets one to three of the oval, elongated or irregular apothecia may develop. Much of the early infection seems to take place near the clefts between the leaflets, and is usually first seen in April, with an abundant development of the fructifications in May and June (Pawsey, 1960) and again in the autumn.

Pawsey also found evidence that incipient apothecia were formed in autumn, and these completed their development in the following spring. Hence, they could be a means by which the fungus survived the winter and continued its spread within the crop. He considered, however, that over-wintering was mainly by means of spores shed in autumn and remaining attached by their mucilage to the surface of *Thuja* leaves.

In severe attacks, much of the lower foliage may be destroyed, dry up and fall, and the plants remain small and stunted, and many may die. *D.thujina* is important only on *Thuja plicata*, although it was first recorded on *T. occidentalis* (Durand, 1913), which is not usually affected (Boudier, 1983a), and it has been reported in Germany on *Chamaecyparis lawsoniana* (Burmeister, 1966). *Thuja standishii* and its f1 hybrids with *T. plicata* are resistant to the disease (Soegaard, 1969). In *T.plicata*, the plants become more resistant with age, and cuttings struck from adult trees are little affected (Soegaard, 1969). Among the forms of *T. plicata*, Boudier (1983) found that the desirable variety *atrovirens* was very susceptible, but the var. *excelsa* often remained healthy when exposed to infection.

Little is known about the factors that control the incidence of needle blight. Pawsey (1960) found that at 20 °C infection became visible about three weeks after inoculation, but at 13 and 5 °C no effects could be seen at the end of 12 weeks. Porter (1957), in Canada, noted that heavy spore discharge took place when relative humidity exceeded 90 per cent, and the temperature was about

12 °C. Phillips and Burdekin (unpublished) also found that spores were discharged in damp periods following rain. Boudier (1983a) pointed out that attacks of needle blight tended to be unpredictable, though damage in deep valleys might occur year after year. He noted that the disease was favoured above all by high humidity, which in nurseries may be encouraged by frequent watering, and in general by shading by trees and by heavy weed cover; incidence of the disease is also increased by root damage in heavy water-logged soils and by reduced vigour of plants after pricking out and transplanting. He found that in France the disease caused most damage in the maritime northwest, and was much less troublesome in parts of the country with a drier, more continental climate. Similarly in Britain the disease causes more destruction in the wetter west than in the drier east (Burdekin and Phillips, 1970).

Much can be done to prevent *D. thujina* from entering and building up in a crop, if *Thuja plicata* is grown only in nurseries isolated by 2 or 3 kilometres from other *Thuja* plants. Isolation alone is rarely effective for long, however, unless a rotation sowing system is also practised (Pawsey, 1963). Such a system can be used only if a number of scattered isolated sites are available that can be linked into groups each designed to produce plants for a large area. Only one nursery in the group supplies plants for the whole district in any one year, and is then completely cleared of all *Thuja* stock before resowing. In the succeeding year a second nursery supplies the plants, is cleared of *Thuja* and resown. Clearance of the *Thuja* must be carefully carried out, otherwise any infected stock remaining will provide a source of infection for the next sowing. Sufficient nurseries are maintained in the group to provide a supply of stock every year. No *Thuja* transplants must be brought into the rotation nurseries, but other tree species can be freely moved in and out.

Even in isolated nurseries run on a rotation, *D. thujina* may sometimes enter and begin to build up to epidemic proportions. It is therefore very important to keep a careful watch on *Thuja* beds.

It may be possible to avoid the disease by growing the resistant species or varieties of *Thuja* mentioned above, if these are arboriculturally or silviculturally suitable, or by raising cuttings from mature trees.

If the disease appears in a nursery, some fungicidal sprays can be used to give a measure of control. Cycloheximide gives excellent results with only 1 to 3 applications over the season (Pawsey, 1965; Burdekin and Phillips, 1970), but this chemical is no longer available or approved. In France, Mathon (1982) gained control by applying triadimefon, benomyl or mancozeb 12 times between April and October. In further trials, Boudier (1983a, 1986) obtained very good results with triadimenol, propiconazole and triadimefon (though he found the latter was phytotoxic). Some other materials, including benomyl and prochloraz, gave good but less effective results. He found that in nurseries where risk of infection was high, at least four applications were needed between June and October, especially in damp periods, and in container nurseries spraying needed to start by the end of April.

**Leaf and shoot death caused by *Coniothyrium fuckelii***
Humphreys-Jones (1980) found the fungus *Coniothyrium fuckelii* causing the death of leaves and shoots of *Thuja orientalis* cv. Aurea Nana, and making 30 per cent of them useless for sale. The fungus is described below (chapter 9) under '*Juniperus*'.


## Post-nursery diseases

STEM DISEASES

**Cambial damage by spring frosts**
Young plantation trees before the thicket stage are liable to cambial injury by spring frosts.

**Cankers caused by *Seiridium cardinale***
The canker fungus *Seiridium cardinale* described above under '*Cupressus* and *Chamaecyparis*' has been found in Britain on western red cedar (Redfern *et al.*, 1984; Strouts *et al.*, 1987), and in Italy it is known on *Thuja orientalis* (Benetti and Motta, 1979; Parrini and Panconesi, 1981). Damage to *Thuja* trees has been less than that to *Cupressus* and *Chamaecyparis*.


ROOT AND BUTT ROTS

Western red cedar is susceptible to butt rot by *Heterobasidion annosum*, and *Armillaria* spp. may kill young trees and cause butt rot in older ones (chapter 3). *Thuja* spp. may also be attacked by *Stereum sanguinolentum* (chapter 20).


# Hemlock: *Tsuga* spp.

Western hemlock (*Tsuga heterophylla*) is another of the minor forest species in Europe, and because of its ability to withstand shade is sometimes used to underplant other conifers. Other species of *Tsuga*, especially *T.canadensis*, may be found in parks and gardens.


GENERAL

The canker and dieback fungi *Ascocalyx abietina* (*Gremmeniella abietina*) (which mainly affects pine, and is described in chapter 5) and *Phacidium coniferarum* (which affects many conifers, and is described in chapter 3) and the usually saprophytic *Sydowia polyspora* (anamorph *Sclerophoma pythiophila*) (also described in chapter 3) have all sometimes been found on species of *Tsuga*.

## NEEDLE DISEASES

**Needle blight caused by *Fabrella tsugae* (Farlow) Kirschst. (*Didymascella tsugae* (Farlow) Maire, *Keithia tsugae* (Farlow) Durand)**
Needle blight caused by *Fabrella tsugae* has occasionally been found on *Tsuga canadensis* (eastern hemlock) in Scotland (Wilson, 1937; Foister, 1948), but so far has been of no importance. Its rounded or elliptical apothecia develop below the epidermis on the undersides of the leaves, and measure 0.3–0.5 mm across. The epidermis eventually splits to form a scale. The hymenium consists of paraphyses with brown, club-shaped tips, and asci that each contain four oval, greenish-brown (at first colourless) ascospores with two unequal cells. The spores measure 13–16 × 6–8 μm (Stevens, 1925).

## ROOT AND BUTT ROTS

*T.heterophylla* is very susceptible to butt rot by *Heterobasidion annosum* and *Armillaria* spp., which may also kill young trees. It may also be attacked by *Rhizina undulata*, the cause of group dying in many conifers (chapter 3).

# 9 Diseases of other conifers

## Monkey puzzle (*Araucaria araucana*)

The only species of *Araucaria* hardy in northern Europe is *A. araucana*, the monkey puzzle. This tree, introduced from South America, is common in parks and gardens.

The monkey puzzle is singularly little affected by disease, though it is susceptible to damage by *Armillaria* spp., which sometimes kill quite large trees (see chapter 3).

The Chilean needle rust *Micronegeria fagi* Diet. & Neger (the alternate host of which is *Nothofagus*) has been found once on *Araucaria* in Britain (Butin, private communication).

## Cedar (*Cedrus* spp.)

Three cedars are common in parks and gardens. They are the cedar of Lebanon (*Cedrus libani*), from Asia Minor and the Lebanon, the Atlas cedar (*C. atlantica*), from North Africa, and the deodar (*C. deodara*), from the Himalayas.

Cedars are sometimes affected by needle fall, the cause of which is unknown.

The common rot fungi *Armillaria mellea* (*sensu lato*) and *Heterobasidion annosum* have both been recorded on cedars, and so has the 'Phomopsis disease' caused by *Phacidium coniferarum* (anamorph *Phacidiopycnis pseudotsugae*) (Peace, 1962). These diseases are described in chapter 3. Cedars may sometimes be decayed by *Phaeolus schweinitzii* and *Sparassis crispa*, which are described in chapter 20.

## *Sugi* or Japanese cedar (*Cryptomeria japonica*)

The '*Sugi*' or Japanese cedar, *Cryptomeria japonica*, a native of Japan and China, is grown as an ornamental in parks and gardens.

Nursery plants are sometimes damaged by autumn frosts (Macdonald *et al.*, 1957), and by the grey mould, *Botrytis cinerea*, which is described in chapter 3.

Japanese cedar is also sometimes attacked by *Phomopsis juniperovora* (described below under '*Juniperus*') and by *Seiridium cardinale* (described above in chapter 8, under '*Cupressus* and *Chamaecyparis*'). These diseases are generally of minor importance on *Cryptomeria*.

# Maidenhair tree (*Ginkgo biloba*)

This tree, a native of China, is common in Britain and other parts of Europe (and in North America) in parks and gardens. It is the only representative of an ancient group related to the conifers.

The fungus *Phoma exigua* Desm. has been found in Britain causing a severe stem dieback and bud and leaf rot of *Ginkgo biloba* in Cambridge and on container-grown nursery stock of the same tree in several nurseries in East Anglia. The spores of the fungus measure 4–8.5 × 2–3.5 μm (Humphreys-Jones, 1982).

# Juniper (*Juniperus* spp.)

The common juniper, *Juniperus communis*, a bush or sometimes a small tree, is one of the few native British conifers, and it occurs widely throughout northern Europe. In addition, many introduced species of *Juniperus* are grown in parks and gardens, but most are low-growing shrubs.

## Nursery diseases

### Grey mould (*Botrytis cinerea*) and black snow mould (*Herpotrichia juniperi*)

Nursery plants of juniper may sometimes be attacked by the grey mould, *Botrytis cinerea* (chapter 3). They may also be affected by the black snow mould, *Herpotrichia juniperi*, which, because of its much greater importance on spruce, is described in chapter 4.

### Shoot blight and canker caused by *Phomopsis juniperovora* Hahn

Shoot blight and canker caused by *Phomopsis juniperovora* is of most importance in North America, but it is also found in Europe, Africa and New Zealand (CMI Descr. F. 370). In Europe it is known in Britain, France and Denmark (CMI Descr. F. 370), Germany (Butin and Paetzholdt, 1974) and the Netherlands (Hahn, 1943).

The fungus was first described by Hahn (1920). He found (Hahn, 1931) that the colourless A-spores were commonly 7.5–10 × 2.2–2.8 μm. The long, narrow, colourless B-spores are curved and often strongly hooked. They measure 20–30 × 0.5–1 μm (CMI Descr. F. 370). Isolates of the fungus typically (but not invariably) produce a yellow coloration and flaming orange crystals in culture in a variety of media (Hahn, 1931). The spore measurements overlap closely with those of the weak pathogen *Phomopsis occulta* (the anamorph of *Diaporthe conorum*), which, however, never produces yellow colours or orange crystals in culture. The spores exude from the pycnidia in whitish or yellowish tendrils.

Peterson (1973) found that the spores germinated and their germ tubes developed most rapidly at a temperature of 24 °C. Spores could withstand temperatures as high as 43 °C and as low as 22 °C. They also withstood desiccation. On dead host material they could survive for up to two years (Hodges and Green, 1961). They are spread by air movements and rain-splash.

*P. juniperovora* most characteristically causes a shoot blight, attacking the young tissues at the shoot tips, which turn brown and die. Spread downwards into older tissues then occurs. If it spreads into larger branches or main stems, perennial cankers may be formed. These may quickly girdle small stems up to about half an inch (12.7 mm) in diameter (Hahn, 1926) and in exceptional cases whole trees may be killed.

The host range of *P. juniperovora* is not entirely clear because of possible confusion with *P. occulta*, a fungus with a very wide host range. There is no doubt that *P. juniperovora* at least mainly affects members of the Cupres-saceae, including species of *Juniperus*, *Chamaecyparis*, *Cupressus* and *Thuja* (Hahn, 1943).

Schoeneweiss (1969), who in North America collected and tabulated the most extensive data on specific and varietal susceptibility to the fungus, found that among the junipers, which were most affected, blight was especially severe on *J.sabina tamarissifolia* and many cultivars of *J.scopulorum* and *J.horizontalis*. Disease was generally slight on *J.communis*, *J . sabina* and *J. squamata*, while there was substantial variation between cultivars of *J. chinensis* and *J. virginiana* (on which the disease is best known in North America). Species of *Cupressus* and *Chamaecyparis* were generally less affected than the junipers. In the case of *Thuja*, *T. orientalis* was severely damaged, but effects were slight on *T. occidentalis* and *T. plicata*.

Infection is most rapid at 100 per cent RH, and disease development is encouraged by high temperatures (32 °C) (Peterson, 1973). The disease is therefore favoured by long spells of wet weather, especially in spring and autumn (Caroselli, 1957), and it may be increased by overhead irrigation. Overcrowding in the seedbed may also affect it (Hahn, 1926). Davis and Latham (1939) found that nitrogen manuring to improve growth could also increase damage by *P. juniperovora*. Though the fungus can readily invade healthy young growth (Peterson, 1973), damage by cultivation, transplanting, pruning or grafting may predispose plants to attack (Hahn, 1926).

*P. juniperovora* causes damage mainly in the nursery. If affected nursery plants are planted out, few survive, and any that recover are poor and stunted (Hodges and Green, 1961). At a later stage, older plants are not usually killed, but death of small branches renders the trees unsightly (Peterson, 1973). In Europe the fungus has as yet caused only sporadic loss. In Britain it has caused the death of shoots and twigs and less often of larger branches of *Juniperus communis* and of various ornamental junipers in widely separated localities (Young and Strouts, 1971). In nurseries in France it has caused considerable damage to many of the Cupressaceae (Morelet, 1982). In Germany it has attacked junipers in nurseries in Holstein (Butin and Paetzholdt, 1974), and it has also affected nurseries in the Netherlands (Hahn, 1943).

Balanced manuring and avoidance of overcrowding in the seedbed may help to reduce the effects of this disease. Seedbeds should be cultivated only in dry weather, and they should not be irrigated overhead (Morelet, 1982). Many chemicals have been tested, and the best control has been obtained by spraying with organomercurials, but the use of these is no longer approved (CMI Descr. F. 370). Good results can be obtained with benomyl (Morelet, 1982), though spraying may need to be frequent to ensure that new young growth is protected. If the disease proves troublesome in any area, the danger of loss may be reduced by growing varieties or cultivars known locally to be little affected.

### Leaf and shoot death caused by *Coniothyrium fuckelii* Sacc.

Humphreys-Jones (1977) has described serious damage to cuttings of *Juniperus communis* var. *compressa* caused by *Coniothyrium fuckelii*. Sometimes only a few leaves were destroyed, but many of the affected plants were killed. Pycnidia of *C. fuckelii* (which is better known as the cause of cane blight of the raspberry) were present on the diseased cuttings. The brownish, elliptical spores measure about 4–5 × 3 μm.

## Post-nursery diseases

### NEEDLE DISEASES

### Needle cast caused by *Lophodermium juniperinum* (Fr.) de Not.

*Lophodermium juniperinum* has been known on both living and dead needles of *Juniperus communis* for about 150 years, and has also been recorded on *J. nana* (Moore, 1959; Dennis, 1968). It is a fairly common fungus, but of little importance as a cause of disease. Its prominent black, elliptical, blister-like hysterothecia develop under the cuticle, usually on the upper sides of the leaves, and may be up to 1 mm long by 0.4 mm wide. They open by a slit to show a whitish hymenium of paraphyses that curl at the tips, and long asci each containing eight thread-like ascospores measuring 60–100 × 2 μm (Dennis, 1968).

### Needle blight caused by *Didymascella tetraspora* (Phillips & Keith) Maire (*Keithia tetraspora* (Phillips and Keith) Sacc.)

In 1880, needle blight caused by *Didymascella tetraspora* was recorded in Scotland on the upper leaves of *Juniperus communis*, but the fungus appears to be uncommon (Moore, 1959). Dark-brown apothecia are embedded in the upper surfaces of the leaves, and measure about 1 mm across. The hymenium is made up of paraphyses with swollen, olive-coloured tips, and asci each

containing four oval, olive-brown, unequally two-celled spores that measure
21–24 × 13–16 μm (Dennis, 1968).

### Needle rusts: species of *Gymnosporangium*

Four species of *Gymnosporangium* sometimes produce pustules on juniper
needles in various parts of Europe. They are *G. gaeumannii* Zogg, *G.
cornutum* Kern, *G. torminalis-juniperinum* Fisch. ex Kern. and *G. dobroznakovii* Mitrophanova (table 9.1).

*Gymnosporangium gaeumannii* is uncommon and of no economic importance. It occurs on the alpine subspecies of the common juniper, *Juniperus
communis* ssp. *nana*, in the Alps in Switzerland (Gäumann, 1959) and on
*Juniperus communis* in Yugoslavia (Mijuskovic and Vujanovic, 1989); it has
been introduced into the eastern Rocky Mountain area of Alberta, Canada
(Ziller, 1974). It forms inconspicuous cushion-shaped, pale- to dark-brown
pustules on the upper sides of the juniper needles. These pustules contain
one- (sometimes two-) celled dark yellow-brown urediniospores mostly
measuring 21–25 × 20–23 μm, and also two- (sometimes one-) celled brown
teliospores mostly measuring 33–46 × 25–33 μm (Gäumann, 1959). Both the
urediniospores and teliospores have long colourless pedicels. The urediniospores each have 8–12 germ-pores (Ziller, 1974). No other stages of this rust
are known.

The more widespread rust *Gymnosporangium cornutum* may also sometimes produce brown pustules on the upper sides of the needles of common
juniper, but the pustules in this case contain only two-celled teliospores
(figure 50d). As *G. cornutum* more characteristically forms its telia on swellings on the stems, it is described below under 'stem rusts'.

The third of these rusts, *G. torminalis-juniperinum*, which is also of minor
importance, has sometimes been recorded as a form species of *G. cornutum*
(*G. juniperinum* (L.) Fries) (Gäumann, 1959). It forms brown pustules about
1 mm high, mostly on the upper sides of the needles of *Juniperus communis*
and other junipers of the *Oxycedrus* group. The pustules contain brown,
rather broadly ellipsoid teliospores measuring 21–30 × 35–49 μm (Kern,
1973).

The aecial stage is on *Sorbus torminalis* (Gäumann, 1959) and the fungus
occurs in Austria, France, Germany, Hungary, Sweden, Switzerland and
Turkey, and in Morocco (Kern, 1973).

*G. dobroznakovii*, which occurs in the USSR in the Crimea and the
Ukrainean Peninsula, and in parts of Asia, produces its horn-like telia mainly
on small twigs, but they may occasionally appear on the leaves and fruits
(Kern, 1973).

# DIEBACKS AND TWIG BLIGHTS

## Shoot dieback caused by *Kabatina thujae* Schneider and von Arx var. *juniperi* (Schneider and von Arx) Morelet

A shoot dieback of many species and cultivars of juniper has been found in southern Britain and (more often) in Germany and the Netherlands. The causal fungus was described as *K. juniperi* together with *K. thujae* (of which it is now considered to be only a variety) by Schneider and von Arx (1966). It has more recently been found in North America (Ostrofsky and Peterson, 1981; Perry and Peterson, 1982; Ostrofsky and Ostrofsky, 1984). It produces erumpent black acervuli with colourless, one-celled ellipsoid conidia measuring 3–5 × 2 μm (Sutton, 1980).

In studies in Europe, Hoffmann and Fliege (1967) concluded that the fungus was a wound parasite, and this was confirmed in North America by Ostrofsky and Peterson (1981). Infection occurs in the autumn (Hoffmann and Fliege, 1967). Ash-grey cankers form below the shoot tips, which may be girdled. The tips of the affected shoots become discoloured and die back in the spring. In Nebraska, USA, Ostrofsky and Peterson (1981) found that acervuli first appeared on the diseased tissues in February, becoming abundant in April and May, then decreasing until the autumn. Spore germination took place at temperatures between 16 and 21 °C whenever the relative humidity rose above 95 per cent (Perry and Peterson, 1982).

The fungus may sometimes cause severe damage to both wild and cultivated junipers. In nurseries, affected plants should be destroyed. Spraying with mancozeb may give control, as Morelet (1982) found the closely related *K. thujae* was sensitive to this chemical.

## Shoot dieback caused by *Seiridium cardinale*

Junipers are among the plants known to be susceptible to attack by *Seiridium cardinale* (Strouts, 1973; Parrini and Panconesi, 1981), though they are rarely if ever affected in the field. *S. cardinale* is further discussed under '*Cupressus*' (chapter 8), on which it causes severe damage.

## Shoot blight and canker caused by *Phomopsis juniperovora*

This disease may sometimes affect large trees, but it is far more important in the nursery, and is therefore considered above under 'nursery diseases'.

# STEM RUSTS

## Species of Gymnosporangium (table 9.1)

About six stem rusts, all species of *Gymnosporangium*, occur on juniper in northern Europe. They are *G. clavariiforme* (Pers.) DC, *G. confusum* Plowr., *G. fuscum* DC (*G. sabinae* (Dicks.) Wint.), *G. cornutum* Kern (*G. juniperinum* (L.) Fries), *G. tremelloides* Hartig and *G. amelanchieris* E. Fischer ex Kern (which is sometimes considered to be a form species of *G.*

Table 9.1  *Gymnosporangium* spp. on *Juniperus* spp.

| Telia | Aecia | Distribution |
|---|---|---|
| 1. On spindle-shaped swellings on stems and branches | | |
| *Gymnosporangium clavariiforme* On *Juniperus communis* (and others of Oxycedrus group); tongue-like, orange | On *Crataegus* (and sometimes *Pyrus* and other Rosaceae) | Europe (including Britain), Asia, N. Africa, N. America |
| *G. confusum* On *Juniperus sabina* (and others of both Oxycedrus and Sabina groups); cushion-shaped, chocolate-brown | On *Crataegus* (and sometimes other Rosaceae) | Europe (including Britain), Asia, Africa, N. America |
| *G. fuscum* On junipers of Sabina group; cushion-shaped, becoming conical, yellow-brown | On *Pyrus* (European pear rust) | Europe (including Britain), Asia, N. America |
| *G. amelanchieris* (stem swelling slight) On junipers of Oxycedrus group; flattened, chestnut-brown | On *Amelanchier* | Europe (*excluding* Britain), Asia, N. Africa |
| *G. fusisporum* On junipers of Sabina group; cushion-shaped to conical, dark brown | On *Cotoneaster* | Switzerland, Cyprus, Iran |
| 2. On spindle-shaped swellings on stems and branches and sometimes on needles | | |
| *Gymnosporangium cornutum* On *Juniperus communis* (and others of Oxycedrus group); cushion-shaped, orange | On *Sorbus aucuparia* | Europe (including Britain), Asia, N. America |

Table 9.1 – *continued*

| Telia | Aecia | Distribution |
|---|---|---|
| 3. On spindle-shaped swellings on branches and associated with witches' brooms | | |
| *Gymnosporangium gracile* — On junipers of Oxycedrus and sometimes Sabina groups; cylindrical, orange | On *Amelanchier, Crataegus, Cydonia* | Mediterranean Europe, N. Africa, N. America |
| 4. On globoid swellings on stems and branches | | |
| *Gymnosporangium tremelloides* — On *Juniperus communis* (and sometimes others of Oxycedrus group); flattened, chocolate-brown | On *Sorbus, Cydonia* (and a form on *Malus*: apple rust) | Europe (*excluding* Britain), Africa, China, N. America |
| 5. On twigs and sometimes on needles | | |
| *Gymnosporangium dobroznakovii* — On junipers of Oxycedrus group; horn-like | On *Pyrus* | USSR and into Asia |
| 6. On needles only | | |
| *Gymnosporangium gaeumannii* — On *Juniperus communis*; small, cushion-shaped, brown | No aecial stage known | Swiss Alps and Yugoslavia; N. America |
| *G. torminalis-juniperinum* — On *Juniperus communis* (and others of Oxycedrus group); pustular, brown | On *Sorbus* | Europe (*excluding* Britain), N. Africa |

* Found in Britain.

*tremelloides*). Of these, the first four species are known in Britain. In addition, *G. fusisporum* E. Fischer (regarded by some as a form species of *G. confusum*) occurs in Switzerland, Cyprus and Iran. *G. dobroznakovii* Mitrophanova occurs in the Crimea, and the Ukrainean Peninsula in the USSR and in parts of Asia, and *G. gracile* Pat. is Mediterranean in its distribution (Kern, 1973). They are of little importance on junipers (though sometimes spectacular in appearance) but some are of greater significance on their alternate hosts.

In all of them the mycelium is perennial in gall-like swellings on the juniper stems (though the swelling is slight in *G. amelanchieris*). In most species the galls are spindle-shaped, but those caused by *G. tremelloides* are globoid in form. The telia of the rusts form on these galls, usually in April and May. When dry, the sori are hard, brownish, cushion-shaped, cylindrical or tongue-like bodies formed of masses of two-celled teliospores with long pedicels. In moist conditions, the pedicels and the mycelium of the fungus swell and gelatinise to form yellow or orange mucilaginous masses, on the surface of which the two-celled teliospores germinate as soon as they mature, and produce basidiospores on four-celled basidia. All these rusts are heteroecious, and the basidiospores therefore infect the alternate hosts, on which spermagonia and aecia arise from July to September or October. The aeciospores reinfect juniper plants. No uredial stages are formed.

*G. clavariiforme* is frequent in Britain and Ireland, occurring throughout the northern hemisphere. In Europe it is known from Spain eastwards through Britain to Scandinavia and the Middle East, into Asia to India, Korea and Japan, in North America (CMI Descr. F. 542), and in North Africa (Kern, 1973; Ziller, 1974). It causes large elongated swellings on the branches of *Juniperus communis* and sometimes on other members of the *Oxycedrus* group. On these it produces cylindrical or tongue-like telia 5–10 mm long (figure 49). These are yellow-brown and horny when dry, but in damp weather gelatinise to a soft, orange-yellow mass. They contain long, narrow teliospores (figure 50a), some of which are pale yellow, and measure 100–120 × 10–12 μm, and some of which are brown and smaller, measuring only 50–60 × 15–21 μm (Wilson and Henderson, 1966). According to Bernaux (1956), the pale spores appear first, in cold weather at the start of the season. The spermagonial and aecial stages of *G. clavariiforme* occur mainly on hawthorns (*Crataegus monogyna* and *C. laevigata*), but sometimes also on pear, *Amelanchier* and some other rosaceous hosts (Gäumann, 1959; Kern, 1973; Ziller, 1974).

*Crataegus monogyna* and *C. laevigata* (as well as *Cydonia*, *Mespilus* and occasionally *Pyrus*) are also aecial hosts of *G. confusum*. This is rare in Britain, but has been found on a few occasions in England on cultivated *Juniperus sabina* and its var. *prostrata*, and very locally in Sussex and Kent on *J. communis* (Moore, 1959). It is widespread in the rest of Europe, and occurs also in Asia, Africa and the USA (CMI Descr. F. 544) on various junipers of both the *Oxycedrus* and *Sabina* groups.

Its telial masses when dry are chocolate-brown and cushion-shaped and

**Figure 49** Gelatinous orange-yellow masses of teliospores of *Gymnosporangium clavariiforme* on swollen stem of *Juniperus communis* (D.H. Phillips)

swell when moist. Again two kinds of spore are present, some dark brown with thick walls and orange contents, and some colourless with thin walls. The brown spores measure 30–50 × 20–25 μm, and the colourless ones are rather longer and narrower (figure 50b) (Wilson and Henderson, 1966).

Also uncommon in Britain, on cultivated *J. sabina*, is *G. fuscum* DC (*G. sabinae* (Dicks.) Wint.), which also occurs widely in other parts of Europe, as well as in Asia and North Africa. It has also been introduced into North America (CMI Descr. F. 545). Its telia, which form on junipers of the *Sabina* group, are at first cushion-shaped, later conical and up to 10 mm high, and yellow-brown when they gelatinise. The teliospores (figure 50c) have pale-brown walls, and measure 34–40 × 20–30 μm (Wilson and Henderson, 1966). This is the European Pear Rust, the aecial stage occurring on *Pyrus*.

*G. cornutum* has been seen occasionally on *J. communis* in England and Wales, and much more often in Scotland. It occurs throughout the northern hemisphere, in Europe, Asia and North America (Gäumann, 1959; Kern, 1973) on junipers of the *Oxycedrus* group. Its teliospore masses are flat or cushion-shaped, chocolate-brown when dry, but orange and gelatinous when moist. The spores may be thick-walled and dark brown, or thin-walled and

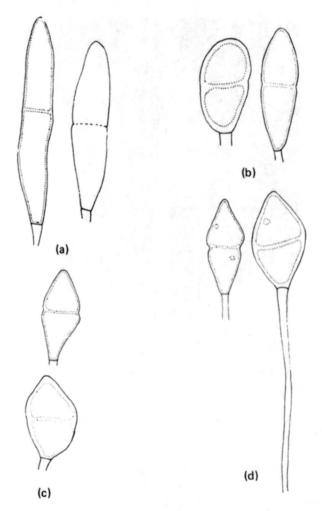

**Figure 50**  Teliospores of the *Gymnosporangium* spp. found on *Juniperus* in Britain (M. Wilson and D.M. Henderson): (a) *G. clavariiforme*, (b) *G. confusum*, (c) *G. fuscum* and (d) *G. cornutum*

yellowish, measuring 32–52 × 18–28 μm (figure 50d). The alternate host is *Sorbus aucuparia* (Wilson and Henderson, 1966).

*G. tremelloides* does not occur in Britain, but has long been well-known in the rest of Europe as well as in Africa, China and North America (Kern, 1973; CMI Descr. F. 549). It forms woody, globoid galls on branches of *Juniperus communis* and sometimes on other junipers of the *Oxycedrus* group. Its flattened, chocolate-brown telia develop on these galls. The general form of the teliospores is ellipsoid, like those of *G. confusum* and *G. fuscum*, and they measure 18–28 × 49–61 μm (Ziller, 1974). The alternate hosts are mainly species of *Sorbus*, but one of its form species (f. sp. *mali*

Erikss.) may sometimes give rise to a disease of apple similar to that on pear caused by *G. fuscum*.

*G. amelanchieris* also has a wide distribution, in Europe, Asia and North Africa (Kern, 1973). Its flattened, chestnut-brown telia form on slight spindle-shaped swellings on the branches of junipers of the *Oxycedrus* group. They measure 20–30 × 36–56 μm (Kern, 1973).

The other three *Gymnosporangium* species have a much narrower distribution. *G. fusisporum* occurs in Switzerland, Cyprus and Iran. Its darkbrown, conic or cushion-shaped telia form on spindle-shaped swellings on branches of junipers of the *Sabina* group, and its teliospores, which are spindle-shaped, sickle-shaped or S-shaped in form, measure 16–24 × 46–90 μm (Kern, 1973). Its aecial stage is on *Cotoneaster*.

*G. dobroznakovii* is found in the USSR (in the Crimea and the Ukrainean Peninsula) and in Asia. It forms its brown, horn-like telia mostly on small twigs, but sometimes on the leaves and fruits of junipers of the *Oxycedrus* group. Its teliospores measure 23–26 × 38–57 μm, and its aecial stage is on pear (Kern, 1973).

*G. gracile* in Europe is confined to the Mediterranean region, and it occurs elsewhere in North Africa and in parts of Texas and Mexico in North America. It forms branch swellings and witches' brooms on junipers of the *Oxycedrus* and *Sabina* groups (most often on the former), and its orange, cylindrical telia form on the affected parts. The teliospores usually measure 15–20 × 45–90 μm. The aecial stage is on species of *Amelanchier, Crataegus* and *Cydonia* (Kern, 1973).

### ROOT AND BUTT ROTS

The root and butt rot fungus *Heterobasidion annosum* has been found on juniper plants in Scotland (Moore, 1959). This fungus is described in chapter 3.

## The incense cedars: *Libocedrus sensu lato*

Various species of *Libocedrus* (some of which are now sometimes separated and placed in *Austrocedrus* and *Calocedrus*) are grown in parks and gardens. *Libocedrus* spp. are known to be susceptible to attack by *Seiridium cardinale*, which is described in chapter 8.

## The dawn redwood: *Metasequoia glyptostroboides*

The dawn redwood, *Metasequoia glyptostroboides*, is a Chinese tree introduced into Britain in 1948. It is now fairly common in parks and gardens in Europe.

*Armillaria mellea sensu lato* (which is described in chapter 3) was found attacking a young tree in Scotland in 1956 (Moore, 1959), and killed another in Surrey in 1970. This tree has also been attacked by the root and butt rot fungus *Heterobasidion annosum* (Strouts *et al.*, 1983).

# Redwood: *Sequoia sempervirens* and Wellingtonia: *Sequoiadendron giganteum*

The redwood, *Sequoia sempervirens*, and the Wellingtonia or big tree, *Sequoiadendron giganteum*, both of which were introduced from the west coast of North America, are grown in Britain and other parts of Europe in parks and gardens.

## General

Wellingtonias are considered to be among the trees most often affected by lightning damage, which may destroy their tops (plate 2)

## Nursery diseases

### Grey mould: *Botrytis cinerea*
Nursery stock of both redwood and Wellingtonia is very susceptible to damage by grey mould (*Botrytis cinerea*), which is described in chapter 3.

## Post-nursery diseases

### NEEDLE DISORDERS

### Winter browning and drought damage
Redwood trees are very liable to a winter browning of the foliage caused by winter cold. Browning may also follow drought in hot dry summers.

### DIEBACKS

### Phomopsis disease caused by *Phacidiopycnis pseudotsugae*
According to Peace (1962), *Sequoiadendron giganteum* is one of the hosts of *Phacidiopycnis pseudotsugae* (the anamorph of *Phacidium coniferarum*, described in chapter 3), though this fungus has so far caused it little damage.

## BRANCH GALLS

**Branch galls caused by *Agrobacterium tumefasciens* (Smith & Townsend) Conn**

Large galls up to 20 cm across have sometimes been found on the branches of *Sequoia sempervirens*, both in Britain (Bull, 1951; Peace, 1962) and in France (Dufrenoy, 1922b) and Germany (Martin, 1957) (figure 51). There is evidence that at least in some cases the crown gall organism, *Agrobacterium tumefasciens*, is associated with these tumours. *A. tumefasciens* has a very wide host range, and is further described in chapter 3.

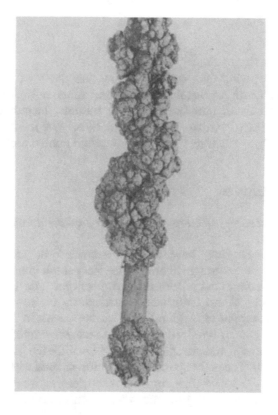

**Figure 51** Crown gall caused by *Agrobacterium tumefasciens* on *Sequoia sempervirens* (Forestry Commission)

## ROOT AND BUTT ROTS

The most important cause of damage to Wellingtonia is the honey fungus, *Armillaria mellea sensu lato*, which is described in chapter 3.

# Yew: *Taxus baccata*

Like Scots pine and the common juniper, the yew is one of Britain's few native conifers. It is widespread in England and Wales, but rare in Scotland. It is common also throughout much of the rest of Europe. Yew woodland and scrub occurs especially on chalk and limestone soils, and yew is also grown as an ornamental in churchyards, parks and gardens, and as a hedge plant.

## GENERAL

Winter cold and wind may cause bronzing and death of yew foliage, and yellowing and death of needles may follow summer droughts (Rose and Strouts, personal communication). Sooty moulds, including *Capnobotrys dingleyae* Hughes (described by Ellis and Ellis, 1985), often follow infestations of the yew scale insect, *Parthenolecanium pomeranicum* (Kawecki).

## NEEDLE DISEASES

### Needle casts caused by *Dothiora taxicola*, *Botryosphaeria foliorum* and *Diplodia taxi*

These three needle fungi have all been recorded on yew in Britain and elsewhere in Europe, though all are of only very small importance.

*Dothiora taxicola* (Peck) Barr (syn. *Sphaerulina taxi* (Cooke) Massee) causes a leaf scorch and twig blight, particularly on yew hedges (Anon., 1951). It is widespread in Scotland, and is also known in southern England (Moore, 1959; Dennis, 1968). Its black perithecia are embedded in the upper sides of the leaves, and measure 140–186 × 125–165 μm, with a short projecting ostiole. The eight-spored asci are club-shaped with a thickened tip, and measure 57–75 × 10–13.5 μm. The colourless, narrowly elliptical ascospores are 3- (sometimes 5-) septate when mature, and measure 20–37.5 × 6.5–9 μm.

The imperfect state is a multilocular, rounded or irregular pycnidium (*Cytospora taxicola* Cooke and Massee) measuring 418–435 × 243–352 μm, and containing colourless, rod-shaped conidia measuring 3–5 × 1 μm (Callen, 1938; Dennis, 1968).

*Botryosphaeria foliorum* (Sacc.) von Arx & E. Muller (*Physalospora gregaria* Sacc. var. *foliorum* Sacc.) was found in the Forth and Clyde area in Scotland by Callen (1938), and has also been recorded in Scotland, at Insch,

**Plate 1.** Spring frost damage

**Plate 2.** Lightning damage

**Plate 3.** Damage by fluorine

**Plate 4.** Grey mould (*Botrytis cinerea*)

*For fuller captions, see pp. xvi–xviii.*

**Plate 5.** Nitrogen deficiency

**Plate 6.** Phosphorus deficiency

**Plate 7.** Potassium deficiency

**Plate 8.** Healthy plant
(cf. plates 5, 6, 7)

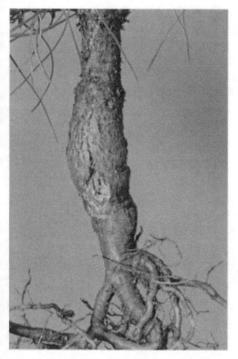

**Plate 9.** Damage by the herbicide chlorthiamid

**Plate 10.** Damage by the herbicide glyphosate

**Plate 11.** Wetwood and slime flux

**Plate 12.** Aecidia of *Peridermium pini*

**Plate 13.** *Thelephora terrestris*

**Plate 14.** Coral spot (*Nectria cinnabarina*)

**Plate 15.** Damage by *Phytophthora cactorum*

**Plate 16.** Top dying of Norway spruce

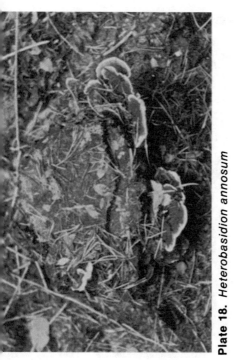

**Plate 17.** Fruit bodies of *Armillaria mellea*

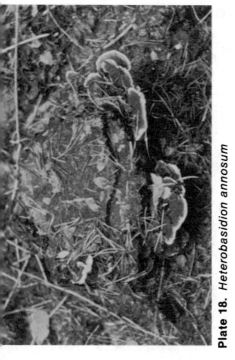

**Plate 18.** *Heterobasidion annosum*

**Plate 19.** *Chrysomyxa abietis*

**Plate 20.** Spermogonia and aecidia of
*Chrysomyxa rhododendri*

**Plate 21.** *Cucurbitaria piceae* on spruce

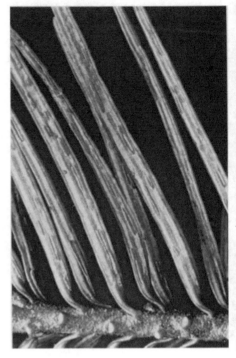

**Plate 22.** Perithecia of *Cucurbitaria piceae*

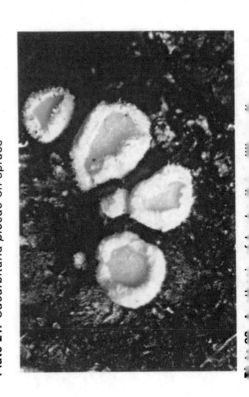

**Plate 26.** Tar spot (*Rhytisma acerinum*)

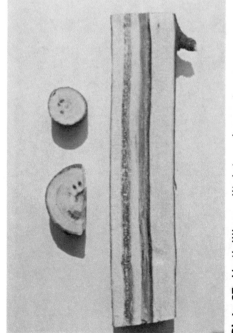

**Plate 28.** Poplar leaf blister

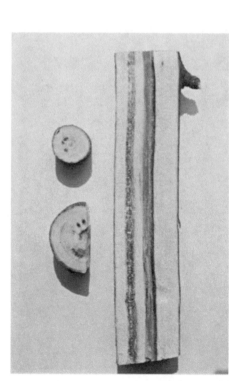

**Plate 25.** *Uncinula aceris*

**Plate 27.** Verticillium wilt: internal symptoms

**Plate 29.** Oak mildew

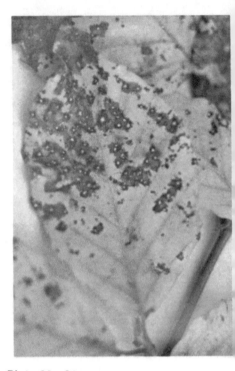

**Plate 30.** *Gloeosporium fagi:* leaf spots on beech

**Plate 31.** Bacterial canker of ash

**Plate 32.** Sooty bark disease of sycamore

**Plate 33.** Beech bark disease: 'tarry spots'

**Plate 34.** Beech bark disease: beech scale infestation

**Plate 35.** Beech bark disease: 'dimpling'

**Plate 36.** *Platychora ulmi* on elm

**Plate 37.** *Gremmeniella abietina*
on pine

**Plate 38.** *Seiridium cardinale:*
damage on *Cupressus*

**Plate 39.** Winter bronzing
of *Thuja plicata*

**Plate 40.** *Didymascella thujina*
on *Thuja plicata*

**Plate 41.** Symptoms
of Dutch elm disease

**Plate 42.** Dutch elm disease:
wilting of shoots

**Plate 43.** Anthracnose of London plane

**Plate 44.** *Sclerotinia laxa*
on *Prunus avium*

Plate 45. *Venturia tremulae* on *Populus*

Plate 46. *Melampsora larici-populina* on poplar leaf

Plate 47. *Marssonina brunnea* leaf spots on poplar

Plate 48. *Glomerella miyabeana* on willow

**Plate 49.** Poplar canker: wilting of foliage

**Plate 51.** Box damaged by *Pseudonectria rousseliana*

**Plate 50.** Poplar canker: association with agromyzid flies

**Plate 52.** *Podosphaera leucotricha* on apple bud

**Plate 53.** Watermark disease of willow: crown symptoms

**Plate 54.** Fireblight on *Sorbus aria*

**Plate 55.** Leaf curl on almond

**Plate 56.** Leaf blotch of horse chestnut

**Plate 57.** Watermark disease of willow: stain in wood

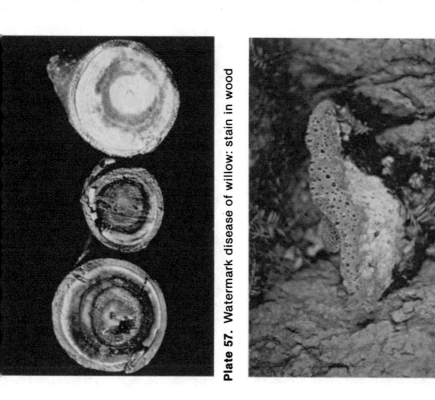

**Plate 58.** *Fomes fomentarius*

**Plate 59.** *Inonotus dryadeus*

**Plate 60.** *Inonotus hispidus*

**Plate 61.** *Ganoderma resinaceum*

**Plate 62.** *Phaeolus schweinitzii*

**Plate 63.** *Coriolus versicolor*

Plate 64. *Laetiporus sulphureus*

on golden yew (*Taxus baccata* var. *stricta*) by Moore (1959). The black, globose pseudothecia are embedded in the leaves, mostly in the upper sides, and measure 183–234 × 138–202 μm, with a small projecting ostiole. The eight-spored asci are club-shaped, and measure 52–90 × 13–25 μm. The colourless, one-celled, ovate-oblong or egg-shaped ascospores are narrower at one of the two rounded ends, and measure 15–23 × 7–12 μm.

The dark-brown pycnidia of the imperfect state (*Phyllostictina hysterella* (Sacc.) Petr.) are embedded in the leaf, mostly on the upper side, and are spherical or ovoid bodies measuring 185–208 × 140–179 μm, containing colourless, one-celled, egg-shaped or elliptical, round-ended conidia measuring 10–16 × 8–10 μm (Callen, 1938).

*Diplodia taxi* (Sow.) de Not. is widespread on the leaves of yew (Moore, 1959), and according to Grove (1937), in Germany it has been found on branches as well as leaves, and has there been known to kill trees. The black pycnidia are embedded in the leaves, and contain elliptical spores with rounded ends. The spores are at first one-celled and colourless, but finally become two-celled and smoky brown, when they measure 20–25 × 8–10 μm (Grove, 1937).

## BRANCH CANKER AND DIEBACK

### Canker and branch girdling of unknown cause
The cause of a sometimes extensive cankering and girdling of twigs and branches found in southern England on *Taxus baccata* and its cultivars is unknown. The symptoms suggest a parasitic bark-invading organism (Strouts, personal communication).

## ROOT DISEASES

### Phytophthora root disease
Phytophthora root disease is the commonest cause of death of yew in nurseries, and has also killed established ornamental trees. The species concerned is most often *Phytophthora cinnamomi*, but *P. cryptogea* and *P. citricola* have also been involved (Strouts, personal communication).

## DECAY AND HEART ROT

The commonest and most damaging heart rot fungus of standing yews is *Laetiporus sulphureus* (*Polyporus sulphureus*), which causes a slow, red-brown, cubical rot. Other, less common decay fungi on yews are *Ganoderma lucidum* and *G. resinaceum*. These three fungi enter via roots and through pruning wounds. Another decay fungus, *Amylostereum laevigatum*, has been found in yew hedges topped to reduce their height (Rose and Strouts, personal communication). These fungi are described in chapter 20. Yews seem to be resistant to attack by species of *Armillaria* (honey fungus).

# 10 Diseases of oak (*Quercus* spp.)

The two British oaks are *Quercus robur* and *Q. petraea,* the pedunculate and sessile oak, respectively. Both are widespread in woodland and in hedgerows, parks and gardens throughout the country. Hybrids between the two are common. The introduced species *Q. cerris* (the Turkey oak) and *Q. ilex* (the evergreen or holm oak) are both common and naturalised in many areas in the south of the country. *Q. robur* and *Q. petraea* also occur widely throughout the rest of Europe, where *Q. cerris*, *Q. ilex* and other oaks such as *Q. suber* (the cork oak) and *Q. pubescens* (the white oak) have a more restricted distribution.

Many other introduced oaks are grown in parks and gardens, though most are uncommon or rare. *Q. borealis*, the red oak, and *Q. coccinea*, the scarlet oak (both from North America) are frequent, however, though more so in the south than in the north. The Lucombe oak, one of the forms of *Q.* × *hispanica*, a hybrid between *Q. cerris* and *Q. suber* (the European cork oak) is also commonly planted in parks in southern England.

Oaks in Britain are generally relatively free of serious diseases. Their fungus diseases have been reviewed by Murray (1974).

## General

### Spring frost damage
In some seasons spring frosts defoliate oaks, which then usually produce a new crop of leaves. This second crop may itself then be especially susceptible to attack by powdery mildew (*Microsphaera alphitoides*), which is further described below under 'nursery diseases'.

### General oak dieback and decline
Oak forests in various parts of Europe (including Britain) (Day, 1927b; Osmaston, 1927b; Young, 1965; Marcu, 1966) and North America (Staley, 1965) have from time to time been affected by a dieback that has often killed many trees over large areas. Students of this phenomenon have all concluded that this dieback, which mainly affected *Q. robur*, was the result of a complex of factors varying in detail from time to time and place to place, but generally including drought, sometimes accentuated by drying winds (Young, 1965), together with attacks by various defoliating insects and a range of fungi,

particularly the oak mildew *Microsphaera alphitoides* and *Armillaria* spp. Oak mildew is further discussed below. *A. mellea* (chapter 3) is common on the roots of oak, and on its own may infest the roots for many years without causing serious damage.

In comparatively recent years, particularly since the 1950s, an oak decline with certain special features has received much attention in eastern Europe, in Bulgaria, Czechoslovakia, Poland, Romania, the USSR and Yugoslavia. In these countries a number of species of *Ophiostoma* (*sensu* Hoog, 1984, and Weijman and Hoog, 1975) have been isolated, and associated with a disease known as vascular mycosis or tracheomycosis of oak. The disease has been reviewed by Gibbs (1981), Delatour (1986b), Urosevic (1987), Oleksyn and Przybyl (1987), EPPO (1990) and others. It may take a chronic or an acute form. In chronic cases the leaves become yellowish in parts of the crown. They then turn brown and die. Some may remain hanging on the tree, but they gradually fall, and the tree dies over a period of up to ten years. Affected trees often develop many epicormic shoots. In acute cases, leaves on individual branches and then on the whole tree wilt suddenly and the tree soon dies, even the roots being affected, so that no coppice regrowth occurs (Oleksyn and Przybyl, 1987). If cuts are made into twigs and branches of diseased trees, they reveal staining in the sap wood similar to that found in elms affected by Dutch elm disease (Delatour, 1986b).

The most important of the *Ophiostoma* spp. involved appear to be *O. roboris* Georgescu and Teodoru and *O. kubanicum* Scherbin-Parfenenko, but others include *O. quercus* (Georgevitch) Nannf., *O. valachicum* Georgescu, Teodoru and Badea, and two fungi usually known as *Ceratostomella merolinensis* Georgevitch and *Ceratocystis longirostellata* Bakshi. These fungi have been rather briefly described, and have not been studied in any detail. Hence, they have become subject to some confusion, and there seems reason to believe that *O. roboris*, *O. quercus* and *Ceratocystis longirostellata* are the same as the common saprophyte *Ceratocystis piceae* (Munch) Bakshi (Gibbs, 1981; Delatour, 1986b; Oleksyn and Przybyl, 1987). More precise studies of the taxonomy and pathogenicity of these fungi are still required. Oleksyn and Przybyl (1987) conclude from their study that the current oak decline in the USSR (and apparently that in the other parts of eastern Europe) has been caused primarily by climatic factors (especially lengthy droughts) but intensified by insects and fungi, of which the most important are species of *Ophiostoma*.

According to Urosevic (1987), in Bohemia *O. valachicum* and *O. roboris* are seed-borne on acorns, and can destroy young seedlings in nurseries.

A study of a decline of *Quercus robur* in Romania indicated that mycoplasma-like organisms were involved (Ploaie *et al.*, 1987).

## Stag-headedness

Old, mature oaks on English clays often become stag-headed, as do others subjected to malpractices such as the covering of their roots with asphalt or concrete.

## Shakes

Oaks may also be subject to the internal cracking of the stem known as 'shakes'. Some of these ('star shakes') radiate outwards from the centre of the tree, others ('ring shakes') run round the tree along the annual rings, and they appear as serious defects in sawn timber. Sometimes, but by no means always, the presence of shakes may be indicated by ribs or grooves in the bark of the standing tree. The development of the shakes is partly associated with genetic factors, but site condition seems to be of greater importance, shakes occurring most often in crops grown in stony, sandy or gravelly soils (Henman and Denne, 1984, 1986, 1988). Work by Savill and Mather (1990) suggests that trees of *Q. robur* and *Q. petraea* with large earlywood vessels are more likely to develop shake than those with smaller xylem elements. They found further that trees with large vessels tend to flush later than the others. They indicate, therefore, that trees liable to shake may be identified and removed if late-flushing trees, which seem predisposed to the disorder, are marked for thinning at flushing time.

## Miscellaneous diseases

The shoot dieback caused by the bacterium *Pseudomonas syringae* pv. *syringae* may affect oaks. It is also found on many other hosts and is therefore considered in chapter 3.

Oaks in continental Europe are sometimes attacked by the chestnut blight fungus, *Cryphonectria parasitica*. This is of much greater importance on chestnut, and is therefore described in chapter 12.

# Nursery diseases

## GENERAL

Young oaks in nurseries are sometimes affected by Verticillium wilt, which is found on a wide range of hosts, and is therefore considered in chapter 3.

## LEAF DISEASES

### *Microsphaera alphitoides* Griffon and Maubl., and other powdery mildews

Several powdery mildews affect oaks in Europe, but the only one of wide-spread and major importance is *Microsphaera alphitoides*. This may be very damaging, particularly on nursery stock, on young trees not long planted out and on young coppice regrowth. The fungus occurs throughout Europe and across Asia as far as India (Srivastava and Kumari, 1983) and China (Yu and Lai, 1982), and in South Africa (Gorter, 1984).

#### *The fungus and the symptoms*

The conidial stage of a powdery mildew described as *Oidium quercinum* was first recorded in Portugal in 1877, but the first undoubted record of the fungus now generally called *Microsphaera alphitoides* was made in France in 1907. The fungus was at least almost unknown until the French report, after which it spread rapidly throughout Europe (Woodward *et al.*, 1929). The perfect stage was first described by Griffin and Maublanc (1912), who considered it to be a new species, to which they applied the name *M. alphitoides*. However, the fungus has been regarded by some (Foëx, 1941) as a form of *M. alni* DC ex Wint. or of *M. quercina* (Schw.) Burr. (which is itself considered by Gardner *et al.*, 1972, to be synonymous with *M. alni*).

*M. alphitoides* attacks the young leaves and sometimes the young shoots of the plants, and appears first from May onwards as small cinnamon-coloured spots. These increase in size, and soon the whole leaf may be covered by a layer of white mycelium and a thick white powder of conidia (figure 52). In severe attacks the leaves shrivel, become brown and fall. Occasionally (usually after hot, dry summers), cleistothecia of the perfect stage may be found in the autumn, mainly on the upper sides of the leaves. They are yellow at first, but when mature become dark brown or black (plate 29). In Britain they were first found in 1945 (Robertson and Macfarlane, 1946; Batko, 1962).

The conidia are borne singly on three-celled conidiophores, and are ellipti-cal or barrel-shaped, measuring 25–37 × 15–22 μm. The cleistothecia measure about 180–200 μm across. They contain several asci, and bear up to about 20 appendages, the ends of which carry several repeatedly dichotomous branches. The colourless, ellipsoid ascospores, of which there are up to eight in each ascus, measure 18–24 × 6–13 μm (Robertson and Macfarlane, 1946). The development of the cleistothecia has been described by Turnau and Czerwonka (1988).

#### *Host range and host susceptibility*

*M. alphitoides* is found occasionally on beech (*Fagus sylvatica*), and (on the European mainland) on the sweet chestnut (*Castanea sativa*) as well as on oak. It is an important pathogen only on the latter, however. Of the oaks, *Q. robur* is especially susceptible; under field conditions *Q. petraea* is less so, though it may be badly affected in the nursery (Anon., 1956b). When young, the leaves of both *Q. robur* and *Q. petraea* are very susceptible, but they

**Figure 52** Leaves of young oak covered by the white mycelium and white powdery spore masses of the oak mildew *Microsphaera alphitoides* (D.H. Phillips)

become more resistant with age (Edwards and Ayres, 1982). In Italy, Luisi and Grasso (1973) listed a number of oaks in decreasing order of susceptibility as follows: *Q. pedunculata* (*Q. robur*), *Q. pubescens*, *Q. frainetto*, *Q. cerris*, *Q. ilex*, *Q. suber*, *Q. coccifera*. Edwards and Ayres (1981) found that *Q. robur* was highly susceptible until mature, while *Q. cerris* and *Q. borealis* were, respectively, moderately and highly resistant. *Q. cerris* and *Q. castaneifolia* both showed resistance in work by Grigor'ev (1984) in the Crimea, where *Q. longipes* and *Q. erucifolia* were susceptible. In Bulgaria, Mirchev and Alam (1985) found that *Q. robur* and *Q. hartwissiana* were susceptible and *Q. cerris* and *Q. rubra* were resistant, while *Q. petraea* and *Q. macranthera* were intermediate in susceptibility. Savvin (1984) also found that *Q. rubra* was resistant.

Provenance trials with *Q. robur* and *Q. petraea* have shown that different provenances may vary somewhat in resistance to the mildew, but that these differences are too small to have much practical importance (Leibundgut, 1969; Rack, 1957a).

*Overwintering of the fungus*
The conidia spread the disease rapidly throughout the summer, but are short-lived. Overwintering is by means of mycelium in a small percentage of the buds (Woodward *et al.*, 1929; Kerling, 1966). Petri (1923) found evidence that overwintering in Italy was by means of chlamydospores, but this has not been demonstrated elsewhere. Likewise so far there seems no evidence that the cleistothecia play any part in overwintering.

*Factors affecting mildew*
Hewitt (1974) found that the oidia germinated best at temperatures between 20 and 30 °C, at a relative humidity between 76 and 96 per cent. In the USSR the disease is severe and spreads rapidly with much leaf infection when warm weather in May, with a near normal relative humidity and rainfall greater than 35 mm, is followed by a cool, moderately wet July (Basova, 1987). Many authors have suggested that there is an association between hot, dry summers and the production of the cleistothecia (Batko, 1962), though Dobbs (1951) found the fruit bodies in North Wales after one notably cool, wet summer.

   Observations (supported in part by experiment) have suggested that ultraviolet light at high elevations (Bolli, 1943) and gaseous sulphur compounds in industrial areas (Köck, 1935) may play some part in mildew control.

   The parasitic fungus *Cicinnobolus cesati* De By. may to some extent suppress mildew attacks in late summer (Woodward *et al.*, 1929). The parasitic fungus *Septoria quercina* may also suppress the development of *M. alphitoides* (Savvin, 1984).

*Importance of mildew*
In the nursery if mildew is left uncontrolled it may kill young seedlings or by the destruction of shoots reduce the young tree to a bush-like shape. Nimmo (unpublished), from the results of spray trials on newly planted stock, concluded that the disease could reduce height growth by 30 per cent over three years, and reduce the number of straight, single leaders by about the same percentage. Viney (1970) considered that in France and Belgium mildew was the main reason for the lack of natural regeneration of stands of *Q. robur*.

   On older trees mildew generally causes less injury, though at times, particularly in the earlier part of this century, it has formed part of a complex involving also drought and frost, and various insects and fungi, which has given rise to dieback and death of many trees over large areas. This complex is briefly considered above under 'general oak dieback and decline'.

*Control of oak mildew*
Viney (1970) has suggested that in France and Belgium it may be necessary to replace stands of *Quercus robur* with the less susceptible *Q. petraea* where mildew is troublesome. *Q. petraea* in any case is in other respects a more satisfactory forest tree than *Q. robur*. This is not always easy, however, as many seed lots contain a mixture of the two species and the hybrid between them. In appropriate situations, however, it may be possible to grow some of

the more resistant species mentioned above under 'host range and host susceptibility'.

Other methods of control can usually be applied only in the nursery. Under conditions in Eastern Europe it has been found that shading seedlings with oats and sunflowers or lupins protects them from mildew (Kartavenko, 1958; Orlos, 1951).

Control in the nursery is generally achieved by the use of fungicides. Bordeaux mixture and other copper sprays have been found to be effective, but generally the best control has been obtained with colloidal and wettable sulphur and dinocap (Woodward *et al.*, 1929; Müller-Kögler, 1954; Rack, 1957b; Jagielski, 1969; Mamedov, 1974). Guseinov (1976) and Marchetti and D'Aulerio (1980) obtained good results with benomyl, while Boudier (1987a) gained good control with flutriafol (preferably preceded by a sulphur spray in heavily infested nurseries). Beyond the nursery stage, Lesovskii and Martyshechkina (1985) in the Ukraine recommend the aerial spraying of medium-aged oak stands with colloidal sulphur, which greatly reduced damage to the trees.

Spraying should be begun at the first signs of infection, and to keep the disease under control it may be necessary to spray several times at intervals of two to three weeks. It might be possible to improve the effectiveness of spray programmes if attacks of oak mildew could be accurately forecast. Work on such forecasting has been done in Bulgaria (Minkevich and Stoyanov, 1987; Shirnina, 1987).

### Other powdery mildews on oak

A number of other powdery mildews have been recorded on oaks in Europe. They include *Microsphaera hypophylla* Nevodovskij. This fungus has been observed in Norway, Sweden, Austria, Switzerland and the USSR (Roll-Hansen, 1966; Luisi and Grasso, 1973), but studies by Zahorovska (1985) indicate that it is the same as *M. alphitoides*.

Another mildew, *Phyllactina guttata*, which is much better known on hazel and ash, has also been recorded on oak (Batko, 1962), on which, however, it causes no significant damage. It is described under 'hazel' in chapter 19.

## DISEASES OF BARK AND SHOOTS

### Bark canker and dieback caused by *Fusicoccum quercus* Oudem. (*F. noxium* Ruhl.)

*Fusicoccum quercus* occurs on oaks in Britain, the Netherlands and Germany. It is most often a saprophyte on dead branches, but especially in Germany, where it has been studied and reviewed by Butin (1981, 1983), it may be an important cause of bark canker and dieback. It occurs mainly on nursery plants of from two to three years of age, but may affect young forest trees up to the pole stage.

In spring and summer the fungus forms a 'summerform' with

*Phomopsis*-like pycnidia up to 1.5 mm across. Later, in autumn, it forms a larger 'winterform' with *Fusicoccum*-like pycnidia in multilocular stromata up to 3 mm in diameter. Both forms produce colourless spindle-shaped to elliptical A-spores mostly measuring 12–14 × 3.5–4.5 μm and colourless short-elliptic to cylindric B-spores mostly measuring 5.5–6.5 × 2.5–3.0 μm.

Infection first becomes visible in the spring, when elliptical, reddish-yellow discoloured areas appear on the bark, usually round a dead bud or dead lateral branch. The small, pimple-like pycnidia of the 'summerform' and later the larger stromata of the 'winterform' appear on the lesions, which become somewhat sunken cankers separated from healthy tissues by a barrier of callus. Commonly the diseased tissues remain thus sealed off, but sometimes in the following growing season the fungus breaks through the callus layer and extends the canker area, causing dieback, and sometimes the death of young plants.

The disease seems to have been found mainly on *Quercus robur* and the red oak *Q. borealis*, but it has also been known to affect the burr oak, *Q. macrocarpa*, and the chestnut oak, *Q. prinus*.

Infection by *Fusicoccum quercus* appears to occur only in certain years preceded by one or two seasons with droughts, and it is associated with sandy soils which readily dry out.

In nurseries the disease has been known to kill young trees or through dieback induce a stunted, bushy growth.

*Fusicoccum* canker is best controlled by avoiding dry, sandy soils when planting or raising oaks.


## ROOT DISEASES

**Root rot caused by *Rosellinia quercina* Hartig and *R. thelena* (Fr.) Rabenh.**
Two species of *Rosellinia*, *R. quercina* and *R. thelena*, occur widely in Europe in association with a root rot which occasionally destroys patches of oak seedlings in nurseries.

*Rosellinia quercina* has been found occasionally on *Quercus robur*, *Q. petraea* and *Q. cerris* in all parts of Britain, causing a root rot of young seedling oak in nurseries. It appears to be commoner and more damaging on the European mainland (Woeste, 1956; Urosevic and Jancarik, 1957). It has been studied in Britain by Waldie (1930) and in Germany by Hartig (1894) and Woeste (1956). The fungus now found in Britain, however, appears to be *R. thelena* (Francis, quoted by Cannon *et al.*, 1985).

*Symptoms and development of the disease*
It usually shows as spreading patches of seedlings which yellow downwards from the tops. Fine hyphae growing through the soil in spring, and arising directly from resting mycelium, from small black sclerotia or from germinating ascospores, enter the young roots mainly through lenticels. The roots (especially the tap roots) are then rotted (figure 53). The mycelium develops

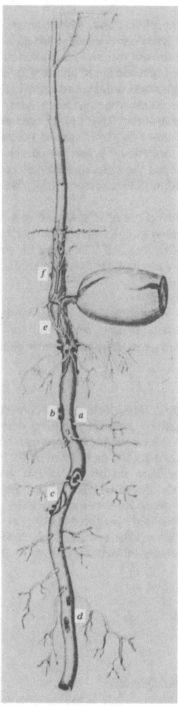

**Figure 53** Seedling oak attacked by *Rosellinia quercina*. Infection tubercles can be seen at a and b, and the cortical tissue around them has been destroyed. Near the acorn, strands of the fungus ramify over the surface of the plant, at e (R. Hartig)

into anastomosing rhizomorphic strands which form a dense mass. This is at first white but finally becomes dark brown or black, and spreads under and over the bark, over the root surface, and on the soil surface around the root collar. Small verticillately branched conidiophores form on the mycelium and produce small ovoid, colourless conidia, and small black sclerotia about the size of a pin's head arise on the root surfaces. Later, in September, short-necked black perithecia measuring 0.8–1 mm across develop in the mycelium on the plant stems just above the soil, and on the soil surface itself. The perithecia contain asci separated by septate paraphyses, and each ascus contains eight spindle-shaped, brown, one-celled ascospores each measuring 17–32 × 7.5–10 μm (Waldie, 1930; Brooks, 1953). These ascospores are shed in the following spring.

Like *R. quercina*, the mycelium of *R. thelena* forms dense brown mats. On these it produces colourless conidia and black perithecia. Its dark-brown ascospores measure 20–26 × 7.5–8.5 μm and at each end bear slender colourless appendages 5–9 μm long (Ellis and Ellis, 1985).

### Factors affecting Rosellinia root rot
The disease appears to be favoured by damp, cool conditions and poor drainage. The sclerotia may persist in the soil for a long time, though infection of a nursery is not necessarily followed by persistent attacks (Peace, 1962).

### Control
As the disease is so sporadic, little work has been done on its control. Waldie (1930) reported that he gained control by burning the soil to a depth of 6 in (approximately 15 cm) with a powerful blowlamp, and then watering with a 5 per cent solution of carbolic acid. Other, more recent, soil sterilants such as dazomet and metham sodium might be tried if necessary. On general grounds, diseased plants should not be transplanted, but should be destroyed. To avoid the transfer of sclerotia, apparently healthy plants from the same beds should not be transplanted elsewhere in the same nursery or moved to other nurseries. As the disease does not attack plants more than two or three years old, however, they may be planted out in the forest or in parks and gardens.

## Post-nursery diseases

### LEAF DISEASES

### Oak mildew: *Microsphaera alphitoides*
The oak mildew, *Microsphaera alphitoides*, may occur on trees of any age, and in woodland it is most important on young trees not long planted, and on coppice shoots. It is most important, however, on nursery stock and therefore is covered above under 'nursery diseases'.

**Leaf blister caused by *Taphrina caerulescens* (Mont. & Desm.) Tul.**

*Taphrina caerulescens* has been found occasionally in Great Britain and Ireland (Henderson, 1954; Brooks, 1953), across the mainland of Europe from France to the Caucasus, and in North Africa and Japan (Viennot-Bourgin, 1949) and in North America (Goode, 1953). In most of these areas it is of only minor significance, but in the southern states of the USA it has caused occasional epidemics following cool, wet springs. Some trees have then been killed by repeated loss of leaves, while the growth of others has been reduced (Wedgeworth, 1926; Goode, 1953).

The fungus causes slightly swollen, pale-green, yellowish or pinkish, rounded or irregular spots up to 12 mm across and concave below. A layer of asci forms in the concave surface, which, hence, becomes silvery grey and later velvety brown. The blisters may cause curling and twisting of the leaves of narrow-leaved oak species (Weber, 1941; Viennot-Bourgin, 1949).

The ascospores produce primary conidia as soon as they are formed, and these conidia themselves bud off secondary spores, so that the asci quickly fill with small 'sprout cells' (Brooks, 1953; Goode and Parris, 1954; Henderson, 1954). The asci vary in size and form with the oak species on which the fungus is growing. They are further described by Mix (1949).

The conidia infect the young leaves as the buds open. Older leaves are not susceptible (Goode and Parris, 1954).

In Great Britain *T. caerulescens* seems to have been recorded only on *Quercus petraea* and the introduced *Q. borealis* (*Q. rubra*) (Henderson, 1954), while in Ireland it has been found on *Q. robur* (Brooks, 1953). In the USA Weber (1941) noted it on almost all the many North American oak species, with *Q. borealis* and *Q. nigra* being especially susceptible, but *Q. phellos* showing resistance to the disease. Also in the USA, Goode and Parris (1954) found almost all the oaks they studied were severely damaged by *T. caerulescens*, only *Q. stellata*, *Q. palustris* and *Q. shumardii* being resistant.

*Control of the disease*
Control measures against leaf blister are necessary only in the southern states of the USA, where spraying the dormant buds before they swell and burst may be required to kill overwintering spores. Good control has been obtained with one application of Bordeaux mixture, zineb, maneb, manzate, captan and several other fungicides (Goode, 1953).

**Leaf spot caused by *Ciborinia candolleana* (Lév.) Whetzel**
**(*Sclerotinia candolleana* (Lév.) Fuckel)**
*Ciborinia candolleana*, which is common and usually regarded as no more than a saprophyte, has sometimes been found in Britain and elsewhere in Europe associated with a leaf blotch on *Quercus robur*, *Q. petraea* and *Q. borealis* (*Q. rubra*). In the summer of 1925 and in 1926 it gave rise to·an epidemic all over Scotland south of Inverness, and over much of England and Wales (Wilson and Waldie, 1927).

From July onwards, affected oak leaves develop circular brown spots, which sometimes spread over the whole leaf. Small, black, hemispherical or lens-shaped sclerotia up to 3 mm across are produced over the winter on the fallen leaves. From these sclerotia stalked apothecia arise in the following summer from May to July. The slender apothecial stalk grows until the apothecium is clear of the litter surface, and so its length varies with the depth of the sclerotium in the litter. The yellowish-brown apothecial disc is up to 4 mm across, and is flat when mature. The outer wall of the apothecium is smooth. The eight-spored asci measure up to $100 \times 7$ μm, and narrow paraphyses occur between them. The ascospores are colourless, more or less elliptical, and measure $6–10 \times 3–4$ μm (Dennis, 1968).

The same fungus is sometimes said to occur on *Castanea* (Brooks, 1953), though Dennis (1968) states that the fungus on sweet chestnut is *C. hirtella* (Boud.) Batra and Korf (*Sclerotinia hirtella* Boud.), which has long hairs on the stalk and cup of the apothecium.

### Leaf spot caused by *Mycosphaerella punctiformis*

*Phyllosticta maculiformis*, an anamorph of *Mycosphaerella punctiformis*, has been found from time to time causing yellow-brown spots on fading and dead leaves (and once on living leaves) of *Quercus robur* and *Q. cerris* in England and Scotland (Moore, 1959; Ellis and Ellis, 1985). The fungus also occurs in other parts of Europe, and in North Africa (Grove, 1935). It is of negligible importance on oak, but is better known and rather more important on the European mainland on sweet chestnut, under which it is therefore described in chapter 12.

### The oak rust: *Uredo quercus* Duby

The rare and inconspicuous leaf rust, *Uredo quercus*, has been found a few times in southern England on *Quercus petraea*, as well as on *Q. ilex* in Devon and Guernsey (Moore, 1959). It occurs sporadically in other parts of Europe, most often in the Mediterranean area.

The yellow uredosori appear as pustules on the undersides of the leaves, especially of sucker shoots (Ellis and Ellis, 1985), and the orange-yellow, obovoid or broadly ellipsoid urediniospores measure $15–25 \times 10–17$ μm; they are covered by minute echinulations. No other stage is known in Britain, though in France, where the rust is also known, telia have been found (Wilson and Henderson, 1966).

## DISEASES OF LEAVES AND SHOOTS

### Leaf spot and dieback caused by *Apiognomonia errabunda* (Roberge) Höhnel *sensu lato* (anamorph *Discula umbrinella* (Berk. & Broome) B. Sutton, syn. *Gloeosporium umbrinellum* Berk. & Broome)

*Apiognomonia errabunda* is generally regarded as a complex of species, formerly included in *Gnomonia* but transferred to *Apiognomonia* by Barr (1978). These species cause leaf spot and dieback (sometimes called anthracnose) on various broadleaved trees in Europe and North America. One, *A. quercina* (Kleb.) Höhnel (*Gnomonia quercina* Kleb.), the anamorph of which is *Discula quercina* (Westend.) von Arx, attacks oak, and sometimes beech (Monod, 1983) and hornbeam.

The *Discula* stage of *A. quercina* causes a widespread leaf spot of oaks, and sometimes (especially in North America) a shoot dieback (Grove, 1937; Neely and Himelick, 1967). On oaks in Europe the fungus usually produces irregular or angular, brownish-black leaf spots with a brown margin, and measuring up to about 15 mm across (Grove, 1937). In most years the disease is not especially damaging, but in the northern parts of Britain, in the summer of 1980, when the weather was exceptionally wet, it was so severe that many oak trees appeared withered (Redfern *et al.*, 1981). In North America, symptoms on the very numerous oaks vary with the species, the leaf spots on some species being light brown with a yellow border (Parris and Byrd, 1962). In severe attacks, the tips of the leaves may die and wither, and the tissues curl and break into irregular shot-holes. Sometimes the disease may spread to the shoot tips, which then curl, shrivel and die.

Rounded, brown conidial pustules, about 250 μm across, form on the spots on both sides of the leaves. The one-celled, more or less colourless, oblong-spindle-shaped spores measure 10–15 × 3–5 μm (Grove, 1937). The perithecia of the perfect stage develop on the fallen leaves, and can be found in the spring following leaf fall. The colourless ascospores are unequally two-celled and mostly measure 13–15 × 3–4 μm (Barr, 1978).

The disease appears to be favoured by cool, wet weather (Neely and Himelick, 1967; Shishkina, 1969), rainfall being the most important single factor (Fergus, 1953).

## DISEASES OF BARK AND SHOOTS

**Dieback caused by *Caudospora taleola* (Fr.) Starb. (*Diaporthe taleola* (Fr.) Sacc., *Hercospora taleola* (Fr.) E. Muller) and *Diaporthe leiphaemia* (Fr.) Sacc. (*Amphiporthe leiphaemia* (Fr.) Butin)**

*Caudospora taleola* and *Diaporthe leiphaemia* (figures 54 and 55) are both common saprophytes on small branches of oak in both Britain and the rest of Europe and in parts of North America (Wehmeyer, 1932; Reid and Cain, 1960). Occasionally, on trees weakened by drought or other causes they may become minor parasites, giving rise to bark cankers, and even to the death of small trees (Moore, 1959) (figures 54, 55).

The perithecia of *C. taleola* are embedded in a whitish stroma under the bark. The stroma is surrounded by a distinct black zone which is often visible on the bark surface (Munk, 1957) or may be seen if the outer bark is cut away. The asci contain eight (or sometimes four) ascospores, which are colourless, two-celled and constricted at the septum and measure 17–25 × 7–9.5 μm. At first the ascospores bear colourless, cylindrical appendages at the apex and

**Figure 54**  Cankers caused by *Caudospora taleola* (*Diaporthe taleola*) on an oak stem
(Forestry Commission)

radiating from around the septum, but these growths soon disappear (Ellis
and Ellis, 1985). Various conidial stages also occur (Wehmeyer, 1932).

The stroma of *D. leiphaemia*, which is also embedded under the bark
surface, is at first orange-yellow or brown, but may later blacken. The
perithecia develop in the stroma. The colourless, two-celled ascospores are
rather spindle-shaped to ellipsoid, slightly curved or inequilateral, and constricted at the septum. They measure 15–20 × 2.5–5.5 μm. Pycnidial locules of
its anamorph *Phomopsis quercella* (Sacc. & Roum.) Died. (*P. quercina* (Sacc.)
Höhn.) may occur round the edges of the stroma; they contain colourless,
spindle-shaped A-spores measuring 8–12 × 2–3 μm, and cylindrical, rather
curved B-spores measuring 20–30 × 1.5–2 μm (Ellis and Ellis, 1985).

In Czechoslovakia, *P. quercella* has been found damaging acorns, and also
attacking young oak seedlings at the root collar, and on older oaks it has
caused cankers associated with a brownish dieback and a wilt (Urosevic,
1987).

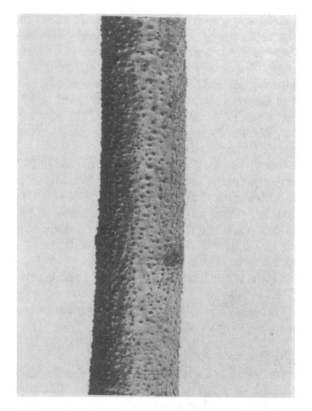

**Figure 55**   Stromata of *Diaporthe leiphaemia* on an oak stem (Forestry Commission)

## Dieback caused by *Colpoma quercinum* (Pers.) Wallr.
### (*Clithris quercina* (Pers.) Rehm)

*Colpoma quercinum* is a common fungus on dead twigs and branches of oak trees, and at times it may cause a minor dieback (Twyman, 1946). It has been said to be a wound parasite, often following insect damage (Brooks, 1953). Its dark, narrow, leathery ascocarps develop below the bark, and at maturity split open by a longitudinal slit to reveal a light-yellow hymenium measuring up to 15 × 2 mm. The club-shaped asci each contain eight thread-like, colourless ascospores that measure 80–95 × 1.5 μm and finally become multiseptate. Slender paraphyses with curled tips occur between the asci. The ascospores are found in May, June and July (Dennis, 1968). An anamorphic stage, *Conostroma didymum* (Fautr. & Roum.) Moesz also occurs (Cannon *et al.*, 1985). Its colourless conidia measure 5–7 × 1.5 μm (Ellis and Ellis, 1985).

## Canker and dieback caused by *Botryosphaeria stevensii* Shoemaker
### (*Physalospora mutila* N.E. Stevens: anamorph *Diplodia mutila* (Fr.) Mont.)

*Botryosphaeria stevensii* is known in Britain and other parts of Europe and in

North America as a saprophyte on various woody plants (Sutton, 1980). In some places it causes cankers of apples, pears and grapevines.

In North America it has been found as the cause of a canker and dieback of oak. In Europe, Vajna (1986) observed it attacking the sessile oak in Hungary, while Ragazzi and Mesturino (1987) recorded it in Italy as the cause of a dieback of *Quercus cerris* and *Q. robur*. In northeast Spain it has been known at least since 1934, causing canker, dieback and wilting of the cork oak, *Q. suber*, isolates from which have been shown experimentally to be highly pathogenic (Luque and Girbal, 1988).

The latter found that the *Diplodia* stage of the fungus formed unilocular and multilocular pycnidia in a stroma. The pycnidia contained ellipsoid macroconidia which were at first colourless and one-celled but finally brown and two-celled, measuring 25–39 × 12.5–16 μm; colourless, ellipsoid, one-celled microconidia (first found by Vajna, 1986) were sometimes also present, and measured 3.9–5.9 × 0.7–1.5 μm. The perithecia of the fungus also sometimes developed. Their ascospores were ellipsoid, unicellular and colourless, but some of them later darkened and became two- or three-celled, measuring 22–35 × 9–18 μm.

Luque and Girbal (1988) considered that *B. stevensii* entered cork oak trees following harvesting of the bark.

### Bark damage caused by *Bulgaria inquinans*
The ascomycete *Bulgaria inquinans* often produces its black, rubbery fruit bodies on the bark of felled or fallen trees. It has also been found causing large lesions on the bark of standing trees of the scarlet oak (Strouts *et al.*, 1986).

### Cankers associated with *Stereum rugosum*
Liese (1930) described a canker on oak associated with the basidiomycete *Stereum rugosum*, and similar cankers on *Quercus robur* in Scotland, apparently caused by the same fungus, were found by Banerjee (1956). More recently *S. rugosum* has been isolated from large perennating cankers on the stems of 25-year-old red oak plants in southern England (Strouts *et al.*, 1982). *S. rugosum* is described and further discussed in chapter 20.

### Bark canker and dieback caused by *Fusicoccum quercus*
*Fusicoccum quercus* may attack young oaks up to the pole stage, but is of most importance on nursery plants. It is therefore considered above under 'nursery diseases'.


## DISEASES OF ROOT AND BUTT

### Phytophthora root disease ('ink disease') caused by *Phytophthora cinnamomi*
Species of *Phytophthora*, which are treated generally in chapter 3, may

sometimes destroy seedling oaks in nurseries. On older trees, *P. cinnamomi* and *P. cambivora* are well known as causes of the Phytophthora root rot of sweet chestnut often called 'ink disease'. This is described in chapter 12. In southwest France, however, between Bayonne and the Spanish border, *P. cinnamomi* causes an ink disease of the red oak, *Quercus rubra*, and occasionally *Q. robur*, *Q. pyrenaica* and their hybrids. It affects trees from a height of 1 to 2 m upwards, and differs somewhat from ink disease of the sweet chestnut. Information on the disease on oaks has been reviewed by Delatour (1986a).

On oak the damage shows mainly at the base of the trunk, where a black discharge oozes from small cracks in the bark. Within, the trunk shows a black, inky stain, which may extend upwards from 1 to 2 m. Healing takes place from the cambium on the edges of the disease area, and alternates with further extension by the fungus, so that many islands of damaged tissue appear in the wood, and the lower part of the butt is rendered useless. Wood-destroying fungi often follow, and increase the damage caused. Some damage and discoloration of the wood may also be found in the roots of the affected trees, which may also develop many epicormic shoots on their trunks.

The disease seems to be encouraged by waterlogging of the soil. *P. cinnamomi* grows most rapidly at relatively high temperatures, its optimum temperature for growth *in vitro* being between 24 and 27 °C. It is discouraged by cold winters. The climate in the Bayonne area is warm, with mild winters, and since 1948, when 'ink disease' was first found on oaks in southwest France, little or no spread seems to have occurred. This may be because other oak wood areas in France have a less equable climate, with colder winters.

The disease is important because red oak is being increasingly grown in the area concerned, and 'ink disease' leads to the loss of the most valuable part of the butt. Often only a few scattered trees are affected, but in some stands up to 40 per cent of the trees may be diseased.

It may in time be possible to combat the disease by the selection of resistant clones. Meanwhile oak should not be planted in areas liable to waterlogging unless drainage work can be carried out.

## DECAY FUNGI

Many rot fungi have been recorded on standing oaks. The most damaging appear to be *Inonotus dryadeus* (which causes a soft white butt rot, and which is important as a cause of root decay leading to windthrow or dieback and death), *Laetiporus sulphureus* (the cause of a brown cubical butt rot), *Ganoderma applanatum*, *G. lucidum*, *G. pfeifferi* and *G. resinaceum*, and *Meripilus giganteus* (another white rot fungus of the roots and the butt). Others include *Bjerkandera adusta*, *Fistulina hepatica* (the beefsteak fungus), *Inonotus hispidus*, *Phellinus robustus*, *Polyporus squamosus*, *Stereum gausapatum*, *S. rugosum*, *S. hirsutum* and *S. frustulatum*.

The agaric *Collybia fusipes*, which is usually at most only a weak parasite, has sometimes been known to cause an extensive rot in oak roots. These decay fungi are described and further discussed in chapter 20.

The honey fungus *Armillaria mellea* may attack oak, and the weak parasite *A. bulbosa* may sometimes attack oak roots. The *Armillaria* species are considered in chapter 3.

## VIRUS AND VIRUS-LIKE DISEASES

Oaks in Germany, Hungary and Czechoslovakia may be infected by the tobacco mosaic virus (TMV), which may remain latent or produce chlorotic flecking, or star-shaped leaf lesions and leaf distortion, or witches' broom-like symptoms with mosaic-like symptoms on the leaves (Cooper, 1979; Nienhaus, 1985). The species usually affected have been *Quercus robur*, *Q. petraea* and *Q. cerris*. Trees with some of these symptoms have also been seen in southern England and Sweden (Cooper, 1979). Mosaic-like symptoms have also been seen on the red oak *Q. borealis* (*Q. rubra*) in Denmark (Kristensen, 1963).

Nienhaus (1985) has described a ring-speckling of leaves of *Q. petraea* in the Rhineland area, associated with spherical particles. Ploaie *et al.* (1987) have also described a disease of *Q. robur* and *Q. petraea* in Romania, in which the annual shoots were stunted, and the leaves were small, and wilting occurred. They found round or oval mycoplasma-like bodies in the phloem cells of the mid-ribs and shoots.

## DECAY IN ACORNS

Acorns may be decayed by *Ciboria batschiana* (Zopf) Buchwald (*Sclerotinia pseudotuberosa* Rehm, *Stromatinia pseudotuberosa* (Rehm) Boud.). This fungus produces its stalked brown apothecia on the fallen acorns, and its ovoid, one-celled, colourless ascospores measure about 7–10 × 4–5 μm (Ellis and Ellis, 1985). In eastern Europe, acorns may also be damaged by *Discula quercina* (the anamorph of *Apiognomonia quercina*).

# 11 Diseases of beech (*Fagus sylvatica*)

The common or European beech, *Fagus sylvatica*, grows widely throughout Europe. It still makes up about 5 per cent of the forest area in Britain, where among the hardwoods it is exceeded in area only by oak. It is one of the few timber trees that grows successfully on chalk and limestone soils. Various copper, cut-leaved, and weeping varieties are common in parks and gardens, and beech is also used for hedging.

## General

### Spring frost damage
Beech leaves are sometimes destroyed by spring frosts. Thus frosts in early and mid-May 1984 caused widespread and severe damage to newly flushed beech shoots in southern Scotland and northern England and in Wales (Strouts *et al.*, 1985; Redfern *et al.*, 1985).

### Drought damage
Beech trees are among those liable to damage by drought. Thus the hot, dry summers of 1975 and 1976 were followed throughout Britain by many cases of fluxing and necrotic patches on beech trunks, and many beech trees eventually died (Young and Strouts, 1977, 1978; Redfern *et al.*, 1979). Similarly, a very dry spring and summer in 1984 led to drought damage to beech (especially dieback of hedgerows) in parts of Scotland (Redfern *et al.*, 1985).

### Damage by sea winds
The foliage of trees near the coast may also be scorched by salt-laden gales.

### Sunscorch
Like other thin-barked trees, beeches may be affected by sunscorch, which causes the death of areas of bark on the south-facing sides of the trunks.

Further information on various forms of climatic damage is given in chapter 2.

# Nursery diseases

## SEEDLING BLIGHT AND DAMPING OFF

### Beech seedling blight caused by *Phytophthora cactorum*

Beech seedling blight is caused by a fungus now usually considered to be *Phytophthora cactorum* (Lebert and Cohn) Schröter, though it has also been described as *P. fagi* Hartig and *P. omnivorum* de Bary (Fitzpatrick, 1930). It is soil-borne, and may give rise to both pre- and post-emergence damping off (Hartig, 1900). Characteristically, in its post-emergence phase, it causes dark green lesions on the stems of young seedlings, and large blotches on their cotyledons and first leaves (figure 56). In damp weather it spreads rapidly, and the affected plants become brown and shrivelled. According to Hartig, the fungus may persist as oospores in the soil for at least four years.

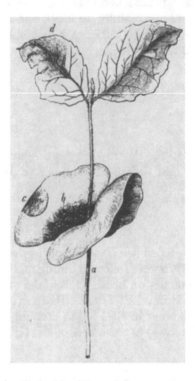

**Figure 56**   Beech seedling attacked by *Phytophthora cactorum*. The cotyledons at b and c and the first foliar leaves at d show blotches caused by the fungus (R. Hartig)

Attack on young seedlings takes place in spring following germination of the oospores, but older, more woody plants are not affected. The disease is encouraged by poor drainage, so good drainage of nursery beds is therefore important. The fungus may also enter nurseries through infected water supplies (Wardlaw and Palzer, 1985). If the disease appears in seedbeds, the plants may be sprayed with Bordeaux mixture (Manshard, 1927); other fungicides effective against *Phytophthora* spp., such as captan, zineb and maneb, may also be tried. Promising results on other nursery trees attacked by *P. cactorum* and other *Phytophthora* spp. have been obtained with metalaxyl and fosetyl-aluminium (Wardlaw and Palzer, 1985; Matheron and Mirchetich, 1985). Infested land should be fumigated with formaldehyde. Avoiding the resowing of beech in the infested soil for five or more years to give time for the disappearance of live oospores of the fungus has also been suggested, and may be an effective control measure, though *P. cactorum* has a wide host range and it may therefore be insufficient to confine the avoidance to beech.

### Other forms of damping off: *Phytophthora citricola* and *Pythium mammilatum*

In one case of severe post-emergence damping off in Britain, the cause was found to be *Phytophthora citricola*, and in another, both pre- and post-emergence damage was caused by *Pythium mammilatum* (Strouts, personal communication).

### Seedling blight caused by *Mycocentrospora acerina* (Hartig) Deighton (*Cercospora acerina* Hartig)

The hyphomycete *Mycocentrospora acerina* is best known as a cause of damage to herbaceous crop plants such as celery and carrots (CMI Descr. F. 537). It was first described by Hartig (1880), however, as the cause of a disease of *Acer* seedlings resembling Phytophthora seedling blight of beech. The same fungus has been found in northern Britain causing a similar disease on seedling beech (Redfern *et al.*, 1982).

The diseased plants are covered by a grey, later black coating, and some may die. The fungus produces long, narrow, curved, tapering, multiseptate conidia. These are colourless, or with the lower, wider cells pale brown, and they measure up to $150 \times 2$–$3$ µm (Ellis and Ellis, 1985). Large brown, resistant chlamydospores carry the fungus over the winter, and it may live as a saprophyte in the soil (Hartig, 1900).

Control of the disease on celery and carrots has been obtained with a benomyl dip after harvesting (Derbyshire and Crisp, 1978).

## Post-nursery diseases

### LEAF DISEASES

### Powdery mildews

The powdery mildews *Microsphaera alphitoides* (the oak mildew) and *Phyllactinia guttata* (*P. corylea*) (the hazel mildew) have been recorded on beech. They are generally of little importance on this host (Burchill, 1978), though they may sometimes defoliate beech hedges.

**Gloeosporium leaf spot caused by *Apiognomonia errabunda* (Rob. and Desm.) Höhnel (anamorph *Discula umbrinella* (Berk. and Br.) B. Sutton, syn. *Gloeosporium fagi* Westend., *G. fagicolum* Passer.)**

The fungus formerly called *Gloeosporium fagi* has long been known as the cause of a leaf spot of beech. In Britain it has been widely noted (Strouts *et al.*, 1986, 1988; Gregory *et al.*, 1988), and has been specifically recorded in Essex, Buckinghamshire, Kent, Surrey and Yorkshire as well as in many parts of Scotland (Moore, 1959). Elsewhere it is known in Belgium, Holland, Denmark, Germany, Austria and Italy (Grove, 1937; Mesturino and Mugnai, 1986).

*G. fagi* causes rounded or irregular spots that are brownish on the upper side of the leaf and olive green below (plate 30). On the undersides of the spots, in summer and autumn it produces small, honey-coloured or brown pustules with colourless oblong-ovoid spores measuring $10–13 \times 4–6$ μm (Ellis and Ellis, 1985). *G. fagicolum* differs from *G. fagi* only in its smaller spores, that measure $12.5 \times 4$ μm (Allescher, 1903). These fungi are in fact anamorphic stages of the ascomycete *Apiognomonia errabunda* (see also chapter 3); these stages are highly variable, and therefore appear variously as species of *Gloeosporium* (a name now rejected), *Gloeosporidium*, *Discula* and *Sporonema* (Monod, 1983). The black perithecia of *A. errabunda* occur on fallen leaves in the spring, and their colourless, unequally two-celled ascospores measure $14–17 \times 3–4$ μm (Ellis and Ellis, 1985).

*G. fagi* is often abundant on leaves infested by the beech woolly aphid (*Phyllaphis fagi*). This combination may cause spectacular browning, especially on the windward side of beech crowns, and the resultant damage is sometimes wrongly attributed to salt-laden winds, windchill and the like (Young, personal communication).

Gloeosporium leaf spot seems to be most common and most severe in cool, wet seasons, but it usually causes only minor damage.

## CANKER, DIEBACK AND BARK DISEASES

Eight or nine species of *Nectria* have been recorded on beech in Britain (Booth, 1959). Most are known to be no more than saprophytes, but three, *N. ditissima* Tul., *N. coccinea* (Pers. ex Fr.) Fr. and the coral spot fungus *N. cinnabarina* (Tode) Fr., are associated with diseases, the first with a canker and dieback the second with a dieback and bark necrosis, and the third (which is further discussed in chapter 3) sometimes causes a dieback of beech hedges. These and other *Nectria* spp. have been much confused, however, and the literature on these diseases is sometimes difficult to interpret.

**Canker caused by *Nectria ditissima* Tul. (anamorph *Cylindrocarpon willkommii* (Lindau) Wollenw.)**

The canker caused by *Nectria ditissima* Tul. was described and figured by Hartig (1900). Hartig considered that this fungus was the cause of canker of many hardwood trees. Later studies have suggested that *N. ditissima* as it is now interpreted is virtually confined to beech (Booth, 1959), although in 1972 it was found on *Sorbus aucuparia*. In Britain the fungus is widespread, but it occurs especially on the South Downs in Hampshire (Peace and Murray, 1956), in the Cotswolds (Brown, 1953) and in the Forest of Dean. Elsewhere it is known in Germany (Langner, 1936), France (Perrin, 1980a), Hungary (Györfi, 1957), Italy (Capretti and Mugnai, 1987), the Netherlands (Gerwen, 1980) and Yugoslavia (Lazarev, 1985).

Its smooth-walled, dark-red perithecia form clusters of 5 to 30 on the bark, and measure about ¼ to ⅓ mm across. The club-shaped asci have a thin, undifferentiated apex, and measure 85–95 × 15–18 μm, and each contains eight oval or elliptical, colourless, two-celled ascospores that measure 14–21 × 5–8 μm.

Pustules of colourless microconidia and macroconidia (*Cylindrocarpon willkommii*) are also formed. The microconidia are oblong to sausage-shaped, 2–2.5 × 0.8–1 μm, and the macroconidia are long, cylindrical, four- to seven-septate, 50–90 × 3–6 μm (Booth, 1959; Dennis, 1968).

At temperatures of 22 °C and above, the fungus produces only mycelium and conidia; maximum perithecial production takes place at 18 °C in the presence of light (Dehorter and Perrin, 1983a). Dehorter and Perrin (1983b) also studied nutrient factors affecting growth and perithecial production. Perrin and Gerwen (1984) found four strains of *N. ditissima* which differed in aggressiveness as well as in their cultural and morphological characters. The fungus produces various pectinases which play a part in the early stages of attack (Perrin, 1984a).

Perrin (1985) found evidence that individual beeches and their progeny differed in susceptibility to canker.

The fungus attacks twigs, branches and stems of beeches, and causes death of buds, discoloration of the leaves, bark necrosis, and deeply sunken cankers that may lead to severe twisting and distortion of the stem (figure 57). Girdling of the trunk may result in dieback of the crown (Peace, 1962; Capretti and Mugnai, 1987).

*N. ditissima* may gain entry through natural openings such as leaf scars, or following damage by insects or honeysuckle, or by rubbing (Peace, 1955a; Peace and Murray, 1956), or by pruning (Brown, 1953).

As a rule the disease causes little damage, though in occasional stands it may affect many trees. As far as possible the affected trees should be removed when thinning. The work of Perrin (1985) suggests that it may be possible to develop clones of beech resistant to the disease.

### Bark disease associated with *Nectria coccinea* (Pers.) Fr.

*Nectria coccinea* causes cankers on some broadleaved trees, and occurs, mainly as a saprophyte, on many others, but it is best known and most important as the fungal component of beech bark disease which in Europe affects the common beech, *Fagus sylvatica*. Much information on beech bark

**Figure 57**   Cankers on beech caused by *Nectria ditissima* (Forestry Commission)

disease can be found in the publications of Houston and Wainhouse (1983) and Lonsdale and Wainhouse (1987).

Beech bark disease is a necrotic disorder which is initiated on the main stem and has a complex aetiology. Various forms of this necrosis are known in parts of eastern North America (there affecting *Fagus grandifolia*), Britain, France and Germany (Houston *et al.*, 1979) and other European countries, including Belgium (Piraux, 1980), Luxemburg (Lies, 1980), Denmark (Thomsen *et al.*, 1949), Switzerland (Leibundgut and Frick, 1943), Hungary (Szanto, 1948) and Yugoslavia (Marinkovic and Karadzic, 1985).

*Symptoms*
Descriptions of the symptoms include those given by Shigo (1963), Perrin (1977), Schütt and Lang (1980) and Lonsdale and Wainhouse (1987). The most characteristic early symptom is the production of necrotic zones in the bark and outermost sapwood.

The development of necrosis is often made externally visible by the exudation of sap which soon darkens and forms a black congealing patch ('tarry spot') on the bark surface (plate 33). Excision of the bark reveals orange-brown necrotic tissue which has a characteristic smell. Old necroses tend to become dark brown or black and friable (figure 58). The necrosis may be confined to localised patches or vertical strips which are delimited by wound periderm and/or callus (Ehrlich, 1934; Shigo, 1963). Rapid, unchecked necrosis may, however, frequently occur, followed by death of the tree. Trees which do not rapidly die show cankering and/or bark roughening of a variable appearance, often together with current exudation of sap from recently formed necroses (Ehrlich, 1934; Lonsdale, 1980b; Perrin, 1980a).

**Figure 58**  Beech bark disease: breaking up of the bark in the final stages of the disease (D.H. Phillips)

Another symptom is the development of foliar chlorosis and/or the production of small, sparse leaves. Although these symptoms are characteristic of the disease in North America, they are not in themselves diagnostic in Europe, where they may also be caused by disorders other than bark disease. Thus in Britain many beeches grow on chalky soils, on which such symptoms may be confused with those of lime-induced chlorosis (Lonsdale and Pratt, 1981).

Although sap exudation from necrotic bark is the main outward symptom of beech bark disease, there is another feature of many diseased trees which may be more conspicuous. This is the presence of a white, fibrous wax on the bark surface, the secretion of a minute sap-sucking insect, *Cryptococcus fagisuga* Lind. (beech scale or felted beech coccus) (plate 34). The intensity of infestation may range from a few isolated colonies to complete coverage of much of the stem. Trees may thus appear whitewashed, although the wax tends to blacken with age. The stems of young trees attacked by *C. fagisuga* often show many shallow pits or 'dimples' (Thomsen *et al.*, 1949) (plate 35), which arise because the xylem below the sites of much insect activity is reduced and abnormal (Lonsdale, 1983a). The apparent significance of *C. fagisuga* in the development of beech bark disease is discussed below.

Once the bark tissues have become necrotic, they usually yield a species of *Nectria* on isolation. This fungus frequently fruits over older lesions, in which the tissues are dark and friable. In Europe the species of *Nectria* is usually *N. coccinia* (Lyr, 1967; Parker, 1974a; Perrin, 1977), the anamorph of which is *Cylindrocarpon candidum* (Link) Wollenw., while in North America the fungus isolated is almost always *N. coccinea* var. *faginata* (Ehrlich, 1934). *Nectria galligena* has been isolated from necroses in limited areas in North America (where it has been shown capable of causing bark disease)(Cotter, 1977; Mielke *et al.*, 1982). A similar report concerning *N. galligena* from beech bark disease lesions in Denmark (Thomsen *et al.*, 1949) appears to have been made in error (Perrin, 1983b).

*Nectria coccinea* has bright-red, oval or subglobose perithecia $\frac{1}{4}$ to $\frac{1}{3}$ mm in diameter, scattered or in groups of 5–30 on a stroma. Its asci are cylindrical with an apical pore surrounded by a 'chitinoid' ring. They measure 75–100 $\mu$m and each contains eight colourless (becoming pale-brown), elliptical, two-celled ascospores that measure 12–15 × 5–6 $\mu$m. Colourless cylindrical micro-conidia and macroconidia are also formed. The microconidia measure 4–9 × 1.5–3 $\mu$m and the macroconidia, which are non- to seven-septate, measure 20–80 × 5–7 $\mu$m (Booth, 1959; Dennis, 1968). Ascospore discharge takes place following wet weather (Parker, 1974b).

Other fungi soon invade the wood beneath the dead bark, sometimes on live as well as on completely dead stems, and these, particularly white-rotting polypores such as *Bjerkandera adusta*, quickly rot the trunk and often lead to snap (Shigo, 1964; Thomsen *et al.*, 1949; Parker, 1974a; Perrin, 1977). Hence, at this stage the disease is sometimes called 'beech snap' (figure 59).

*Suggested causes of beech bark disease*

The cause of beech bark disease has been much debated. In North America Ehrlich (1934) concluded that the primary cause of the disease was attack by *Nectria coccinia* var. *faginata*, following infestation by *Cryptococcus fagisuga*. In Ehrlich's experience, whenever the disease appeared in the field, both these organisms were present and in the absence of either, beech bark disease

**Figure 59** Beech bark disease: in the final stages, secondary decay fungi have caused 'beech snap' (Forestry Commission)

failed to develop. The fungus could always be isolated from the affected tissues and it could only attack tissues previously damaged by the insect. These views have generally been accepted in Canada and the USA (Forbes *et al.*, 1965; Shigo, 1964) and Houston *et al.* (1979) considered that they could be applied equally to the disease in Europe.

The North American situation is more clear-cut than that in Europe, since it involves a distinct front of invasion of the insect which overshadows local variations in disease development. The insect was accidentally introduced from Europe around 1890 to the eastern seaboard. The fungus which appears in its wake is of unknown geographic origin. In Europe, where both insect and fungus are endemic, in some affected stands abiotic factors such as drought and extreme frost have sometimes been considered at least as important as major causes of the disease. Some workers (for example, Perrin, 1977; Parker, 1974b; Houston *et al.*, 1979) have confirmed the insect/*Nectria* association, while others have variously favoured the fungus without the insect, the insect without the fungus and abiotic factors as primary causes.

Ehrlich (1934) found that *N. coccinea* var. *faginata* could cause the disease in the absence of the insect if artificially introduced into bark by wound-inoculation. He suggested that the role of the insect was limited to producing wounds through which fungal entry could occur. However, only small, rapidly delimited lesions were produced in inoculation experiments by Leibundgut and Frick (1943) with *N. coccinea* on European beech and similar results were obtained by Parker (1974b) and Perrin (1979) in their inoculation of young trees. Zycha (1951a) cited the results of Leibundgut and Frick and suggested that *N. coccinea* was merely one of a variety of saprophytic colonists of lesions produced not by any organism (save possibly a virus) but by drought or extreme frost. His evidence for this abiotic origin of the lesions was the correlation between the timing of their appearance and the occurrence of the abnormal weather conditions (Zycha 1951a,b, 1959a). Schwerdtfeger (1963) and Peace and Murray (1956) have also found correlations between the disease and extremes of weather, thus supporting Zycha's view that the beech coccus and *N. coccinea* are secondary colonists and part of a succession that finally includes the white rot fungi that cause decay of the wood. Dimitri (1967) and Paucke (1968) have lent support to the 'abiotic theory' by artificial induction of necroses through the application of extreme heat or cold.

Other factors thought to be important have included sunscorch (Leibundgut and Frick, 1943), shallowness of the soil and neglected thinning (Peace, 1955), closeness of planting (Parker, 1983) and predisposition of the trees by lime-induced chlorosis (Day, 1946). Lonsdale and Pratt (1981) studied an area in which soils over a chalky drift differed in that in places they were 'chalky', with finely divided chalk in the upper horizon, while in other places they contained no such chalk and were 'non-chalky'. Beech at age 29 on both these soil types exhibited similar infestations of *Cryptococcus fagisuga*, but trees on the 'chalky' soils showed more *Nectria* lesions, even trees with only moderate infestations of beech scale. Another factor may be the age of the crop, several authors (Leibundgut and Frick, 1943; Zycha, 1951b; Peace, 1955; Schütt, 1979) having found that most damage occurs in older trees. Perrin (1979) was able to produce undelimited, exuding lesions only on older trees following wound inoculation with *N. coccinea*. Ehrlich (1934), on the other hand, found trees of all ages to be susceptible in North America, while in Britain severe outbreaks have occurred from the 1960s onwards on pole-stage crops (Parker, 1974a).

Combinations of *C. fagisuga* and/or *Nectria* together with abiotic or unknown factors have been suggested by other authors. Thus, Lyr (1967) found that the fungus could produce typical beech bark disease on beech logs, but much more effectively at 15–25 °C than at 3–5 °C, this effect being largely attributable to water stress at the high temperature. Rhumbler (1922) claimed that the insect and the *Nectria* (then included in *N. ditissima*), together with wood-boring insects, were both involved, but that the first stage in disease development was a slime-flux of unknown origin.

The insect on its own has been regarded as capable of causing death of the entire tree (Hartig, 1878) or extensive bark necrosis (Kunkel, 1968) and this idea was supported by Braun (1976, 1977), who described a process whereby the peculiar anatomical structure of beech bark leads to the progressive development of necrosis following initial, rather superficial, damage by the insect. Braun suggested that *N. coccinea* was not essential in this necrotic breakdown of the tissues, but could accelerate the process. Perrin (1983a) has shown that there is a close and specific relationship between *C. fagisuga* and *N. coccinea* on beech. *C. fagisuga* produces pectinases which predispose beech bark to attack by *N. coccinea*, which then produces further pectinases which play a part in the development of beech bark disease (Perrin, 1983c).

The above views on the aetiology of beech bark disease seem largely contradictory, but there is some evidence that they all possess some measure of validity. Lonsdale (1980a) found that wound inoculation of pole-stage beech with *N. coccinea* resulted in extensive necrosis only when a heavy coccus infestation was present. He also found typical *Nectria* lesions on drought-stressed older trees which had little or no insect infestation. And on young infested trees he found a positive correlation between the incidence of sap exudation and the presence of chalky soil conditions (Lonsdale, 1980b). From these data Lonsdale suggested that the invasive ability of *N. coccinea* depended largely on stress in the host and that heavy coccus infestation, drought and nutritional disorders were examples of potentially interchange-able stress factors. He pointed out that Hartig (1900) had long ago observed the role of the coccus in increasing the rate of invasion of bark tissues by '*N. ditissima*', the taxon which then included *N. coccinea*. Variations in the source and degree of stress cause variations in the form of the disease.

*Interactions with other organisms*
Some beetles and other invertebrate predators may attack and reduce beech scale populations (Mayer and Allen, 1983). The fungus *Verticillium lecanii* Viegas can attack both *C. fagisuga* and *N. coccinea* (Lonsdale, 1983b; Lonsdale and Sherriff, 1983). Another fungus, *Nematogonum ferrugineum* (Pers.) Hughes, which occurs in England (Houston, 1983c) and France (Perrin, 1977) as well as in North America, is also capable of parasitising *N. coccinea* on trees affected by beech bark disease (Houston, 1983c).

The presence of bark-encrusting lichens and fungi may also affect colonisa-tion of beech bark by *C. fagisuga*. Thus the lichen *Lecanora conizaeoides* Nyl. ex Cromb. encourages colonisation (Ehrlich, 1934; Houston *et al.*, 1979). In North America, Houston (1983b) found that some lichens encouraged growth by *C. fagisuga*, while others adversely affected it. Among the bark fungi, *Ascodichaena rugosa* Butin inhibits colonisation by the insect (Houston, 1976).

There is, however, little evidence that these organisms have any major effect on beech bark disease in the field.

*Variations in host resistance*
In European beech forests, trees vary in the degree to which they become infested by the beech scale, largely because individual trees vary in susceptibi-lity, and populations of the insect differ in aggressiveness (Lonsdale and Wainhouse, 1987). Similar variations have been found in North America

(Houston, 1983a). Isolates of *Nectria coccinea* also vary somewhat in pathogenicity (Lonsdale, 1980a). Perrin (1984b), who inoculated seedlings raised from diseased and healthy beech trees, found that beeches also varied in susceptibility to *N. coccinea*, and that resistance was genetically controlled.

*Losses caused*

It is difficult to assess the losses caused by beech bark disease. In eastern North America it has sometimes killed more than half the trees (Ehrlich, 1934) and many of the remaining stems have been made almost useless (Shigo, 1963). Miller-Weeks (1983) has reported heavy losses varying from 30 to 50 per cent of the trees in parts of New England, while in Canada Magasi and Newell (1983) have suggested that in the Maritime Provinces the future of beech as a forest crop is bleak because of this disease.

Similar heavy losses have sometimes been recorded in Europe (Thomsen *et al.*, 1949). In France, however, Perrin (1983a) has pointed out that although losses have increased in recent years in northern and northeastern forests in areas ecologically least suited to beech, in the country as a whole, losses have declined. In Germany, Lang (1983) has stated that although the disease is present in many areas in stands of all ages, it cannot be said to threaten the existence of beech plantings. In Britain, outbreaks tend to fall into an 'early' phase (at about 15 to 25 years), a 'peak' phase (from 20 to 35 years) and an 'aftermath' phase (from about 30 to 40 years or beyond) (Lonsdale and Wainhouse, 1987). At the peak period, losses may seem so great that the whole beech crop is in jeopardy. Nevertheless in most cases even in severely affected young stands, enough trees remain to form a mature beechwood. As the woods recover after the peak phase, compensatory growth by healthy trees tends to fill the gaps caused by the death of diseased stems, often leaving little evidence of losses (Parker, 1983).

Valentine (1983) has produced a model of forest growth which attempts to evaluate the consequences of beech bark disease.

*Control*

Infestation of stems by *C. fagisuga* can be controlled by physical removal (for example, scrubbing) or by insecticide sprays (tar-oil winter wash in December to January or diazinon in May or November). There is some evidence that this may quite rapidly reduce susceptibility to the formation of *Nectria* lesions (Perrin, 1980b). As far as forest trees are concerned, these controls are uneconomic and they can be used only on small amenity plantings of special value. At present it can only be recommended that salvage fellings of infected trees should be made quickly before rot fungi can enter and destroy the wood. Only heavily infested trees should be removed, however, as moderately and slightly affected stems may recover (Lonsdale and Wainhouse, 1987).

Lonsdale and Wainhouse (1987) have also reviewed the use of replacement species for beech following severe beech bark disease. They point out that beech can be replanted on most soils, but the findings of Lonsdale and Pratt (1981) suggest that it is best replaced by some other species on sites with much finely divided chalk in the upper soil horizon. However, few species other than beech grow well on very calcareous soils. Ash, sycamore, small-leaved lime and Italian alder can be used on the deeper, more fertile soils over chalk or limestone. Norway maple is a further choice, and this tree will also tolerate more shallow soil conditions on such sites. The mixture of beech with a conifer nurse in the earlier stages of growth may also both reduce disease and improve over-all productivity.

## VIRUS AND VIRUS-LIKE DISEASES

Tomato black ring virus (one of the nepovirus group) has been isolated from leaves of beech in Germany and Scotland. Affected leaves showed chlorotic rings and flecks, mottling and spotting (Cooper, 1979; Nienhaus, 1985).

Nienhaus (1985) found flexuous particles of what appeared to be a virus of the potyvirus group in the Rhineland in beech trees the leaves of which were chlorotic and mottled and reduced in size. At a later stage of the disease the trees showed a branch dieback.

Nienhaus *et al.* (1985) studied a decline in beech trees in the Rhineland and Westphalia. They isolated a range of viruses from several groups from about 40 per cent of the sampled trees, and thought that these viruses might be playing some part in predisposing the trees to the disease complex concerned.

Also in Germany, Parameswaran (1980) has described a bark necrosis of beech trees in the phloem of which he was able to detect mycoplasma-like bodies.

## DECAY FUNGI

Many decay fungi affect beech, among them the honey fungus, *Armillaria mellea*, and occasionally *Heterobasidion annosum*, which more often attacks conifers. These are both described in chapter 3. *Bjerkandera adusta* (*Polyporus adustus*) is common on trees damaged by beech bark disease. It causes a white rot which breaks up the stem and so leads to the syndrome called 'beech snap'. Beech trees, especially those affected by beech bark disease, may also be attacked by *Inonotus nodulosus*, *Pseudotrametes gibbosa* (*Trametes gibbosa*) and (mainly in continental Europe) *Fomes fomentarius* (Schütt, 1979). The latter causes a white or yellowish-white rot, most often in the upper parts of the tree. *Perenniporia fraxinea* may sometimes cause a basal white rot. *Ganoderma applanatum* and *G. adspersum* both give rise to a butt and branch rot of over-mature beeches. Another *Ganoderma* sp., *G. pfeifferi*, may decay roots, stems and branches and kill the whole tree. *Pleurotus ostreatus* (the oyster mushroom) may also cause extensive decay of stems and branches, while *Meripilus giganteus* (*Polyporus giganteus*) decays and kills the smaller roots. Beech is also occasionally attacked by *Laetiporus sulphureus*, which causes a brown cubical rot. The ascomycete *Ustulina deusta* rots the stem base and major roots, which often leads to windblow. These fungi are described in chapter 20.

# 12 Diseases of sweet chestnut (*Castanea* spp.)

The only species of *Castanea* of any importance in Britain and the rest of Europe is *C. sativa*, the sweet or Spanish chestnut. This tree, which is native in the Mediterranean area, is grown in more southerly parts of Europe for its nuts as well as for its timber, though in some areas, especially in Italy, it has declined in importance because of damage by the chestnut blight fungus, *Cryphonectria parasitica*. In Britain it was probably introduced by the Romans, and it is now widespread and often naturalised in large parts of the south and east. It is still grown there as coppice for fencing, hop poles, etc., and if grown as a standard it produces valuable timber. It is important as an ornamental in parks and gardens. Though sweet chestnut fruits ripen in southern England, they remain small, and nearly all the chestnuts sold for eating are imported.

As might be expected of a tree of Mediterranean origin, *C. sativa* is rather susceptible to damage by spring frosts.

## General

### Powdery mildews
In mainland Europe (but not so far in Britain), sweet chestnut leaves, both in the nursery and at a later stage, may sometimes be attacked by the powdery mildew *Microsphaera alphitoides*. This produces a white coating on the leaf surface. It is most important on oak, and is therefore described in chapter 10. Also sometimes found on sweet chestnut is another powdery mildew, *Phyllactinia guttata* (Burchill, 1978), which is described in chapter 20 under 'Corylus', its most important host. These two powdery mildews are not of major importance on *Castanea*.

## Leaf diseases

### Leaf spot caused by *Mycosphaerella punctiformis* (Pers.) Starb.
(*M. maculiformis* (Pers.) Schröter: anamorphs *Phyllosticta maculiformis* Sacc. and *Septoria castaneicola* Desm.)
*Phyllosticta maculiformis* (*P. betulina* Sacc.) and *Septoria castaneicola* (*Cylindrosporium castaneicola* (Desm.) Berl.), the anamorphs of *Mycosphaerella punctiformis*, produce small spots on sweet chestnut leaves (Klebahn, 1934),

usually in mid or late summer. The fungus attacks sweet chestnut in Europe (especially on the continent, where it is rather more important than it is in Britain), and is known also in North Africa and North America; it sometimes also affects oak, and may be found on the leaves of many other broadleaved trees.

The affected chestnut leaves bear small yellow-brown spots on their upper sides, and the anamorphic stages of the fungus both appear on the undersides of the spots in summer and autumn. The black pycnidia of *P. maculiformis* contain colourless, cylindrical, straight or curved spores which measure 4–6 × 1–1.5 μm (Grove, 1935), while the long, narrow spores of *S. castaneicola* measure 30–40 × 3–4.5 μm (Ellis and Ellis, 1985). *Mycosphaerella puncti- formis* has many synonyms (Cannon *et al.*, 1985). Its pseudothecia develop over the winter on the undersides of the fallen leaves. Their colourless, two- celled ascospores measure 8–14 × 2–3 μm (Ellis and Ellis, 1985).

The fungus is usually of little significance, but in parts of mainland Europe in cool, damp summers it may sometimes spread and cause local premature defoliation (Cambonie, 1932; Biraghi, 1949; Bazzigher, 1956b). The Japanese chestnut, *Castanea crenata*, is said to be resistant (Bazzigher, 1956b).

# Diseases of shoots and stems

Diebacks and cankers on sweet chestnut caused by the two fungi *Cryptodiaporthe castanea* and *Diplodina castaneae* have both been recorded in Britain, and more often found in the rest of Europe and in North America. They are, however, little known, and have caused only minor loss. What little is known about them is summarised below. The two diseases are sometimes considered to be different aspects of only one.

**Dieback and canker caused by *Cryptodiaporthe castanea* (Tul.) Wehmeyer (anamorph *Discella castanea* (Sacc.) von Arx, syn. *Fusicoccum castaneum* Sacc.)**
A dieback and canker of sweet chestnut associated with *Cryptodiaporthe castanea* is known in France as *javart* disease. This popular name is due to a resemblance between the small cankers on the chestnut shoots and the ulcers sometimes formed on horses' feet between hoof and hair, a disorder known as *javart* in French, or quitter or quittor in English (Day, 1930). The disease has been found in Switzerland (Défago, 1937) and in the USA (Fowler, 1938) as well as in France and England.

*Symptoms*
The symptoms of the disorder show as a wilting and defoliation of coppice shoots, often on only part of a stool. Small greyish or reddish-brown fungal stromata up to 2 mm across appear in cracks in the bark of the affected shoots, from which the bark eventually peels (Moreau and Moreau, 1953).

Sunken cankers may girdle the stem and cause it to collapse (Défago, 1937) (figure 60).

### The causal fungus

The perithecia of the fungus are embedded usually in groups in the central part of the stroma. They measure 240–800 × 200–400 μm and their openings scarcely protrude above the stromatic surface. The perithecia contain asci that measure 50–60 × 7–8 μm. Each ascus contains eight colourless, more or less cylindrical, two-celled ascospores measuring 11–16 × 2–3 μm, each usually with a colourless, short cylindric appendage at each end (Wehmeyer, 1932). The pycnidia of the anamorph (*Discella castanea*) form as irregular cavities usually round the edges of the stroma. The pycnospores are colourless, oval or spindle-shaped, measuring 4–9 × 3–5 μm. Ellis and Ellis (1985), however, consider the imperfect stage to be a *Diplodina* with long one-celled (but finally two-celled) conidia measuring 8–12 × 2–3 μm (see below, *Diplodina castaneae*). In very damp weather the outer layers of the pycnidium may break open and reveal the spores within as a gelatinous mass (Défago, 1937).

### Entry into the host and factors affecting the disease

*C. castanea* appears to be a rather weak wound parasite, which may also enter through dead buds. In strongly growing trees, callus formation arrests the growth of the fungus, which is able to enlarge the canker only if the host is weakened by adverse conditions (Défago, 1937).

### Damage caused

Though this dieback has been recorded in Britain, it is little known there, and causes no serious damage. Elsewhere it seems at least at times to have been of greater importance. Thus, in the USA, Fowler (1938) reported that it often caused damage in both nurseries and plantations, killing branches, reducing growth and deforming the trees of the American chestnut. In Switzerland it has also killed young trees (Défago, 1937).

### Control

Control measures for this disease have never been worked out in Britain because the disorder has been so rarely noted. Défago (1937) in Switzerland suggested that the disease was best avoided by careful site selection and soil preparation. If the disease appears, infected shoots should be cut back at least 15 cm below the cankers.

### Dieback and canker caused by *Diplodina castaneae* Prill. and Delacr. (*Cytodiplospora castaneae* Oud.)

In their study of diseases of sweet chestnut in the Forest of Marly, west of Paris, Moreau and Moreau (1953) isolated *Diplodina castaneae* from wilted trees also invaded by *Cryptodiaporthe castanea*. In Britain Day (1930) found *D. castaneae* on a sweet chestnut pole from a coppice crop much of which

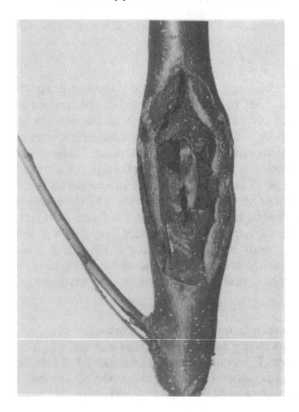

**Figure 60**   Canker on sweet chestnut coppice shoot caused by *Cryptodiaporthe castanea* (Forestry Commission)

showed severe dieback; he regarded it as a form of *javart* disease, as did Lanier *et al.* (1976).

*Symptoms*
Bright-brown patches showed up clearly in contrast to the purplish-grey bark of healthy parts of the shoots.

*The causal fungus*
Small, black stromata formed on the brown areas. Pycnidia were produced as cavities in these stromata. They contained colourless, two-celled, spindle-shaped spores measuring 6–7 × 1–1.5 μm (Grove, 1935) or 9–12 × 2.5–3 μm (Day, 1930). It is of interest that Ellis and Ellis (1985), who consider the anamorph of *Cryptodiaporthe castanea* (above) to be a *Diplodina*, describe its finally two-celled conidia as measuring 8–12 × 2–3 μm. The fungus appears to be a weak wound parasite.

*Control*
Day (1930) suggested that the disease could be controlled by cutting out and destroying all the diseased shoots. He also recommended trenching round an

infected area to isolate it and prevent the possible spread of the fungus from diseased to healthy roots, though there appears to be no evidence that the disease can be transmitted in this way.

### Chestnut blight caused by *Cryphonectria parasitica* (Murrill) Barr (*Endothia parasitica* (Murrill) Anderson and Anderson)

The literature on the blight of sweet chestnut caused by *Cryphonectria parasitica* has been reviewed by Griffin (1986) and Anagnostakis (1987), who both supply extensive bibliographies. Chestnut blight does not occur in Britain, and is considered to be a possible threat to sweet chestnut plantings there.

The disease, which is thought to have originated in China or Japan, was first found in 1900 in eastern North America (Anagnostakis, 1987), where it spread rapidly and virtually destroyed the American chestnut, *Castanea dentata*. Chestnut blight reached Europe in 1938, when it was found in Italy, and the effect on the European sweet chestnut was almost as serious as it had been on *C. dentata* in America (Pavari, 1949). Spread rapidly occurred westwards into France, Spain and Portugal, east into Yugoslavia, Greece, Hungary and Turkey and the Ukraine, USSR, and north into Switzerland, and apparently as far north as Belgium (CMI Map No. 66).

### *The causal fungus*

A standard description of the causal fungus, *Cryphonectria parasitica*, is given by Sivanesan and Holliday (CMI Descr. F. 704). The fungus may sporulate at any time, but the pycnidia of its *Endothiella* stage are produced most abundantly in spring and summer. They form in small, orange to brownish stromata, and in damp weather their spores exude as orange tendrils. The individual spores are colourless, one-celled and rod-shaped, and measure 3–5 × 1–1.5 µm. The perithecia also form in the stromata, one to many being embedded in each stromatal pustule, from which only the tips of their long necks emerge. The ascospores are shed mainly in summer and autumn. They are colourless and equally two-celled, measuring 7–12 × 3.5–5 µm. They are shed following rainfall, and may be carried long distances by the wind. The sticky pycnospores are spread mainly by birds, small mammals, and insects and mites, and perhaps also by slugs (Turchetti and Chelazzi, 1984). Uscuplic (1983) found evidence that the fungus could spread by the colonisation of stumps after felling. Jaynes and DePalma (1984) found it on the outer shells of a varying proportion of the nuts of *Castanea dentata*. They succeeded in reducing but not eliminating the infection by a hot water treatment.

### *Symptoms*

The fungus enters through wounds, and spreads through the bark and into the cambium and outer sapwood, producing greyish to buff-coloured mycelial fans in and under the bark. It gives rise to swollen, reddish girdling cankers on stems, branches and coppice shoots (figure 61). These cankers are at first smooth (and most conspicuous when the bark is wet) but they eventually become cracked and fissured. As the fungus interferes with the conducting system of the tree, it also causes wilting and death of leaves beyond the cankers, and the resulting yellowed and brown foliage is therefore the most conspicuous symptom in the growing season. Below the cankers, affected trees commonly develop epicormic shoots.

**Figure 61**   Lesion on sweet chestnut caused by *Cryphonectria parasitica* (Forestry Commission)

*Host range*
In Europe, *C. parasitica* causes disease mainly in the sweet chestnut, *C. sativa*, though it sometimes attacks the sessile oak, *Quercus petraea*, the evergreen oak, *Q. ilex*, and the white oak, *Q. pubescens* (Biraghi, 1951). In North America, as noted above, the American chestnut has been more or less eliminated by the disease; other chestnuts there, including *C. pumila*, the chinkapin, are also susceptible. In addition, the post oak, *Q. stellata*, and the live oak, *Q. virginiana*, may be affected, and *C. parasitica* has been associated with cankers in the pecan *Carya illinoensis* (Hepting, 1971); it is also known as a saprophyte on various other species of *Quercus* and *Carya*, and also on

species of *Acer* and *Rhus* (Boyce, 1961). In Japan it has been known to attack *Eucalyptus* (Old and Kobayashi, 1988).

Within the genus *Castanea*, the Chinese chestnut, *Castanea mollissima*, and the Japanese chestnut, *C. crenata*, are both resistant (Hepting, 1971), though not immune. In a survey of *C. mollissima* plantings in the eastern USA, Jones *et al.* (1980) found that a good many trees showed cankers, but almost none were killed. Even within the very susceptible *C. dentata* and *C. sativa* resistant populations or individuals occur (Griffin *et al.*, 1983; Govi and D'Aulerio, 1982).

*Breeding for resistance*
Much work has been done on the selection and breeding of resistant chestnut varieties, especially in the USA and Switzerland. This work has been reviewed by Hartline (1980), Burnham *et al.* (1986) and Burnham (1988). The latter has given an account of attempts to breed resistance into *Castanea dentata*, while Bazzigher (1981) describes the Swiss programme in which crosses between *C. sativa* and resistant Asian species were made, producing after some 25 years about 40 000 trees for further testing.

*Hypovirulence*
Not much more than ten years after the discovery of chestnut blight in Italy, Biraghi (1951) noticed that healing and recovery was taking place in a coppice area formerly severely affected by the disease. The fungus in the trees appeared to be in some way abnormal, and it had failed to spread beyond the outer layers of the bark. Further work in France by Grente (1965) and others, and later in North America and elsewhere, showed that if what Grente described as the hypovirulent strains were inoculated into affected trees, the disease in the latter regressed, and a large proportion of the mycelium of the formerly virulent fungus within the cankered tissues was changed into the hypovirulent form. The production of pycnidia on the cankers was much reduced, and perithecial development was suppressed. Continued studies indicated that the hypovirulent strains were affected by a disease which could be transmitted by hyphal anastomosis. The disease was associated with various forms of double-stranded RNA (dsRNA), the presence of which was genetically controlled by determinants in the fungal cytoplasm (Anagnostakis, 1987).

*Control*
In Yugoslavia Uscuplic (1983) has recommended the treatment of fresh stumps with creosote because he found an association between spread of the disease and spore infection of the stumps. Conventional treatments with fungicides have proved of little effect. The fungus quickly acquired resistance to benomyl, carbendazim and thiophanate-methyl (Delen, 1980). To some extent the standard methods of disease control designed to minimise the available inoculum can be applied, by felling diseased trees, pruning out infected branches, and burning the resulting debris.

Where appropriate, it may be possible to grow resistant chestnut species, selections and hybrids. Thinking on disease control has been greatly changed by the discovery of the hypovirulent strains of the fungus. The presence of these strains in Italy has decreased the severity of the disease there. Though there are still some problems to be solved, it is hoped that the extensive work done on the hypovirulent strains may make it possible to spread them artificially in areas affected by chestnut blight, in both Europe and North America, and so use these strains as a form of biological control (D'Aulerio *et al.*, 1987). Hypovirulence and its use in the control of the disease has been reviewed by Fulbright *et al.* (1988).

## DISEASES OF ROOTS

**Phytophthora root rot ('ink disease') caused by** *Phytophthora cinnamomi* **Rands and** *P. cambivora* **(Petr.) Buism.**
Probably the best-known of the diseases of sweet chestnut in Britain is the Phytophthora root rot caused by *Phytophthora cinnamomi* and *P. cambivora*. It is also of importance in other parts of Europe. The disease is often called 'ink disease' because a blue-black inky stain may often be found round the damaged roots. This name is not a good one, however, as this stain is not associated only with this disease, but a similar inky fluid will also seep even from fresh, undiseased chestnut wood or bark in contact with the ground (Peace, 1962); the disease, on the other hand, also occurs on other hosts (including beech), in which no such staining is induced (Day, 1938). Inky staining is, however, a characteristic of a somewhat similar disease of oak, caused by *P. cinnamomi* in southwest France. This disease is described under 'oak' in chapter 10.

*Distribution and the causal organism*
On *Castanea*, Phytophthora root rot is known to occur throughout most of Europe, in Asia (in India and Turkey), in Australia and New Zealand, in Canada and the USA, and in Argentina (Urquijo Landaluze, 1942).

The causal fungus in Britain is most often *Phytophthora cinnamomi*, but *P. cambivora* may sometimes be involved. In Czechoslovakia (Juhasova and Bencat, 1981) and Switzerland (Reichard and Bolay, 1986) it seems possible that *P. cambivora* is most often associated with the disease. These fungi are further discussed in chapter 3. Both occur on other hosts, and both have often been misidentified in the past. The world distribution of *P. cambivora* is given in CMI Distribution Map No. 70, and that of *P. cinnamomi* in the corresponding Map No. 302.

The host range and geographical distribution of *P. cinnamomi* are both very wide. The host range of *P. cambivora* is much narrower. Indeed, Crandall (1950) considered that the fungus occurred only on *Castanea*, but according to Waterhouse and Waterston (1966: CMI Descr. F. 112) it is now known to affect also *Acer, Fagus, Juglans, Nothofagus, Quercus* and *Ulmus*.

*Symptoms on sweet chestnut*

Symptoms of the disease on chestnut may vary, and sometimes quite severely affected trees may show no obvious symptoms above ground (Grente and Solignat, 1952). In the case of coppiced chestnut only some of the shoots may show symptoms, while others remain apparently healthy.

The disease primarily affects the roots, and symptoms in the aerial parts follow and are consequent upon the root infection. At first the fine roots are destroyed, and the fungus spreads inwards into the larger roots and so to the root collar. The tap roots are usually especially severely affected, and as they are destroyed the tree may react by producing further lateral roots. The diseased roots may exude a violet or blue-black inky slime, and cankers may form on the larger roots and at the bases of the trunks (Day, 1938; Dufrénoy, 1922; Milburn and Gravatt, 1932). These basal lesions are the most characteristic and useful symptoms for diagnostic purposes in the field.

The first symptoms to be observed, however, are usually those in the crown, and they arise as a result of the root damage and often the girdling of the trunk by the causal fungi. Sometimes the first visible sign of disease is the falling of unopened male catkins early in the season (Schell, 1922). Leaf symptoms may vary, and with regard to these the disease seems to take two main forms (Day, 1938; Milburn and Gravatt, 1932); these, however, are probably only stages in the same disorder, depending, for example, on the time of attack or its severity. In the first form, the leaves are very small, and after flowering the following year the tree dies. In the second form, the leaves are of normal size, but they quickly become yellow, and finally brown. On such trees fluid may exude from affected bark at the bases of their trunks. The trees may then recover, or die slowly in a few years (Day, 1938). In France, where the trees may be grown for their fruit, the nuts on affected trees are small, and may abort before they can mature; they then hang on the branches for several years (Dufrénoy, 1922).

*Host variation in sweet chestnut*

It has been known at least since 1922 that the Japanese chestnut *Castanea crenata* and the Chinese chestnut *C. mollissima* are immune or at least highly resistant to Phytophthora root rot (Dufrénoy, 1922). Hence, these trees have been planted to some extent, and used in crossing programmes with *C. sativa* in France (Dufrénoy, 1922), Spain (Urquijo Landaluze, 1944, 1963) and Portugal (Guerreiro, 1949). In the USA, Milburn and Gravatt (1932) found that the Chinese chestnuts *C. henryi* and *C. seguinii* also showed high resistance.

*Factors affecting the disease*

It is the general view that the disease is encouraged by various soil factors. Day (1938) found that severe attacks occurred only on shallow, heavy, compact clay soils with poor drainage below. There is no doubt that the disease is worst on wet, poorly drained soils. According to Schell (1922) death often takes place in less than two years on poor soils, but on rich soils the affected trees may persist for four or five years or more. Dufrénoy (1922) noted that the disease developed most rapidly in those affected trees with the most abundant foliage.

*Losses caused*
Though Phytophthora root rot has often been found in sweet chestnut crops in Britain, in Europe it haş done most damage in France, Spain and Switzerland. In France in the earlier part of this century, it caused heavy losses in the forests of the Pyrenees, and was also damaging in other areas (Schell, 1922). In Spain, in the better areas half the trees were killed in 25 years; in others in which the trees were a valuable source of income, almost all were destroyed (Del Cañizo, 1942). In Switzerland, Reichard and Bolay (1986) examined a forest in the canton of Geneva in which nearly all the chestnut trees were affected and many killed. Crandall *et al.* (1945) considered the disease was also the main cause of the loss of the American chestnut *Castanea dentata* in a large part of the eastern USA.

*Control*
In Britain, Day (1939) suggested that the disease could sometimes be arrested in its early stages on individual trees by cutting out patches of diseased bark and then applying a sterilant to the wound and to the surrounding soil. In general, however, it is best prevented by avoiding wet, ill-drained soils when planting chestnut.

More attention has been paid to chemical treatments in other countries in which the disease is of more importance. Sulphate of iron and various copper compounds, applied to wounds on trunks or roots, or to the soil, have been tried with some success in some areas (see, for example, Schell, 1922; Del Cañizo, 1942; Urquijo Landaluze, 1949; Fernandes, 1955). Dufrénoy (1922) recommended soil disinfection with carbon bisulphide or calcium carbide. He also noted that newly affected trees might be saved by severe pruning, which slowed down the disease and also encouraged the formation of new roots. Such methods are likely to be of limited application.

Following the work on selection and breeding for resistance, mentioned above, Cristinzio (1938) in Italy advocated replacing the susceptible trees with the Japanese chestnut, *Castanea crenata*. This tree, however, does not grow well in Britain.

**Other *Phytophthora* spp. attacking sweet chestnut roots**
Among other *Phytophthora* spp. (described further in chapter 3) known to attack sweet chestnut roots are *P. cryptogea*, which is associated with a dieback, and *P. cactorum*, associated especially with a collar rot.

DECAY FUNGI

A number of decay fungi attack the sweet chestnut. Among them are the *Armillaria* spp. (probably mainly *A. mellea*) which invade the roots. The *Armillaria* spp. are described in chapter 3.

*Laetiporus sulphureus* is important as the cause of a brown cubical rot which may render the trees dangerous. Another brown cubical rot is caused by *Fistulina hepatica*, the beefsteak fungus. *Phellinus robustus* occurs locally on sweet chestnut, mainly in continental Europe; it causes a yellowish-white decay, chiefly in the sapwood. *Ganoderma lucidum* occasionally causes a white rot at the base, while *Stereum gausapatum* may give rise to a yellowish-white pipe rot in chestnut stems and branches. These fungi are described in chapter 20.

## NUT ROTS

Chestnut fruits may be attacked by the ascomycete *Ciboria batschiana*, which turns them into a black stroma (Neergaard, 1979). This fungus is described under 'oak' in chapter 10.

The hyphomycete *Acrospeira mirabilis* Berk. and Br. sometimes partially converts the nuts into a brown, powdery mass. The globose, warty terminal cells of its three-celled spores are darker brown than the rest, and measure 20–30 μm across (Ellis and Ellis, 1985).

# 13 Diseases of ash (*Fraxinus* spp.), birch (*Betula* spp.) and alder (*Alnus* spp.)

## Ash (*Fraxinus* spp.)

The common ash (*Fraxinus excelsior*) occurs throughout most of Europe, and is widespread in Britain, especially on calcareous sites. In other parts of Europe, the manna ash, *F. ornus* (often planted in British parks and gardens), is native in southern Europe and Asia Minor, and the narrow-leaved ash, *F. angustifolia*, and the Caucasian ash, *F. oxycarpa*, both from southern Europe, are occasionally planted in parks and gardens in Britain. Also found in European parks and gardens are various species from Asia and North America.

### GENERAL

**Crown dieback (ash decline)**
Older ash trees, especially on clay soils in the southern midlands of England, often show a severe dieback of the crowns, parts of which have few or no leaves. The cause of this damage is unknown, but most of the affected trees are in hedgerows bordering fields with intense arable farming, while few are to be found in towns or woodlands (Pawsey, 1984; Hull and Gibbs,1991)

### LEAF DISEASES

**The powdery mildew *Phyllactinia guttata* (*P. corylea*)**
The powdery mildew *Phyllactinia guttata* sometimes produces its sparse mycelium on ash leaves, as well as on those of hazel (under which it is described in chapter 19), and other trees and shrubs.

**Minor leaf spots caused by *Phyllosticta fraxinicola* Curr., *Ascochyta metulispora* Berk. and Br. and *Venturia fraxini* (Fr.) Aderh. (anamorph *Spilocaea fraxini* (Aderh.) Sivan., syn. *Fusicladium fraxini* Aderh.)**
*Phyllosticta fraxinicola* has been recorded as fairly widespread and common on ash, as a cause of rounded or irregularly oval brown leaf spots with a reddish-brown margin. A few black pycnidia occur on the upper leaf surface,

and may contain mostly ellipsoid, colourless, one-celled spores measuring 4–6 × 1.5 μm (Grove, 1935).

Another member of the Deuteromycotina, *Ascochyta metulispora*, causes brown leaf spots up to 1 cm across on ash leaves. The brown pycnidia are embedded in the upper sides of the spots, and their colourless conidia, which eventually become two-celled, measure 8–11 × 2.5–3 μm (Ellis and Ellis, 1985)

*Spilocaea fraxini* (formerly called *Fusicladium fraxini*), the anamorph of *Venturia fraxini*, causes purplish or violet, later brown areas on ash leaves. The two-celled brown conidia measure 15–26 × 5 μm (Batko, 1974). The brown ascocarps, which bear brown setae on their tops, are embedded in the fallen leaves, and their brown, two-celled ascospores measure 11–14 × 3–4.5 μm (Ellis and Ellis, 1985). The fungus occurs in Britain, and in Czechoslovakia, Germany, Italy, Romania and Switzerland (Sivanesan, 1977).

## DISEASES OF BARK AND SHOOTS

### Bacterial canker or bacterial knot caused by *Pseudomonas syringae* van Hall subsp. *savastanoi* (Smith) Janse pv. *fraxini* Janse

Bacterial canker, which may greatly reduce the value of the wood, is the most damaging disease of ash in Britain, and it has been recorded in almost the whole of Europe, including the USSR (CMI Map No. 134).

The symptoms of the disease, which Janse (1982) prefers to call bacterial knot because of the form of the lesions, have been described in detail by Boa and Preece (1981), Dowson (1957), Janse (1981a), Riggenbach (1956) and van Vliet (1931). The cankers (plate 31) are very variable in size, from about 0.5 cm to over 30 cm across (Oganova, 1957), with most of the cankers on the smaller branches, but the largest, which may persist for many years, on the older branches and occasionally on the trunks.

Cankers are first noticeable as small longitudinal cracks. These may progress only to form small rough protuberances, but usually widen and develop into sunken cavities. The affected tissues are at first filled with layers of corky tissue parallel to the cambium, but in time the corky material is extruded, and replaced by mucilage containing bacteria. The edges of the cankers are raised, rough and irregular, and often yellowish or reddish in colour. Janse (1981a) found two types of excrescence, the first swollen and wart-like, the second more canker-like, with the wood exposed and the wound often surrounded by callus tissue. Janse (1982) made a full study of the anatomy and development of the cankers, which he found mainly affecting the bark.

The bacterium present in the cankers, *Pseudomonas syringae* subsp. *savastanoi* pv. *fraxini* (Janse, 1981b) exudes from the affected tissues as a yellowish slime in humid weather in spring and early summer, when a pure culture is most easily obtained (Janse, 1981a,b). The pathovar *fraxini* will generally attack only *Fraxinus excelsior* and some of its cultivars (Janse,

1981a). Methods for the identification of *P. syringae* and the separation of its many subspecies and of the pathovars of the subspecies *savastanoi* are given by Lelliot and Stead (1987) and Janse (1981b).

Various fungi have been found associated with bacterial cankers. Thus Riggenbach (1956) in Switzerland found *Fusarium lateritium*, *Pleospora herbarum* and *Plenodomus rabenhorstii* as well as *Pseudomonas syringae* subsp. *savastanoi*; in the presence of the latter, the growth of the fungi, particularly *F. lateritium*, was stimulated. Janse in the Netherlands (1981b) also found *Fusarium lateritium* and *Plenodomus rabenhorstii* in association with the cankers. In the USSR Oganova (1957) isolated *Auricularia mesenterica* from rotting tissues in bacterial cankers. In Britain a species of *Nectria* often occurs with the bacteria in the larger cankers, which Dowson (1957) considered could not be produced by the bacteria alone.

In many countries, in both Europe and North America, however, a distinct fungal canker of many broadleaved trees (including ash), caused by *Nectria galligena*, also occurs. This is separately discussed below and in chapter 3.

According to Dowson (1957), *P. syringae* ssp. *savastanoi* pv. *fraxini* enters young shoots through wounds caused by hailstones. It may also enter through leaf scars, as well as through wounds made by insects, and through frost cracks, lenticels and perhaps bud-scale scars (Janse, 1981a).

Skoric (1938) considered that the bacteria were spread by rain, and insects played no important part in dissemination. However, Oganova (1957) in the USSR and Tubeuf (1936) in Germany believed that the ash bark beetle *Leperisinus fraxini* (*L. varius*) damaged the trees and transmitted the disease. Janse (1981a) concluded that the ash bark beetle played little part in transmission, though it had been shown experimentally that it could carry the disease.

Though ash canker causes considerable deformation and damage to trees in Britain, it seems to cause rather greater loss on the mainland of Europe. In Yugoslavia, Skoric (1938) found that young trees affected by bacterial canker grew to no more than 1 to 2 m. According to Tubeuf (1936) canker sometimes caused severe damage in Germany, where many diseased trees had to be felled. In the Netherlands, Janse (1981a) found that the growth of young trees could be much reduced, and affected plants might die.

*Factors affecting the disease*

Janse (1981a) found that bacterial canker caused most damage on soils unsuited to ash, including acid peats, sites on which root growth was restricted, and in some roadside plantings.

*Control*

Little can be done to control bacterial canker of ash, though Oganova (1957) in the USSR recommended the cutting out of affected branches, and the complete removal of severely diseased trees. She also suggested that wounds should be avoided, and any found should be treated with an antiseptic. As she considered that *Leperisinus fraxini* played a part in the spread of the disease, she recommended the disinfection of the soil around the trees to kill bark beetles. Spraying with copper compounds has some controlling effect. These measures, apart from the removal of diseased trees when thinning, are not practicable on a forest scale.

**Fungal canker caused by *Nectria galligena* Bres.**
In the Netherlands van Vliet (1931) found two distinct types of canker on ash. The cankers of the first type were bacterial in origin, like those already described, and occurred on both trunks and branches. The second, caused by *Nectria galligena* var. *major* (the perithecia of which were often present on them), were verrucose, and occurred only on the trunks of young trees. The diseased cortex on these cankers became reddish-brown. D'Oliveira (1939) in Portugal also found cankers of both bacterial and fungal origin, and on the latter found *Cylindrocarpon mali* or its variety *flavum*, the imperfect stage of *Nectria galligena* and its variety *major*, respectively. Similar bacterial and fungal cankers, the latter associated with *Nectria ditissima* or *N. galligena*, were described in Germany by Tubeuf (1936), who also found cankers, mainly on small shoots, caused by frost. Janse (1981a, 1982) also compared cankers on ash caused by *Nectria galligena* with the bacterial cankers. He found that the fungus caused typical target cankers, and affected trees suffered extensive dieback of the branches, and much leaf fall. Cankers on the trunk were rare. The fungus penetrated the bark and cambium and entered the xylem. Cankers formed mainly round buds, leaf scars, branch stubs and dead twigs.

These cankers have been extensively studied on apple, especially in Europe and North America, and on many broadleaved trees, mainly in eastern North America. Characteristically, as the fungus in the cankers grows and causes damage when the tree is dormant and is delayed and partly overgrown by callus tissue in the growing season, the cankers show a target-like form. As these cankers occur (especially in North America) on a wide range of broadleaved trees, they are discussed further in chapter 3, and under '*Malus*' in chapter 18. Booth (1959) considered that the fungus on the ash cankers was *N. galligena* (from which the variety *major* was not separable). The fungus forms bright-red perithecia, usually in groups of two to five, on a stroma. The two-celled, colourless ascospores may be oval, ellipsoid or spindle-shaped, and measure 14–22 × 6–9 μm.

**Minor cankers associated with *Phomopsis* spp. and *Hysterographium fraxini* (Pers. ex Fr.) de Not. (*H. elongatum* (Wahlenb.) Corda)**
Grove (1935) stated that *Phomopsis controversa* Traverso occurred as a saprophyte on dry branches of *Fraxinus excelsior*, while *P. scobina* Höhnel was rather common in England and Scotland on twigs and leaf stalks of *F. excelsior* and *F. ornus*. He considered it likely, however, that these two *Phomopsis* species were one and the same fungus. Wehmeyer (1932) took the same view, regarding them as conidial stages of *Diaporthe eres* Nitschke.

The two fungi were studied in Scotland by Macdonald and Russell (1937), who concluded that they should be regarded as two species, on both their cultural characteristics and their pathogenicity. They found that *P. controversa* grew rapidly in culture to form a white woolly mycelium, produced no pycnidia when inoculated into plum and apple fruits, and scarcely entered the tissues when inoculated into ash twigs. Its colourless fusiform A-spores measured 5.5–9 × 1.5–4 μm, and no B-spores were to be found. There were dark zones round the pycnidia.

*P. scobina* grew slowly in culture to form a white, sometimes later pinkish, tufted mycelium, produced many pycnidia when inoculated into plum and apple fruits, and entered the tissues when inoculated into ash twigs. Colourless A- and B-spores occurred, the fusiform A-spores measuring 7–13 × 2–4 μm and the hair-like B-spores measuring 16.5–25 × 1 μm. No zone lines were present round the pycnidia.

Parallel studies in the field indicated that *P. scobina* could cause lesions up to 25 cm long, completely encircling large branches, and many pycnidia occurred in the brown sunken bark, which cracked and fell away on older lesions. Young twigs might be induced to form small witches' brooms. This fungus was found in 1985 fruiting abundantly on dead bark and on the margins of large perennating cankers of a green-barked form of common ash, *Fraxinus excelsior*, in Avon, England (Strouts *et al.*, 1987).

Macdonald and Russell (1937) found that *P. controversa* was generally only weakly parasitic. It sometimes killed small twigs but on larger branches and trunks it was associated only with small cankers, which soon ceased to develop.

*P. controversa* and *P. scobina*, respectively, appear to be the anamorphs of *Diaporthe controversa* (Desm.) Nitschke ex Fuckel and *D. scobina* Nitschke, which form part of the complex centred around *D. eres* Nitschke, which is briefly discussed in chapter 3.

*Hysterographium fraxini* is known in Britain, though it appears to be rare except in West Wales (Dennis, 1968). It is normally only a saprophyte on the bark of ash and sometimes other deciduous trees and shrubs, but Zogg (1943) concluded that in Switzerland in damp conditions and on waterlogged soils it could act as a wound parasite. It could then cause elongated, canker-like depressions in stems and branches, and produce dieback of shoots. It could then do considerable damage, though without reinfection the lesions did not persist from year to year.

The black, rather elliptical hysterothecia occur on the lesions, and open by a slit. The club-shaped asci contain up to eight ascospores. These are ellipsoid, golden or chestnut-brown, divided into many cells by about five to ten transverse and one to three longitudinal septa, often constricted at the central septum, and measuring 30–50 × 12–20 μm (Dennis, 1968).

### The violet root-rot fungus *Helicobasidium brebissonii* (*H. purpureum*)

The perfect stage of the violet root rot fungus, *Helicobasidium brebissonii*, has been found on the bark of ash (Moore, 1959). This fungus is further considered in chapter 3.

## VIRUS AND VIRUS-LIKE DISEASES

Two virus diseases of the foliage of ash have been recorded in Britain (Cooper, 1979). In the first the symptoms are erratic, and may be only partially expressed. They show as a chlorotic vein banding, oak leaf patterns and chevrons, or a mild mottle. Arabis mosaic virus was found in leaves with symptoms, but may not be the cause of the disease. In the second of these diseases, which has been found on root stocks of *Fraxinus americana* and *F. excelsior*, affected leaves are distorted, and show chlorotic ring and line patterns. In this case, Arabis mosaic virus was isolated and produced the symptoms of the disorder in inoculated plants.

Also in Britain, Cooper *et al.* (1983) found Arabis mosaic virus and a tabovirus resembling tobacco rattle virus in a branch of *Fraxinus mariesii*, the Chinese flowering ash. The leaves showed yellow vein patterns and blotches of dead tissue.

In Italy, tobacco necrosis virus appears to be the cause of small chlorotic leaf spots surrounding a central dead area (Casalicchio, 1965).

A disease of *Fraxinus americana*, the American ash, is known in the USSR as well as in North America (Cooper, 1979). The affected trees develop witches' brooms, on which the normally pinnate leaves become simple and chlorotic. The disease is transmissible by dodder, and a mycoplasma-like organism has been found in the phloem of the diseased trees.

## DECAY FUNGI

Ash is usually resistant to attack by the *Armillaria* species, which are described in chapter 3. Of the other decay fungi, the commonest on ash is *Inonotus hispidus*, and *Daldinia concentrica* often causes a branch rot. *Perenniporia fraxinea* (*Fomitopsis cytisina*) may cause a severe white basal rot, and *Phellinus igniarius* a soft, white, spongy rot, mainly of the heartwood. These rot fungi are further discussed in chapter 20.

# Birch (*Betula* spp.)

Of the three birches native to Britain, one, *Betula nana*, is only a shrub that grows locally on moorlands in the Scottish mountains. The other two, *Betula pubescens*, the downy or white birch, and *B. pendula*, the silver birch, both occur commonly throughout the country and in most of Europe, the former generally on damper soils, the latter on the drier ones. A good many introduced birches are occasionally to be found in parks and gardens, and the variety 'Dalecarlica' of *B. pendula* is quite common.

# General

## Summer heat and drought

Birch is resistant to winter cold and frost and tolerates a wide range of site conditions. However, it may be badly affected (though less so than beech) by summer heat and drought, and in Britain was the first tree to show conspicuous damage in the phenomenally hot dry summers of 1975 and 1976. Affected trees showed varying degrees of premature browning and defoliation, death of bark with dark exudations, and ultimately death.

# Nursery diseases

## LEAF DISEASES

### The leaf rust *Melampsoridium betulinum* (Fr.) Kleb.

The leaf rust *Melampsoridium betulinum* occurs widely throughout Europe, Asia and North America. It is common on the leaves of birches of any age, but it causes significant damage only on nursery stock and occasionally on young natural regeneration. In Britain (and in Europe as a whole) it is known mainly on *Betula pubescens* and *B. pendula*, but it has been found on various introduced species, though Peace (1962) observed *B. ermanii* and *B. utilis* to be very resistant.

The abundant orange uredinia appear in late summer as very small orange pustules up to about 0.5 mm in diameter on the undersides of the leaves, showing as pale-yellow spots on the upper leaf surfaces. The urediniospores are ellipsoidal or subclavate, with a spiny wall, and measure 22–38 × 9–15 μm. They are formed in long chains (Roll-Hansen and Roll-Hansen, 1980b). Dooley (1984) in North America found that their optimum germination temperature *in vitro* was 10 °C. Orange to yellow-brown telia appear on the undersides of the leaves in late summer and early autumn, and may cover the whole leaf surface. The prismatic teliospores form below the epidermis, have smooth walls, and measure 30–52 × 8–16 μm (Wilson and Henderson, 1966).

There is evidence that the fungus may overwinter in birch buds (D'Oliveira and Pimento, 1953), as urediniospores under dormant bud scales (Dooley, 1984), and in fallen leaves, on which new uredinia may arise in the spring to reinfect the young expanding leaves (Weir and Hubert, 1918).

The spermagonia and aecia, which occur on larch, appear to be rare, and the fungus can certainly be found on birches far from any affected larches.

Control measures for birch rust have not been worked out. If the disease should prove troublesome in the nursery, general control measures could be applied, including the clearing up and destruction of any fallen leaves and the early application of one of the fungicides known to be effective against other leaf rusts.

# Post-nursery diseases

## LEAF DISEASES

### Leaf spot caused by *Taphrina betulae* (Fuckel) Johansson

The uncommon fungus *Taphrina betulae* is known as the cause of pale-green to yellow, later reddish-brown leaf spots on birch over much of northern Europe (Mix, 1949), including Scotland (Henderson, 1954; Henderson and Watling, 1978). The spots may be up to 1 cm across, and in midsummer bear a layer of asci on either or both leaf surfaces. The asci measure 20–24 × 11–13 µm, and contain eight elliptical or ovate ascospores. These measure 4–6 × 3.5–5 µm, but may bud in the ascus to produce many smaller spores.

In Britain the fungus has been recorded on *B. pendula* and *B. pubescens*, and in other parts of Europe also on *B. intermedia*, *B. medwediewi* and *B. turkestanica* (Mix, 1949).

A second species of *Taphrina*, *T. betulina* Rostrup, also attacks birch and fruits on its leaves. Since it also gives rise to witches' brooms, it is described below under 'diseases of shoots'.

### Minor leaf spots caused by various fungi

A number of fungi have been associated with leaf spots on birch. They have generally caused little damage, and in many cases little is known about their pathogenicity. They include *Phyllosticta betulina* Sacc., *Discula betulina* (Westend.) v. Arx, *Gloeosporium betulae* Fuckel, *Marssonina betulae* Magn., *Melanconium betulinum* Schm. and Kunze, *Asteroma microspermum* (Peck) Sutton and *Fusicladium betulae* (Rob. and Desm.) Aderh.

Grove (1935) reported that the black, globose pycnidia of *Phyllosticta betulina* were not uncommon in Britain scattered on living leaves of birch, usually on the upper side. The colourless, one-celled spores measured about 4–7(–10) × 1–2.5 µm. Affected leaves sometimes became brown and withered.

Grove (1935) also described two species of *Gloeosporium* associated with leaf spots on birch in Britain and throughout the rest of Europe. *G. betulinum* (now *Discula betulina* (Westd.) Arx, syn. *Myxosporium devastans* Rostr.) formed rounded brown spots on both sides of the leaves, but its small pustules were on the undersides only. Its colourless, one-celled spores generally measured 4–10 × 2.5 µm. It is common on birch leaves, and in some seasons may cause premature defoliation, moving upwards from the base of the tree (Young and Strouts, personal communication; Redfern *et al.*, 1981).

*G. betulae* (*Ceuthospora betulae* (Fuckel) von Arx) produces pustules on spots on the upper or on both sides of the leaves. Its colourless, one-celled spores measure 13–16 × 2 µm.

*Marssonina betulae* has been found forming pustules on irregular or stellate, brownish-black spots with a dentate margin (Grove, 1937). The colourless spores, which are finally two-celled, and then occasionally constricted at the septum, measure 17–22 × 5–6 µm (Batko, unpublished).

*Melanconium betulinum*, which was recorded by Grove (1937) on dead birch bark, was reported on birch leaves by Redfern *et al.* (1973). The dark, smoky-brown, one-celled, narrowly ellipsoid or almond-shaped spores measure 13–18 × 5–6.5 μm. *M. betulinum* is an imperfect state of *Melanconis stilbostoma* (Fr.) Tul.

*Asteroma microspermum* (*Gloeosporium betulae-albae* Sacc. and Dearn.) may also cause brown spots on birch leaves. Its colourless conidia measure 6.5–9.5 × 2–2.5 μm (Ellis and Ellis, 1985).

*Fusicladium betulae*, another fungus which may cause leaf spots on birch, is the anamorph of the ascomycete *Venturia ditricha* (Fr.) P. Karsten (Cannon *et al.*, 1985). Its pale-brown, two-celled conidia develop in summer, and measure 15–23 × 5–8 μm. The brown, setose pseudothecia of the perfect stage are embedded in the tissues of the fallen leaves, and form over the winter and in the following spring and early summer. Their pale-brown, unequally two-celled ascospores measure 15–23 × 5–8 μm (Ellis and Ellis, 1985).

### Powdery mildews: *Microsphaera penicillata* and *Phyllactinia guttata*

The powdery mildew *Microsphaera penicillata*, which is described below under '*Alnus*', may also sometimes produce its white or greyish mycelium and conidia on birch leaves. The hazel mildew *Phyllactinia guttata* (described under '*Corylus*', in chapter 19) may also sometimes affect birch (Hamacher, 1987).

## DISEASES OF SHOOTS

### Witches' brooms caused by *Taphrina betulina* Rostr. (*T. turgida* (Sadeb.) Giesenhagen)

Witches' brooms, formed of dense masses of adventitious twigs (figures 62, 63), are very common in Britain on *Betula pubescens* and *B. pendula* (especially the former), and occur elsewhere in Europe also on *B. nana*, *B. aurata*, *B. carpatica* and *B. intermedia* (Henderson, 1954; Mix, 1949) and in Japan on *B. maximowicziana* (Koike and Tanaka, 1986). They are caused by the fungus *Taphrina betulina* (though very similar brooms are caused by the mite *Eriophyes rudis*). The leaves on the affected shoots may become yellow, and on their undersides they bear a layer of asci that measure 41–75 × 15–21 μm. The ascospores, which measure 4.5–6.5 × 4–5.5 μm, quickly bud in the ascus to produce many smaller spores. The fungus becomes perennial in the brooms. The infected twigs are susceptible to frost, and many of them may be destroyed in autumn and winter (Blomfield, 1924; Henderson, 1954).

### Black knot caused by *Anisogramma virgultorum* (Fr.) Theiss. and Sydow (*Plowrightia virgultorum* (Fr.) Sacc.)

Massee (1914) described growths on birch somewhat like witches' brooms but associated with the fungus *Anisogramma virgultorum*. Small black lesions

**Figure 62**   Witches' brooms on birch caused by *Taphrina betulina* (D.H. Phillips)

**Figure 63**   Witches' broom on birch caused by *Taphrina betulina* (D.H. Phillips)

formed round the points of infection, sometimes girdling the stems. Below the girdled point adventitious shoots like those of witches' brooms grew out. Some of these shoots were themselves killed and replaced by others. The fungus forms a black stroma on the branches, erumpent from the wood. The perithecia are embedded in the stroma, and contain eight-spored asci measuring 45–66 × 12 µm. The unequally two-celled, egg- or club-shaped ascospores are colourless or greenish, and measure 8–12 × 4–6 µm (Ellis and Ellis, 1985)

### Cankers caused by *Nectria galligena*
*Nectria galligena*, which is well known as the cause of apple canker, may also cause cankers on twigs and branches of birch. As it has a wide host range, it is described in chapter 3.

## DISEASES OF FRUITS AND SEEDS

The small, pale-brown, stalked apothecia of *Ciboria betulae* (Woronin) White are often found on fallen, mummified fruits of birch. Their colourless, one-celled ascospores measure 10–14 × 3.5–5 µm (Ellis and Ellis, 1985).

## VIRUS DISEASES

### Cherry leaf roll virus (CLRV) and 'flamy' birch
Cherry leaf roll virus (CLRV), which has been found on *Prunus avium*, *Sambucus nigra* and other hosts (CMI/AAB Descr. V. 80), has been isolated from *Betula pendula* (*B. verrucosa*) by Cooper and Atkinson (1975). Schmelzer (1972) found the same virus on *B. pendula* in East Germany.

Cooper and Atkinson (who fully describe the virus) found that birches affected by the virus became bushy, with small leaves on affected branches. From July onwards the leaves showed mild chlorotic blotches, rings and line patterns. These symptoms became more clearly defined with time and by October the lines and rings were yellow. Cuttings from diseased trees and seedlings raised from infected trees grew more slowly than those from healthy material (Cooper *et al.*, 1984). The virus may be transmitted by pollen and by seed (Nienhaus, 1985), though when Schimalski *et al.* (1980) raised seedlings from naturally infected trees they found that only 0.2–4.3 per cent were infected. Cooper *et al.* (1984) found that many of the embryos in seeds from infected trees aborted.

A stem grooving and pitting and figuring of wood known as 'flamy' birch (Gardiner, 1965) is also thought to be caused by a virus (Cooper, 1979).

## DECAY FUNGI

Birch is very frequently attacked and killed by the root-rot fungus *Armillaria mellea* and occasionally by *A. ostoyae*, and sometimes by *Heterobasidion annosum*, which are described in chapter 3. Birch is one of the least durable of woods, and the bracket fungus *Piptoporus betulinus* is common as the cause of a brown cubical rot of standing trees. Other decay fungi may also affect it, among them *Fomes fomentarius* (which causes a white rot, especially in the upper parts of the tree), *Inonotus obliquus* (which causes a white ring rot) and a form of *Phellinus igniarius* (which causes a soft spongy rot, mainly of the heartwood). Birch is also one of the many hosts of the silverleaf fungus *Chondrostereum purpureum*. These fungi are described in chapter 20.

# Alder (*Alnus* spp.)

The native common alder, *Alnus glutinosa*, is widespread in damp woods and by lakes and streams throughout Britain (and most of the rest of Europe). The introduced grey alder, *A. incana*, a native of eastern Europe, is sometimes planted, especially in small woods in Scotland and on reclaimed tips. The American red alder, *A. rubra*, is occasionally found in parks and gardens, and so far to a very limited extent in forest plantings.

## GENERAL

### Dieback of unknown cause
Alders, especially *A. rubra* but sometimes also *A. glutinosa*, may be affected by dieback. In Britain this disorder has not been investigated, though it is suspected that scale insects may be involved. In Hampshire such dieback of *A. rubra* followed an infestation by the willow scale (*Chionaspis salicis*), which had passed unobserved because it was rendered inconspicuous by algae and lichens (Young and Carter, unpublished). In Germany and Holland various fungi that appear to be at most weak parasites have been found associated with it (Truter, 1947).

## LEAF DISEASES

### Powdery mildews: *Microsphaera penicillata* (Wallr.) Lev. and *Phyllactinia guttata*
The powdery mildew *Microsphaera penicillata* (*M. alni* var. *extensa* (Cooke and Peck) Salmon) may produce whitish colonies on alder leaves, especially on the undersides. Its small, dark-brown fruit bodies bear dichotomously branched, colourless appendages, and their ascospores measure 18–24 × 10–12 μm (Ellis and Ellis, 1985).

Another powdery mildew, *Phyllactinia guttata* (*P. corylea*) may also form a sparse mycelium on alder leaves. The appendages on its fruit bodies are unbranched, with bulbous bases. It is commoner on hazel, under which it is described in chapter 19.

### *Taphrina* spp. on leaves

Alders in Europe are attacked by five species of *Taphrina*, four of which affect the leaves. These are *T. tosquinetii* (the commonest), which causes a distortion of the leaves; *T. sadebeckii*, which causes a leaf spot; *T. viridis*, which also causes a leaf spot (and may be only a form of *T. sadebeckii*); and *T. epiphylla*. The last causes both a leaf spot and the production of witches' brooms, and is discussed below under 'diseases of shoots'. A further *Taphrina* sp., *T. amentorum*, gives rise to outgrowths on the catkin scales.

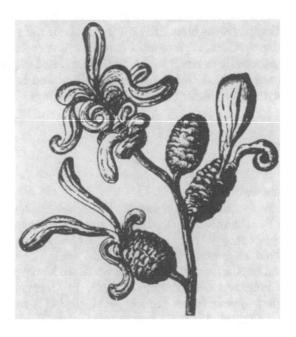

**Figure 64**   Catkin blight of alder caused by *Taphrina amentorum* (R. Hartig)

### Leaf curl caused by *Taphrina tosquinetii* (Westend.) Magnus
### (*T. alnitorqua* Tul., *Exoascus alnitorquus* (Tul.) Sadeb.)

The ascomycete *Taphrina tosquinetii* causes a distortion of the leaves (usually called leaf curl) and of the twigs of *Alnus glutinosa* in Britain, and of other species of alder, including *A. crispa* and *A. hybrida*, in other parts of Europe (Mix, 1949).

Leaf curl as it affects alders in Britain has been described by Henderson (1954) and Bond (1956). The fungus is most common on sucker shoots at the bases of trees, though sometimes on trees in crowded, damp conditions, twigs in the crown up to 20 feet (6.5 metres) from the ground may also be affected.

The cortex of diseased shoots is thick and wrinkled, and on these affected shoots the leaves, which are late in unfolding, at first appear reddish and crumpled. Parts of an affected leaf may become swollen, or the whole leaf may become enlarged, thickened, brittle and distorted. A bloom of developing asci appears on the swollen areas of one or both leaf surfaces.

The asci of the fungus are more or less cylindrical, and arise from basal cells which form below the cuticle and partly between the epidermal cells. They measure 17–40 × 7–13 μm (Mix, 1949). At first they contain eight ovate or elliptical, colourless ascospores which measure 2.5–5.5 × 2.5–5 μm. The ascospores may be discharged through an apical slit or may bud to form many smaller cells which are passively liberated (Bond, 1956).

The infected shoots may die after one or two seasons, but may persist for as long as ten or twelve years.

### Leaf spot caused by *Taphrina sadebeckii* Johansson
In Britain, *Taphrina sadebeckii* causes bright-yellow, unthickened spots up to 1 cm across on the leaves of *Alnus glutinosa* (Henderson, 1954). Elsewhere in Europe and in Japan it occurs also on *A. hirsuta* and its variety *sibirica*, and on *A. rugosa* (Mix, 1949).

The cylindrical asci arise from the subcuticular mycelium and measure 17–65 × 10–21 μm. They contain eight ovate or elliptical ascospores which measure 4–6 × 3.5–5 μm and often bud in the ascus (Mix, 1949).

### Leaf spot caused by *Taphrina viridis* (Sadeb.) Maire
*Taphrina viridis*, which does not occur in Britain, greatly resembles *T. sadebeckii* (of which it may be a form), but it causes small, pale-green or yellow, unthickened spots only on the leaves of the European green alder, *Alnus viridis*. It is found in mountainous areas in Scandinavia, Germany and Austria, and the Italian Alps. The ascospores, which bud in the ascus, measure 4.5–5 × 4–4.5 μm (Mix, 1949).

### The leaf rust *Melampsoridium betulinum*
The leaf rust *Melampsoridium betulinum*, described above under 'birch', may also attack alder (Roll-Hansen and Roll-Hansen, 1981).

## DISEASES OF SHOOTS

### Witches' brooms and leaf spots caused by *Taphrina epiphylla* Sadebeck
The perennial mycelium of *Taphrina epiphylla* may cause witches' brooms on the grey alder, *Alnus incana*. In summer, the leaves on the brooms show yellow, unthickened spots which bear the palisades of asci. These measure 20–60 × 10–23 μm. The ascospores measure 4–7 × 4–7 μm. The fungus

occurs in Norway and Sweden, Germany and Poland (Mix, 1949). It is exceedingly similar to *T. sadebeckii* (Hartig, 1894), of which it may be a form.


## DISEASES OF CATKINS

### Catkin blight caused by *Taphrina amentorum* (Sadeb.) Rostrup (*T. alni-incanae* (Kuhn.) Sadeb.)

*Taphrina amentorum* sometimes affects the female catkins of *Alnus glutinosa* and *A. incana* in Britain (Henderson, 1954), and elsewhere in Europe and in Japan and Alaska occurs also on *A. hirsuta*, *A. hybrida* and *A. rubra* (Mix, 1949). The fungus causes conspicuous red tongue-like outgrowths up to about 2 cm long and 0.5 cm wide to develop on the catkin scales (figure 64).

The cylindrical asci emerge from between the epidermal cells on both sides of the outgrowths. Mix (1949) found that the asci measured 26–53 × 10–23 μm. Each contains eight colourless, ovate or elliptical ascospores. These measure 4.5–6 × 4–5 μm, but may bud in the ascus to form other smaller spores.


## DECAY FUNGI

The commonest cause of rot in standing alders is *Inonotus radiatus* (*Polyporus radiatus*), which causes a white, flaky rot, especially towards the top of the tree. Among the other decay fungi affecting alder is another white rot organism, *Daedaleopsis confragosa*. These fungi are described in chapter 20.

# 14 Diseases of elm (*Ulmus* spp.)

Only one species of elm, *Ulmus glabra*, the wych elm, seems to be truly native in Britain. It occurs widely in northern Europe, and a number of weeping and other varieties, grown in parks and gardens, are also found. Other elms, including the English or common elm, *U. procera*, and the smooth-leaved or European field elm, *U. minor* (*U. carpinifolia*), are thought to have been introduced to Britain around the Roman period. In the rest of Europe *U. minor* is widespread over a large part of the continent, while *U. procera* occurs chiefly in Spain. Varieties of the smooth-leaved elm include *U. minor* var. *cornubiensis*, the Cornish elm, and var. *sarniensis*, the Guernsey or Wheatley elm. The European white elm, *U. laevis*, grows across Europe from France to Finland. Many elms were widely grown in town and country, in woodlands, in hedgerows, in parkland and in suburban streets. All the elms in Europe (and in North America) have been severely affected by the epidemic of Dutch elm disease which began in the late 1960s. As a result, in large areas most of the mature trees have been lost.

## General

### Wetwood
Elms are among the broadleaved trees sometimes affected by wetwood. High water and gas pressure in the trees then causes sap to ooze from the bark and flow down the trunks. The water-soaked bark is killed, and the sap flow often becomes very evil-smelling through the growth in it of various bacteria and yeasts. At least in some cases wetwood appears to be associated with bacteria within the trunks, and affected trees may also become liable to dieback. Wetwood is further discussed in chapter 3.

## Leaf diseases

### MINOR LEAF FUNGI

**Platychora ulmi (Schleicher) Petrak (*Euryachora ulmi* (Schleicher) Schröter, *Systremma ulmi* (Schleicher) Theiss. and Sydow; anamorph *Piggotia astroidea* B. & Br.)**
Several minor leaf fungi attack elm, though none is of economic importance.

The best-known is *Platychora ulmi*. From midsummer into the autumn and the early part of winter this forms small rounded black bodies on living and moribund leaves, and it is therefore sometimes called tar spot (Moore, 1959) (plate 36). These bodies, which are about 2–3 mm in diameter, are the stromata of the fungus, and at first contain pycnidial cavities (*Piggotia astroidea*) producing clear brown cylindrical spores measuring 9–11 × 4.5–6 μm (Ellis and Ellis, 1985). Later in the winter, perithecial cavities also arise, and produce cylindrical, eight-spored asci with ovoid, colourless, very unequally two-celled ascospores that eventually turn pale olive-brown (Dennis, 1968). These spores measure 10–12 × 4–4.5 μm.

### *Mycosphaerella ulmi* Kleb. (anamorph *Phloeospora ulmi* (Fr.) Wallr., syn. *Septogloeum ulmi* Died., *Septoria ulmi* Fr.)

*Phloeospora ulmi*, the anamorph of the ascomycete *Mycosphaerella ulmi*, causes brown (at first yellow) spots on elm leaves, and sometimes on their branches. Its small pustules arise on the spots on the undersides of the leaves, and bear white masses of long, cylindrical or sausage-shaped spores with rounded ends. At first the spores are one- or two-celled, but when mature they are three- (or occasionally four-) septate, and then measure 30–58 × 5–6 μm (Grove, 1937). This fungus is common on *Ulmus glabra* and *U. procera* (Moore, 1959). The ascospores of the perfect stage, which occurs on the fallen leaves, are two-celled and colourless, and measure 26–28 × 2.5–4 μm (Ellis and Ellis, 1985).

### *Cylindrosporella inconspicua* (Cav.) von Arx (*Gloeosporium inconspicuum* Cav.)

*Cylindrosporella inconspicua* also produces small, inconspicuous brown spots on the undersides of living and moribund leaves of English elm (*U. procera*). Pustules on the spots bear small, colourless, ellipsoid spores measuring 1–2 (–4) × 0.5–1 μm.

### *Taphrina ulmi* (Fuckel) Johansson (*Exoascus ulmi* Fuckel)

Also inconspicuous and easily overlooked is *Taphrina ulmi*. So far in Britain this fungus has been recorded only in Kent, and on hedgerows in Somerset, where it is known to be common (Bond, 1956). Elsewhere in Europe it is known across the continent from Britain through the Netherlands, Germany and Poland, to Sweden, and it also occurs in North America (Mix, 1949). It causes only a diffuse blotching of the leaves of the suckers and adventitious shoots. The blotches turn brown and the dead tissues fall out so that holes remain in the leaves. The small, naked asci form palisades on the leaf surfaces, and measure 12–20 × 8–10 μm. They each contain 4–8 spherical ascospores that often bud in the ascus (Bond, 1956) and measure 3–6 × 3–6 μm (Mix, 1949).

### Powdery mildews : *Phyllactinia guttata* and *Uncinula clandestina*

The powdery mildew *Phyllactinia guttata* attacks many broadleaved trees, including elms, on which it produces a sparse white growth. The trees are little affected. This fungus is described under '*Corylus*' (on which it is more important) in chapter 19. Another powdery mildew, *Uncinula clandestina* Schroet., has been recorded on elm in Spain (Burchill, 1978), and is listed by Ing (1990a) as occurring on elm in Britain. The small brown cleistoscarps of this fungus are about 95 μm across, and bear a ring of colourless appendages coiled at the tips. They contain from 4 to 20 asci, each of which contains from 2 to 3 ascospores (Salmon, 1900).

## DIEBACKS AND CANKERS

### Diebacks and cankers caused by *Nectria* spp.
Elms are among the trees most often attacked by the coral spot fungus, *Nectria cinnabarina*, on which it causes dieback and bark necrosis. *N. cinnabarina* is described in chapter 3. Elms may also be affected by cankers caused by *N. coccinea*. This fungus is more important as a cause of beech bark disease, and is therefore described under '*Fagus*' in chapter 11.

### Canker and dieback caused by *Plectophomella concentrica* Redfern and Sutton
An inconspicuous and apparently minor canker and dieback of elm caused by *Plectophomella concentrica* has been described by Redfern and Sutton (1981). The disease is widespread on wych elm (*Ulmus glabra*) in northeastern parts of Britain from Scarborough in northeast England to Stonehaven in east Scotland. It has also been found on the same host in west Scotland and in Orkney, and (on English elm) across southern England from Surrey and Hampshire to Worcestershire (Greig and Redfern, 1987). On one occasion the fungus was also found in Edinburgh causing a canker on lime (*Tilia* sp.). In many respects *P. concentrica* resembles *P. ulmi* (Verral and May) Redfern and Sutton (*Dothiorella ulmi* Verral and May), which causes a wilt disease of elm in North America.

*Symptoms*
*P. concentrica* causes swollen cankers mainly on one-year-old shoots, though infection of two- and three-year-old material may also take place (figure 65). Young affected shoots usually die during the winter after infection. The fungus then appears to die out, and in the ensuing growing season healthy recovery shoots grow out from the live shoot below the canker. When older and larger shoots are affected, the cankers usually heal, though the healing process may take several years to complete.

Inside affected shoots, black streaks similar to those caused by *Ophiostoma ulmi* may often be found in the xylem up to 5 or 6 cm above and below the cankers.

**Figure 65**    Canker on elm caused by *Plectophomella concentrica* (D.B. Redfern)


The causal fungus, *P. concentrica*, forms rings of epidermal or subepidermal, dark-brown or black stromata, in which the conidia are formed in locules. These locules open by several circular ostioles. The colourless, one-celled conidia are ellipsoid to cylindrical, straight or slightly curved, with smooth walls, and measure 2.4–4 × 1–1.5 μm.

The optimum temperature for growth of the fungus in culture was found to be 20 °C. Growth was still good at lower temperatures down to 5 °C, but at 25 °C it abruptly declined.

*Infection*
Infection appears to take place in the dormant season, usually on the internodes, but occasionally associated with buds and leaf scars.


WILT DISEASES


**Dutch elm disease caused by *Ophiostoma ulmi* (Buism.) *Nannf.***
**(*Ceratocystis ulmi* (Buism.) Moreau, *Ceratostomella ulmi* Buism.,**
**anamorph *Graphium ulmi* *Schwarz*) and by *Ophiostoma***
***novo-ulmi* Brasier**
The taxonomy of the fungus causing Dutch elm disease has been the subject of debate for many years; in particular whether the genus *Ophiostoma* is preferred to *Ceratocystis* and more recently on the most appropriate specific

names. The reasons for the current choice of *Ophiostoma ulmi* and *O. novo-ulmi* will emerge as the history of the disease is revealed below.

This disease is one of the most serious tree diseases known and has caused extensive losses in Europe, North America and western Asia. The first record was made in Picardy in France in 1918 and over the next decade it spread to a number of other European countries. It was first identified in Britain in 1927 at Totteridge in Hertfordshire. The status and development of the disease in Britain up to 1960 has been fully reviewed by Peace (1960). In 1930 the disease was first reported in the United States, which it probably reached as a result of the shipment of infested logs from Europe and it was found for the first time in Canada in 1944. In Asia the disease has been found in Turkey, Iran, Kashmir and Uzbekistan (Gibbs, 1978a).

There was a general decline in the level of the disease in Britain during the 1940s and 1950s. However, in the late 1960s there was renewed concern about the disease and several locally severe outbreaks were reported. At first it was thought that these were local flare-ups similar to those which had occurred from time to time in the previous two decades. By 1970 it was clear that a new and more severe epidemic had started and by the late 1970s over three-quarters of the original elm population of 23 million in southern England were dead or dying. At the same time serious losses were reported from many other parts of Europe.

Brasier and Gibbs (1973) demonstrated that this epidemic probably originated from the importation of diseased elm logs from North America containing a hitherto unidentified aggressive strain of the causal fungus. Subsequent research has indicated that there may have been two simultaneous epidemics present in Europe, one developing in western Europe from North American sources and the other spreading into Europe from the east (Brasier, 1979). The disease continues to spread and the majority of mature elms in North America, Europe and central and southwest Asia are likely to be killed.

There are several hundred references in the literature to research on Dutch elm disease. A few of the most important of these will be noted in this section but readers may find it useful to refer to review articles, including those of Peace (1960), Brasier and Gibbs (1978), Gibbs (1978a), Sinclair and Campana (1978), Stipes and Campana (1981) and Brasier (1986, 1990).

Following the onset of the severe outbreak of the disease in the late 1960s, it was established that two distinct forms of the fungus were present in both Europe and North America. The two forms were initially termed the aggressive and non-aggressive strains (Gibbs and Brasier, 1973). The non-aggressive subgroup is now believed to have been responsible for the earlier pandemic of the disease in Europe and North America in the 1920s–40s. The highly pathogenic aggressive subgroup has been responsible for the second pandemic. This subgroup itself comprises two races, the North American and the Eurasian races (Brasier, 1979) (figure 66).

Brasier (1991) has made a detailed study of the biological differences between the aggressive and non-aggressive subgroups. They differ, for example, in cultural characteristics such as appearance, growth rate and

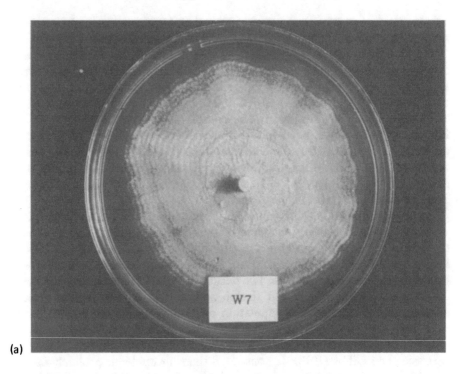

(a)

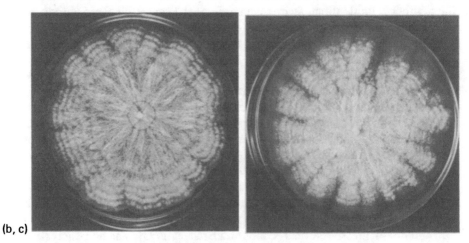

(b, c)

**Figure 66**   Cultures of the non-aggressive (a) and aggressive (b,c) forms of the Dutch elm disease fungi. The aggressive strain (*Ophiostoma novo-ulmi*) occurs in two well-marked forms, the North American (NAN) (b) and the Eurasian form (EAN) (c) (C.M. Brasier)

temperature optima and at the same time have markedly different pathogenicities in elms of moderate resistance to the disease. There is also a particularly significant reproductive barrier between the two subgroups exemplified both in culture and in the field. As a consequence of these morphological, physiological and genetic characteristics Brasier (1991) has proposed that the old non-aggressive subgroup is retained as *O. ulmi* and the aggressive subgroup be a new species, *O. novo-ulmi.·*

This new classification puts the historical and biological role of *O. ulmi*, *sensu stricto*, during the first pandemic in the early part of the century in a clearer perspective.

The fungus is carried by bark beetles that emerge in spring and summer from galleries (figure 67) under the bark of infested elm wood and feed in the tops of elms, where they form grooves in the crotches of the twigs (figure 68). The beetles bear the spores of the fungus on and in their bodies, and so infect many of the wounds they make. They later breed under the bark of dead and dying elms and in fallen or felled elm material (Gibbs *et al.*, 1977). The beetles mainly responsible for transmission in Britain are *Scolytus scolytus* and *S. multistriatus* (figure 69), but *S. laevis* is the main carrier in Sweden (Heybroek, 1967). The European *S. multistriatus* is also the most important carrier in North America, but there the native beetle *Hylurgopinus rufipes* can also transmit the disease.

**Figure 67** Dutch elm disease: undersurface of infected bark showing maternal and larval galleries of *Scolytus scolytus*, one of the insect vectors of the disease (Forestry Commission)

The fungus produces its fructifications in sheltered moist places, mainly under dead bark when it begins to lift and separate from the wood, and in the galleries made by the bark beetles. It has long been known to be heterothallic, so perithecia are formed only when two compatible mating types are present. The perithecia of *O. ulmi* and *O. novo-ulmi* have been described by

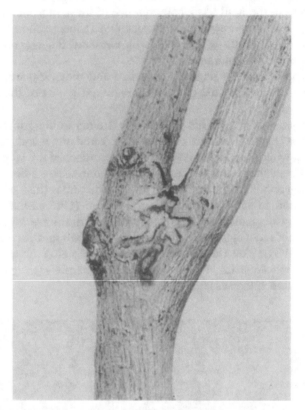

**Figure 68**   Dutch elm disease: elm twig showing feeding grooves made by the bark
beetle *Scolytus scolytus* (Forestry Commission)

Brasier (1991). The perithecial dimensions of *O. ulmi* are similar to those
given by Buisman (1932). Brasier (1991) recorded that the black globose
perithecial base of *O. ulmi* was 100–150 μm wide and the necks were 280–420
μm long. On the other hand, the perithecial base of *O. novo-ulmi* was 75–140
μm wide and the necks 230–640 μm long. Brasier (1991) noted that there was
a smaller overlap in the neck length to base width ratios of the two species.
The eight-spored asci soon break down to a jelly in which are embedded the
colourless, slightly curved ascospores that measure 4.5–6 × 1.5 μm. The
spores ooze out in the jelly, and are commonly to be seen in a round droplet
surmounting the perithecial neck.

There are three asexual spore forms of the fungus (Brasier, 1981). The
coremiospores or synnematal spores are borne at the tips of black coremial
stalks 1–2 mm in height. The single-celled spores are hyaline and measure 2–5
× 1–3 μm. Conidia of the *Cephalosporium* stage (more recently referred to

**Figure 69**   *Scolytus scolytus* (actual length 5–6 mm), one of the beetles that carry
Dutch elm disease (Forestry Commission)

*Sporothrix*: de Hoog, 1974) are produced on short mycelial branches or
conidiophores, and the hyaline single-celled spores range in size from 4 to 6 ×
2–3 μm. Mucilaginous droplets containing a cluster of these spores can often
be found. The third asexual form is the yeast stage, and budded yeast cells,
very variable in size, can be found in liquid state cultures and also on agar
plates and in the xylem vessels of infected trees.

The coremia appear on newly dead material and, in Britain, coremial
production takes place in all except the coldest weather. The formation of
perithecia is more restricted, and is most abundant from November to
February (Brasier, 1981).

Once within the tree the fungus increases mainly by a yeast-like budding
process, and the bud spores so formed are distributed in the sap stream, and
so spread rapidly throughout the current xylem tract. The fungus rarely
succeeds in growing laterally through the living summer wood from one
growth ring to the next, of either the preceding or succeeding year (Peace,
1960), though it may cross in this way in the roots and lower part of the trunk
(Banfield, 1968).

In inoculation experiments, Kerling (1955) found the fungus had moved 5 cm above the inoculation point in two days, and 75 cm two days later. When Banfield (1941) inoculated trees early in the season, when transport of the spores is most rapid, he found the fungus moved from the foot to the top of tall trees in between 20 minutes and 48 hours. He also made inoculations 40 feet (12.5 metres) up the trunks of trees, and the spores then moved down to the base in 15–20 minutes.

In Britain, the vector beetles emerge from May to October, so that they begin to transmit the disease in May, and the first external symptoms are usually visible in June. Parts of the crown then show wilting of the leaves, which turn yellow and eventually brown (plates 41, 42) and dieback of twigs and branches may follow. Fast-growing twigs may curl to form 'shepherd's crooks', which are useful in the detection of diseased trees in winter (figure 70). In severe attacks, the entire tree is killed before the end of the summer; if it survives, the tree is likely to die in the following season, as a result of the fungus passing from one annual ring to the next (probably in the roots). In the earlier epidemic of the disease in Britain (Peace, 1960) trees frequently recovered from the disease, but since the arrival of *O. novo-ulmi* recovery has become relatively rare.

**Figure 70**  Dutch elm disease: diseased twigs twisted into 'shepherds' crooks' (D.H. Phillips)

Internal symptoms soon begin to appear in infected trees. Dark droplets of gum become visible in the living cells and also in some of the vessels, in which colourless bladder-like tyloses also form and fill the conducting cavities. Fungal hyphae may also sometimes be found in the vessels, tracheids, fibres and parenchyma cells (Kerling, 1955; Wilson, 1965), and some breakdown of pits and cell walls may occur as a result of their activities (Oellette, 1961). The affected tissues show in transverse sections of the twigs and branches as rings of dark-brown spots in the spring wood (figure 71). If the bark and outer wood are pared away they show as discontinuous streaks (figure 72). These spots and streaks give a good indication of years in which active infection was present.

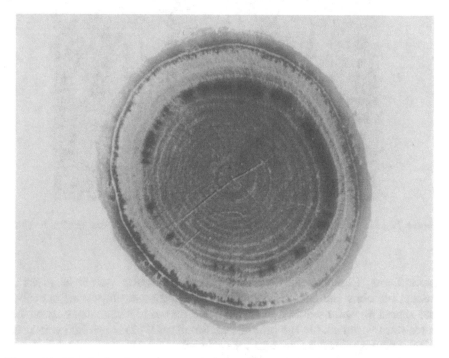

**Figure 71**  Dutch elm disease: a cross-section of a diseased twig, showing two rings of darkly stained, infected vessels (Forestry Commission)

Wounds are necessary for primary infection to take place (Smucker, 1937), and though it has been shown (Smucker, 1935) that direct spread by airborne spores is possible, almost all transmission is by the feeding of vector beetles. Secondary spread from tree to tree along root grafts and the common root systems of suckers is especially important in hedgerows or in closely planted avenues (figure 73).

The fungus is usually considered to cause wilting and death partly by the plugging of the conducting system and partly by the production of toxins (Peace, 1960). There is no doubt that materials that can be used to induce wilting artificially are present in culture filtrates of *C. ulmi* (Kerling, 1955). In

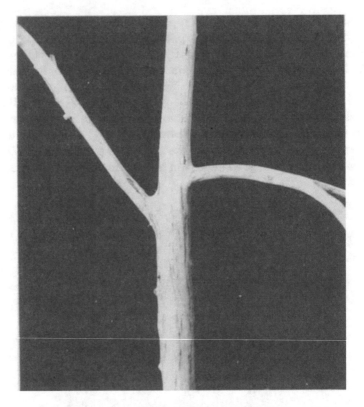

**Figure 72**  Dutch elm disease: infected stem stripped of the outer tissues to show
longitudinal streaks (Forestry Commission)

experiments, Elgersma (1967) found that the vascular system in stems of
susceptible elms had become completely blocked seven days after infection,
and almost no water would pass through it, whereas blockage was only partial
in resistant varieties. On the other hand, Smalley (1962) controlled wilting by
injecting trees with 2,3,6-trichlorophenylacetic acid (TCPA), which induced
tylosis formation, and he considered that the tyloses at least in part were
responsible for the limitation of the spread of infection. Pomerleau (1968) has
suggested that blockage of vessels cannot be the direct cause of wilting, but
that before blockage occurs the small spores of the fungus reach the leaves,
and there directly give rise to the death of the leaves and shoots.

*O. ulmi* and *O. novo-ulmi* are the only important fungi that cause epidemic
wilting of elms in Britain, though on very rare occasions Verticillium wilt
(caused by *Verticillium albo-atrum* and *V. dahliae*) is said to have been
recorded and to have caused somewhat similar symptoms, including streaking
of the wood (Peace, 1960).

For practical purposes it may be said that *O. ulmi* and *O. novo-ulmi* attack
only *Ulmus* and the related genus *Zelkova*, and it has been suggested that
some tropical members of the Ulmaceae, including *Trema*, *Celtis* and

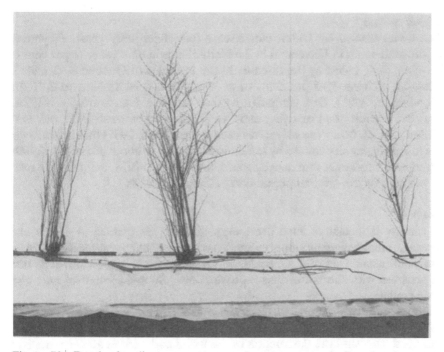

**Figure 73** Dutch elm disease: an excavated root system of elm, showing young suckers. The causal fungus spreads from sucker to sucker along the interconnecting roots, to give rise to secondary infections (Forestry Commission)

*Holoptelea*, which are very resistant, might be used in the production of resistant hybrids (Smalley and Riker, 1962).

All the species and hybrids of elm commonly grown in Britain are susceptible to Dutch elm disease, though the hybrid Huntingdon elm, *Ulmus* × *hollandica* 'vegeta', appears to show some resistance. *U. americana* and other American elms are even more susceptible than those of Europe (Beattie, 1937). Some Asian elms, including *U. pumila*, *U. japonica* and *U. parvifolia*, are resistant.

*Breeding for resistance*
Work on the selection and breeding of elms resistant to Dutch elm disease has been undertaken in several countries, including the Netherlands and the United States. In the Netherlands, a number of clones have been developed which incorporate crosses between European elms and *U. wallichiana* from India. In the United States, selections have been made from *U. pumila* or hybrids including this as one parent (Lester, 1978). All these selections show at least moderate resistance to *O. novo-ulmi*. The resistant selections differ in form and vigour from the native species of North America and Europe and cannot be considered as replacements for them. However, several may be suitable for planting as windbreaks or as urban trees.

*Losses caused*

The losses caused by Dutch elm disease have been very great. Heybroek (1966) states that in Hilversum in the Netherlands in nine years 71 per cent of the elms were killed by the disease. In the Province of Quebec in Canada it killed an estimated 16 per cent, or at least 600 000 white elms in 20 years (Pomerleau, 1961). In a campaign to eradicate the disease from the USA, over 2.5 million dead or dying elms were removed between 1933 and 1937 (Anon., 1938). More recently it has been stated (Hart, 1965) that in Michigan the losses in amenity caused by the disease amount to about 30 million dollars per annum. Losses in southern England, from 1970 to 1978, amounted to over 75 per cent of the original population of 23 million elms.

*Control*

A variety of measures have been suggested for the control of Dutch elm disease, ranging from attempts to eradicate the disease by chemical control by means of insecticides and fungicides and more recently to slow down its rate of progress with the aid of insect pheromones. Biological control may also have a place.

Eradication programmes were tried in the United States soon after the disease arrived there, but none succeeded. However, the felling of diseased and dead trees and the destruction of the wood and bark at an early stage of an epidemic can markedly slow down its rate of progress. The object of such sanitation felling is to eliminate the breeding grounds of the insect vector and thus reduce the beetle populations. These measures, sometimes in conjunction with other control measures such as insecticidal spraying, have been carried out over many years in parts of the United States (Neely, 1967, 1972). In the outbreak during the 1970s in Britain, many local authorities were unable to contain the spread of the disease but a few, including East Sussex and Brighton, have succeeded in markedly slowing down its progress by sanitation felling (Gibbs, 1978a).

Sanitation felling is considered to be the basis of any control programme against Dutch elm disease. Other control methods, many of which involve the use of chemicals, have little hope of success unless the beetle population is brought under control.

Insecticides have long been associated with Dutch elm disease control campaigns in the United States. DDT and later methoxychlor have been used to protect trees from attack by beetles (Doane, 1962) but have subsequently been banned.

More recently interest has centred on possible control of the vector beetles by use of pheromones (Peacock, 1975). The chemical constituents of the pheromone released by the virgin female of *S. multistriatus* have been synthesised and tested in the field. It remains to be seen whether traps incorporating this pheromone can be used for mass-trapping or whether their use is limited to monitoring beetle populations within an area.

Webber and Hedger (1986) studied the interactions between *O. ulmi* and various saprophytic fungi in the bark of elms dead or dying after attack by the disease. *Phomopsis oblonga* was found to be a strong competitor of *O. ulmi* in the bark. On the basis of these observations there is clearly potential for natural biological control of the source of inoculum for further spread of the disease.

Many fungicides have also been tested and among the most promising are carbendazim and thiabendazole. Techniques for injecting solutions of these chemicals into trees have been developed as either curative or preventative treatments (Smalley, 1978) (figure 74). However, such treatments are costly and can only be justified on valuable specimen trees or groups of trees and even then there is no guarantee of success.

**Figure 74** Dutch elm disease: injecting a tree with a curative or protective fungicide (Forestry Commission)

Root transmission of the disease is common in urban areas through root grafts and in hedgerow English elms through the common root system (figure 73). Where sanitation programmes are adopted it may be necessary to employ measures to prevent root transmission. This can be achieved by physically breaking the root connections using mechanical diggers or trenching machines or by the use of soil sterilants such as methyl bromide or metham sodium (Epstein, 1978).

**Verticillium wilt**
Elms may very occasionally be attacked by *Verticillium albo-atrum* and *V. dahliae*, the causes of Verticillium wilt. This disease is described in chapter 3.

## VIRUS AND VIRUS-LIKE DISEASES

**Elm mottle virus**
Elm mottle virus (E mot V) has been found in Britain, Bulgaria, the USSR and Germany, causing leaf mottling and ring-spot and line-pattern symptoms on the leaves of *Ulmus glabra*, *U. minor* and *U. vulgaris* [*sic*] (Cooper, 1979; Nienhaus, 1985). Its characteristics have been described by Jones and Mayo (1973), who found that it could be seed-borne. It can also be mechanically transmitted, and occurs in the pollen (Cooper, 1979; Nienhaus, 1985). The same virus has been found in lilac, but its economic importance is unknown.

**Elm witches' broom**
Available information on elm witches' broom, which seems to be caused by a mycoplasma-like organism, has been summarised by Cooper (1979) and Nienhaus (1985). The disease has been found on *Ulmus minor*, the smooth-leaved elm, in Czechoslovakia, the United Kingdom and Italy. On affected trees twigs with short internodes bear small chlorotic leaves. Sometimes the leaves at the proximal ends of affected twigs show 'cowl-formation', in which the leaf blade twists round the midrib to form a cone. Transmission has been achieved by budding. Elm witches' broom disease in Italy is described by Pisi *et al.* (1981).

## ROOT ROT AND DECAY FUNGI

Elm is one of the many trees the roots and butt of which are sometimes attacked by *Armillaria mellea* (which is described in chapter 3). The common-est decay fungi of elm, however, are *Pleurotus ulmarius* and *Polyporus squamosus*. The first of these causes a brown heart rot, while the second causes a fibrous white rot of trunks and large branches, especially near the top of the tree. *Rigidoporus ulmarius* may cause a brown cubical rot at the butt, and occasionally *Perenniporia fraxinea* produces a severe white rot, also at the base of the tree. The ascomycete *Ustulina deusta* also attacks the butt, and destroys the roots. These fungi are described in chapter 20.

# 15 Diseases of sycamore and maple (*Acer* spp.)

The only species of *Acer* native in Britain is the field maple, *A. campestre*, a small tree of woodland edges, hedgerows and boundaries. The sycamore, *A. pseudoplatanus*, introduced to Britain four or five hundred years ago, and now naturalised, is the only *Acer* species grown there as a forest tree, as well as for shelter and amenity planting. Also long introduced in Britain is the Norway maple, *A. platanoides*, much grown in streets, parks and gardens. These three *Acer* spp. are all native and widespread over most of the mainland of Europe (Evans, 1984). A number of other maples are native in the more southerly and southeasterly parts of Europe, and many introduced species from the Far East and North America are extensively grown in Britain and the rest of Europe as ornamental trees.

## SEEDLING BLIGHT

A seedling blight resembling Phytophthora seedling blight of beech, and caused by *Mycocentrospora acerina*, was first described by Hartig (1880) on *Acer*. It is not of major importance on this host, and appears rather more common on beech, under which it is described in chapter 11.

## LEAF DISEASES

**The powdery mildew *Uncinula bicornis* (Fr.) Lév. (*U. aceris* (DC) Sacc.)**
In Britain the common powdery mildew *Uncinula bicornis* forms a usually sparse, inconspicuous white covering on one side (mostly the lower) or occasionally on both sides of the leaves of sycamores and field maples (plate 25). Elsewhere in Europe it has also been recorded on Norway maple and *Acer rubrum* (Winter, 1887) and on the Tartar maple, *A. tataricum*, and the Oregon maple, *A. macrophyllum* (Burchill, 1978).

The mycelium first produces colourless, ellipsoid or oval conidia that measure 25–35 × 15–20 μm. Later, black, sub-globose ascocarps (cleistothecia) measuring up to about 0.2 mm across are formed. On their upper halves they bear many short, colourless, spreading appendages that fork at the tip to two (sometimes three or four) recurved hooks (figure 75). Within are four to twelve pear-shaped asci measuring 70–95 × 45–55 μm, each containing from six to eight colourless, elliptical ascospores that measure 17–27 × 7–14 μm (Dennis, 1968; Ellis and Ellis, 1985). The ascocarps are formed towards the end of the season, and shed their ascospores in the following spring, and these cause the primary infections on the young foliage.

**Figure 75**   Cleistocarps of *Uncinula bicornis*, showing the bifurcated appendages, and (in a dissected cleistocarp) the asci and ascospores (K.F. Tubeuf, after Tulasne)

The fungus is usually of no practical importance, and often passes unnoticed. If it causes concern, as it may in some nurseries in France, it may be controlled by collecting and burning infected leaves at the end of the season, and spraying the plants with flutriafol, which can be used to both prevent and eradicate infections (Boudier, 1987a), or with dinocarp or with colloidal or wettable sulphur.

**Other powdery mildews**
The powdery mildew *Phyllactinia guttata,* which is best known on hazel, has also been recorded on *Acer*, on which it is of no importance. It is described under 'hazel' in chapter 19.

A third powdery mildew, *Unicinula tulasnei* Fuckel, is known on leaves of *Acer* spp. in parts of mainland Europe (Viennot-Bourgin, 1956). It has often been regarded as a variety of *U. bicornis* (*U. aceris*) (Salmon, 1900) but has hooked but usually unbranched appendages and smaller spores.

**Giant leaf blotch disease of sycamore caused by *Pleuroceras pseudoplatani* (von Tubeuf) Monod (*Ophiognomonia pseudoplatani* (von Tubeuf) Barrett and Pearce, *Gnomonia pseudoplatani* von Tubeuf; anamorph *Asteroma pseudoplatani* Butin & Wulf)**

The giant leaf blotch disease of sycamore caused by *Pleuroceras pseudoplatani* (*Ophiognomonia pseudoplatani*) was described from Germany some sixty years ago, and has more recently been studied there by Wulf (1988). It has also been found in Denmark. In Britain it has been known in the Oxford area since 1973 (Barrett and Pearce, 1981), and has since been found in Gloucestershire (Strouts *et al.*, 1982) and apparently in some more northerly areas (Redfern *et al.*, 1987).

Infection of young leaves takes place in damp weather in May from ascospores shed from perithecia in dead leaves on the soil surface. Hence, infection tends to be most severe in branches close to the ground (Wulf, 1988). Symptoms of the disease first appear in mid-June. The fungus quickly causes large, radially spreading lesions which most often originate at the junction with the petiole and appear dark brown above and pale greyish-brown below, with infected leaf veins blackish in colour.

By early September the lesions may occupy up to 70 per cent of the leaf surface, and on some trees as many as 75 per cent of the leaves may be affected. Such trees are severely disfigured and their amenity value is much impaired. Affected leaves fall prematurely, from the early part of September.

The perithecia of *P. pseudoplatani* are found on the lesions on over-wintered infected leaves, beginning in March. Tubeuf (1930) and Wulf (1988) found these perithecia mainly on the upper sides of the leaves, but Barrett and Pearce (1981) found them chiefly on the leaf surface closest to the ground. The numerous necks of the perithecia project from the leaf surface like small black bristles. The perithecia embedded in the leaf tissues measure 150–320 μm across. Their necks are laterally attached, and measure 260–330 μm, but of this length only 160–220 μm projects above the leaf surface. The asci, which measure 60–90 × 6–10 μm, each contain eight elongated, unequally two-celled ascospores measuring 45–65 × 0.5–1 μm (Barrett and Pearce, 1981) or 40–50 × 1.5–3 μm (Wulf, 1988). From August onwards, inconspicuous acervulus-like pustules of an anamorphic stage, *Asteroma pseudoplatani*, may be found on the lesions. They produce club-shaped, straight or slightly curved spermatia which measure 6–7.5 × 2–3 μm (Wulf, 1988).

Giant leaf blotch affects the common sycamore and at least one of its purple-leaved garden varieties (Wulf, 1988). In areas in which the disease is troublesome, it can best be controlled by collecting and burning autumn leaves, which form the inoculum source in the following season.

**Tar spot caused by *Rhytisma acerinum* (Pers.) Fr. (anamorph *Melasmia acerina* Lév.)**

By far the commonest cause of tar spot of sycamore and some other *Acer* spp. in Britain and the rest of Europe, as well as in North America, is *Rhytisma acerinum* (plate 26). In Britain this fungus is common wherever the sycamore is grown, except in the polluted atmosphere of large industrial towns (Jones, 1944) and in very remote upland areas (Maxwell, 1933a,b). A fungus which causes a very similar tar spot disease of Norway maple in New York State, USA, may be a different though closely related species (Hudler *et al.*, 1987).

### Development of the disease

The disease has been described by Bracher (1924) and Jones (1923, 1925). Ascospores produced on fallen leaves in April and May infect the new foliage, their germ tubes entering through the stomata on the upper side of the leaves. Water-soaked spots appear at the sites of the infection, and enlarge and become yellow. The fungus grows in the leaf, and within and just below the epidermal cells on the upper leaf surfaces gives rise to a thick stroma with a shining black exterior, so that the yellow areas become transformed into black, slightly raised spots, usually up to about 15 mm across.

In summer, what appear to be spermagonial cavities (*Melasmia acerina*) form in the stroma, and open by round pores to extrude numerous colourless, slender, slightly curved spermatia that measure 6–9 × 0.5–1 μm (Grove, 1937), and the function of which remains uncertain (Jones, 1925).

Over the winter, on the fallen leaves, ascocarps form in cavities in or around the edges of the spermagonial stromata. The ascocarps are elliptical, and open by long, narrow, sometimes forked slits. They contain club-shaped asci that measure up to 130 × 10 μm, each containing a bundle of eight long, narrow, colourless ascospores slightly thicker at the upper end, measuring 60–80 × 1.5–2.5 μm. Between the asci are slender paraphyses with curled and sometimes forked tips (Dennis, 1968). The young ascocarps may be found in February, and the ascospores are shed in April and May as the host leaf buds open and the leaves expand (Grove, 1937). The ascospores have a gelatinous sheath (Bracher, 1924) and stick to the leaves after ejection. As they germinate they become bicellular before the production of a germ tube. The ascospores are ejected for only a short distance, after which they are further dispersed by air currents (Jones, 1925).

### Host range

In Britain *R. acerinum* is commonest on sycamore, but is occasional on the field maple. Elsewhere in Europe it has been found also on the Norway maple and the box elder (Grove, 1937; Laubert, 1933). In North America it affects many *Acer* spp., and the related *R. punctatum*, which is discussed briefly below, attacks the bigleaf maple, *Acer macrophyllum* (Anon., 1967).

### Variation in R. acerinum

Cross-inoculation studies have suggested that *R. acerinum* exists in several strains. One of these, which affects only sycamore, has sometimes been separated as *R. pseudoplatani* Muller (Muller, 1912). This has been listed as occurring in Britain (Ramsbottom and Balfour-Browne, 1951).

*Factors affecting tar spot: the effect of air pollution*

The asci of *R. acerinum* are exposed only when there is sufficient water to cause the swelling and curling back of the stomatal covering above them. The shedding of their ripe ascospores is then brought about by small changes in humidity. The spores shed on to the abaxial surfaces of the leaves will only germinate and cause infection in the presence of a film of water (Greenhalgh and Bevan, 1978). Infection in the spring therefore depends mainly on the presence of inoculum in the form of infected fallen leaves from the previous year, and suitably damp conditions for ascospore release and germination when young leaves are developing on the host plants.

In addition, however, it has been known for a long time that *R. acerinum* is sensitive to damage by atmospheric sulphur dioxide and perhaps by other air pollutants (Jones, 1944). The effect of sulphur dioxide on the incidence of tar spot has been studied by Bevan and Greenhalgh (1976), Greenhalgh and Bevan (1978), Oszako (1985) and Leith and Fowler (1988). Bevan and Greenhalgh (1976) found that in the areas in which they worked, tar spot failed to develop where the annual average sulphur dioxide concentration was greater than 85–90 $\mu g/m^3$ air. Further studies (Greenhalgh and Bevan, 1978) showed that at levels of pollution below this cut-off figure, the number of tar spots per unit area increased as pollution levels fell. Transplant experiments indicated that the fungus was vulnerable to the air pollutant at the time of infection. High levels of pollution prevented infection, but once infection had taken place, later pollution did not stop the development of the disease. In this connection, Leith and Fowler (1988) found that in Edinburgh, though pollution was present, it was not at levels high enough to affect the incidence of tar spot, and they found that the degree of infection appeared to be wholly dependent on the distance and amount of the available inoculum source, and was unaffected by the sulphur dioxide levels in the city. Oszako (1985) similarly found that in Warsaw tar spot occurred only when sulphur dioxide levels were low, the critical concentration appearing to be 0.025 $mg/m^3$.

*Damage caused*

Though in severe attacks *R. acerinum* causes some premature defoliation, and some may regard its black stromata as unsightly on trees grown for ornament, its effects are rarely serious. It has sometimes been known to cause appreciable damage to seedlings and young trees, especially in Germany (Wagner, 1927), but has so far been of little significance in Britain.

*Control of tar spot*

As the disease is of so little importance, control measures are not usually employed against it, but sources of infection may be reduced by the collection and burning or burial of infected leaves in autumn and winter. Young stock can be sprayed at intervals from bud burst with copper, dithiocarbamates or other fungicides (Wagner, 1927; Anon., 1967c; Wittmann, 1980). In North

America, Hudler *et al.* (1987) found that the same or a closely similar fungus could be controlled by spraying with benomyl, mancozeb or triadimefon.

**Tar spot caused by *Rhytisma punctatum* (Pers.) Fr. (anamorph *Melasmia punctata* Sacc. and Roum.)**

Less common, but also recorded on sycamore leaves in Britain (Cannon *et al.*, 1985) and elsewhere in Europe and in North America, is *Rhytisma punctatum*. In North America this fungus also occurs, especially in the Pacific Northwest, on the bigleaf maple, *Acer macrophyllum*, on which its life history and cytology have been studied by Woo and Partridge (1969). The morphology and life history of *R. punctatum* much resemble those of *R. acerinum*, and again it causes tar spots on the upper surfaces of the leaves, but the main spots are made up of smaller black stromata each up to about 1–5 mm across upon a light green background. Their asci measure 70–80 × 9–10 μm, and each contains 8 colourless ascospores wider near the apex than at the base, and measuring 30–60 × 1.5–2 μm (Rehm, 1896). In their studies in North America, where the fungus is common especially in the more westerly areas, Woo and Partridge (1969) found that each individual stroma contained one apothecium. Infection takes place in late spring and early summer by ascospores shed from apothecia produced in fallen leaves. The ascospores are ejected only a short distance and are then apparently dispersed by air currents, as in the case of *R. acerinum*. Light-green lesions appear on the leaves from early July, and the black stromata then form and develop their apothecia on fallen leaves over the winter. A spermagonal stage, *Melasmia punctata*, occurs in the stromata in Europe, but Woo and Partridge found no sign of spermagonia on bigleaf maple. In inoculation experiments they obtained infection only on the upper leaf surfaces.

Control measures for this disease are the same as those for tar spot caused by *R. acerinum*.

**Leaf spot caused by *Venturia aceris* (Höhnel) Sivanesan (*Phaeosphaerella aceris* Höhnel)**

*Venturia aceris*, the olive-green, unequally two-celled ascospores of which measure 9–10.5 × 3.5–4.5 μm, causes brown spots on the leaves of Norway maple in Germany (Sivanesan, 1977).

**Leaf spot caused by *Cristulariella depraedans* (Cooke) Höhnel (*Polyactis depraedans* Cooke, *Illosporium diedickeanum* Sacc.)**

*Cristulariella depraedans*, which occurs in Europe and North America, causes spotting and withering of the leaves of *Acer* spp. In Britain and Germany it affects the sycamore and the Norway maple (Batko, 1974; Redhead, 1975), and in Canada and the USA it has been found on the Norway maple and on *Acer palmatum*, *A. rubrum*, *A. saccharum*, *A. saccharinum*, *A. spicatum* and *A. circinatum*, and also (in British Columbia) on the rosaceous *Aruncus sylvester* (Bowen, 1930; Redhead, 1975).

The fungus, one of the Deuteromycotina (Fungi Imperfecti) causes grey spots on the green leaf. The spots are at first water-soaked and only about 1 mm across, but enlarge somewhat, and become greyish brown, and may join together to cover large areas (figure 76). Fungal fruit bodies arise on the spots on one or both sides of the leaf, sometimes particularly on the leaf veins (Bowen, 1930). Under a hand lens they look like minute white pinheads. Each consists of a simple, erect, colourless, septate hypha about 100–270 μm long and 8–16 μm wide, bearing at its top a compact, flattened-globose sporing head about 100–150 μm across. The sporing head is made up of one-celled, repeatedly di- or trichotomously dividing branches that radiate from a central rounded cell. At their outer ends the outermost branches carry two- or three-lobed cells somewhat resembling basidia, that produce small, globular, colourless conidia singly on the lobes. These conidia measure 2–3 or 4 μm across (Bowen, 1930; Waterman and Marshall, 1947). After his study of the fungus, however, Redhead (1975) took a different view, regarding the sporing heads as complex 'propagules' or 'macroconidia', which germinate to produce infective hyphae. The fungus may also form small black sclerotia (Cooke, 1885). With time, but often quite rapidly, the affected leaves wilt and wither, and fall early from the tree.

**Figure 76** Leaf spots on sycamore leaf caused by *Cristulariella depraedans* (D.H. Phillips)

In Britain, epidemics caused by *C. depraedans* on sycamore were recorded in Norfolk about 1879, in Perthshire about 1935, on the Clyde in 1945 (Moore, 1959), and on sycamore and Norway maple in east and southeast England from August to October 1968 (Burdekin, 1969). In August and September 1980 it was widespread in South Wales and southern England and in Argyll, and outbreaks in various parts of the country were also reported in 1981 and 1986.

Attacks appear to be associated with cool, wet summers. The disease is often most severe on the lower branches, on which affected leaves are usually first found.

As *C. depraedans* causes epidemics only infrequently, and when it does it attacks only late in the season, it is of no importance as a cause of damage, and no control measures have so far been required to deal with it.

**Other minor leaf spots caused by species of *Phyllosticta*, *Diplodina*, *Phloeospora* and *Leptothyrium***

Various species of *Phyllosticta*, *Diplodina*, *Phloeospora* and *Leptothyrium* have been recorded in Britain associated with spotting of the leaves of *Acer* spp. Several of these fungi may occur together on the same leaf, and they form a very confused group that needs further study (Moore, 1959).

*Phyllosticta aceris* Sacc. produces more or less rounded, brownish spots up to about 10 mm or more across on living leaves of the field maple. Its black pycnidia, which are embedded in the leaf tissues, open by a pore usually on the upper side of the leaf, and contain colourless, ovoid spores measuring 5–7 × 2.5–3 μm. The fungus may be found from July to September in southern England and the Midlands, and occurs also in continental Europe as far south as Italy (Grove, 1935). *P. campestris* Pass. in litt. apud Brun. has also been recorded in Britain on field maple seedlings, but Grove (1935) considered that it was no more than a young stage of *P. aceris*.

*P. platanoidis* Sacc. has been recorded on sycamore in parts of southern England, South Wales and southern Scotland (Moore, 1959) and on young seedlings of Norway maple in Oxfordshire (Grove, 1935). It causes rather indistinct spots, and the globose, black pycnidia, which are embedded in the leaf with their openings on the underside, contain straight, cylindrical, colourless spores measuring 4–6 × 0.5–1 μm. *P. platanoidis* is often accompanied by *Phloeospora aceris*, and also by *Leptothyrium platanoidis*, which Grove (1935) considered was a form of the *Phyllosticta*.

*Diplodina acerum* Sacc. & Br., the pycnidia of which contain colourless, two-celled spores measuring 10–20 × 3–4.5 μm, was found by Batko (unpublished) causing considerable but very local damage to leaves of Norway maple at Auchencastle, Scotland, in early September, 1957. This fungus appears to be one of the many synonyms of *D. acerina* (Pass.) Sutton (*Septogloeum acerinum* (Pass.) Sacc.), the spores of which measure 13.5–15.5 × 3–3.5 μm (Sutton, 1980).

*Phloeospora aceris* (Lib.) Sacc. is common in summer and autumn on leaves, especially of seedlings, of sycamore, and sometimes of field maple, and is probably the best known of these minor leaf fungi (Redfern *et al.*, 1973, 1981, 1987). It causes brown, rounded spots, and embedded in these are large pycnidia with a poorly developed wall. The pycnidia contain more or less cylindrical or rather narrowly club-shaped, eventually 4-celled, colourless spores that measure 20–30 (–45) × 2–(4–5) μm. *P. pseudoplatani* Bubak, which has also been recorded on *Acer*, was considered by Grove (1935) to be the same fungus as *P. aceris*.

Grove (1935) also described *P. platanoidis* Petr., found causing small spots on the leaves of young plants of Norway maple in Oxfordshire. The pycnidia, on the upper side of the leaf, contain long hair-like, curved, four-celled spores measuring 60–70 × 1.5–2.5 μm.

*Leptothyrium platanoidis* Pass. apud Brun. was also recorded by Grove (1937) causing brown spots on leaves of seedling sycamores in South Wales and Staffordshire. The pycnidia open on the undersides of the leaves, and contain straight or slightly curved spores measuring 4.5 × 1 μm. As noted above, in Grove's opinion *L. platanoidis* was part of the life cycle of *Phyllosticta platanoidis*.


## DISEASES OF THE BARK AND TRUNK

### Sunscorch and other climatic damage

Like other thin-barked trees, *Acer* spp. may be affected by sunscorch, which leads to the death of areas of bark on the south-facing sides of their trunks.

After the hot dry summer of 1976, Murray (1978) described stem lesions and crown dieback of sycamores. He isolated various fungi from the affected trees, but concluded that the damage was due primarily to drought stress.

At about the same time, Bevercombe and Rayner (1978, 1980) and Rayner *et al.* (1981) described some curious diamond-shaped cankers on sycamore trees. Sometimes the cankers were occluded by a covering of callus tissues, while in other cases no healing took place, and some trees died. Often, but not in all cases, the fungus *Dichomera saubinettii* (Mont.) Cooke was isolated from the cankers. The disease was traced back over some 40 years, but was again especially prevalent after the 1976 hot dry summer. Like Murray, these authors also concluded that drought stress appeared to be the primary cause of the disorder.

Gregory (1982) examined sycamore trees with severe cankers and bark necrosis. He found that *Nectria coccinea* and *Diplodina acerina* could cause some extension of these wounds, and considered that these fungi were weak parasites able to cause such damage following summer droughts (see above under 'other minor leaf spots').

Climatic damage is further discussed in chapter 2.


### Diebacks, cankers and bark necrosis caused by *Nectria* spp.

The extension of drought-induced cankers and bark necrosis by *Nectria coccinea* has been mentioned above (sunscorch and other climatic damage).

*N. cinnabarina* (coral spot) is another wound parasite. It may cause a dieback of *Acer* spp., in which it produces a brown or green stain in the wood. *N. galligena*, well-known as the cause of canker in apple and pear, has also been associated with cankers in *Acer* spp., as well as in many other trees. *N. coccinea* is described under '*Fagus*' (chapter 11), on which it is involved in beech bark disease; the other two *Nectria* spp. are described in chapter 3.

**Sooty bark disease caused by *Cryptostroma corticale* (Ell. and Everh.) Gregory and Waller**

The imperfect fungus *Cryptostroma corticale* has been found in the Great Lakes region in Canada and the USA, and in England, France and Germany (CMI Map No. 272; Plate and Schneider, 1965). In North America and in Germany it is known almost entirely as a saprophyte and as a cause of asthma (Towey *et al.*, 1932; Plate and Schneider, 1965). In Britain it was first noticed in 1945 in Wanstead Park in northeast London (Waller, 1952). Within a few years it appeared in many adjacent parts of Greater London, where strong circumstantial evidence suggested that it was the cause of a devastating wilt and dieback of sycamores (Gregory *et al.*, 1949). Surveys soon revealed its presence outside Greater London in Surrey, Essex and Hertfordshire, and further afield in Norfolk (Peace, 1955), and later as far north as Kettering, Northants (Pawsey, 1962b). Beyond the London area, however, it was almost invariably a saprophyte on trees killed by some other agency (Peace, 1955). In a very severe outbreak in 1976, however, it caused damage as far afield as 160 miles from its original centre in London (Young, 1978). Not long after the discovery of the fungus in London, it was also found to be destroying sycamores in France, first in Paris, and later in Grenoble (Moreau and Moreau, 1954).

How the fungus reached Europe is unknown, though Gregory and Waller were of the opinion that it came to Britain in timber imported into London docks.

*The causal fungus*

*C. corticale* has been most fully described by Gregory and Waller (1951). These authors considered it to be mainly an invader of the phloem and cambium, but it has more recently become apparent (Young, 1978) that the fungus first invades the woody tissues and then moves out to the cambium and bark at a later stage. It forms a thin, black stroma in the cambium, and one, or sometimes up to three, sporing stromata in the bark. The sporing stroma is at first a white mycelial sheet, but it splits parallel to the surface, to form a roof and a floor. These are separated by the growth of vertical columns, so that a shallow space arises, up to about 1 mm deep. Hence, the affected area of bark is pushed up to form a blister. The roof, floor and columns blacken as they mature. The floor is lined by a bluish-grey layer of conidiophores from the ends of which long fragile chains of spores are formed. These conidia are one-celled, smooth, oval, brown when mature, and measure 4–6.5 × 3.5–4 μm. They fill the stromatic cavity in a sooty mass, and between them run sticky, unbranched capillitial threads attached to the floor. With time, the bark breaks up and falls from the affected trees, and the spores are exposed as a velvety layer that is gradually dispersed by the wind (plate 32). In France, Moreau and Moreau (1954) found the ascomycete *Eutypa acharii* associated with *C. corticale* on affected trees. In Britain, *E. acharii* has been found growing as a saprophyte on wood under the dead bark of sycamore, and may be mistaken for the cambial stroma of *C. corticale*. However, the perithecia of *E. acharii*, about 0.5 mm in diameter and embedded in the wood beneath a black layer, are very distinctive.

Alcock and Wheeler (1983) found that different isolates of *C. corticale* grew at different rates, both on agar and in young sycamore plants. In further studies, Bevercombe and Rayner (1984) examined isolates from Wales and a large part of southern England, and found that they fell into 11 different somatic-compatability groups.

*Symptoms of the disease*
Careful observation has shown that the first symptom of the disease is a wilting of the leaves of part of the crown of the tree. This wilting has been seen at varying times betwen May and September, and is followed by dieback of the affected branches (Gregory and Waller, 1951). The wilting leaves dry and their petioles droop, but they remain attached to the tree. The woody tissues inside a wilted branch are stained dark yellow or green. Later, at any time of the year, sporing lesions may be found, either as blisters or as open spore masses. The affected trees usually die within a few years. The disease is often first noticed only when the bark lesions are visible, and the affected tree fails to come into leaf in the spring. Inspection then shows dark vertical blisters on the trunk and branches. If the bark is then stripped off the blisters, a characteristic crackling sound may be heard as the stromatic columns break, and the brown spore mass is exposed.

On dead smooth-barked sycamores, the fungus may spread within the bark over the whole trunk, and as the bark breaks up and falls, the spores form a sooty deposit on the vegetation around (Gregory *et al.*, 1949). Sometimes, however, the sporing lesions may be high up in the crown, when they are difficult to see, and on thick-barked sycamores only small stromata a few inches across are formed (Gregory and Waller, 1951; Peace, 1955).

If a partly affected tree is cut up, a yellow or dark greenish-brown stain like that in the wilted shoots may be found in the heartwood extending up and down the trunk, sometimes for the whole length of the stem and into the branches. Both the stain and the fungus may also be traced downward into the root system (Moreau and Moreau, 1951; Townrow, 1954). In places the stain extends into the sap wood and reaches the bark, and most of the stem blisters appear to form where the stain touches the cambium. Once the tree is dead, the stain disappears, leaving the wood a uniform grey colour (Gregory *et al.*, 1949).

*Host range*
C. *corticale* is known as a pathogen chiefly in Britain and France. There it is restricted almost entirely to sycamore, though it has also been found on field maple (*Acer campestre*), the Norway maple (*A. platanoides*) and (in France only) the box elder (*A. negundo*) (Peace, 1955).

In North America it has occurred mainly as a saprophyte on maple logs (probably on *Acer saccharum*), though it was also recorded by Towey *et al.* (1932) on dying maples, hickories (*Carya* spp.) and basswoods (*Tilia* spp.).

*Factors affecting the disease*
The most severe outbreaks of sooty bark disease have occurred in and around Greater London, in summer the warmest region in Britain, and have developed in years following particularly hot summers. Thus in the forty years up to 1980 very high summer temperatures occurred in 1947, 1948, 1955, 1959, 1975 and 1976 and disease outbreaks developed in the subsequent season or seasons. Young (1978) studied the meteorological data related to these years and concluded that the disease was likely to become acute in seasons following those where any summer month (June, July or August) had a mean daily maximum temperature of 23 °C or more.

Observations by Townrow (1954) indicated that the fungus grew three times as fast in culture at 25 °C as it did at 10 °C and that spore germination also increased with temperature. These observations provide some evidence to support the hypothesis that this is a temperature-regulated disease.

Dickenson and Wheeler (1981) showed that the fungus grew more rapidly in the wood of sycamore plants held at 25 °C rather than 15 °C, and greater growth also occurred in trees affected by water stress. It seems likely that high temperatures and often associated water stress play a very important role in the rate of fungal growth in the host and therefore in the development of serious outbreaks of the disease.

In Wanstead Park when it was first found there, C. *corticale* killed between 15 and 20 per cent of the sycamores each year from 1948 to 1951 (Gregory and Waller, 1951). Since that time there has been a series of outbreaks, associated with hot summers, culminating in a severe outbreak in 1976 and 1977 following the hot summer of 1975 and the record temperatures in the summer of 1976. Although no formal surveys were undertaken at this time, death of sycamore occurred over many parts of southern Britain and was particularly severe in the Greater London area.

*Control*
If the relationship between summer temperatures and the disease is maintained, in Britain the disease should largely be confined to the southern areas, and Greater London in particular. Serious outbreaks are likely to occur only infrequently in seasons following hot summers (though such summers may increase in number if current theories on global warming are borne out). In summers with average or below average temperatures the disease is likely to remain quiescent. Unless there is a dramatic change in the summer tempera-

tures in Britain, active measures to control the disease would therefore seem to be unnecessary.

## WILT DISEASES

### Verticillium wilt

*Acer* spp. are among the trees most often and most severely affected by Verticillium wilt, which is described in chapter 3 (plate 27). Young and Strouts (1975) described unusual symptoms of this disease on sycamores, in which slime fluxing took place from longitudinal cracks in the bark. These symptoms resemble those described by Gravatt (1926) on red and silver maples in the USA.

## ROOT DISEASES

### Phytophthora root rot

A root rot of snake-bark maple (*Acer pensylvanicum*), which in one case killed a number of trees in southern England, appeared to be caused by *Phytophthora cambivora*, which was quickly followed by *Armillaria mellea* (Brasier, 1971). *Acer* spp. are also known as hosts of the ubiquitous *P. cinnamomi*. These *Phytophthora* spp. are further described in chapter 3.

## VIRUS DISEASES

### Stunt and leaf spot of *Acer negundo* and *A. pseudoplatanus*

Cooper (1979) has summarised information on a virus disease of *Acer negundo* and *A. pseudoplatanus* known from Hungary and the United Kingdom. Affected plants show a shortening of the internodes and failure of the terminal bud. The leaves change in outline, and show chlorotic spots.

### Maple leaf perforation

Subikova *et al.* (1985) in Czechoslovakia have described a maple leaf perforation disease which can be mechanically transmitted. Spherical virus particles, and sometimes mycoplasma-like organisms, appear to be associated with the disease.

## ROOT ROTS AND DECAY FUNGI

*Acer* spp. are among the hosts of one or more of the *Armillaria* species, which are described in chapter 3.

Among the commonest decay fungi of sycamore are *Inonotus hispidus* and *Polyporus squamosus*, both of which are wound parasites causing white or yellowish rots, mainly in the top of the tree. Also sometimes found on sycamore are *Rigidoporus ulmarius*, the cause of a dark-brown cubical rot, mainly at the base of the tree, and *Ganoderma applanatum*, which may attack the roots, butts and large branches and cause a soft, spongy white rot. These fungi are described in chapter 20.

# 16 Diseases of poplar (*Populus* spp.)

Of the many species of poplar, the aspen, *Populus tremula*, is native in Britain and throughout most of Europe. The black poplar, *P. nigra*, is less common in Britain but is also native, and likewise occurs elsewhere in Europe. Many introduced poplars are also grown in Britain. They include the white poplar, *P. alba*, which is common and more or less naturalised in southern England, and is native in France, Belgium and Holland and eastwards. The Lombardy poplar, *P. nigra* var. *italica*, from northern Italy, is also much planted. The forms grown in Europe for landscaping and timber (for basket and match making, but now chiefly for pulpwood and veneer material) are mainly clones of the balsam poplar, *P. trichocarpa*, from North America, and of the hybrid black Italian poplar, *P* × *euramericana*. Over a large part of continental Europe these poplars are an important crop.

Poplar diseases have been reviewed by the Groupe de Travail des Maladies de la Commission Internationale du Peuplier (Anon., 1979), and Pinon (1984a) has discussed the management of poplar diseases.

## Nursery diseases

### GENERAL

Poplars are among the hosts of the crown gall organism, *Agrobacterium tumefasciens*, which is briefly described in chapter 3.

### DISEASES OF BARK AND SHOOTS

**Dothichiza bark necrosis or trunk scab caused by *Cryptodiaporthe populea* (Sacc.) Butin (anamorph *Discosporium populeum* (Sacc.) B. Sutton, syn. *Dothichiza populea* Sacc. and Briard, *Chondroplea populea* (Sacc.) Kleb.)**
The ascomycete *Cryptodiaporthe populea* is found throughout Europe and into Turkey, in the eastern parts of North America, and perhaps in Argentina (CMI Map No. 344, 1968). It is now known also in China (Zhong, 1982). It attacks both plantation trees and nursery plants, and in some areas from time to time causes a severe bark necrosis. In Britain it is common, but has so far been of minor importance, and has caused significant damage only in nurser-

ies. In continental Europe, however, it has severely affected large trees and even caused their deaths (Schönhar, 1952; van der Meiden and van Vloten, 1958). In recent years severe attacks have occurred in northern Italy (Anselmi, 1986).

The perithecia of the fungus are uncommon in Britain; they were first found there on Lombardy poplar in Gloucestershire in January 1969 (Burdekin, 1970), and mature in February and March (Ellis and Ellis, 1985). They are long-necked and wholly embedded in the host tissues with only the ostioles visible at the surface; they contain asci measuring 75–85 × 12–16 μm, each with eight two-celled, colourless, elliptical ascospores measuring 16–24 × 6–9 μm (Butin, 1958a).

The anamorphic stage, now called *Discosporium populeum*, but formerly known as *Dothichiza populea*, is very common, and consists of pycnidia up to 2 mm across, embedded in the host tissues but wide open when mature, and forming waxy masses of colourless, sub-globose or egg-shaped conidia measuring 8–12 × 7–9 μm (Butin, 1958a; Goidanich, 1940).

The pycnidia are formed throughout much of the year, from March till late autumn, and the pycnospores can survive for five years (Gremmen, 1958; Taris, 1959), though usually under natural conditions they persist at most for not more than ten months (Butin, 1962). The pycnidia cease to produce spores only in very dry weather (Goidanich, 1940).

In culture, *C. populea* grows well in media containing malt or yeast extracts (Butin, 1958b) or in poplar extracts (Hubbes, 1959; Zhao *et al.*, 1983). It produces an oily, partly crystalline toxin, which is partly responsible for the damage it causes (Braun and Hubbes, 1957; Butin, 1958b; Hubbes, 1959). The fungus grows best at a temperature of about 20 °C, but growth can still take place at temperatures only a few degrees above freezing (Viennot-Bourgin and Taris, 1957). The conidia germinate when the relative humidity is 90.5 per cent or higher (Butin, 1962), at an optimum temperature of 20 °C and at an optimum pH between 5.0 and 9.0 (Zhao *et al.*, 1983).

Various strains of *C. populea* exist. They differ in host specialisation, and change the colour of a poplar extract medium to various shades of yellow and brown (Braun and Hubbes, 1957; Brendel, 1965), the deepest brown colours being produced by the most pathogenic strains (Hubbes, 1959). Anselmi (1986) suggests that hypovirulent strains of the fungus may be developing in France, Italy and Yugoslavia.

*Mode of entry*
*C. populea* is a facultative parasite that enters the tree mainly through wounds (including leaf scars and bud scale scars, and damage by insect pests as well as other mechanical injuries) (Viennot-Bourgin and Taris, 1957; Gremmen, 1958; Schmidle, 1953; Zhong, 1982), but also through lenticels (Hubbes, 1959), and through shoot tips and buds (Braun and Hubbes, 1957; Viennot-Bourgin and Taris, 1957). It has been known to enter through leaf scars following premature defoliation caused by *Melampsora larici-populina* (van der Meiden and van Vloten, 1958).

*Symptoms*
The fungus gives rise to dead, oval, generally sunken, greyish, brown or black patches on the stems round the bases of twigs, particularly at the junction of one- and two-year-old growth. Similar lesions may occur on branches and around buds (Butin, 1956a). On areas of active infection the black pycnidia of *C. populea* protrude through the dead bark, though older parts of lesions are often colonised by other fungi, such as *Cytospora chrysosperma* (*Valsa sordida*), *Glomerella miyabeana* and *Chondrostereum purpureum*, and by bacteria (Gremmen, 1958; Breuel and Börtitz, 1966). As a reaction to the attack, the host may produce wound periderm, and this may sometimes prevent further spread of the pathogen (Butin, 1956b).

*Time of infection and factors affecting the disease*
There is some disagreement concerning the time of infection and the factors affecting the disease, perhaps partly because the fungus may remain latent for as long as two years after infection (Gremmen, 1958).

Most writers have agreed that infection occurs mainly in the winter months. Thus Schönhar (1953) concluded that infection took place mostly in winter, and in spring and autumn affected only drought-weakened trees. Müller (1953b) also considered that infection took place mainly in winter, and that the trees were resistant in the growing season. Braun and Hubbes (1957) found that callusing of wounds took place only above 12 °C, and wounded plants became infected only below this temperature. Later, Hubbes (1959) failed to infect plants at temperatures above 16 °C. On the other hand, Aerts *et al.* (1959) found in their experiments that infection occurred only in spring and summer. According to Viennot-Bourgin and Taris (1957) there are three main periods of infection. Most infection takes place in winter (when the tree is most susceptible) and in April and May (when the buds open, and spores are first available). To a small extent, infection may also take place through wounds in the early part of the summer. In experiments with spores, Schmidle (1953) succeeded in inoculating plants at all seasons, but the percentage of successful inoculations between April and July was relatively low, and was at its highest in September and October. More recently, Gremmen (1978) has partly resolved the question of time of infection by observing that infection can occur both in the spring, through bud scale scars, and in the autumn, through leaf scars.

Factors thought to affect the disease include temperature conditions (which have been already mentioned), water loss (Müller, 1953a; Butin, 1956b), frost (Franken, 1956; Kalandra, 1962), excess nitrogen fertiliser (Gremmen, 1978), leaf diseases (Gremmen, 1978) and waterlogging (Schmidle, 1953). In Hungary, Szontagh (1981) found that the severity of disease was increased if trees were irrigated in autumn and winter with waste water. Most workers have concluded that these and other factors operate mainly by weakening the trees and predisposing them to attack, and that plants checked, for example, by transplanting are especially vulnerable. Peace (1962) notes that in Britain plants with tops two or more years old have been especially liable to attack.

In the USSR, Shirnina (1986) and Shirnina and Nechaeva (1986) found evidence that various fungi, including *Trichoderma lignorum* (*T. viride*), *Alternaria tenuis* (*A. alternata*) and *Trichothecium roseum*, had an inhibiting effect on the growth of *Cryptodiaporthe populea*, and reduced its development on poplar cuttings.

*Host susceptibility*
Results of many tests of species and clones for resistance have been published, but in most cases they show little agreement one with another. As a general rule it appears that poplars of the *Leuce* section are least susceptible (Bavendamm, 1936; Kalandra, 1962). Of the common hybrid black poplars, *P.* 'Robusta' is particularly susceptible, and *P.* 'Gelrica' is resistant (Kalandra, 1962). *P. nigra* var. *italica* is very susceptible (Veldeman and Welvaert, 1960). Werner and Siwecki (1978) have demonstrated that the mechanism of resistance may be related to the ability of the host to form lignin and cork in the tissues below the infection court. Other differences in histology, physiology and biochemistry may also be involved (Anselmi, 1986).

A survey of the most important clones grown in Italy, including the newly introduced *P × euramericana* Pittori group, has been made by Anselmi (1986). This showed that clones of *P. deltoides* and *P. trichocarpa* and the hybrids between them were highly resistant, while those of *P. nigra* were generally susceptible. The *P. × euramericana* hybrid black poplars showed wide variations, some being highly resistant, others highly susceptible, and others of intermediate susceptibility.

*Losses caused*
Among the poplar diseases found in Britain, bark necrosis is second in importance to bacterial canker. Occasional severe attacks have occurred on dry sites, on overlarge transplants, and on plants crowded in the nursery, though the over-all damage caused has been small.

In some other countries severe damage has been reported. Thus in Italy up to 95 per cent of the plants have sometimes been killed by this disease (Goidànich, 1940). More recently all the plants of some clones have been attacked in some areas in northern Italy (Anselmi, 1986). Similar losses on five- to eight-year-old trees were reported in Germany by Kampfer (1931). Extensive damage has also been found on young plantation trees in the USA (Waterman, 1957).

*Control*
To avoid the disease, good rooting and growing conditions should be provided (Waterman, 1957; Peace, 1962; Magnani, 1963b). Overcrowding in the nursery and the use of overlarge plants should both be avoided, and any pruning of nursery plants should be done before mid-August to allow some healing of pruning wounds before growth ceases (Peace, 1962).

If such steps are taken, the disease should rarely be a problem in Britain. If it is (as it may well be in some other parts of Europe), attempts should be made to improve growing conditions, avoiding overcrowding and drying out, and ensuring a balanced fertiliser regime. Steps should also be taken to select suitable resistant clones. This may present difficulties because these clones need to be culturally suitable, and resistant not only to *C. populea* but also to other major diseases (Anselmi, 1986). As a temporary measure, spraying may be considered, since various fungicides, especially copper compounds, have been found to reduce the spread of the disease (Bavendamm, 1936; Wettstein and Donaubauer, 1958), and its carry-over into the following season (Dahte, 1960).

## Cytospora dieback and bark necrosis caused by *Valsa sordida* Nits. (anamorph *Cytospora chrysosperma* Fr.)

*Valsa sordida*, an ascomycete whose pycnidial state is *Cytospora chrysosperma*, is a saprophyte and weak wound parasite that sometimes causes a dieback and bark necrosis of poplars, mostly of nursery plants and young trees. It is a common fungus throughout most of Europe, in North America, North and South Africa, Australasia and Chile (CMI Map No. 416). So far it has proved of no importance in Britain.

### The causal fungus
The fungus grows in the bark of twigs, branches and stems, and its pycnidial stage, which may be found from May to September, erupts as scattered greyish-black stromata about 2 mm across. Masses of sausage-shaped conidia measuring $4-5 \times 1$ μm form in cavities in the stromata, and in damp weather emerge in numerous long, yellow or golden tendrils (Grove, 1935). The perithecia, which are seldom found in Britain, are also embedded in the stromata; they are black and flask-shaped, and measure 0.5 mm across. Their necks are 0.5–0.6 mm long, and their ostioles open at the stroma surface. The club-shaped asci are with or without a short stalk. They measure $26-43 \times 4.3-6.4$ μm, and each contains eight colourless, sausage-shaped ascospores measuring $6.6-10.9 \times 1.3-1.7$ μm. These perithecia occur in autumn (Schreiner, 1931b).

The fungus grows readily in culture, and resists temperatures well below freezing (Taris, 1956).

### Entry into the plant
As noted above, *V. sordida* enters through wounds. It is common on the bark even of healthy trees (Christensen, 1940), and so is always available to enter if conditions favour it. It can also grow on dead tissue and spread thence to live parts of the plant (Stahl, 1967). It may also be able to enter through lenticels (Persson, 1955).

### Symptoms and development of the disease
Within the plant, the fungus grows in the bark, kills the cambium and enters the wood. Small twigs and branches may be girdled and dieback ensue, but on larger branches and stems sunken lesions are produced (Schreiner, 1931a),

and on these the scattered stromata of the fungus are found. Stahl (1967) in Australia noted infection only on strips of bark about 10–15 cm wide on the sunny side of the trunk. The fungus often spread from these strips, and killed the crown above the lesions.

Viennot-Bourgin and Taris (1957) found that, like *Cryptodiaporthe populea*, *V. sordida* invaded the trees in the winter, but was then able to progress in its attack over the growing season.

The histopathology of infection has been studied and described by Biggs *et al.* (1983).

### Host range and variations in resistance

As a cause of disease, *V. sordida* is of importance almost solely on poplar, though it has been once reported in Britain as a cause of the death of young willows (Moore, 1959), and elsewhere has been recorded on species of *Juglans* and *Ulmus* (Christensen, 1940) and *Sorbus* (Peace, 1962).

There is some evidence that various poplar species and hybrids vary in resistance to attack by the fungus, though much of the information is derived only from field observations. Species or hybrids found very resistant by one worker may be recorded as highly susceptible by another, and even the same workers have sometimes noted some species as apparently almost immune in one season that in another year were the only ones attacked (Müller-Stoll and Hartmann, 1950).

Bloomberg (1962b), however, studied a species and two hybrids that covered a range of resistance, and found evidence that the more resistant the tree the greater its water storage capacity, and resistance to water loss. Tsyplakova (1967) found that chemical extracts from bark and wood of resistant trees inhibited the growth of cultures of *V. sordida*.

### Factors affecting the disease

It is generally agreed, however, that environmental factors are more important than those inherent in the tree. Those most often considered significant are moisture stress, temperature, the quality of the site and the resulting vigour of the trees, and closeness of planting.

Thus Bloomberg (1962a) found that he could inoculate plants successfully only when he subjected them to drought and low atmospheric humidity. Further, if he inoculated cuttings that had been allowed to dry out, necrosis developed, but ceased if the cuttings were put in water, while cuttings inoculated when in water only developed necrosis if removed from the water and allowed to dry out.

Stahl (1967), who found the disease developed only on the sunny side of the trunk, concluded that sunshine on the bark raised its temperature high enough at times to interfere with the formation of antibiotic substances (phytoalexins) produced by the plant in response to infection. The fungus, which is normally unable to attack, was then able to gain a hold. Tao *et al.* (1984) also found evidence that the fungus invaded the bark following injury by solar radiation.

Schreiner (1931a,b) found that trees of poor growth on unfavourable sites were most affected, and on vigorous trees the necrotic patches often healed over. Similarly, when he transplanted diseased trees from a poor site to a good one, they recovered, but those left in the original site were killed by the fungus. He also found that cuttings stored over the winter were very liable to infection.

Treshow and Harward (1965) reported a correlation between necrosis and earlier leaf blight caused by a *Marssonina* sp.

*Losses caused by the disease*
Damage and loss caused by this disease in Britain have so far been negligible. Elsewhere, however, it has sometimes given rise to heavy losses. Thus in one season it killed many thickly planted poplars in the Bashkir Republic (USSR) (Ibragimov, 1957). Schreiner (1931a) regarded it as the most important disease of poplars in New England, where in one case in which 10 000 cuttings were stored over the winter, only 20 or 30 survived its attack. More recently, Worrall (1983) has reported the loss of more than half the poplar plants in some Californian nurseries. Also in the USA, Treshow and Harward (1965) found that the main damage caused by the disease was indirect and resulted from suckering and excessive branching, which led to misshapen boles. Similarly, in Australia, Stahl (1967) noted that few affected trees were killed but many lost all commercial or aesthetic value and had to be removed. Epidemics were rare, but when they did occur, whole plantings could be badly damaged.

*Control*
To prevent Cytospora dieback and bark necrosis, care should be taken to provide good growing conditions in nurseries, with sufficient water and nutrients. Damage to the plants, particularly in winter, should be avoided (Müller-Stoll and Hartmann, 1950), and so should overcrowding.

Other means of control are likely to be rarely needed in Britain, though Schreiner (1931b) recommended that cuttings stored over the winter should be kept at 2 °C to avoid infection. Trunk injection with ferric sulphate and spraying with 1 per cent Bordeaux or Burgundy mixture gave promising results in the USSR (Ibragimov, 1964), where Kleiner (1965) also recommended spraying with 1 per cent DNOC in early spring and late autumn to prevent attack. In North America, where the fungus attacks cuttings, causing 'blackstem disease', Walla and Stack (1980) recommended the dipping of cuttings in thiram before winter storage, finding that this more than doubled the subsequent number of saleable plants.

**Leaf blister caused by *Taphrina populina***
Leaf blister caused by *Taphrina populina* is described below under 'post-nursery diseases', but is common on trees at all stages, including those in the nursery.

**Leaf rusts:** *Melampsora* **spp.**
The leaf rusts included in the genus *Melampsora*, and considered below under 'post-nursery diseases', are also common in the nursery.

## Post-nursery diseases

### LEAF DISEASES

**Powdery mildew:** *Uncinula adunca* **(***U. salicis***)**
The powdery mildew *Uncinula adunca* is known in Britain only on *Salix* spp. (under which it is described in chapter 17), but in continental Europe it occurs also on poplars.

**Leaf blister caused by** *Taphrina populina* **Fr. (***T. aurea* **(Pers.) Fr.)**
*Taphrina populina* occurs widely in Europe, and is known also in eastern North America and in India, China and Japan (Mix, 1949). It is an asco-mycete that causes large, conspicuous blisters on poplar leaves (plate 28). The blisters appear from May and June until the autumn as convex rounded or irregular projections, sometimes on both sides of the leaf, but more often on the upper side only. The upper sides of the blisters remain the same green colour as the rest of the leaf surface, but the concave underside soon becomes lined by a bright golden-yellow layer of asci. The asci vary in size on different poplar species and varieties (Mix, 1949) but generally measure about 70–90 × 18–22 μm, and they contain numerous bud spores that measure about 2 × 1 μm (Dennis, 1968).

The fungus overwinters as spores on the bud scales, and in damp weather attacks the young shoots when the buds burst in spring (Schneider and Sutra, 1969).

Servazzi (1935) has suggested that various physiologic forms of the fungus exist, and attack different poplar species.

*T. populina* is found especially on clones of *P* × *euramericana*, but occurs also on *P. nigra* and some of its varieties, *P. deltoides* (Magnani, 1960; Servazzi, 1935) and sometimes other poplar species (Mix, 1949). Taris (1970) studied a range of cultivated poplars, including varieties and clones of *P. deltoides* and *P* × *euramericana*, and found variations in susceptibility in both these groups. Thus among the forms of *P.* × *euramericana* the clones 'Regenerata' and 'Serotina' were very susceptible, while a number of forms of the clone 'Robusta' were resistant.

Leaf blister is common and often striking , but usually does little damage to the trees. As a rule no control measures need be applied against it, but if required, nursery plants and young trees may be sprayed with captan or a copper fungicide as the buds begin to swell and infection is likely from the overwintered bud spores.

**Leaf spot diseases caused by species of *Drepanopeziza***
The anamorphic stages of the following *Drepanopeziza* species cause leaf spots on poplars in much of Europe:

*Drepanopeziza punctiformis* Gremmen (anamorph *Marssonina brunnea* (Ell. and Everh.) Magnus
*D. populorum* (Desm.) Höhnel (anamorph *Marssonina populi-nigrae* Kleb., syn. *M. populi* (Lib.) Magnus)
*D. populi-albae* (Kleb.) Nannf. (anamorph *Marssonina castagnei* (Desm. and Mont.) Magnus)

A further less important species, *D. tremulae* Kleb. (anamorph *M. tremulae* Kleb.), occurs mainly in Italy.
The apothecial stages of these species form on the fallen leaves.

*Drepanopeziza punctiformis* appears to be the most damaging of these fungi. It occurs widely in Europe, in Britain (Phillips, 1967a), France, Belgium, the Netherlands, Germany, Austria, Italy, Yugoslavia and Hungary (Donaubauer, 1967; Gergaez, 1967; Vucinic, 1977), Bulgaria, Greece, Spain and Portugal (Anon., 1979) and also in Japan (Donaubauer, 1967), China (Cao, 1987), Canada (Donaubauer, 1967), the USA (Ostry and McNabb, 1986) and New Zealand (Spiers, 1984).

The fungus causes very small, dark-brown spots up to about 1 mm across on leaves and petioles and sometimes on the young green twigs (plate 47); when infection is severe the spots coalesce over the whole leaf surface. The macroconidia of *Marssonina brunnea* form in pustules on the spots, and are colourless, unequally two-celled, often sickle-shaped, $15-18 \times 4-6$ μm. Sometimes, colourless, one-celled, elliptical microconidia measuring $2-4 \times 1-3$ μm may be found in the same pustules (Anon., 1979). The apothecia of the perfect stage are dark brown, and occur in May on both sides of the leaves. The asci are clavate, measuring $45-84 \times 9-12$ μm, and each contains eight colourless, ellipsoidal ascospores measuring $6-17 \times 4-7$ μm (Byrom and Burdekin, 1970).

*D. punctiformis* is made up of a number of strains which vary greatly in their pathogenicity (Spiers, 1983a).

Infection takes place in spring, from conidia produced in acervuli overwintered on infected shoots and from ascospores from perithecia on fallen leaves. The relative importance of these sources of inoculum differs. Thus, in the Netherlands primary infections arise from ascospores, which are produced as the young leaves unfold, but in Italy conidia appear to form the main primary infection source (Anon., 1979).

As may be expected in a fungus which grows over a wide geographical and climatic range, conidia germinate over a large range of temperature, from 9 to 25 °C. There is a marked interaction between temperature and rainfall, infection taking place most readily in warm, damp periods, with little infection occurring when the weather is either cool or dry (Anon., 1979). Poplar leaves are most easily attacked when young, becoming more resistant to infection with age (Spiers and Hopcroft, 1984).

*D. punctiformis* usually attacks poplars of the *Aigeiros* and *Tacamahaca* sections (Magnani, 1965), but after a study of isolates from New Zealand and elsewhere, Spiers (1984) concluded that the anamorph *M. brunnea* could be divided into two form species, one of which, f.sp. *trepidae*, attacked *P. tremulae* and perhaps *P. tremuloides*, of the *Leuce* group. Within the poplar groups, different clones differ considerably in resistance (Donaubauer, 1965; Ostry and McNabb, 1986). Melchior *et al.* (1983) also found differences in resistance in seed provenances of *P. deltoides* from various parts of the USA. Clones of *P.* × *euramericana* also differ greatly in resistance (Anon., 1979), and the fungus has caused much damage to some of these.

Spiers and Wenham (1983a) have shown that conidia of *D. punctiformis* may be carried on seed of *Populus deltoides*, and diseased seedlings may arise from the infested seed.

This fungus may cause serious premature leaf fall and reduction in growth, particularly in Italy, where it has been especially damaging on sites of low fertility (Castellani and Cellerino, 1964; Cellerino, 1966). It has also caused significant damage in the Netherlands and Japan. So far in Britain it has not been of major importance, although locally serious outbreaks have occasionally been reported.

Leaf spot disease caused by *D. punctiformis* is clearly affected by many subtle interactions between variations in the susceptibility of host species and clones, pathogenicity of strains of the causal fungus, climate and weather, and site characteristics.

*Control*

As the disease has not so far been of major importance in Britain, few if any measures have been taken to control it there. Elsewhere, because it may be seed-borne, and probably carried by plants, Cao (1987) has suggested a need for quarantine measures to prevent its long-distance spread. Spiers and Wenham (1983b) found that the dusting of infested seed with benomyl, captafol, chlorothalonil, thiophanate and thiram did not damage the seed and prevented seedling infection.

In areas where the disease is known to cause loss, poplars should not be grown on infertile sites, and care should be taken to avoid excess nitrogen or a deficiency of potash, and poplar clones resistant to the fungus should be chosen. On a small scale, diseased shoots can be pruned out, and steps taken to remove, dig in or rot down infected leaves, to reduce the inoculum available in the spring. Spraying from the beginning of the season with various chemicals has also shown promise. Among such fungicides have been those in the dithane group, including dithane M45 (Castellani and Cellerino, 1967) and maneb (Guldemond and Kolster, 1966), copper-based materials and difolotan and dodine (Gojkovic, 1970; Castellani, 1967).

*Drepanopeziza populorum* occurs throughout Europe, including Britain and Ireland (O'Riordain and Kavanagh, 1965), as well as in Canada and the USA (Donaubauer, 1967) and in China (Li, 1984).

The anamorphic stage is *Marssonina populi-nigrae*, of which *M. populi* is considered to be a synonym (Zycha, 1965; Donaubauer, 1967), and it is in this stage that the fungus produces brown spots on the upper sides of poplar leaves, and sometimes brown or black spots on their twigs and branches. The leaf spots have irregular edges, and are larger than those caused by *Marssonina brunnea*, reaching up to about 5 mm or more across (Anon., 1979; Ellis and Ellis, 1985) and often coalescing to form blotches, on which from July to September small tawny pustules may be found. From these pustules, whitish tendrils of macroconidia emerge in damp weather. The spores are rather pear-shaped, straight or curved, constricted at the septum that divides them irregularly well below the middle, and measure 17–25 × 6–11 μm (Ellis and Ellis, 1985). Microconidia measuring 2.5–8 × 1–2.5 μm may also occur (Anon., 1979).

The perfect stage is a brown apothecium with club-shaped asci measuring 80–100 × 13–14 μm, each containing eight colourless, ellipsoid, one-celled (later two-celled) ascospores measuring 10–18 × 5–9 μm (Gremmen, 1962; Anon., 1979; Ellis and Ellis, 1985).

*D. populorum* attacks poplars in the *Aigeiros* section, including various clones of *P.* × *euramericana* (Gremmen, 1964), though it is less damaging on these than *D. punctiformis*. It also attacks *P. nigra* (Magnani, 1964), especially the Lombardy poplar, *P. nigra* 'Italica', on which it causes most damage in Britain, as well as in Italy (Anon., 1979) and in France (Boudier, 1983b). On the Lombardy poplar it causes premature loss of leaves from the base upwards, so that only a tuft of foliage at the top of the crown may be left alive (figure 77). Trees thus defoliated over several seasons may be killed. *P. deltoides* is also sometimes attacked (Anon., 1979).

The same fungus sometimes also affects the balsam poplar, *P. balsamifera* (*P. tacamahaca*) in the *Tacamahaca* section (Laubert, 1936), and in China, Li (1984) has described a form which attacks poplars in the *Leuce* section.

*Control*

Control measures for *D. populorum* are broadly the same as for *D. punctiformis*. In China, Wan and Wang (1987) have found differences in resistance to this fungus between various clones of the *Aigeiros* section, so further studies may show that resistant clones can be used in the control of the disease.

*Drepanopeziza populi-albae* occurs throughout much of Europe (including Britain), and has also been recorded in Iran, Canada and the USA (Anon., 1979).

The anamorphic stage, *Marssonina castagnei*, causes spots and blotches much like those of *M. populi-nigrae* (but with no well-defined darker margin) on the upper sides of the leaves. Small pustules of macroconidia arise on the blotches in moist weather in summer and early autumn. The conidia, which ooze out in short whitish tendrils, are more or less pear-shaped, curved, with

**Figure 77** A Lombardy poplar almost entirely defoliated by *Marssonina populi-nigrae*, so that the only remaining foliage forms a tuft at the top of the crown (D.H. Phillips)

no constriction at the septum that divides them below the middle. They appear to be very variable in size: according to Grove (1937), they measure 25–26 × 9 μm, though the figures given by Magnani (1964) are 18.4–23.6 × 6.5–7.5 μm, those by Zycha (1965) are 15.5–19 × 8.0–9.5 μm, and those by Ellis and Ellis (1985) are 18–20 × 5–9 μm.

*Drepanopeziza populi-albae* appears to be almost confined to *Populus alba* (Laubert, 1936; Gremmen, 1964), though Magnani (1964) recorded it in Italy on *P.* × *euramericana*, and Spiers (1983b) found that it was capable of attacking poplars in the *Tacamahaca* and *Leuce* sections. Naidenov (1985) in Bulgaria found that it was quite common on white and grey poplars and aspens (all in the *Leuce* section). Among the forms of *P. alba* he found the Dutch clone 'Raket' was very susceptible. Spiers (1988) found some forms of *P. alba* very susceptible, but others, including some Italian selections, showed high resistance in the field.

So far, *D. populi-albae* has proved of only minor importance. If any steps are needed against it, measures suggested for *D. punctiformis* should prove effective.

*Drepanopeziza tremulae* occurs mainly in Italy, where it seems to be confined to the aspen, *Populus tremula*, on which its anamorph *Marssonina tremulae* causes spots up to about 3 mm across on the undersides of the leaves (Anon., 1979).

The colourless, oblong or oval to rather pear-shaped conidia measure 13–17 × 4–5 μm (Rimpau, 1962). The apothecia of the perfect stage form on the fallen leaves, and their one-celled, colourless ascospores measure 8–9 × 1.8–2 μm (Anon., 1979).

The disease caused by *D. tremulae* is of minor importance. If required, the control measures suggested for *D. punctiformis* should be effective against it.

**Leaf spots, leaf blights and diebacks caused by *Venturia tremulae* Aderh. (*V. macularis* (Fr.) Muller and von Arx) and *V. populina* (Vuill.) Fabric.**

Two species of *Venturia* are of some pathological significance on poplars in Europe. The first is *V. tremulae* and the second is *V. populina*.

Leaf blight and shoot dieback caused by *V. tremulae* (anamorph *Pollaccia radiosa* (Lib.) Bald. & Cif.) is known throughout Europe and North America. In Britain it has been found only a few times (Sutton, 1961; Sutton and Pirozynski, 1963; Redfern *et al.*, 1973, 1974; Young and Strouts, 1978).

In its imperfect stage (*P. radiosa*), the fungus produces small tufts of conidiophores, each of which bears one conidium. The conidia are ellipsoid or club-shaped with a flattened base, almost colourless or brownish, smooth, two- or three-celled, and measure 14.5–19 × 6–7.5 μm (Sutton, 1961). The pseudothecia of the perfect stage (*V. tremulae*) are embedded in the leaf tissues, with a short, projecting neck furnished with stiff bristles, and the asci within each contain eight two-celled, olivaceous, ellipsoid ascospores septate above the middle and measuring 8–14 × 4–6 μm (Sivanesan, 1977).

Morelet (1985) has described three varieties of *V. tremulae*:

(1) var. *tremulae* (anamorph *Pollaccia radiosa* var. *radiosa*). This is the best-known form in Europe, attacking the European aspen and sometimes the white poplar, *P. alba*, and the hybrids between them.

(2) var. *populi-albae* (anamorph *P. radiosa* var. *populi-albae*). This relatively uncommon variety causes leaf spots on the white poplar in Europe. It is also present in North America.

(3) var. *grandidentata* (anamorph *P. radiosa* var. *lethifera*). This is the most important variety in North America, on *P. tremuloides* and *P. grandidentata*.

From this it will be seen that these varieties occur on poplars of the *Leuce* section, though poplars of the *Aigeiros* section are also sometimes attacked (Sivanesan, 1977).

The variability of the fungus, which exists in at least two pathotypes, needs to be taken into account when breeding poplars for resistance (Morelet, 1986). Dance (1961b), in Canada, carried out spore-trapping experiments and found most spores in the air in June, with a small later peak in August and September. The number of conidia in the air was directly related to rainfall, but the duration of the rain was more important than the total amount.

The disease caused by *V. tremulae* in Europe has been studied by Gremmen (1956) and Weisberger (1969). It first appears as rounded or irregular buff-coloured leaf spots up to 1 cm across, with a black margin. Deep olive-coloured pustules of *Pollaccia radiosa* occur in the spots, mostly on the upper side of the leaf. The pustules are made up of the tufts of conidiophores and conidia. Soon the spots enlarge and coalesce, and the distal part of the leaf blackens (plate 45). Much of the foliage may be affected, and the fungus may spread down into the shoots, which die back and become characteristically hook-shaped, resembling the 'shepherd's crooks' produced in elms affected by Dutch elm disease (Gremmen, 1956; Dance, 1961b).

The fungus overwinters as mycelium in the diseased shoots and as pseudo-thecia in the withering and fallen leaves (Shavrova, 1967).

It causes most damage to seedlings, suckers and young trees, which become malformed, while nursery plants are stunted, and natural regeneration is delayed (Dance, 1961b; Ginzburg, 1961; Anderson and Anderson, 1980).

Where control measures are needed, plant sanitation measures by the pruning out of diseased shoots and the destruction of infected leaves, and spraying with Bordeaux mixture and other copper fungicides, have been suggested (Pospisil, 1960; Shavrova, 1967). In North America Anderson and Anderson (1980) have obtained control by applying benomyl.

Leaf blight and dieback caused by *Venturia populina*, the anamorph of which is *Pollaccia elegans* Servazzi, is also widespread throughout much of Europe (Sivanesan, 1977) as well as in North America (Dance, 1961a), but it is usually of little importance. *V. populina* much resembles *V. tremulae*, and the two fungi cause very similar diseases. Indeed, they have often been confused with one another, but they differ in the size and details of the septation of their ascospores, and (though with some overlap) in their host range, *V. populina* affecting poplars in the *Tacamahaca* and *Aigeiros* sections.

The disease has been studied by Gremmen (1956) and Dance (1961a). Symptoms resemble those caused by *V. tremulae*, but are usually less severe. The mainly 3-celled conidia of the *Pollaccia* stage, formed on olive-coloured pustules on the irregular leaf spots, mostly measure 26–35 × 9–13 μm. The pseudothecia of the perfect stage develop on the withered or fallen leaves. Their two-celled ascospores are pale olivaceous brown, with the septum in or near the lower third, and measure about 20–23 × 11–13 μm (Sivanesan, 1977).

The disease occurs mainly on young trees, and Dance (1961a) found that the foliage is susceptible only when young. The disease is rare in Britain, but it does occur there from time to time; Redfern *et al.* (1974) recorded it as a cause of dieback in Scotland and northern England in 1973. In France it seems to have increased in recent years (Taris, 1980). It has also reduced the volume of poplar trunks in Italy, causing considerable damage (Ciccarone, 1988).

Control measures are the same as those for leaf blight and shoot dieback caused by *V. tremulae*.

**Poplar leaf rusts: *Melampsora* spp.**
In Britain, three species of *Melampsora* attack the leaves (and to some extent the shoots) of poplars. They are *M. allii-populina* Kleb., *M. larici-populina* Kleb. and *M. populnea* (Pers.) P. Karsten. The latter, however, is divided into three main *formae speciales* which have themselves sometimes been regarded as species. All are heteroecious, producing their often numerous and conspicuous uredinia and telia on poplars and their spermagonia and aecia on various alternate hosts. They are described by Wilson and Henderson (1966) and Ellis and Ellis (1985). They are all exceedingly similar in their stages on poplar, and can be separated with some difficulty, and then only by the use of minute microscopical differences (Gremmen, 1954; Hennebert, 1964: Pinon, 1973). Kurkela (1980), however, has suggested that the various species included under *M. populnea* can be separated by serological tests. These rusts occur in both the nursery and the plantation, and are more important in the former than in the latter. In Britain, though very common, they appear to be of minor importance. In parts of the European mainland, where they also occur, they may cause substantial defoliation and greater damage.

Also of some importance in parts of continental Europe, but absent from Britain, is *M. medusae* Thüm., another heteroecious rust, with its uredinia and telia on poplars and its spermagonia and aecia on various conifers.

These rusts are listed in table 16.1.

**Table 16.1** The most important poplar leaf rusts

| Poplar hosts (uredinia and telia) | Rusts | Alternate hosts (spermagonia and aecia) |
|---|---|---|
| Black and balsam poplars (*Aigeiros* and *Tacamahaca* sections) | 1. *M. allii-populina*<br>2. *M. larici-populina* | *Allium* and *Arum*<br>*Larix** |
| White poplars and aspens (*Leuce* section) | 3. *M. populnea* including:<br>a. *M. larici-tremulae*<br>b. *M. aecidioides*<br>c. *M. pinitorqua* | <br><br>*Larix*<br>*Mercurialis*<br>*Pinus* |
| White poplars and aspens, black and balsam poplars (*Leuce*, *Aigeiros* and *Tacamahaca* sections) | 4. *M. medusae* | Various conifers[†] |

* Not certainly known on larch in Britain.
† Not on conifers in Europe.

*M. allii-populina* Kleb. occurs throughout much of Europe and the Middle East, and in North Africa and South America. In Belgian nurseries, Steenackers (1988) found that it was of some significance, causing more infections than *M. larici-populina*. In general, however, at least in France, it tends to become more common south of the Loire (Pinon, 1986b). It produces its small uredinia and telia on the undersides of the poplar leaves. The uredinia are up to about 1 mm across, round and bright orange in colour, with more or less elliptical, oblong or broadly club-shaped urediniospores measuring 24–38 × 11–18 μm and with a uniformly thickened wall. The dark-brown telia, which form on the undersides of the leaves under the epidermis, are about the same size as the uredinia, and are made up of groups of column-shaped teliospores measuring 35–60 × 6–10 μm. The spermagonia and aecia of the fungus form on species of *Allium* and *Arum* and sometimes of *Muscari*.

Somda and Pinon (1981) found that the optimum temperature for germination of the urediniospores was 16 °C.

*M. larici-populina* Kleb. occurs widely in Europe and in Asia, Africa, South America, Australia and New Zealand. It appears to be the commonest of the *Melampsora* species in Britain and the northern parts of Europe. It produces small orange uredinia up to 1 mm across on the lower sides of poplar leaves (plate 46), with elliptical urediniospores measuring 30–40 × 13–17 μm; their walls are more thickened at the equator than elsewhere. The reddish-brown telia are also small, and occur on the upper sides of the leaves. They are black when mature, and the columnar spores, which arise below the epidermis, measure 40–50 × 7–10 μm.

The spermagonia and aecia of the fungus are formed on larch, but in many areas it can overwinter by means of urediniospores as well as by teliospores (Chiba and Zinno, 1960), and it often occurs on poplars when larch is absent from the vicinity. Indeed, in Britain it is not certainly known on larch, though elsewhere in some areas it seems to appear on poplars only when larches are planted nearby (van der Meiden and van Vloten, 1958).

The uredinia require a relative humidity of at least 95 per cent for germination, the optimum temperature for which is 16 °C (Krzan, 1981a,b).

*M. larici-populina* exists in Europe in at least two pathotypes, known as E1 and E2 (Dam, 1985; Pinon, 1986c; Pinon *et al.*, 1987), which differ in their host ranges on poplar clones. There is now evidence of yet a third pathotype in Belgium (Steenackers, 1988).

In Australia, Sharma and Heather (1988) found that various fungi, the most important of which is *Cladosporium tenuissimum* Cooke, may act as hyperparasites, destroying the uredinia of *M. larici-populina*.

*M. populnea* (Pers.) P. Karsten (*M. tremulae* Tul.) is a common rust throughout much of Europe and into the Middle East, as well as in Kenya and South Africa and in Uruguay (CMI Map No. 389). Its uredinia and telia are produced on the undersides of poplar leaves, and are very small, with the uredinia measuring only up to about ½ mm across and the telia about twice this diameter. The uredinia are usually first seen in August and are yellowish orange, and the urediniospores are subglobose, measuring 15–25 × 11–18 μm.

The dark-brown, subepidermal telia develop soon after the uredinia, and contain columnar teliospores measuring 22–60 × 7–12 μm. In the USSR, Minkevich and Bazova (1984) found that the disease on poplars was favoured by warm wet winters. Infection by aeciospores from the alternate hosts takes place on the young poplar leaves.

*M. populnea* is a complex with three main *formae speciales*, f. sp. *pinitorqua* R. Hartig, f. sp. *laricis* R. Hartig and f. sp. *rostrupii* Wagner ex Kleb., often regarded as separate species, named *M. pinitorqua* Rostrup, *M. larici-tremulae* Kleb. and *M. rostrupii* Wagner ex Kleb. The spermagonia of *M. pinitorqua* form on pines and those of *M. larici-tremulae* on larches, and these species are therefore described further under '*Pinus*' and '*Larix*' in chapters 5 and 6, respectively. The spermagonial and aecial stages of *M. rostrupii* (syn. *M. aecidioides* Plowr.) occur on *Mercurialis*.

*M. medusae* is widespread in North America, and occurs also in Argentina and Japan (Ziller, 1974) and in Australia (Singh and Heather, 1982) and New Zealand (Spiers, 1978) and southern Africa (Trench *et al.*, 1987). It is also known in Europe, in Spain (where it was found in 1925) and France (Pinon, 1986b).

In North America, Australia and southern Africa the fungus has caused considerable defoliation and damage to poplars. In extreme cases most of the leaves may fall, leaving only a few tufts at the ends of the shoots (Trench *et al.*, 1987). In Europe, however, so far its effects and its spread have been limited, and it still occurs only in the Iberian peninsula and southwest France (Pinon, 1986b).

On poplars it produces its uredinia and telia on the undersides of the leaves, as yellow spots. The ellipsoid or obovoid urediniospores measure 23–35 × 15–23 μm, and the columnar teliospores measure 29–45 × 10–15 μm (Ziller, 1974). The spermagonia and aecia form on the needles of species of *Abies*, *Larix*, *Pinus*, *Picea*, *Pseudotsuga* and *Tsuga* (Ziller, 1974) but these stages appear to be unknown in Europe. In North America, Prakash and Thielges (1987) found that the rust on poplars existed as a number of races which differed in their pathogenicity.

*Host range*

The broad host range of these rusts is given in table 16.1, and in greater detail by Anon. (1979). Within species, varieties and hybrids, different poplar clones vary in resistance to the rusts (Donaubauer, 1964; Chiba, 1964), but clones resistant in one area may be susceptible in another (Schreiner, 1959), as the rusts exist as a range of physiological races each capable of attacking a different range of clones (van Vloten, 1944; Magnani, 1966). Thus Pinon (1986c) has drawn together the reactions of various important European poplars to the races E1 and E2 of *Melampsora larici-populina*. Some clones show resistance or susceptibility to both these races, while others are resistant to one but susceptible to the other. Pinon (1986b) (from data of Van Kraayenoord) also indicates the marked variability in susceptibility to *M. medusae* of many major European clones. Gremmen (1980) points out that poplar breeders and growers are faced with serious problems because when new poplars come into large-scale use, new races of these rusts may soon arise to attack them. He therefore recommends the growing of mixtures of genetically varied poplars interspersed with other broadleaved trees.

## Losses caused

Although by the end of the season these rusts are often very conspicuous, covering the leaves with sori and causing premature defoliation, their effects in Britain generally appear to be small, because their attack usually comes late, when growth of the trees has been completed. In parts of continental Europe, however, *Melampsora* spp. cause serious damage, and in Nordbrabant in Holland, in the years following 1948–1950, after extensive planting of larch in the area, it became impossible to establish *P.* 'Serotina', *P.* 'Marilandica' and *P.* 'Heidemij' because of severe attacks by *M. larici-populina*. Financial losses were estimated to be up to 400 Dutch florins per hectare (van der Meiden and van Vloten, 1958). Further, extensive damage by *Cryptodiaporthe populea* (anamorph *Dothichiza populea*) often followed the rust attacks, because the rust caused premature leaf fall, and the spores of *C. populea* were then able to enter the fresh leaf scars.

## Control

In Britain it is rarely necessary to take any steps to control these leaf rusts. In areas where *Melampsora* spp. (especially *M. larici-populina*) are troublesome, a search should be made for suitable resistant clones to replace those affected by the disease. Control of the alternate hosts of the rusts (*Larix* in the case of *M. larici-populina*) should be undertaken in the vicinity of the poplars. Poplar plantings should be suitably fertilised, avoiding excess nitrogen and any deficiency of potash (Anon., 1979). Spraying with fungicides has given good results in some places. Usually this is possible only in nurseries and on small trees, but in New Zealand control over larger areas has been obtained by aerial spraying (Sheridan, 1981). Bordeaux mixture and other copper fungicides control *Melampsora* spp. (Saric and Milatovic, 1960; Nohara *et al.*, 1961). Pei and Shang (1984) in China tested various fungicides against *M. larici-populina*, and found Bordeaux mixture the most effective, but other useful materials included chlorothalonil, zineb, thiram and thiophanate-methyl. Mancozeb is also said to be effective (Anon., 1979).

## Leaf spots caused by *Phyllosticta populina* Sacc. and *Septoria populi* Desm.

*Phyllosticta populina* and *Septoria populi* have both been recorded as the cause of leaf spots on poplar in Britain (Grove, 1935; Moore, 1959), though neither has so far proved to be of major importance.

*Phyllosticta populina*, which occurs on *Populus nigra* and *P.* × *euramericana*, produces small rounded spots that later join to become angular or sinuous brown or grey blotches with a darker margin. Pycnidia about 150 μm across appear on the blotches, on the upper side of the leaf, and contain ellipsoid or oval conidia measuring 6–8 × 2.5–3.5 μm (Grove, 1935).

*Septoria populi* has also been found in the field, mainly on *Populus nigra*, and its varieties and on *P.* × *euramericana* (especially cv. 'Serotina') (Grove, 1935), though Magnani (1962) in Italy succeeded in inoculating *P. alba*, *P. deltoides*, *P. tacamahaca* (*P. balsamifera*), *P. simonii* and *P.* × *berolinensis*. In Britain it caused considerable damage to aspen at one site in Surrey in 1977 (Young, private communication) but it seems to be of most importance in the area around the Mediterranean (Anon., 1979).

It produces dry, pale spots with a brownish-black margin, up to about 2 mm across, scattered over the leaf. Pycnidia about 225 μm in diameter appear on the spots, on top or on both sides of the leaf, and contain curved, worm-like, two-celled spores measuring 30–45 × 3–3.5 μm.

*S. populi* is considered to be the conidial stage of *Mycosphaerella populi* Schroet.

## DISEASES OF TWIGS, BRANCHES AND STEMS

### Cankers associated with *Nectria coccinea* and *N. galligena*
*Nectria coccinea*, which is most important as the fungus involved in beech bark disease (under which it is therefore described in chapter 11), has also been known to cause cankers on poplars. Another canker sometimes found on poplars is caused by *N. galligena*. This fungus is best known as the cause of apple canker, but has many other hosts, and is therefore described in chapter 3.

### Dothichiza bark necrosis or trunk scab caused by *Cryptodiaporthe populea*
As already noted, Dothichiza bark necrosis or trunk scab may attack plantation trees as well as nursery plants, and in continental Europe may severely damage or even kill large trees. In Britain, however, it has caused serious losses only in the nursery. It is therefore described above under 'nursery diseases'.

### Cytospora dieback and bark necrosis caused by *Valsa sordida*
Cytospora dieback and bark necrosis, which may be found on trees in plantations, is nevertheless mainly a disease of nursery plants and young trees. It is therefore described in the 'nursery diseases' section.

### Cankers caused by *Hypoxylon mammatum* (Wahlenb.) Miller
*Hypoxylon mammatum* is an important cause of canker in poplars in North America, where it chiefly affects *Populus tremuloides*, but may also attack *P. grandidentata* and more rarely the balsam poplars (Hepting, 1971). It is also known there on species of *Acer* and *Betula* and many other broadleaved trees, though often only as a saprophyte (Sinclair *et al.*, 1987). It also occurs in Europe, where its significance has been discussed by Pinon (1986a).

Hypoxylon cankers form on branches and trunks, and young trees, which are those usually most affected (Falk *et al.*, 1989), may be girdled and killed, either by *H. mammatum* itself or by other, secondary fungi and by insects. On older trees, girdling of branches may lead to the death of the ends of the shoots, and the resulting dead leaves, which remain long on the tree, draw attention to the disease (Sinclair *et al.*, 1987). On trunks, elongated cankers develop, often round a branch stub.

Infection, apparently by ascospores, takes place through wounds, including those made by insects (Ostry and Anderson, 1983), provided that they reach as far as the wood. The fungus then spreads to the cambium and into the bark, which becomes cracked and broken. In the first year after infection the bark is raised in blisters, and if it is peeled off, the blistering is found to be due to the presence of a grey mycelial mat and numerous small coremial pillars which force up the bark above them. These pillars and the mycelial mat bear many colourless, unicellular conidia of a *Nodulisporium* stage (Sinclair *et al.*, 1987; Ellis and Ellis, 1985). Later, whitish but finally black perithecial stromata form on the cankered tissues, each containing about 5–30 perithecia which open to the surface by small papillae. The asci in these perithecia each contain eight dark-brown, unicellular, oblong or ellipsoid ascospores each measuring about 20–36 × 6–14 μm (Ellis and Ellis, 1985).

The fungus produces a toxin which may be responsible for some of the damage it causes (Pinon, 1984b). Isolates differ in pathogenicity (Griffin *et al.*, 1984).

Some severe attacks of Hypoxylon canker have been observed in *P. tremula* × *P. tremuloides* hybrids in Sweden, but generally in Europe the disease mainly affects the common aspen in areas from the alps southwards. It becomes rare in the plains further north (Pinon, 1986a), though the fungus occurs fairly widely. In Britain it is rather rare, and seems to be saprophytic only.

In southern Europe it may kill branches or entire trees (Capretti, 1983), and Pinon (1986a) estimates that losses in the hotter, drier areas may reach 10 per cent or more. Moisture stress appears to render trees more liable to the disease (Bélanger *et al.*, 1989).

This canker is at present of small consequence in Europe because the common aspen is of little economic importance there, but Pinon (1986a) points out that in several European countries, France being prominent among them, there is an increasing interest in the selection, breeding and cultivation of poplars in the *Leuce* section, especially hybrids between *P. tremula* and *P. tremuloides* and clones of the grey poplar *P. canescens*. Hence, if the breeding programme fails to take account of Hypoxylon canker, the disease could increase in importance in the future. To deal with the problem, resistance known to occur in *P. alba* can be utilized.

### Cankers caused by *Ceratocystis fimbriata* Ell. and Halst. (*Endoconidiophora fimbriata* (Ell. and Halst.) Davidson

*Ceratocystis fimbriata* occurs mainly in the warmer parts of the world, causing perennial target cankers on a number of trees, and a tuber rot of sweet potato

(CMI Descr. F 141 and CMI Map No. 9). A form of the fungus causes canker stain of plane, and is described under '*Platanus*' in chapter 19.

C. *fimbriata* has long been well known as the cause of target cankers on the aspen *Populus tremuloides* in North America. Its effects on this tree there are described by Sinclair *et al.* (1987). More recently it has been found causing similar cankers in Poland (Gremmen and de Kam, 1977; Przybyl, 1984a,b,c,d, 1988), where it has affected clones of *P. trichocarpa* and hybrids between poplars in the *Tacamahaca* and *Aigeiros* groups.

The concentrically zoned, sunken cankers are ellipsoid and up to about 50 cm long (Gremmen and de Kam, 1977) and the trunk becomes swollen around them. On *P. tremuloides* they ooze a yellow-orange liquid and later when the bark sloughs off they become dark brown or black. They form on the trunk around wounds and at the origins of small branches (Sinclair *et al.*, 1987).

Fruit bodies of the fungus develop on the cankers. They include asexual forms which produce colourless, cylindrical endoconidia measuring 8.5–32.8 × 2.7–7.5 μm, usually colourless but sometimes light-brownish, barrel-shaped endoconidia measuring 6.4–13 × 4–9.6 μm, and thick-walled oval conidia measuring 7.0–24 × 5.1–15.2 μm. The perithecia of the perfect stage are black and rounded, with long necks fimbriate at the tips. The ascospores emerge to form a sticky droplet at the end of the neck, and are bowler-hat-shaped, measuring 4–7.2 × 2.5–7.5 μm (Gremmen and de Kam, 1977; Przybyl, 1984a).

Infection takes place through wounds, including those made by sap-feeding insects. The cankers cause reduction of the wood quality, and weakness which may lead to stem breakage (Sinclair *et al.*, 1987)

## Bacterial canker caused by *Xanthomonas populi* (Ridé) Ridé and Ridé (*Aplanobacterium* (*Aplanobacter*) *populi* Ridé)

The bacterial canker caused by *Xanthomonas populi* (Ridé and Ridé, 1978), which is much the most important disease of poplars in Britain, is also of major importance in France, Belgium, Holland, Germany, Poland, the USSR and Hungary, and appears to be present also in Romania and Yugoslavia (CMI Map No. 422). The name *X. populi* is not a legitimate one under the international code of nomenclature, but it is employed here because it is the only one at present in general use for this bacterium.

*X. populi* was first isolated in 1956 by Ridé, who proved that it produced typical cankers on inoculation into susceptible poplars (Ridé, 1958, 1963). It is a Gram-negative rod with a single polar flagellum observable by immuno-fluorescence techniques (Ridé and Ridé, 1978). It has stringent nutrient requirements, but if grown on a standard medium with the addition of yeast extract, glucose and peptone, it forms creamy mucoid colonies. It produces one of the xanthomonadin pigments characteristic of *Xanthomonas* species. It occurs mainly on *Populus* spp., but de Kam (1978, 1981) and Jenkins and Starr (1982) have distinguished a separate pathovar which attacks only species of *Salix*. Different isolates of *X. populi* appear to differ in pathogenicity (Ridé *et al.*, 1986; Steenackers, 1989).

### Symptoms of the disease

The various symptoms of the disease have been described by Sabet and Dowson (1952), Ridé (1963), Jobling and Young (1965) and others. The first visible symptom is seen in early spring, when a dense, whitish slime that swells in damp weather exudes mainly from cracks in young branches from one or two (or sometimes up to four) years old. The tissues below the cracks are greenish and translucent, because their intercellular spaces are filled with bacterial slime, and it is at this time that the bacterium is most easily isolated from the infected wood. These young branches may be girdled, and a dieback results (plate 49), but on older and larger branches and stems cankers may arise, and develop for many years (figure 78). Slime exudes also from these cankers, and sometimes from the bases of buds, and is usually easy to find on diseased trees of most (but not all) susceptible clones in spring and early summer. It becomes progressively scarcer as the season advances, however, and also more and more contaminated by secondary organisms, and less and less effective as an inoculum (Ridé, 1966).

The cankers may be 'closed', rough excrescences, usually up to 3 cm across, or 'open', elongated lesions with exposed wood surrounded by a raised rim of callus tissue (Sabet and Dowson, 1952). Jobling and Young (1965) have also described series of elongated and sometimes sinuous cankers arising over the tunnels made by the larvae of agromyzid flies.

When Ridé (1959) inoculated poplar leaves with *X. populi*, it produced diffuse blotches that became grey in the centre, but similar inoculations by Whitbread (1967) resulted in no symptoms.

Once cankers have arisen, they become colonised by other organisms, particularly *Pseudomonas syringae*, which was at one time considered to be the cause of the disease (Sabet and Dowson, 1952), though inoculations with pure cultures of this bacterium always fail to produce the characteristic symptoms of canker, but cause only a local necrosis of the bark, with subsequent recovery (Ridé, 1963).

Inoculation tests with *X. populi* have shown that trees infected from April to July produce cankers in the same year. If infection takes place between August and early October, slime production and necrosis of the affected branch takes place at the time of bud burst in the following season. Inoculations made at other times have given variable results (Ridé, 1959).

### Entry into the plant

*X. populi* appears to enter the plant through wounds of various kinds. Infection succeeds only if the bacterium reaches the cambium (Ridé, 1966). It has been suggested that under natural conditions entry is gained through bud scale, stipule and leaf scars, and wounds made by insects, small cracks caused by virus diseases, and perhaps by frost, as well as through pruning (Ridé,

**Figure 78** Canker on a poplar stem, caused by *Xanthomonas populi* (D.H. Phillips)

1959, 1963, 1966; Jobling and Young, 1965; Whitbread, 1967; Burdekin, 1972).

Bud scale scars and stipule scars are both present at the beginning of the season, when bacterial slime is most abundant. Fresh stipule scars were successfully inoculated by Ridé (1966). Leaf scars remain open to infection for only a few days (Ridé, 1963), and they occur mainly at the end of the season when little slime is present. Nevertheless, some slime may be found at times throughout the growing season (Burdekin, in Phillips, 1967b), and leaf scars have been successfully inoculated (Whitbread, 1967). Careful field observations have also shown them to be important infection sites (Burdekin, 1972).

When Jobling and Young (1965) found elongated cankers over the tunnels made by the larvae of agromyzid flies (plate 50), they considered that the bacterium entered the exit holes made by the insects, and grew in the parenchyma filling the tunnels. Burdekin (1966) found that the bacterium could be introduced into the tunnels, and multiplied there. Openings leading directly into tunnels within the cambium are clearly ideal entry points for *X. populi*, which, as noted above, can invade the plant only if it reaches cambial tissue. Ridé and Viart (1966) were also of the opinion that a *Dendromyza* sp. played some part in the transmission of the disease, and Ridé (1963) also noticed bacterial slime emerging from galls formed by the moth *Gypsonoma aceriana* (*Semasia aceriana*).

Sabet (1953) carried out freezing experiments with poplars, but they provided no evidence that frost damage led to the establishment of bacterial canker. Cankers of purely physical origin may sometimes arise near the bases of poplars (and other trees) as a result of frost cracking, but *X. populi* is not involved in these.

Little is known about the mode of spread of the bacterium from tree to tree, though rain splash is probably involved, and carriage by insects has been suggested. In France, Ridé and Viart (1966) noticed that the disease spread rapidly in the direction of the prevailing wind.

*Susceptibility of poplar species, varieties and clones*
Many tests have been made to test the relative susceptibility of various species, varieties and clones of poplar, but the results as a whole are difficult to interpret because until the isolation of *X. populi* they were carried out by inoculation with natural slime, which is a variable material, containing different concentrations of *X. populi* and other organisms (Ridé, 1963). Bacterial slime may also be difficult to obtain, particularly as late as September, the best month for inoculation tests (Ridé, 1959). To be reliable, tests must be carried out over several seasons by standard methods, using virulent cultures of *X. populi*, and parallel tests should be made with plants exposed to natural infection in the field (Ridé, 1963). De Kam and Heisterkamp (1987) studied two methods used in the comparison of the susceptibility of poplar clones. Leaf scars were inoculated in September and stipule scars in May to June. The very susceptible and very resistant clones were easily identified using either method, but the stipule scar inoculations gave a better indication of the relative susceptibility of the intermediate clones than did the leaf scar method. Kechel and Böden (1985) have developed a promising test procedure using tissue cultures to examine clonal resistance.

Broadly, within the *Leuce* group, the aspen (*P. tremula*) is susceptible (and indeed was formerly a main source of bacterial slime for resistance tests), while the white and some clones of the grey poplar (*P. alba* and *P. canescens*, respectively) are resistant (Peace, 1962). In the *Aigeiros* group, most clones of *P. nigra* are also resistant, and there are some resistant clones of *P. deltoides* (Anon., 1979). The commercially important hybrid black poplars show great variation between the clones, however. Thus *P.* × *euramericana* 'Brabantica' is very susceptible, but the cvs 'Robusta', 'Serotina', 'Heidemij' and others are reasonably resistant (Anon., 1979). In the *Tacamahaca* group some clones of *P. trichocarpa* (including the cv. 'Fritzi Pauley') are very resistant (Anon., 1979), but some others are highly susceptible (Whitbread,

1967). Before silviculturally desirable clones can be recommended for plant-ing, they therefore need adequate testing for resistance to bacterial canker as well as to other diseases, such as the leaf spot fungus, *Drepanopeziza punctiformis*.

### Factors affecting the disease

Most of the factors known to affect the disease under natural conditions have been already mentioned. Thus of high importance is the susceptibility of the individual poplar species, variety or clone concerned. So is the presence of suitable wounds, such as bud scale, stipule and leaf scars, damage by agromy-zid flies, or pruning, at the time when bacterial inoculum is available. Stipule scars are present when bacterial slime is most abundant (Ridé, 1966). Leaf scars occur mainly when slime is scarce, as the amount of available inoculum declines throughout the season. Hence, the provision of early leaf scars through premature defoliation caused by *Taphrina populina* or *Marssonina brunnea* may also be important (Ridé, 1966). Already-diseased poplars are important sources of *X. populi* when planting new stands.

### Losses caused

The damage caused to susceptible poplars may lead to severe dieback and even death of the tree. Even a small number of cankers on the trunks of the trees may make them valueless. Apart from this direct damage, the disease greatly restricts the use of available clones, as many that are otherwise silviculturally desirable cannot be grown because they are susceptible to canker.

### Control measures

In France (Ridé, 1963) and in some other countries, attempts have been made since 1957 to control poplar canker partly by legislation demanding the removal of diseased trees that act as sources of inoculum, and also by the careful roguing out of diseased plants from poplar stool beds to prevent the use and distribution of infected cuttings.

However, useful as these measures are, the disease is controlled chiefly by the planting of resistant clones. Thus in Britain, where government grants are available to support the planting of certain trees, only a limited number of poplars are eligible for this grant aid. The approved poplars are all forms which grow well and at the same time are reasonably resistant to attacks by *X. populi*. They include *P. trichocarpa* 'Fritzi Pauley', *P.* × *euramericana* cvs 'Eugenii', 'Robusta', 'Heidemij', 'Serotina' and others (Evans, 1984).

### Mistletoe: *Viscum album*

Poplars in Europe (including Britain) are often hosts of the common mistle-toe, *Viscum album*. This hemiparasite is especially frequent in eastern and southeastern France. Most often affected are clones of *Populus* × *eurameric-ana*, whereas *P. nigra*, *P. alba* and *P. tremula* are rarely if ever attacked (Anon., 1979). The bushy growths of the mistletoe grow mainly on the branches, and usually do little damage unless they attack the trunk, when the wood quality is reduced. Indeed, the mistletoe plants are often harvested and marketed as Christmas decorations. *V. album* attacks a good many other trees also, and a short account of it is therefore given in chapter 3.

## DISEASES OF CATKINS

### Catkin blight caused by *Taphrina johanssonii* Sadeb. and *T. rhizophora* Johans.

*Taphrina johanssonii* and *T. rhizophora* are two uncommon ascomycetes that infect poplar catkins and cause swelling of the carpels. The swollen carpels are sterile, and become covered by a golden-yellow layer of club-shaped asci.

In Britain, *T. johanssonii* has been recorded in Scotland on *Populus tremula* (aspen) (Foister, 1961). Elsewhere, where it occurs throughout much of Europe, in North America and in Japan, it has sometimes been found also on *P. tremuloides*, *P. grandidentata*, *P. canescens*, *P. nigra* var. *pyramidalis*, *P. sieboldii* (Mix, 1949) and *P. fremontii* (Viennot-Bourgin, 1949). Its asci measure 60–140 × 12–27 µm, and contain numerous oval or globular bud-spores measuring 4–10 × 1.5–4 µm (Dennis, 1968).

*Taphrina rhizophora* Johans. has also been recorded in Britain (Ramsbottom and Balfour-Brown, 1951) as well as in Poland and Sweden (Mix, 1949). It scarcely differs morphologically from *T. johanssonii*, but its asci measure 120–160 µm in length, and it is restricted to *Populus alba* (Mix, 1949; Viennot-Bourgin, 1949).

## VIRUS AND VIRUS-LIKE DISEASES

### Poplar mosaic

Poplar mosaic is a virus disease common in Britain (Tinsley, 1967; Biddle and Tinsley, 1971a) as well as throughout much of continental Europe in France, Holland, Germany, Denmark, Poland, Czechoslovakia, Switzerland, Italy, Yugoslavia and Bulgaria, and in Canada and the USA (Berg, 1964; Cooper *et al.*, 1986) and in China (Xiang *et al.*, 1984)

The disease and the virus, Poplar Mosaic Virus (PopMV), which causes it have been described in *CMI/AAB Descriptions of Plant Viruses* No. 75, and by Berg (1964), who found that the filamentous particles of the virus measured up to 735 nm. Isolates of the virus may differ somewhat in pathogenicity and in their optimum temperatures for growth (van der Meer *et al.*, 1980; Cooper and Edwards, 1981). The best herbaceous test plant is *Vigna sinensis*, the cowpea (Berg, 1964), though *Nicotiana glutinosa* may also be used (Tinsley, 1967). The virus can also be identified by enzyme-linked immunosorbent assay (ELISA) (Cooper *et al.*, 1986), for which test material is now commercially available. Virus particles in poplar leaves can also be detected by electron microscopy (Atkinson and Cooper, 1976).

*Symptoms of the disease*
In poplars the virus causes vein clearing and spotting, so that star-shaped or diffuse light-green or yellowish spots appear scattered over the leaves (figure 79). Symptoms vary somewhat with the species or clone of poplar, and in some clones necrotic spots may be found on leaf veins and petioles. In severe infections, leaf curling may occur, and swellings appear on twigs (Berg, 1964). Infection may be very distinct on some clones, for example, *P.* 'Heidemij', while on others, particularly at the end of the summer and into the autumn, symptoms are difficult to detect (van der Meiden, 1964). Cooper *et al.* (1986) found that even within the *Aigeiros* section, symptoms on *Populus deltoides* were distinct and common, but on *P. nigra* they were indistinct amd seldom seen. Berg (1964) found that symptoms were especially clear in warm growing seasons.

*Distribution of virus within the tree*
Within the plant, distribution of the virus may be uneven (Cooper and Edwards, 1981), and symptomless cuttings taken from diseased trees usually produce symptomless plants, though cuttings from plants showing no symptoms occasionally give rise to diseased plants (Berg, 1964).

*Means of spread*
The virus is spread mainly by the use of diseased cuttings. Berg (1964) failed to pass the disease from poplar to poplar by pruning shears or aphids, and succeeded in transmitting it only by grafting. He readily passed it to herbaceous hosts by sap transmission, however, and work by Kontzog (1989) supports the view that passage on pruning tools is a possibility. Cooper and Edwards (1981) also failed to transmit the disease by aphids. Biddle and Tinsley (1968) from field observations concluded that subterranean spread of the virus probably took place, but considered that this was not to any large extent through root grafts.

*Host range*
Poplar mosaic affects almost all the poplars in the *Aigeiros* section, including many clones of *P. × euramericana*. Members of the *Tacamahaca* section and crosses between species in this and the *Aigeiros* section are also affected (Berg, 1964).

*Damage caused*
In Britain, Biddle and Tinsley (1971b) found that poplar mosaic could greatly reduce the growth in height and diameter of some clones in the nursery, but it had little effect on the growth of older trees in the plantation. In the Netherlands, the growth of commercial clones is also little affected beyond the nursery stage, though in Italy, where the disease is more severe and may

**Figure 79** Poplar mosaic virus: symptoms on leaves (Forestry Commission)

cause early leaf fall, growth reduction of trees in the plantation may occur (van der Meiden, 1964b). In China, the growth in volume of diseased trees may be 30 per cent below that of healthy ones (Xiang *et al.*, 1984).

Apart from its effect on growth, the virus may reduce the specific gravity of the wood of affected trees, but it may sometimes cause a slight increase in the strength of timber (Biddle and Tinsley, 1968, 1971b).

*Control of the disease*
So far the disease can be controlled only by the destruction of diseased propagating stools to prevent the dissemination of diseased planting stock.

**Poplar witches' broom**
Sharma and Cousin (1986), in France, have described a witches' broom disease of poplars which is especially common along the main roads in the vicinity of Paris. In the sieve-tubes of diseased *Populus alba* var. *nivea*

examined under the electron microscope they found mycoplasma-like organisms (MLOs) which were absent from those of healthy plants.

## DECAY FUNGI

Poplars are among the trees the roots of which may be damaged by *Armillaria* spp. (probably mainly *A. mellea*), which are described in chapter 3. The silver leaf fungus, *Chondrostereum purpureum*, is quite common on poplar trees. Also on poplars *Perenniporia fraximea* and *Ganoderma applanatum* may cause a white basal rot, and *Laetiporus sulphureus* and *Pholiota squarrosa* both cause brown rot in poplar trunks. These are described in chapter 20.

# 17 Diseases of willow (*Salix* spp.)

Many willow species occur in Britain and throughout the rest of Europe. Some are only small creeping forms, like *Salix repens*, others are only shrubs. Some, like *S. triandra* and *S. alba* var. *vitellina*, are grown as osiers. Of the tree species, *S. alba* var. *coerulea* is grown for its timber, which is used to make cricket bats. *S.* × *sepulcralis chrysocoma* is the common weeping willow with golden-yellow branchlets which is much grown in parks and gardens. This tree is also often called *S. alba* 'tristis'. The Japanese *S. matsudana* var. *tortuosa* is among the introduced species and varieties also now planted in gardens.

Butin (1960), in a still useful review, has summarised much of the information on willow diseases.

## LEAF DISEASES

**The powdery mildews *Uncinula adunca* (Wallr.) Lév.) (*U. salicis* (DC) Winter) and *Phyllactinia guttata* (Wallr.) Lév.)**
The powdery mildew *Uncinula adunca* is sometimes found on various *Salix* spp. in Britain (Bisby and Mason, 1940; Thomas, 1974; Greenhalgh, 1976), and in other European countries is said to occur also on *Populus* and *Betula*. It is also present on *Salix* and *Populus* in North America (Hepting, 1971). The white mycelium forms patches on the leaves on one or both leaf surfaces. The very dark brown, rounded cleistothecia are scattered over the white patches. They have colourless, unbranched appendages that may be curved at the top (figure 80) and contain eight to twelve asci, each with four to five spores. The colourless, elliptical spores measure 25–30 × 15–19 μm (Rabenhorst, 1884).

Species of *Salix* are also among the many trees sometimes attacked by *Phyllactinia guttata*. This mildew is most common on hazel, and is therefore described under '*Corylus*' in chapter 19.

**Leaf spot caused by *Rhytisma salicinum* (Pers.) Fr. (anamorph *Melasmia salicina* Lév.) and *R. symmetricum* J. Müller**
The uncommon *Rhytisma salicinum* has sometimes been found producing irregularly rounded, thick, shining black stromatic spots on the upper sides of willow leaves in Britain, mainly in the islands off the west coast of Scotland. It is also found elsewhere in Europe, especially in Scandinavia. Its stromata somewhat resemble the tar spots made on sycamore leaves by the related fungus *Rhytisma acerinum*, but at maturity they measure only 2–5 mm across. In the autumn, one to three pycnidia of the *Melasmia* stage form within each

**Figure 80** Cleistocarps of *Uncinula adunca* (*U. salicis*), showing the curled appendages (K.F. Tubeuf, after Tulasne)

stroma. The pycnidia contain colourless, long cylindrical spores 5–6 μm long
(Grove, 1937). The ascocarps of the perfect stage, also embedded in the
stromata, form in winter on the fallen leaves. They contain long asci each
having 8 colourless, thread-like ascospores measuring 60–90 × 1.5–3 μm
(Viennot-Bourgin, 1949).

The rare *R. symmetricum* has also been found producing many small
stromatic spots on willow leaves in Britain (figure 81) and in other European
countries. The stromata extend from one side of the leaf to the other, and the
spots therefore appear on both upper and lower leaf surfaces. The fruit bodies
are also embedded in the stromata on both sides of the leaves. Their asci
contain thread-like ascospores that measure 30–108 μm in length (Dennis and
Wakefield, 1946).

These *Rhytisma* species are of little pathological importance.

### Leaf spot caused by *Marssonina kriegeriana* (Bres.) Magn.

*Marssonina kriegeriana*, found by Nattrass (1930) to be the cause of a leaf
spot of willows in Egypt, has occasionally been found also on willows in
Britain (Grove, 1937; Cannon *et al.*, 1985), on the leaves of which its greyish
acervuli form on small dark spots.

Its colourless, curved, two-celled conidia appear to be variable, and
measure 17–25 × 5–7 μm according to Ellis and Ellis (1985) or 13–17 × 4–7
μm according to Grove (1937). The apothecia of its perfect state,
*Drepanopeziza triandrae* Rimpau (Rimpau, 1962), have not yet been found in
Britain. Their one-celled, colourless ascospores measure 10.5–12 × 5–5.6 μm
(Holliday, 1989). The fungus also occurs in other parts of Europe, but does
not seem to be common or important.

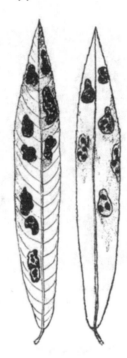

**Figure 81**   Willow leaf with stromatic spots caused by *Rhytisma symmetricum* (K.F. Tubeuf)


**Leaf spot caused by *Drepanopeziza salicis* (Tul. and C. Tul.) Höhnel (*Trochila salicis* Tul. and C. Tul.) (anamorph *Monostichella salicis* (Westend.) v. Arx, syn. *Gloeosporium salicis* Westend.)**

*Monostichella salicis* was recorded (as *Gloeosporium salicis*) in Britain on *Salix alba*, *S. caprea* and *S. fragilis* by Grove (1937), and it has also been found there on *S. amygdalina* and *S. triandra* (Ellis and Ellis, 1985). In Europe as a whole it is known also on *S. pentandra* (Lanier *et al.*, 1978) and on *S. americana* hort. (Butin, 1960). It has also been found on *S. fragilis* in the USA (Nelson, 1965).

*M. salicis* causes dark-brown or black, crowded spots, which often coalesce and may eventually cover the whole leaf. The conidial pustules form on the upper leaf surface in summer and autumn. The ellipsoid or oblong, colourless, one-celled conidia measure 12–16 × 4–6 μm (Grove, 1937).

The dark apothecia of the perfect stage, *Drepanopeziza salicis*, form on both sides of the dead leaves in the spring. Their one-celled, colourless ascospores measure 10–16 × 6–7 μm (Ellis and Ellis, 1985).

The fungus is quite widespread in Europe. In Italy, Anselmi (1977, 1980) found that primary infections were caused mainly by mature ascospores, infection taking place chiefly when temperatures were above 12 °C and the leaves were wet after heavy rain. Anselmi (1977) recommended the use of resistant clones of willow, some of which were already known, and chemical control by spraying with benomyl or thiophanate-methyl, or (to less effect) with maneb or mancozeb.

**Leaf spot caused by *Ramularia rosea* (Fuckel) Sacc.**
Ellis and Ellis (1985) record *Ramularia rosea* as the cause of small brown leaf spots on *Salix caprea*, *S. triandra* and *S. viminalis*. The pale-reddish colonies are on the undersides of the leaves. Bundles of conidiophores emerge from the stomata, and bear long, rather narrow cylindrical, one- to four-celled conidia which measure 15–27 × 2.5–3.5 μm. The fungus, which also occurs on willows in other parts of Europe (Tubeuf, 1895), is of little pathological importance.

## DISEASES OF LEAVES AND SHOOTS

### Rusts of tree willows and osiers
Five species of *Melampsora* affect tree willows. They are *M. larici-pentandrae* Kleb., *M. allii-fragilis* Kleb., *M. epitea* var. *epitea* Thüm., *M. capraearum* Thüm. and *M. salicis-albae* Kleb. *M. epitea* also attacks osiers, as do *M. ribesii-viminalis* Kleb. and *M. amygdalinae* Kleb. Full descriptions of these rusts, as well as of others affecting creeping and bush willows, are given by Wilson and Henderson (1966) and Gäumann (1959). Their host ranges on tree willows and osiers are given in Table 17.1.

The autoecious *M. amygdalinae* Kleb. may cause serious damage to the basket willow *Salix triandra* (the almond willow), making the rods useless. The bright-orange aecial pustules (caeomata) are produced with the orange spermagonia on the stems and young leaves, usually on the undersides. The globoid to ovoid aeciospores measure 18–23 × 14–19 μm. The small orange uredinia are almost all on the undersides of the leaves. The ovoid or oblong urediniospores measure 20–40 × 11–18 μm. The reddish-brown (later almost black) telia occur under the epidermis on the undersides of the leaves. The teliospores, measuring 8–15 × 7–14 μm, are very much shorter than those of any other *Melampsora* sp. on willows.

The life cycle of *M. amygdalinae* was examined by Ogilvie (1932). This rust overwinters either as telia on fallen leaves, or as mycelium in the buds and as perennial uredinia on uncut rods. The aecial stage causes cankers on the young stems in April. Uredinia, also on cankers, may be found from March to October, and spread the disease rapidly, damaging the leaves as well as the stems. The telia occur on the leaves from July onwards. This rust is locally damaging in England; it occurs throughout Europe and into Asia.

All the other *Melampsora* spp. on willows are heteroecious, and their teliospores are more than 20 μm in length. Both the uredinia and telia are produced on the willows. In three species, *M. salicis-albae*, *M. allii-fragilis* and *M. larici-pentandrae*, the urediniospores are smooth at their upper ends, but otherwise echinulate. In *M. salicis-albae* Kleb. (*M. allii-salicis-albae* Kleb.) the urediniospores are oblong, or club- or pear-shaped, measuring

**Table 17.1** Rusts on tree willows and osiers

| Rust | Tree willows and osier hosts (uredinia and telia) | Alternate hosts (spermagonia and aecia) |
|------|---------------------------------------------------|------------------------------------------|
| 1. *Melampsora larici-pentandrae* | *S. fragilis\** *S. pentandra* and their hybrids\* | *Larix* spp. |
| 2. *M. allii-fragilis* | *S. fragilis\** *S. pentandra* and their hybrids\* | *Allium* spp. (not in Britain) |
| 3. *M. epitea* var. *epitea* *Euonymus* form (*M. euonymi-caprearum*) | *S. caprea\** *S. cinerea\** *S. alba†* | *Euonymus europaeus* |
| *Larix* form (*M. larici-epitea*) | *S. alba†* *S. caprea* × *viminalis\** *S. daphnoides\** *S. fragilis\** *S. viminalis* # *S. triandra* # | *Larix* spp. |
| *Ribes* form (*M. ribesii-purpureae*) | *S. purpurea* # | *Ribes* spp. (not in Britain?) |
| 4. *M. capraearum* | *S. caprea\** *S. cinerea\** *S. caprea* × *viminalis* *S. aurita\** | *Larix* spp. |
| 5. *M. salicis-albae* | *S. alba†* *S. alba* var. *vitellina†* *S. pentandra\** | *Allium* spp. (not in Britain) |
| 6. *M. ribesii-viminalis* | *S. viminalis* # | *Ribes* spp. (not in Britain?) |
| 7. *M. amygdalinae* | *S. triandra* # | None; autoecious with all stages on *Salix* |

\* *Shrub or small tree.*
† *Tree.*
# *Osier.*

20–36 × 11–17 μm. In spring they occur in large pustules up to 2 mm long on the young twigs. Later, throughout the summer and autumn they are present in smaller spots on the undersides of the leaves. The dark-brown telia are on both sides of the leaves, under the epidermis, and the teliospores measure 24–45 × 7–10 μm.

Overwintering is by the teliospores on the leaves or by mycelium which perennates in the branches and gives rise to uredinia in spring, before the aecia arise on the alternate hosts, *Allium ursinum* and other *Allium* spp.

Hence, once established, it can persist in *Salix* without the further presence of the alternate host (Weir and Hubert, 1918; Wilson and Henderson, 1966). It does not seem to be among the rusts found on *Allium* spp. in Britain (Ellis and Ellis, 1985), where it is fairly common (as it is elsewhere in Europe) on *Salix alba*, and on its variety *vitellina*, the rods of which it makes too brittle for basket-making. It also occurs throughout Europe and into Asia on *S. pentandra* and on various hybrids between *S. alba* and other willows (Gäumann, 1959).

In *M. allii-fragilis* Kleb., which appears to be rather rare and of little pathological importance, the spermagonia and aecia again occur on *Allium* spp. (though apparently not in Britain), and are indistinguishable from those of *M. salicis-albae*. The small reddish-orange uredinia appear mostly on the undersides of the leaves of *S. fragilis* and *S. pentandra* (Ellis and Ellis, 1985; Gäumann, 1959), and the obovoid urediniospores measure 22–33 × 13–15 μm. The dark-brown, shining telia are mainly on the upper sides of the leaves, and occur under the cuticle. The teliospores measure 30–48 × 7–14 μm. The fungus occurs in Britain and throughout the rest of Europe.

*M. larici-pentandrae* Kleb. is also scarce and of little economic importance. In Britain it is known only in Scotland and Ireland. The bright-orange uredinia are mostly on the undersides of the leaves, and the ovoid or ellipsoid urediniospores measure 23–44 × 12–16 μm. The yellowish-brown (later almost black) telia are under the epidermis, on the undersides of the leaves. The teliospores measure 28–38 × 6–11 μm.

*M. larici-pentandrae* occurs throughout Europe on *Salix fragilis* and *S. pentandra*. Its spermagonia and aecia develop on *Larix* spp.

In the remaining three species, *M. epitea*, *M. capraearum* and *M. ribesii-viminalis*, the urediniospores are spiny over the whole surface.

*M. epitea* Thüm. consists of several varieties or races, sometimes regarded as distinct species. The orange-yellow uredinia are on both sides of the leaves. The ellipsoid or globoid urediniospores measure 12–25 × 10–18 μm. The small, yellowish-brown (later almost black) telia are under the epidermis, mostly on the undersides of the leaves. The teliospores, which are not much thickened at their tips, and have no conspicuous germ pore, measure 20–50 × 7–14 μm. *M. epitea sensu lato* is common in Great Britain and Ireland and occurs throughout Europe and into Asia and North Africa. The forms on tree willows and osiers are all races or form species of the variety *epitea*, with *Euonymus europaeus* (*M. euonymi-capraearum* Kleb.), *Larix* spp. (*M. larici-epitea* Kleb.) and *Ribes* spp. (*M. ribesii-purpureae* Kleb.) as the alternate (spermagonial and aecial) hosts. The fungus is of little economic significance, though it may attack many *Salix* spp. (Gäumann, 1959).

*M. capraearum* Thüm. (*M. larici-capraearum* Kleb.) has its uredinia on the lower leaf surfaces; they show as pale-yellow spots on the other side. The ovoid or ellipsoid urediniospores measure 14–21 × 13–15 μm. The dark reddish-brown telia are under the cuticle on the upper sides of the leaves. The teliospores are thick-walled above, with a clearly marked germ pore, and measure 30–40 × 7–14 μm. *M. capraearum* is common in Britain and throughout the rest of Europe and into Asia, affecting *Salix aurita*, *S. caprea* and *S. cinerea* (Ellis and Ellis, 1985), but it is of minor importance. Its spermagonia and aecia are on larch.

*M. ribesii-viminalis* Kleb. has small, pale orange-yellow uredinia on the undersides of the leaves of *Salix viminalis*. Its rounded urediniospores measure 15–19 × 14–16 μm. The dark-brown, shining telia are under the cuticle on the upper sides of the leaves. The teliospores, like those of *M. epitea*, are unthickened above, with no clearly visible germ pore. They measure 25–40 × 7–14 μm. It is doubtful if the spermagonial and aecial stage, on *Ribes* spp., is present in Britain, and the fungus on *Salix* seems to be rare. The fungus is found elsewhere in Europe, but is of no importance.

Of these seven rusts, it will be seen that *M. allii-fragilis*, *M. larici-pentandrae* and *M. ribesii-viminalis* are uncommon. *M. amygdalinae* is locally common in England, on the basket willow *S. triandra,* and *M. epitea* is common on various willows, including the basket willows *S. viminalis*, *S. triandra* and *S. purpurea*. *M. salicis-albae* is fairly common on the basket willow *S. alba* var. *vitellina*.

Pathological information (apart from the few points on overwintering referred to above) is scanty. Ogilvie and Hutchinson (1933) found that the germination of the urediniospores of *M. amygdalinae* and *M. epitea* was encouraged by cool, damp conditions, and the spread and severity of the disease was much reduced in hot weather.

*Control*

Control measures for rust have been proposed only in the case of basket willows, which locally at one time formed an important crop. Their acreage has now much declined, however. Ogilvie and Hutchinson (1933) suggested that the spread and development of *M. amygdalinae* on *S. triandra* could be slowed and reduced by delaying the cutting of the rods in spring to avoid the production of young, susceptible growth when the basidiospores arise from the teliospores. If cutting cannot be delayed, young growth can be removed by grazing cattle or the use of a tar-oil winter wash to kill the young shoots.

Stott *et al.* (1980) in trials with various fungicides against the same fungus found that fenpropimorph and benodanil gave the most effective control, closely followed by triadimefon and oxycarboxin.

### Anthracnose caused by *Marssonina salicicola* (Bres.) Magn. (anamorph of *Drepanopeziza sphaerioides* (Pers.) Höhnel)

The anthracnose caused by the fungus *Marssonina salicicola* has long been the most damaging disease of the weeping willow, *S.* × *sepulcralis chrysocoma*, on the shoots of which it causes very unsightly cankers. In New Zealand, Murray (1926) found weeping willow trees so badly affected that they lost their weeping habit. The disease also occurs in Britain on the osier *S. purpurea* (Nattrass, 1930). Elsewhere it has been found in Argentina on the cricket bat willow, *S. alba* var. *coerulea* (Jauch, 1952), and in New Zealand on *S. fragilis* (Murray, 1926), as well as on various willows in the Netherlands (van Poeteren, 1938), Germany (Butin, 1960), Italy (Anselmi, 1979), Norway (Reinfeldt, 1979) and Portugal (Costa, 1982).

The fungus causes small black elliptical spots on the developing shoots; these spots spread to become elongated, irregular, sunken cankers with a dark-brown or black raised rim and a paler centre (figure 82). Typically, black or purple-black spots also develop on the leaves, which in severe attacks become distorted and are shed. Similar spots may sometimes also be found on the catkins. The numerous unsightly cankers and leaf spots which cover weeping willows in severe outbreaks much reduce the ornamental value of the trees. Whitish fruiting pustules (acervuli) of the fungus develop on the cankers and leaf spots. The acervuli may be up to 400 μm across (Nattrass, 1930), but generally measure about 150–200 μm. They contain the colourless, often curved, club-shaped or pear-shaped conidia, which measure 11–19 × 3–7 μm (Nattrass, 1930; Murray, 1926). The spores are two-celled, with the lower cell much smaller than the upper.

The perfect stage of the fungus is *Drepanopeziza sphaerioides* (Fr.) Nannf. (Butin, 1960; Rimpau, 1962), which so far does not appear to have been recorded in Britain. Its colourless, one-celled ascospores measure 13–22 × 6–8.5 μm (Holliday, 1989). The apothecia do not seem to be generally common, and Anselmi (1979) in Italy found no mature ascospores.

The fungus overwinters on the cankers, on the edges of which it forms small stromata (Anselmi, 1979), and probably on the bud scales and on dead, fallen leaves. Conidia formed on these early in the year infect the young leaves and the newly elongating shoots. The leaves become more resistant with age, being very susceptible when young (Anselmi, 1979). Nattrass (1930) found some evidence from inoculation experiments that the fungus was a wound parasite. Its attacks are especially severe in wet growing seasons. The optimum temperature range for the germination of the conidia is from 12 to 21 °C (Anselmi, 1979).

Murray (1926) in New Zealand found that *S. fragilis* was less badly affected than the weeping willow. Rose (1970) reported that selection and breeding for resistance against willow anthracnose had begun at Long Ashton.

*Control*

Reinfeldt (1979) has recommended propagating willows only from young green tips, which are likely to be free from the fungus.

In the long term the growing of resistant stock would appear to be the best answer to the disease. Meanwhile, spraying with a suitable fungicide remains the only practicable control method, though it can usually be done only on relatively small trees.

Spraying with a copper fungicide has often been suggested, for example, by Rose (1970), who advised applying copper at weekly intervals from bud-break until midsummer. Reinfeldt (1979) also found spraying with copper oxychloride was beneficial. Gaggini (1970) obtained excellent control with captafol and quinomethionate, the first of which was also found effective by Reinfeldt (1979). Good results have been obtained with benomyl and with

**Figure 82** Willow anthracnose: leaf spots and stem cankers on weeping willow, caused by *Marssonina salicicola* (Forestry Commission)

maneb and mancozeb (Reinfeldt, 1979; Anselmi, 1979). Spraying needs to begin early, when the buds begin to swell, and two or more applications are usually required.

**Black canker caused by *Glomerella cingulata* (Stoneman) Spaulding and von Schrenk (*G. miyabeana* (Fukushi) von Arx and Müller: anamorph *Colletotrichum gloeosporioides* Penz) and willow scab caused by *Pollaccia saliciperda* (All. and Tub.) von Arx, syn. *Fusicladium saliciperdum* (All. and Tub.) Tub.), teleomorph *Venturia saliciperda* Nuesch)**

Black canker, caused by a form of *Glomerella cingulata* commonly known as *G. miyabeana*, and willow scab, caused by *Pollaccia saliciperda* (the anamorph of *Venturia saliciperda*), have often been confused, partly because their symptoms are very similar. Furthermore, inoculation experiments by some workers have indicated that *G. miyabeana* is a major pathogen, and *P. saliciperda* at most only a secondary organism (Nattras, 1928; Dennis, 1931; Rupert and Leach, 1942; Wiejak, 1960). Alcock (1924), Clinton(1929) and Brooks and Walker (1935), however, all found good evidence, at least in the areas and under the conditions in which they were working, that *P. salici-perda* could act as an important primary pathogen in the absence of *G. miyabeana*.

**Black canker**

Black canker, caused by *Glomerella miyabeana* (*Physalospora miyabeana* Fukushi), which is now frequently regarded as a specialised race of *G. cingulata* (von Arx, 1957b) is important mainly on basket willows. On these it may make the rods useless for basket making (Brooks, 1953). In Czechoslovakia, Farsky and Fekete (1965) found that the disease, then recorded there for the first time, had destroyed up to 50 per cent of the first-year setts in the area affected.

As well as in Britain and Czechoslovakia, black canker is known in Japan (Fukushi, 1921), Germany (Diercks, 1959), the Netherlands (van Poeteren, 1935), Canada (Harrison, 1965; Conners *et al.*, 1941) and the USA (Rupert and Leach, 1942).

The first symptoms of the disease, which may be found as early as May, show as reddish-brown, angular spots on the leaves. The spots spread, and the leaves blacken and shrivel, but commonly remain attached to the stem (plate 48). Often the fungus spreads further through the petiole into the stem, where it produces oval, elongated, flattened cankers, which are then generally isolated by a layer of cork (Nattrass, 1928; Dennis, 1931). The bark over these cankered areas dries, contracts and splits.

The fruit bodies of the fungus occur mainly on the shoot cankers. The conidial pustules (acervuli) develop first, in spring, on shoots in which the mycelium has overwintered; they have been described as *Colletotrichum gloeosporioides* Penz (Arx, 1957b). The spore mass is colourless or pale pink (Nattrass, 1928) to light brick-red (Brooks, 1953), and the one-celled, ellipsoid, colourless conidia measure 13–23 × 4–7 μm (Nattrass, 1928; Brooks, 1953). The fungus also overwinters as perithecia which form in autumn. The one-celled, oblong-ellipsoid, colourless ascospores measure 15–17 × 5.5–7 μm, and the asci are interspersed with slender paraphyses (Brooks, 1953).

Rybak-Mikitiuk (1962) found that the optimum temperature for conidial germination and mycelial growth was 26 °C, and in the field the disease spread most rapidly at a temperature of 25 °C when the relative humidity was high.

In Britain, black canker has been recorded on *Salix americana* hort., *S. alba* var. *vitellina* and *S. alba* var. *cardinalis* (and occasionally on a few other ornamental willows) (Nattrass, 1928; Dennis, 1931). Elsewhere it has also been found on *S. alba* × *fragilis* (Farsky and Fekete, 1965) and the grey willows (van Poeteren, 1935).

Control measures are probably necessary and feasible only in basket willow beds and in nurseries, and on young plants not long established. Spraying with Bordeaux mixture and other copper sprays has often been recommended (Brooks, 1953). Trials with more recently developed fungicides suggest that a weekly spraying throughout the growing season with zineb or a mixture of captan and brestan are likely to give control (Diercks, 1959). In the USA,

Barnard and Schroeder (1984/85) found that they could control the fungus on *Acacia* plants by spraying with chlorothalonil.

**Willow scab**
Willow scab, caused by *Pollaccia saliciperda* (*Fusicladium saliciperdum*), may cause considerable damage in damp seasons. In England it has been found especially on *Salix decipiens* (Brooks and Walker, 1935), and in Scotland Alcock (1924) recorded it on *S. alba* var. *vitellina*. It has been found elsewhere on various willow species and varieties in Germany (Janson, 1927), the Netherlands (von Arx, 1957a), Poland (Twarowska, 1968), Canada (Harrison, 1965) and the USA (Steinmetz and Prince, 1938).

In Canada, Bloomberg and Funk (1960) reported that one-third of the crown of some trees was destroyed by the fungus. In Britain, Brooks and Walker (1935) found that in a wet season trees might lose almost all their leaves by the middle of the summer. Very severe damage has also been recorded in Germany (Appen, 1927) and Poland (Twarowska, 1968).

The symptoms of the disease (figure 83) are very similar to those of black canker, and first become visible in early spring. Irregular spots form on the young leaves, and as they spread, the leaves blacken, die and fall. Black lesions may girdle shoots so that twigs and even whole branches may be killed. In conditions favourable to the fungus, the progress of the disease may be very rapid (Brooks and Walker, 1935; Bloomberg and Funk, 1960).

Small olive-brown pustules of *Pollaccia saliciperda* develop on the affected leaves, mainly along the veins on the undersides (Brooks and Walker, 1935; Clinton and McCormick, 1929). The fungus overwinters as similar pustules on affected twigs (Brooks, 1953). The olive-brown, rather slipper-shaped conidia measure 15–23 × 5–10 μm and are eventually two-celled.

The perfect stage of the fungus was once considered to be *Venturia chlorospora* (Ces.) Karst., which, however, appears to cause only minor leaf spots. The pale-brown, two-celled ascospores of *V. chlorospora* measure 11–18 × 5–7 μm, and its conidial stage, which is known only in culture, is regarded by Sivanesan (1977) as a *Cladosporium*. Nuesch (1960) has shown that the very active parasite which causes willow scab is in fact a separate species which he has named *V. saliciperda*, and of which the anamorph is *Pollaccia saliciperda* (formerly known as *Fusicladium saliciperdum*), which is described above. The brown pseudothecia of *V. saliciperda*, which form on the dead leaves, have brown setae around the ostiole, and their olive-green, unequally two-celled ascospores measure 11–14 × 3.5–5 μm (Sivanesan, 1977).

The fungus overwinters on the dead twigs infected in the previous season. In spring, pustules of the *Pollaccia* stage arise on the diseased twigs, and the conidia they produce attack the newly opening buds and young leaves (Brooks and Walker, 1935; Clinton and McCormick, 1929).

In Scotland, Alcock (1924) found *P. saliciperda* on *Salix alba* var. *vitellina*, while Brooks (1953) noted that it occurred in Britain especially on *S. fragilis* var. *decipiens*. *S. alba* var. *vitellina* was also found to be very susceptible in eastern Canada and the USA (Clinton and McCormick, 1929; Steinmetz and

**Figure 83**   Dieback of *Salix alba* caused by *Pollaccia saliciperda* (*Fusicladium salici-perdum*) (Forestry Commission)

Prince, 1938). A great many other willow species are known to be susceptible, the osier *S. americana* hort. being especially badly affected in Germany and the Netherlands (Appen, 1927; von Arx, 1957a). In Poland, Kochman (1929) found that *S. alba*, *S. blanda* and *S. babylonica* were the most liable to damage. On the other hand, some clones of *S. fragilis* have shown resistance (Steinmetz and Prince, 1938), and so have *S. viminalis*, *S. purpurea*, *S. amygdalina*, *S. pentandra*, and *S. alba* var. *calva* (*S. alba* var. *coerulea*) (Appen, 1927; Harrison, 1965; Kochman, 1929)

The development of the disease is greatly encouraged by cool, wet weather (Brooks and Walker, 1935; Davidson and Fowler, 1967). A fertiliser balance high in nitrogen and low in potassium has also been said to favour the disease (Anon., 1927; Janson, 1927).

In general, control measures are probably warranted only in osier beds and in nurseries and on young trees. In view of the evidence that the disease may be spread with infested stock (Anon., 1927), cuttings should be taken only from healthy material. Winter washes of copper sulphate solution, for example, may be used in stool beds (Alcock, 1924; Twarowska, 1968). Avoidance of winter pruning is said to aid in control (Pape, 1925). Various fungicides, including Bordeaux mixture and lime sulphur, have been recommended (Alcock, 1924). More recently, spraying with phygon (dichlone) (Harrison, 1965), chinosol, maneb, zineb, captan and brestan (Diercks, 1959; Twarowska, 1968) has been suggested.

### Canker and dieback associated with *Cryptodiaporthe salicina* (Curr.) Wehm. and *C. salicella* (Fr.) Wehm.

The canker and dieback of willows caused by *Cryptodiaporthe salicina* is generally of minor significance; the fungus is common as a saprophyte on dead twigs, and its pathogenicity has sometimes been doubted (Moore, 1959). Schwarz (1922) and Broekhuijsen (1934) were both able to demonstrate its pathogenicity, however, and Bier (1959) showed that it became pathogenic when the relative turgidity of the plant fell to 80 per cent or less.

The disease is known in Britain (Grove, 1937; Redfern, 1974), the mainland of Europe, especially in the Netherlands (Broekhuijsen, 1934), and Canada (Bier, 1959).

Affected shoots die back to a node, at first appearing brownish-green. The lesions bear the flat, disc-shaped pycnidia of the imperfect stage of the fungus (Broekhuijsen, 1934; Schwarz, 1922), which form just within the bark. This stage is sometimes considered to be *Discella carbonacea* (Fr.) Berk. and Br. (Cannon *et al.*, 1985), now regarded as a *Diplodina* (Sutton, 1980). The bark eventually splits open and releases the spores. These are almost colourless or slightly greenish, spindle-shaped, two-celled, and widest at the septum. They measure 13–18 × 3.5–5 μm (Grove, 1937). The perithecia of the perfect stage later appear scattered singly or in small groups, also just beneath the bark surface, through which their small black ostioles emerge. They contain club-shaped asci each with eight colourless, ellipsoid or inequilateral, blunt-ended, finally two-celled ascospores measuring 15–20 × (3.5)–4.5–7.5 μm (Wehmeyer, 1932).

In Britain the disease has been found especially on *Salix alba*, *S. alba* var. *vitellina* and *S. fragilis* (Grove, 1937). On the mainland of Europe it has been found on *S. alba* var. *vitellina pendula* (the Dutch weeping willow) (Schwarz, 1922), *S. alba* and *S. alba* var. *coerulea* (Mooi, 1948), *S. viminalis* (Broekhuijsen, 1934), and most other willows (Butin, 1960). Records in Canada include *S. hookeriana* and *S. scouleriana* (Bier, 1959).

Another species of *Cryptodiaporthe*, *C. salicella*, is also sometimes regarded as a cause of a minor dieback of willows, although it is of doubtful pathogenicity (Moore, 1959). Its perithecia resemble those of *C. salicina*, but the ascospores measure 15–22 × 2–3 μm (Ellis and Ellis, 1985). Its imperfect stage is said to be *Diplodina microsperma* (Johnston) Sutton (Sutton, 1980) (of which Sutton regards *Discella carbonacea* as a synonym) or *Diplodina salicis* (Westend.) Boerma (Cannon *et al.*, 1985). The one- or two-celled conidia measure 16–20 × 4–4.5 μm (Ellis and Ellis, 1985).

*C. salicina* and *C. salicella* are sometimes considered to be inseparable from one another. They are further discussed by Wehmeyer (1932) and Butin (1958).

**Twig blight caused by *Cryptomyces maximus* (Fr.) Rehm**
On rare occasions *Cryptomyces maximus* (figure 84) has been found causing shining dark-brown or black, irregularly elongated, blister-like stromatic cushions up to 8 in (200 mm) long on the twigs and branches of willows (Alcock and Maxwell, 1925). The edges of the cushions are at first yellow at the edges. At maturity the outer tissues of the stroma break to expose a yellowish-brown hymenium. The cylindrical-clavate, eight-spored asci measure up to 250–330 μm, and contain colourless or pale-yellowish, ellipsoid to ovoid, one-celled ascospores measuring 30–37 × 13–17 μm (Ellis and Ellis, 1985) and covered by a thin gelatinous coating. The asci are separated by septate hyphae which unite at their tips to form a brownish gelatinous layer (Dennis, 1968). In wet weather the stromatic cushions swell and fall off, to leave elongated scars on the bark. Affected twigs may be girdled and die (Alcock and Maxwell, 1925).

*C. maximus* has been found in Scotland damaging *Salix fragilis* (Alcock and Maxwell, 1925), and in Norfolk attacking the cricket bat willow, *S. alba* var. *coerulea* (Anon., 1952). It has also been found in Britain on *S. viminalis* (Ellis and Ellis, 1985). Elsewhere in Europe it occurs also on *S. incana* and *S. amygdalina* (Butin, 1960).

According to Alcock and Maxwell (1925), *C. maximus* may be followed by the weakly parasitic *Scleroderris fuliginosa* Karst. and *Myxosporium scutellatum* (Otth) Petrak, which then cause further damage.

**Cankers caused by *Nectria galligena***
Species of *Salix* are among those on which cankers may be caused by the fungus *Nectria galligena*, which has a wide host range, and is therefore described in chapter 3.

**Bacterial canker caused by '*Xanthomonas populi* ssp. *salicis*'**
The bacterium known as '*Xanthomonas populi*' causes the well-known canker of poplar described in chapter 16. A form of this bacterium was found by de Kam (1978, 1981) attacking *Salix dasyclada*, and has been described as '*X. populi* ssp. *salicis*'.

WILT DISEASES

**Watermark disease caused by *Erwinia salicis* (Day) Chester (*Bacterium salicis* Day, *Pseudomonas saliciperda* Lindeijer)**
Watermark disease, caused by the bacterium *Erwinia salicis*, is particularly important as a cause of stain and degrade in the cricket bat willow, *Salix alba* var. *coerulea*. The wood of trees affected by the disease is unsaleable, partly

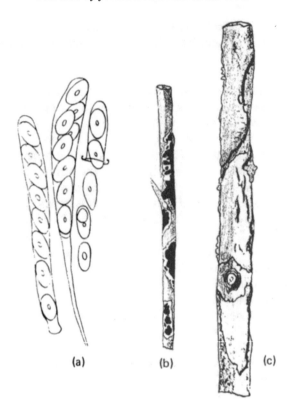

**Figure 84** *Cryptomyces maximus*: (a) asci and ascospores, (b) twig of *Salix incana* with stromata of the fungus, (c) an osier twig, after the stromatic cushions have scaled off in the autumn (K.F. Tubeuf)

because it is badly discoloured (plate 57) and partly because any cricket bats made from it are likely to splinter (Metcalfe, 1940).

The disease was first described in England by Day (1924). It is known to occur there in Norfolk, Suffolk, Cambridgeshire, Essex, Middlesex, Wiltshire, Gloucestershire, Bedfordshire, Hertfordshire and Yorkshire (Dowson and Callan, 1937; Bryce, 1950; Preece, 1978). A similar disease on willow in the Netherlands was originally considered to be caused by a separate bacterium, *Pseudomonas saliciperda* Lindeijer. It is now the view in the Netherlands, however, that *P. saliciperda* and *E. salicis* are the same organism (Gremmen and de Kam, 1970). Watermark disease is also known in Belgium (Rijkaert *et al.*, 1984)

In recent years, especially between 1965 and 1975, the disease has caused much damage in the Netherlands, where trees have had to be felled in large numbers (Gremmen and de Kam, 1981; de Kam and Tol, 1988).

Gremmen and de Kam (1981) have pointed out that *E. salicis* can often be found in trees that show no symptoms of the disease, and that inoculations with pure cultures of the bacterium may result in few infections, whereas better results are obtained by inoculation with sap from diseased trees. They have therefore suggested that watermark disease may not be caused by *E. salicis* alone, but some second, unknown factor may also be involved.

*E. salicis* is described by Dowson (1937, 1957). It is a straight, cylindrical rod, either single, in pairs end to end, or occasionally in chains. The individual cells measure 0.8–1.7 × 0.5–0.7 μm, and move by means of five to seven long, peritrichous flagella. The bacterium produces a light-yellow pigment on sterilised potato, and forms yellowish colonies on potato agar. Further information on isolation and identification is given by Lelliott and Stead (1987) and Preece *et al.* (1979). Wong and Preece (1973) describe a slide-agglutination test for the detection of the bacterium in the field, and de Kam (1982a,b) describes immunofluorescence (IF) and enzyme-linked immunosorbent assay (ELISA) methods which might be used in identification. Isolates of *E. salicis* differ in pathogenicity, only about half of those collected by Gremmen and de Kam (1981) producing symptoms of the disease.

As noted above, *E. salicis* is important especially as the cause of watermark disease of *Salix alba* var. *coerulea*. It also affects other willows, including other varieties of the white willow, *S. alba*, such as var. *vitellina* (Day, 1924; Wong and Preece, 1973). In the Netherlands the disease has been found affecting *S. caprea, S. viminalis, S. amygdalina* and *S. triandra* as well as *S. alba* (Lindeijer, 1931; Burger, 1932). Day (1924) and Wong and Preece (1973) failed to find it on *S. fragilis*. Wong and Preece were also unable to find it on the hybrid *S × rubens (S. viridis)*, though Dowson and Callan (1937) found this hybrid affected. There is therefore evidence that willows vary in susceptibility to the disease, and variations in resistance are found to clonal level (de Kam, 1983). De Kam (1984) has put forward suggestions for the testing of willows for resistance.

Gremmen and de Kam (1981) found that inoculation was most effective at the end of the growing season, and that inoculations made in the spring invariably failed.

External symptoms first appear in late April or early May, when the young leaves on one or more branches turn reddish and wither (plate 53). This phase of the disease is known as the red-leaf stage. A thin, sticky liquid containing bacteria exudes from insect exit holes and other wounds on affected twigs, the wood of newly infected trees being filled with this bacterial slime (Dowson, 1957). Further leaf reddening, wilting and dieback may continue through the summer, though only parts of the crown are affected in any one year, and throughout the season fresh green shoots may appear on the diseased branches. These new adventitious shoots may later be attacked in their turn (Day, 1924; Metcalfe, 1940). The diseased trees rarely die, but soon become stag-headed.

Internally the wood of infected trees shows a watermarked stain. This may form complete or interrupted rings or spread widely through the wood. On exposure to the air it turns bright red and then black (Metcalfe, 1940) (plate 57). Sometimes, in recently affected trees, the staining is confined to a few shoots with withered leaves. In other cases, in long-diseased trees, it occurs almost throughout, reaching even as far as the roots. At times, however, when staining is extensive inside the tree, only a few shoots may show dieback and leaf wilting, and the severity of the external symptoms may be a poor indication of that of the deterioration within. In this connection, Miller-Jones *et al.* (1977) discuss the possible use of electrical resistance measurements in the detection of the disease.

Once inside the tree, the bacteria spread through the xylem, persisting from one year to another, and commonly spreading outwards from one annual ring to the next (Dowson, 1957).

It has been suggested that the wilting and 'red-leaf' symptoms are caused by obstruction of the water-conducting system by the bacteria, and by the effects of the toxic substances and the consequent production of tyloses (Day, 1924; Dowson, 1957). The papers of Wong and Preece (1978a,b,c) provide much information on the histology and biochemistry of wood infected by *E. salicis*.

It has long been known that the disease tends to spread in the direction of the prevailing wind, and studies of the distribution of the bacteria on the leaves of willows and other trees in plantings affected by watermark disease support the view that *E. salicis* is spread mainly by wind and rain (Zweep and de Kam, 1982).

It has often been surmised that transmission is by various insects, and even by the willow tit, *Parus montanus* (Dowson, 1957). Callan, who made a prolonged study of insect transmission (and much of whose work remains unpublished), failed to demonstrate that it definitely took place (Callan, 1939). Gremmen and de Kam (1981) similarly failed to find evidence in support of insect transmission.

Day (1924) observed that the bacteria appeared to gain entry through wounds in the branches, and Dowson (1957) noted that it was through the exit holes made in the bark by adult midges and sawflies that the bacterial ooze emerged in the summer.

Field observations support the view that watermark disease spreads into cricket bat willow plantings from other susceptible willows, especially *S. alba*, growing nearby (Dowson and Callan, 1937; Zweep and de Kam, 1982). Spread may also take place from the stumps of felled infected trees (Wong *et al.*, 1974). Field evidence also suggests that the disease may be distributed with the setts used in propagation (Bryce, 1950; Wong *et al.*, 1974), though studies by de Kam (1983) failed to support this view.

*Control*

The disease is controlled by careful examination and roguing of propagating beds to prevent the distribution of infected setts, and by the felling and destruction of infected trees. De Kam (1984) in the Netherlands also recommends that the branches of pollard willows should be cut back every four years. The stumps of felled diseased trees should be killed by the use of a suitable herbicide. In areas in Britain in which the cricket bat willow is an important crop the local authorities have the power under the Watermark Disease (Local Authorities) Order 1974 (SI 1974 No. 768) to inspect willows and demand the destruction of any that show symptoms of the disease (Preece and Wortley, 1979). Under this and similar earlier orders a vigorous campaign has been maintained against watermark disease for some 20 years, and when Wong *et al.* (1974) carried out an intensive survey in Essex in 1972 they found that only about 0.075 per cent of the trees were infected. Nevertheless the disease is still a cause for much concern, especially as it has done so much damage in the Netherlands in. recent years. Some cultivars of *Salix alba*, including cvs Liempole and Drakenburg, are known in the Netherlands to be very susceptible, and should be avoided (de Kam, 1983). More work on the selection of resistant clones may therefore be justified.

**Twig and branch dieback and wilt caused by a fluorescent *Pseudomonas* sp.**
Hunter and Stott (1978) have briefly described a twig and branch dieback caused by an un-named fluorescent pseudomonad which produces a non-diffusible fluorescent pigment on King's B medium. The bacterium was causing leaf-yellowing and premature leaf fall in seven-year-old cricket bat willows in Essex. Severe, slightly sunken cankers occurred on the main branches, on which narrow, longitudinal cracks were also present. The cracks measured up to 3.6 m long and 25–85 mm wide. The underlying tissues showed a brown stain, mainly in the vessels, for some way beyond the leading edge of the canker. A similar disease caused by a *Pseudomonas* sp. has been reported in Canada.


DECAY FUNGI

The cricket bat and weeping willows are very susceptible to attack by the honey fungi, *Armillaria* spp., which are described in chapter 3, and kill many willow trees. Among the decay fungi which may attack willows are *Ganoderma applanatum* and *G. resinaceum*, which cause a white, often cubical rot. Other fungi which may attack willows, entering through wounds, are the white rot fungi *Inonotus hispidus* and *Phellinus igniarius*, and *Laetiporus sulphureus*, the cause of a brown cubical decay. These fungi are described in chapter 20.

# 18 Diseases of wild and ornamental rosaceous trees

This chapter deals with diseases of wild and ornamental trees in the family Rosaceae. We discuss the many diseases that affect orchard apples, pears, cherries and other fruit trees only if they also attack ornamentals. Diseases of orchard trees are described in books on fruit tree diseases (for example, Wormald, 1955), and in general texts by Scopes and Stables (1989), Smith *et al.* (1988) and Buczacki and Harris (1981).

## Hawthorn (*Crataegus* spp.)

The hawthorns are small thorny trees or shrubs. Of the two native British species, *Crataegus monogyna* is common throughout the country, and *C. laevigata* occurs in the Midlands and the southeast; *C. monogyna* is much used as a hedgerow plant. Various double-flowered and pink- or red-flowered forms of these species are grown as ornamentals, and so are the North American *C. crus-galli* (the cockspur thorn) and hybrids such as *C.* × *prunifolia* (the broad-leaved cockspur thorn).

LEAF DISEASES

**The powdery mildew *Podosphaera clandestina* (Wallr.) Lév.**
**(*P. oxyacanthae* (DC) de Bary)**
The common powdery mildew *Podosphaera clandestina* (CMI Descr. F. 478) forms a white powdery covering over the leaves of hawthorn (and sometimes of other woody rosaceous plants) in Europe and North America.

The colourless, ellipsoid conidia are borne in chains, and measure 18–23 × 9–10 μm. Perithecia (cleistothecia) are found in late autumn. They are brown when mature, and up to 90 μm across. They bear up to 30 stiff equatorial appendages which are dichotomously branched at their tips. They each contain one globose ascus with eight ovoid ascospores measuring 15–25 × 10–12 μm (Ellis and Ellis, 1985).

Khairi and Preece (1978a,b), who studied *P. clandestina* on hedgerow plants of *C. monogyna*, found that clipping of the hedges encouraged the development of lateral buds which became infected, and on which the mildew then overwintered and produced conidia which gave rise to the primary infections in the following spring. In Britain, at least, the cleistothecia failed to mature and shed ascospores, and seemed to play no important part in the epidemiology of the disease.

This mildew is common but does appreciable damage only on soft growth on young trees and on newly planted hedges. Khairi and Preece (1978b) have suggested that to reduce the disease, hawthorn hedges should be clipped only in midwinter. Chemical control may be obtained by spraying with imazalil, benomyl or penconazole (Brooks *et al.*, 1989).

### Leaf spots and blotches caused by *Monilinia johnsonii* (Ell. and Ev.) Honey (*Sclerotinia crataegi* Magn.) and *Phloeospora oxyacanthae* (Kunze and Schm.) Wallr.

The macroconidial stage of the ascomycete *Monilinia johnsonii* may grow over the surface of hawthorn leaves as a sweet-scented, powdery grey to buff-coloured mould, which causes dark-brown or black blotches. In southeast England, following wet springs it has occasionally caused substantial damage, destroying much of the foliage, the blackened remains of the dead leaves at first continuing to hang on the tree (Dowson and Dillon-Weston, 1937; Wormald, 1937; Strouts *et al.*, 1984). The colourless, almost globose conidia of this stage, which is known as *Monilia crataegi* Diet., measure 9–16 μm in diameter, and form in chains separated by narrow connectors, so that they look like beads on a string (Ellis and Ellis, 1985). A microconidial stage may develop on the fallen fruits, and is followed by stalked brown apothecia with a paler light-brown disc. In spring these shed their colourless, one-celled ellipsoid ascospores, which measure 10.5–14 × 5–6 μm, and these spores initiate the primary infections, the macroconidial stage then developing over the summer (Bond, 1961).

*Phloeospora oxyacanthae*, one of the Deuteromycotina, has been known in Europe for many years as the cause of yellowish or brown spots on hawthorn leaves (Tubeuf, 1895). Its brown, open and cup-shaped pycnidia are mostly on the undersides of the leaves, and are up to 150 μm across. The pycno-spores, which ooze out in yellowish masses, are club-shaped, have six to eight septa, and measure 50–80 × 6–8 μm (Grove, 1935). The fungus is widespread and sometimes damaging, as in favourable conditions the spots spread over the leaves, which become brown and scorched.

### Leaf spots caused by *Entomosporium mespili* (the anamorph of *Diplocarpon mespili* (Sorauer) B. Sutton)

Hawthorns are sometimes attacked by *Entomosporium mespili*, the teleo-morph of which is *Diplocarpon mespili*. The fungus, which causes small red spots, has a wide host range in the Rosaceae. Its conidia are complex structures, with two large cells, one basal and one apical, separated by two or three small equatorial cells. The apical and equatorial cells each have one long, unbranched flexuous appendage.

**Leaf curl caused by *Taphrina crataegi* Sadeb.**
*Taphrina crataegi* sometimes attacks hawthorns in Britain and the rest of Europe. It then causes a reddish or yellowish swelling and inrolling of the edges of the leaves, and sometimes a deformation of the shoots. A layer of cylindrical asci forms on the underside of the leaf within the inrolled margin. The ascospores measure 4–5.5 × 3.5–4.5 μm (Mix, 1949).

## DISEASES OF LEAVES AND SHOOTS

**The hawthorn/juniper rusts *Gymnosporangium clavariiforme* (Pers.) DC and *G. confusum* Plowr.**
Two heteroecious rusts widespread in northern Europe, with their telia on *Juniperus*, have *Crataegus monogyna* and *C. laevigata* as the primary hosts on which they produce their spermagonia and aecia (figures 85, 86). These rusts, which are described by Wilson and Henderson (1966), are *Gymnosporangium clavariiforme* and *G. confusum*. Both cause conspicuous red or orange swellings on the leaves, twigs and fruits. The spermagonia of *G. clavariiforme* are at first yellow and finally brown. The white or yellowish or finally brownish aecia are cylindrical, but later torn almost to the base. The aeciospores are pale brown and globose or almost so, measuring 22–30 × 18–26 μm.

**Figure 85** *Gymnosporangium clavariiforme* on hawthorn, showing the aecia on a leaf, fruit and branch. The aecium at a (× 17) shows the fringed top (W.B. Grove)

The spermagonia of *G. confusum* are orange. The yellowish or pale-brown aecia are more or less cylindrical, with an open fringed top, and the globose or ellipsoid, brown aeciospores measure 19–26 × 19–22 μm.
These two rusts are widespread in Europe, into Asia and North Africa, and

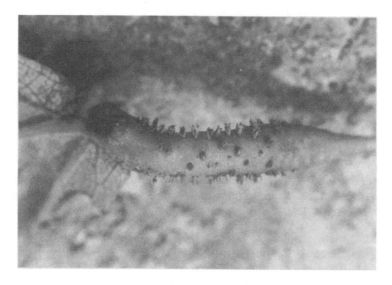

**Figure 86** Aecia of *Gymnosporangium clavariiforme* on a swollen leaf stalk of hawthorn (D.H. Phillips)

in North America. Both occur in Britain, where *G. clavariiforme* is fairly common where hawthorn and juniper grow together. They are of little economic importance.

**Other European *Gymnosporangium* spp.**
Of more limited distribution on *Crataegus* in Europe are *G. gracile* Pat. and *G. malyi* Picbauer (*Roestelia malyi* (Picbauer) Kern). *G. gracile* is found in southern Europe as well as in North Africa and the southern USA and Mexico. Its aecia occur on the undersides of the leaves and on the shoots. They are cylindrical at first, but become deeply lacerate. Their aeciospores are globoid with brown walls, and measure 25–28 μm across. Its telia form on junipers of the *Oxycedrus* and *Sabina* groups. *G. malyi* is known only on *Crataegus* in Yugoslavia. It produces its cylindrical aecia on the fruits, and its brown, subglobose aeciospores measure 18–23 × 23–36 μm (Kern, 1973).

## BACTERIAL WILTS

**Fire blight caused by *Erwinia amylovora* (Burr.) Winsl.**
Hawthorn is one of the hosts of the serious disease fireblight, caused by the bacterium *Erwinia amylovora*, and affected trees may be important sources of infection to nearby commercial pear and apple orchards (Lelliott, 1959; Lelliott and Billing, 1984). Hence, infected trees should be grubbed up and burnt. Hawthorns near apple and pear orchards should be cut back each year to hinder blossoming, as the disease (which is briefly described in chapter 3) enters the plant primarily by infection of the flowers.

VIRUS DISEASES

Cooper (1979) has reported a foliage disease of *Crataegus* widespread in Britain. The symptoms, chlorotic ring and vein banding patterns, seem to be caused by a form of chlorotic ring spot virus.

DECAY FUNGI

Among the few root rot and decay fungi known to attack *Crataegus* is *Phellinus pomaceus*. Others occasionally found on hawthorns include *Armillaria* spp., *Ganoderma adspersum*, *Laetiporus sulphureus* and *Heterobasidion annosum*. These fungi are described in chapters 3 and 20.

# Apple and pear (species of *Malus* and *Pyrus*)

Numerous cultivars of the apple, *Malus sylvestris*, and of the pear, *Pyrus communis*, are grown for their fruit. A number of crab apples (including some introductions from Asia) are grown as ornamentals in streets and gardens, as is also the Asian willow-leaved pear, *P. salicifolia*.

LEAF DISEASES

**Apple scab caused by *Venturia inaequalis* (Cooke) Wint. (anamorph *Spilocaea pomi* Fr., syn. *Fusicladium dendriticum* (Wallr.) Fuckel) and pear scab caused by *V. pirina* Aderh. (anamorph *Fusicladium pyrorum* (Lib.) Fuckel)**
By far the commonest disease of ornamental *Malus* varieties is scab, caused by *Venturia inaequalis* (CMI Descr. F. 401), which is also the most troublesome disease of the apple fruit grower. The fungus causes brown or olive-green spots and sometimes blisters on the leaves (figure 87). The conidia of the *Spilocaea* stage form on the leaf spots, and are brown, slipper-shaped, one- or two-celled, measuring $16$–$24 \times 7$–$10$ μm (Ellis and Ellis, 1985) (figure 88a). If infection is severe, the leaves may wither and fall. The fungus also causes blister-like lesions on young shoots; these lesions burst open to reveal more pustules of the brown conidia.

In orchard apple varieties the fungus spreads from the leaves to the fruits. The affected fruits may then fall before they can enlarge, or become ill-shaped or cracked. On fully developed fruits, later infections cause unsightly dark spots (much like those on the leaves), which reduce the market value of the fruit. Sometimes fruit may appear healthy when picked, but late infections may develop and cause black spots on the apples in store.

The flask-shaped pseudothecia of the perfect stage of *V. inaequalis* are produced in winter on the fallen leaves. They are brown, with apical setae. Ripe asci occur in these perithecia in the spring. The ascospores from these

**Figure 87** Blisters on apple leaf caused by the scab fungus *Venturia inaequalis* (D.H. Phillips)

fruit bodies infect the young leaves, and so complete the life cycle. These ascospores (figure 88b) are pale olive-green, and unequally two-celled (with the septum above the middle), and measure 12–15 × 6–7 µm (Wormald,1955).

A disease very similar to apple scab also affects pears, and the pear scab fungus, *Venturia pirina*, is similar to *V. inaequalis* in all but minor details. It produces dark-brown, velvety patches on the leaves, and sometimes causes leaf distortion. Like apple scab, it may also cause leaf fall, produce blister-like pustules on the shoots, and spread to the fruits, which are then deformed, scabby and split, and so reduced in value. The olivaceous ascospores of *V. pirina* measure 14–20 × 4–8 µm, and are two-celled, with the septum below the middle. The conidia of the *Fusicladium* stage are olivaceous, one- or two-celled, measuring 17–28 × 8–10 µm (Ellis and Ellis, 1985).

*Control of scab*
Some control of apple and pear scab may be obtained by cutting out and burning diseased twigs before the buds burst, and raking up and burning fallen leaves. Protective fungicidal sprays may be applied at about 10-day intervals from bud burst to mid-June. A great many effective fungicides are listed by Ivens (1990), Richter (1988), Alford and Locke (1989) and Brooks *et al.*, 1989). They include benomyl and thiophanate-methyl (continued use of

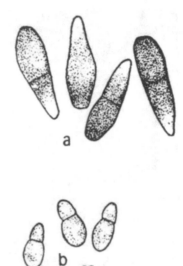

**Figure 88**    Slipper-shaped conidia (a) and two-celled ascospores (b) of *Venturia inaequalis* (H. Wormald)

which may, however, lead to the emergence of resistant strains of the scab fungi), bupirimate + triforine, captan (which may damage some apple varieties), mancozeb and penconazole.

**Powdery mildew: *Podosphaera leucotricha* (Ell. and Everh.) Salmon**
*Podosphaera leucotricha* (CMI Descr. F. 158) distorts the leaves, especially on young shoots, of apples and sometimes pears, covering them with a white, powdery growth (plate 52). The disease can be controlled by cutting out infected shoots in spring, and spraying the plants (especially the developing shoots) from the pink bud stage until extension growth has ceased. The many available fungicides include benomyl, carbendazim, thiophanate-methyl, bupirimate, penconazole, triadimefon and triforine (Brooks *et al.*, 1989).

SHOOT DISEASES

**Canker caused by *Nectria galligena***
Canker, caused by *Nectria galligena*, is troublesome on some ornamental *Malus* and *Pyrus* varieties. It is described in chapter 3.

**Blossom wilt and spur blight caused by *Monilinia laxa***
Blossom wilt and spur blight caused by *Monilinia laxa* may be found on crab apples, though it is commoner on flowering cherries, under which it is therefore described. The form on apple differs slightly from that on cherry, however, and is described as *M. laxa* f. sp. *mali* Wormald *sensu* Harrison.

## DISEASES OF FRUITS

**Brown rot caused by *Monilinia fructigena* Honey and Whetzel (*Sclerotinia fructigena* Aderh. and Ruhl.) (anamorph *Monilia fructigena* Pers.)**
Many diseases affect the fruits of apples, but most are more important to the fruit grower than to the gardener and grower of ornamental trees. Crab apple fruits, however, are often spoilt by brown rot caused by *Monilinia fructigena* (CMI Descr. F. 617), which may also sometimes kill shoots and produce cankers on small shoots. Buff-coloured pustules of the anamorph *Monilia fructigena* form on the fruits, usually in concentric circles, and the fruits eventually become mummified and fall (figure 89). The conidia of the fungus are colourless, ellipsoid to lemon-shaped, and form in chains. They measure 15–20 × 9–11 μm (Ellis and Ellis, 1985).

**Figure 89** Three crab apples—*Malus* 'Frettingham Victoria'—attacked by the brown rot fungus *Monilinia fructigena*. The fruit at top left shows the buff-coloured pustules of the anamorphic stage *Monilia fructigena*, and the blackened fruit at top right is almost mummified (D.H. Phillips)

## VIRUS DISEASES

Many virus diseases affect orchard apples. Among these is the apple chlorotic leaf spot virus (ACLSV) (Virus Descr. 30), a sap-transmissible virus with elongated particles from about 600 to 700 nm. It occurs on many woody rosaceous plants, including apple, on which it produces symptoms only on a few ornamental *Malus* spp. (Smith, 1972; Delbos and Dunez, 1988).

## ROOT AND DECAY FUNGI

Crab apples may often be attacked by the honey fungus, *Armillaria mellea* (described in chapter 3), and by the silver-leaf fungus, *Chondrostereum purpureum*. Other decay fungi sometimes found on crab include *Ganoderma applanatum*, *G. adspersum*, *Inonotus hispidus* and *Phellinus pomaceus*. These fungi are described in chapter 20.

## OTHER DISEASES

Apples and pears (especially the latter) are among the hosts of *Erwinia amylovora*, the fireblight organism, though the disease is rarely found on the ornamental forms of these trees. Fireblight is described in chapter 3.

# Blackthorn (*Prunus spinosa*)

The blackthorn is widespread in Europe, and is much used as a hedgerow shrub or tree. The diseases described below are of minor importance (although some are more damaging on orchard plums in some countries).

## LEAF DISEASES

**Powdery mildews: *Podosphaera tridactyla* (Wallr.) de Bary and *Uncinula prunastri* (DC) Sacc.**
*Podosphaera tridactyla* (CMI Descr. F. 187) and *Uncinula prunastri* may both form a sparse, rather evanescent covering of mycelium and conidia on the leaves of blackthorns. The former is of greater importance on the cherry laurel, *Prunus laurocerasus*, under which it is described below. The appendages on its cleistothecia are several times dichotomously branched at their ends. The dark cleistothecia of *U. prunastri* have equatorially attached, colourless, unbranched appendages curled at their tips. They each contain from 10 to 15 ovate asci each with six to eight ascospores which measure 16–20 × 8–10 μm (Salmon, 1900; Ellis and Ellis, 1985).

**Leaf spot or leaf blotch caused by *Polystigma rubrum* (Pers.) DC (anamorph *Polystigmina rubra* Sacc.)**
*Polystyigma rubrum* produces stromata in thickened, rounded or rather angular, bright-red spots or blotches on the leaves of blackthorns (especially in coastal areas), and sometimes also on the bullace, *Prunus institia*. In some

parts of Europe it also attacks orchard plums (Wormald, 1955). Pycnidia of the anamorphic state form in the stromata, opening on the underside, and containing colourless, walking-stick-shaped conidia measuring 25–30 × 1 μm. Perithecia form on the dead fallen leaves. Their ascospores, which are colourless and single-celled, measure 11–13 × 4–5 μm, are shed in the spring, infecting the young leaves (Wormald, 1955; Ellis and Ellis, 1985). If control is necessary, fallen infected leaves can be collected and destroyed and the trees sprayed with a copper-based fungicide soon after the young leaves emerge (Wormald, 1955).

### The leaf rust *Tranzschelia pruni-spinosae* (Pers.) Dietel
*Tranzschelia pruni-spinosae* var. *pruni-spinosae* occasionally produces small brown uredinia followed by black telia on the undersides of blackthorn leaves, the upper sides of which show small, bright-yellow spots. The single-celled urediniospores measure 20–40 × 10–18 μm and the two-celled telio-spores 30–45 × 10–20 μm. The aecial stage of this rust occurs on cultivated anemones and the wild *Anemone nemorosa* (Wormald, 1955; Ellis and Ellis, 1985).

## DIEBACK

### Dieback caused by *Diaporthe perniciosa* (*D. eres*)
*Diaporthe perniciosa*, one of the many forms of *D. eres*, which is described in chapter 3, may sometimes cause a minor dieback of blackthorns.

## FRUIT DISEASE

### Pocket sloes caused by *Taphrina pruni* Tul.
*Taphrina pruni* (CMI Descr. F. 713) causes a disease of plums in which the fruits become pale, elongated and hollow, without a stone. The affected fruits are known as 'pocket plums' or 'bladder plums'. They later shrink, become covered with a whitish bloom of asci, and eventually fall. The disease persists in infected shoots (Wormald, 1955). The ascospores measure 4–7 × 3–6 μm, but usually bud in the ascus (Mix, 1949). The same fungus occasionally attacks blackthorn, producing pale, hollow 'pocket sloes', and infected black-thorn shoots become swollen, with deformed, strap-shaped leaves (Ellis and Ellis, 1985).

# Almond (*Prunus dulcis*)

The almond, *Prunus dulcis*, is often grown in suburban gardens for its flowers.

## LEAF DISEASES

### Leaf curl caused by *Taphrina deformans* (Berk.) Tul.
### (*Exoascus deformans* (Berk.) Fuckel)

The commonest disease of ornamental almonds is the leaf curl caused by *Taphrina deformans* (CMI Descr. F. 711), which is better known as a disease of the peach. It is known wherever almonds and peaches are grown. The leaves of affected shoots become curled soon after they unfold, and as they grow they become increasingly distorted and thickened, and reddish in colour (plate 55). The fungus produces a layer of asci, mainly on the upper surface of the leaf, and the affected leaves then show a whitish bloom. The diseased leaves usually shrivel and fall, but are often replaced by new ones, which generally remain uninfected. The ascospores of the fungus (and secondary spores budded from them) appear to carry it over the winter in the bud scales (Wormald, 1955).

Leaf curl may be controlled by spraying with a copper spray just before the leaves begin to swell in January or the beginning of February and again two weeks later, and again just before leaf fall in the autumn (Brooks *et al.*, 1989).

## DISEASES OF LEAVES AND SHOOTS

### Freckle or scab caused by *Venturia carpophila* E.E. Fisher
### (anamorph *Cladosporium carpophilum* Thüm., syn. *Fusicladium*
### *carpophilum* (Thüm.) Oudem.)

*Venturia carpophila* (CMI Descr. F. 402) is of most importance on orchard plums, but it sometimes also affects ornamental almonds (and peaches), causing leaf spotting, and gum-exuding lesions at the bases of the spurs. An olivaceous-brown, velvety turf of conidiophores develops on the leaf spots and the spurs, bearing short chains of brown conidia measuring 12–20 × 4–5 μm. The dark-brown pseudothecia of the perfect stage are embedded in the fallen leaves, and their two-celled ascospores measure 12–20 × 4–6 μm (Wormald, 1955; Ellis and Ellis, 1985; Holliday, 1989). The disease can usually be controlled on ornamental trees by cutting out the diseased shoots.

### Leaf spot and blossom wilt caused by *Stigmina carpophila*

*Stigmina carpophila*, which causes a shot-hole disease of peach, and sometimes of related plants, including cherries, plums, apricots and cherry laurels, occasionally causes a spotting of the leaves of ornamental almonds, and a wilting of the blossoms. *S. carpophila* is described below under 'cherry'.

## DISEASES OF STEMS AND BRANCHES

### Bacterial canker caused by *Pseudomonas syringae* pv. *morsprunorum*

Ornamental almonds may sometimes be attacked by *Pseudomonas syringae* pv. *morsprunorum*. This bacterium occurs more commonly on cherry, under which it is therefore described below.

# Cherry

The gean or wild cherry, *Prunus avium*, and the bird cherry, *P. padus*, are both native throughout most of Europe. Both are native in Britain, the gean being commoner in the south and the bird cherry in the north. Cultivars of both these species are grown in gardens, and so are forms of the sour cherry, *P. cerasus*, which originated in Asia. Even commoner are asiatic flowering cherries, especially *Prunus subhirtella, P. × hillierii* 'spire', *P. × yedoensis, P. sargentii* and varieties of the Japanese cherry *P. serrulata*.

## LEAF DISEASES

### Powdery mildew: *Podosphaera tridactyla*
*Podosphaera tridactyla* sometimes forms a sparse, rather evanescent covering of mycelium and conidia on the leaves of flowering cherries. It is usually commoner and more noticeable on the cherry laurel, *Prunus laurocerasus*, under which it is described below.

### Scab caused by *Venturia cerasi* Aderh. (anamorph *Fusicladium cerasi* (Rabenh.) Sacc.)
*Fusicladium cerasi*, the anamorph of *Venturia cerasi*, occasionally causes scab of varieties of the sour cherry, *Prunus cerasus*. It produces small, often sunken, olive velvety spots on the fruits (which may fall), and on the leaves and young shoots. The broadly spindle-shaped conidia are olivaceous brown, usually one- but occasionally two-celled, and measure 16–23 × 5–7 μm. The brown pseudothecia of the perfect stage are setose round the ostiole. They are embedded in the fallen leaves. It is doubtful if they occur in Britain, though they are known in other parts of Europe. Their pale-brown ascospores are two-celled, with a septum below the middle, and measure 10–14 × 4–6 μm (Sivanesan, 1977). The disease can be controlled, if necessary, by the use of a copper-based spray.

### Leaf spot and shot-hole caused by *Blumeriella jaapii* (Rehm.) v. Arx (*Coccomyces hiemalis* Higg.; anamorph *Phloeosporella padi* (Lib.) v. Arx, syn. *Cylindrosporium padi* (Lib.) P. Karst. ex Sacc.)
*Blumeriella jaapii* may sometimes attack ornamental cherries, including forms of *Prunus padus* (the bird cherry) and *P. cerasus* (the sour cherry), as well as plums, apricots and orchard cherries. It is known almost wherever the host trees are grown (but so far does not seem to be present in Britain).

It causes very small brown leaf spots (at first purplish on the upper side), which develop in early summer. The spots may later fall out, to form 'shot-holes', and affected leaves may turn yellow and fall (Wormald, 1955; Seemüller, 1988b).

Small white acervuli (and sometimes, in damp conditions, a white mildew-like growth) develop on the undersides of the leaf spots. The acervuli are of two kinds. One, a *Microgloeum* state, forms small, colourless conidia measuring 7.5–9.5 × 1.5 μm ; the other, known as *Phloeosporella padi*, has elongated, colourless, two-celled spores measuring 60–75 × 3 μm. The dark-brown apothecia of the teleomorphic stage develop on the undersides of the fallen leaves. Their ascospores, which are shed and cause the primary infections in spring and early summer, are colourless and oblong or tear-shaped, measuring about 25 × 2.5 μm (Ellis and Ellis, 1985).

### Leaf rust, *Pucciniastrum areolatum* (Fr.) Otth

The spermagonia and aecia of *Pucciniastrum areolatum* form on spruce, on which this fungus causes a cone rust and a shoot blight (see chapter 4). Its aeciospores from spruce infect the bird cherry (*Prunus padus*), and the cultivated *P. virginiana* and *P. serotina*. Its yellow or brown uredinia are formed on reddish spots on the undersides of the leaves of the *Prunus* spp., followed by red to dark-brown, crust-like telia, which occur on red spots chiefly on the upper sides of the leaves. The teliospores germinate in the following spring, producing the basidiospores that reinfect the spruce flowers. The yellowish urediniospores measure 15–20 × 10–15 μm, and the brown, cylindrical, two- to four-celled teliospores measure 20–30 × 10–15 μm (Ellis and Ellis, 1985).

### Silver leaf caused by *Chondrostereum purpureum* (*Stereum purpureum*)

Silver leaf is not a true leaf disease, but the earliest and most characteristic symptoms appear on the leaves, which exhibit a silvery sheen. Death of branches or of the whole tree then occurs. The causal fungus, *Chondrostereum purpureum*, sometimes attacks flowering cherries, though it has a wide host range (and is best known as the cause of silver leaf of plum trees). It also causes a wood rot, and is further described in chapter 20.

## DISEASES OF LEAVES AND SHOOTS

### Leaf curl and witches' broom caused by *Taphrina wiesneri* (Rathay) Mix
### (*T. cerasi* (Fuckel) Sadeb., *T. minor* Sadeb.)

*Taphrina wiesneri* (CMI Descr. F. 712), which occurs in Europe (including Britain), and in North America, South Africa, Japan and Australasia (Mix, 1949), may attack many cherries, including the flowering ornamental forms, causing leaf curl, with slight thickening of the diseased leaves, or witches' brooms, or both. Bond (1956) also found a form of the disease with stunted, apparently incipient brooms which seemed to link those showing only leaf curl with those in which witches' brooms were fully developed. Brooms on small trees are sometimes as large as the still healthy parts of the plants. The stems of the brooms are usually thickened towards the base. Infected leaves are pinkish or reddish, and smell of new-mown hay (Tubeuf, 1895; Ellis and Ellis, 1985). The undersides of diseased leaves are eventually covered by a whitish bloom caused by a layer of asci. These at first each contain eight colourless, rounded to elliptic ascospores which measure 3.5–9 × 3–6 μm, but often bud within the ascus (Mix, 1949). The disease can be controlled by cutting out the brooms, and spraying as for leaf curl of almond (Wormald, 1955).

## DISEASES OF STEMS AND BRANCHES

### Bacterial canker caused by *Pseudomonas syringae* pv. *morsprunorum* (Wormald) Young *et al.*

The bacterium *Pseudomonas syringae* pv. *morsprunorum* (CMI Descr. B. 125) may cause a canker (and sometimes a shoot wilt) on flowering cherries (Brooks *et al.*, 1989) and also on the ornamental purple-leaved plum, *Prunus pissardii* (Wormald, 1955). In Britain it is especially common on the gean, *P. avium* (Strouts, personal communication). Trees girdled by cankers quickly die. On ungirdled stems, sunken cankers arise, and gum may exude from the lesions. In summer, leaf and shoot infection may take place. Water-soaked spots then appear on the leaves. These spots become brown, and the central dead area falls out to leave 'shot-holes'. Shoot infection may cause the shoots to curl if it remains one-sided, but the tops of girdled shoots wither and die. Separation of the bacterium from related forms is discussed by Garrett *et al.* (1966).

Bacterial canker may be controlled by cutting out diseased parts in the summer, and spraying with Bordeaux mixture or some other copper-based spray, first in mid-August and then three times at intervals of three to four weeks (Brooks *et al.*, 1989).

## DISEASES OF LEAVES, SHOOTS, BLOSSOMS AND FRUITS

### Blossom wilt and twig blight caused by *Monilinia laxa* (Aderh. and Ruhl.) Honey ex Whetzel (*Sclerotinia laxa* Aderh. and Ruhl.) (anamorph *Monilia cinerea* Bonorden, *M. laxa* (Ehrenb.) Sacc.)

Like orchard trees, flowering cherries may commonly be affected, especially in wet seasons, by the blossom wilt and twig blight caused by *Monilinia laxa* (anamorph *Monilia cinerea*) (CMI Descr. F. 619). The same fungus causes a similar disease on plums and other related trees, and its f. *mali* causes blossom wilt and spur canker of apple. These fungi also cause brown rot of the fruits of rosaceous orchard trees (Moore, 1959).

On flowering cherries, leaves may be infected (plate 44), and the fungus then grows down into the shoots; the whole shoot then wilts. More often, flower infection occurs, and the fungus grows through the flower stalk into the spur. The flowering spur wilts, and the flowers and leaves wither.

The greyish, cushion-like pustules of sporophores and conidia of the *Monilia* stage of the fungus are produced on the affected twigs in the following spring. The conidia form in chains, and are ovoid and single-celled, measuring 5–19 × 4–12 μm on the overwintered leaves, but 8–23 × 7–16 μm on leaves newly infected (Wormald, 1955). The perfect stage of the fungus, which is rare in Britain, forms its brown, long-stalked apothecia on mummified fruits. The colourless, one-celled ascospores measure 8–18 × 3–8 μm (Ellis and Ellis, 1985).

The disease may be controlled by cutting out infected spurs, collecting and destroying any diseased fruits, and spraying the trees in the dormant season with tar-oil winter wash; the trees may also be sprayed with carbendazim or vinclozolin just before flowering or as the first flowers open (Alford and Locke, 1989).

**Leaf spot, shot-hole and blossom blight caused by *Stigmina carpophila* (Lév.) M.B. Ellis (*Coryneum beijerinckii* Oudem., syn. *Clasterosporium carpophilum* (Lév.) Aderh.)**

*Stigmina carpophila* is of most importance on orchard cherries, peaches and apricots. Sometimes, however, it also affects flowering cherries, almonds and cherry laurels, producing a variety of symptoms which vary somewhat with the host (Seemüller, 1988a).

On cherries it chiefly causes purplish or brown leaf spots which increase to a diameter of about 10 mm. They may be bordered by a pale-green zone, and the dead tissue eventually drops out to leave a 'shot-hole'. Petiole infections usually cause girdling and the death of the leaf. Sometimes the flowering shoots are infected, and blossom blight ensues. Later infections cause spotting and malformation of the fruits. Infected tissues bear minute spore-bearing stromata (sporodochia). On these, small erect sporophores produce pale-brown, three- to seven-septate conidia which measure about 30–60 × 9–18 μm (Ellis and Ellis, 1985). These conidia are spread by rain splash, and infect the young leaves and blossoms as they open. Further infection of buds and twigs takes place in the autumn.

If control on flowering cherries is necessary, effective results can be obtained by spraying with captan, dithianon and other organic fungicides, which are less damaging to the foliage than the copper-based materials used in the past (Seemüller, 1988a). Benomyl and carbendazim are also effective (Brooks *et al.*, 1989). Spraying should begin when the buds begin to swell, and one or two further applications may be needed.

FRUIT DISEASES

As noted above, the cherry scab fungus, *Venturia cerasi*, and the leaf spot, shot-hole and blossom blight fungus, *Stigmina carpophila*, may attack the fruits, as well as other parts of the plants. Cherry fruits may also be attacked by the blossom wilt fungus, *Monilinia laxa*, also described above, and the brown rot fungus, *M. fructigena*, described above under 'fruit diseases of *Malus* spp.'.

**Pocket cherry caused by *Taphrina padi* (Jacz.) Mix**
*Taphrina padi* sometimes attacks the bird cherry, *Prunus padus*, killing flowers and deforming the fruits, which become elongated and hollow, with persistent styles. The asci produced on the diseased tissues each contain eight colourless, round to elliptic ascospores which measure 5–6 × 4–5 μm, but frequently bud in the ascus (Mix, 1949).

## VIRUS DISEASES

Many virus diseases affect orchard cherries. Among those affecting ornamental flowering cherries are the cherry leaf roll virus (CLRV) and cherry green ring mottle.

Cherry leaf roll virus occurs on many hosts, and on cherries is found especially on *Prunus avium* and its varieties. Diseased trees come late into leaf and flower, and later the edges of the leaves are uprolled and may turn purplish. Growth of the trees becomes poor, and the bark often splits and exudes gum (Smith, 1972). Certain identification involves the use of test plants and serological techniques (Cooper, 1988).

Cherry green ring mottle disease, which is probably caused by a virus, is of importance on some cultivars of the Japanese cherry *P. serrulata*. Symptoms again include bark splitting, as well as epinasty of the foliage, and sometimes a dieback (Smith, 1972; Desvignes, 1988).

## DECAY FUNGI

Cherries may often be attacked by the honey fungus, *Armillaria mellea*, which is described in chapter 3. As noted above, cherries may also be affected by the silver leaf fungus, *Chondrostereum purpureum*. *Laetiporus sulphureus* may decay cherry stems, and *Phellinus pomaceus* sometimes rots their stems and branches. Among other decay fungi found on cherries is *Pholiota squarrosa*. These fungi are described in chapter 20.

# Cherry laurel (*Prunus laurocerasus*) and Portugal laurel (*P. lusitanica*)

These bushes or small trees are mainly used as hedge plants.

## LEAF DISEASES

### Powdery mildews: *Podosphaera tridactyla* (Wallr.) de Bary and *Sphaerotheca pannosa* (Wallr.) Lév.

*Podosphaera tridactyla* (CMI Descr. F. 187) causes marked distortion on the leaves at the shoot tips of cherry laurel, which it covers with a whitish, sparse and evanescent layer of mycelium and colourless conidia which form in chains and measure 22–45 × 14–20 μm. Its brown cleistothecia have appendages which are attached around their apices and branch dichotomously several times. Each fruit body contains one ascus with eight colourless ascospores which measure 18–30 × 12–15 μm. The same fungus also attacks blackthorns and cherries (and orchard plums and apricots).

A second powdery mildew found on the cherry laurel, especially in France (Boudier, 1978), is a form of the rose mildew, *Sphaerotheca pannosa* (CMI Descr. F. 189). This forms a much more dense covering of mycelium and spores, which is at first white, but later grey to buff. The colourless conidia, which form in chains, measure 20–35 × 14–20 μm. The appendages on the brown cleistothecia (which are rarely seen) are few and unbranched. Each cleistothecium contains one ascus, with eight colourless ascospores which measure 20–30 × 14–20 μm.

If necessary, these mildews can be controlled by spraying with such fungicides as benomyl, carbendazim, imazalil, penconazole or triforine (Brooks *et al.*, 1989).

### Leaf spot and shot-hole disease caused by *Stigmina carpophila*

*Stigmina carpophila*, which attacks various *Prunus* spp., especially cherries, under which it is described above, may also attack cherry laurels.

## DIEBACK

### Branch dieback associated with *Cytospora lauro-cerasi* Fuckel

Inoculation experiments by Strouts *et al.* (1986) have indicated that *Cytospora lauro-cerasi*, a fungus which is usually only a saprophyte, is the cause of a branch dieback of cherry laurels which is common in Britain. The black necks of the conidiomata of this coelomycete emerge through small whitish discs. Their slightly curved, sausage-shaped conidia measure 5–6 × 1 μm (Ellis and Ellis, 1985).

## OTHER DISEASES

### Silver leaf caused by *Chondrostereum purpureum*

The Portugal laurel seems to be especially susceptible to silver leaf, caused by *Chondrostereum purpureum* (Strouts, personal communication), which is described in chapter 20.

# Mountain ash or rowan, whitebeam and service trees (*Sorbus* spp.)

The mountain ash or rowan, *Sorbus aucuparia*, is common throughout Europe (including Britain). Two other widespread European *Sorbus* spp., *S. aria*, the whitebeam, and *S. torminalis*, the wild service tree, in Britain occur more locally, mainly in the southern half of England and Wales. Many other very local native species also occur, and various introduced species are grown in parks and gardens.

## LEAF DISEASES

### The powdery mildew *Podosphaera clandestina* (Wallr.) Lév. var. *aucupariae* (Erikss.) U. Braun (*P. aucupariae* Erikss.)

The variety *aucupariae* of the powdery mildew *Podosphaera clandestina* (figure 90), which is well known on hawthorn (under which it is described above) also occurs on mountain ash and other *Sorbus* spp.

### Scab caused by *Venturia inaequalis* f. sp. *aucupariae*

The form species *aucupariae* of *Venturia inaequalis*, described above under 'diseases of apple', causes scab of *Sorbus aucuparia*, *S. aria* and *S. torminalis* (as well as the service tree, *S. domestica*) (Sivanesan, 1977).

### The leaf rusts *Gymnosporangium cornutum* Kern. (*G. juniperi* Link) and *Ochropsora* ariae (Fuckel) Ramsb.

The rust *Gymnosporangium cornutum* (*G. juniperi*), the telia of which develop on *Juniperus communis*, produces its spermagonia and aecia on mountain ash. In Britain, it is common in Scotland, but much less so in England and Wales. Groups of the spermagonia, which are at first yellow, but later turn black, arise on red or orange spots on the upper surfaces of the leaves. The yellowish-brown aecia are cylindrical or horn-shaped (rostelioid), up to about 5 mm high and about 0.5 mm across. They appear on thickened orange spots on the undersides of the leaves. The brown, two-celled aeciospores measure 20–29 × 18–25 μm (Wilson and Henderson, 1966).

*Ochropsora ariae* is widespread but generally uncommon in Europe and Asia. Its uredinia and telia form on spots on the leaves of *Sorbus aria*, *S. aucuparia*, *S. torminalis* and some other *Sorbus* species (though these stages seem to be unknown on *Sorbus* in Britain). The very small uredinia arise on small spots on the undersides of the leaves, and the pale-brown uredinio-spores measure 20–28 × 16–21 μm. The red, waxy, crust-like telia are also on the undersides of the leaves. Their teliospores measure up to 70 × 10–14 μm, and are eventually four-celled. The aecial stage, which occurs on *Anemone nemorosa*, has been found occasionally in Britain (Moore, 1959). Affected

**Figure 90** White mycelium and powdery spores of *Podosphaera clandestina* var. *aucupariae* (*P. aucupariae*) on leaf of mountain ash (D.H. Phillips)

leaves become elongated and pale in colour, with many aecia, the aeciospores of which measure 18–26 × 17–21 μm (Ellis and Ellis, 1985).

These rusts are of little or no importance on ornamental trees.

## CANKERS AND DIEBACKS

**Canker and dieback associated with *Cytospora rubescens* Fr. (anamorph of *Eutypella sorbi* (Alb. and Schw.) Sacc.)**
*Cytospora rubescens* has been shown by MacBrayne (1981) in Scotland to be the cause of a canker and dieback of various *Sorbus* spp. Its sausage-shaped conidia measure 3.5–4 × 1 μm, and emerge in red tendrils (Ellis and Ellis, 1985).

**Bark death associated with *Chondrostereum purpureum***
Young and Strouts (1979) found evidence that a quite common progressive bark dieback and the death of long narrow strips of bark on rowan trees was

caused by the silver leaf fungus, *Chondrostereum purpureum*, which is described in chapter 20.

## WILT DISEASES

### Fireblight caused by *Erwinia amylovora*
Like many other members of the Rosaceae, *Sorbus* spp. are among the hosts of *Erwinia amylovora*, the cause of fireblight. The diseased trees may therefore become a danger to adjacent plantings of apples and pears. Fireblight is very common in Britain on the whitebeam, *S. aria*. Strouts *et al.* (1985) found evidence that the bacterium could enter whitebeam through undamaged bark, probably via the lenticels. Fireblight is described in chapter 3.

# 19 Diseases of other broadleaved trees

The root rot fungi *Armillaria mellea* and *Heterobasidion annosum* (*Fomes annosus*), the silver leaf fungus *Chondrostereum purpureum* and the grey mould *Botrytis cinerea* have been found on most of the trees included in this chapter. These fungi are described in chapters 3 and 20. They are not further considered here, and are mentioned only if, for example, some particular tree is especially susceptible to attack by one of them.

## Horse chestnut (*Aesculus* spp.)

The horse chestnut, *Aesculus hippocastanum*, is native in southeast Europe, and is widely grown as an ornamental in other parts of Europe, including Britain. So are the red-flowered *A.* × *carnea* and sometimes other species from North America and Asia.

### LEAF DISEASES

**Guignardia leaf blotch caused by *Phyllosticta sphaeropsoidea* Ell. and Ev. (anamorph of *Guignardia aesculi* (Peck) Stewart)**
Leaf blotch is an important disease of horse chestnut and related species of *Aesculus* in eastern North America (Stewart, 1916) and in Europe, where it has caused particular damage in Austria (Lohwag, 1962), Italy and Switzerland (Scaramuzzi, 1954) and Yugoslavia (Milatovic, 1956); it also occurs in France, where it is the chief disease of horse chestnut (Boudier, 1985), in Belgium (Grove, 1935), in Germany (Schneider, 1961) and in Portugal (Caetano, 1985), as well as in Britain (Grove, 1935; Moore, 1959).

The fungus usually associated with the disease is *Phyllosticta sphaeropsoidea*, which is common on affected leaves in Britain. Its teleomorphic stage is *Guignardia aesculi* (Hudson, 1987). This has been considered to be at least rare in Britain (Moore, 1959). Elsewhere in Europe it is known to develop on the fallen leaves (Schneider, 1961), and Hudson (1987) obtained it in England by overwintering infected leaves, which produced pseudothecia from late April to June in the following year. *P. sphaeropsoidea* has more or less globose black pycnidia 90–175 μm across. Their colourless, ellipsoid to globose conidia are truncate at the base and fresh spores each bear one inconspicuous tapering appendage at the apex. They emerge in white tendrils, and measure 12–20 × 9–10 μm. The pseudothecia of *G. aesculi* are also black and globose, and up to 150 μm across. Their colourless, ellipsoidal ascospores widen at the centre, and measure 14–23 × 7–9 μm. A spermagonial stage, known as *Leptodothiorella aesculicola* (Sacc.) Sivan., also occurs. Its spermagonia contain small, colourless,ˈmore or less oblong or dumb-bell-shaped spermatia measuring 6–10 × 1–2 μm (Hudson, 1987).

In Britain, the disease first appears in June, July or August. Symptoms on affected leaves (plate 56) show as very conspicuous irregular blotches that later spread over much of the leaf surface, and as a marginal scorch that often runs from the leaf tips along their edges, and extends irregularly inwards. In colour the blotches vary from a dull light-brown to a deep chestnut, and they are eventually bordered by a distinct yellow margin (Grove, 1935; Pawsey, 1962; Stewart, 1916). The pycnidia are scattered on the spots, mainly on the upper surface, and, as noted above, the perfect stage arises on the over-wintering leaves.

In Britain and in the rest of Europe, the disease is found mainly on the horse chestnut, though the first British record was on *Aesculus parviflora* at Kew (Grove, 1935). In eastern North America it occurs mainly on horse chestnut and the Ohio buckeye (*A. glabra*) but it is found also on *A. parviflora* (Stewart, 1916). Neely and Himelick (1963) in the USA inoculated many *Aesculus* spp., most of which were severely affected.

In southern Europe, leaf blotch has sometimes caused considerable damage to nursery plants and mature trees (Scaramuzzi, 1954; Milatovic, 1956; Kispatic, 1962; Lohwag, 1962). In Britain it caused premature leaf fall on nursery stock in Gloucestershire in 1953 (Moore, 1959), and Pawsey (1962) reported scattered cases on mature trees as far north as Edinburgh. In recent years the disease has been common in southern England, though damage has usually been important only on trees in the nursery.

The severity of the damage has not justified the use of control measures on established trees in Britain. Sweeping up the fallen leaves in autumn should reduce the sources of infection. In the case of severe attacks in the nursery, and elsewhere on young established trees, spraying with lime sulphur or Bordeaux mixture or dusting with sulphur (Stewart, 1916), or, more recently, spraying with zerlate, parzate (Davis, 1948) or with ziram or zineb (Kispatic, 1962), have been found to give control. Brooks *et al.* (1989) recommend the use of a prochloraz–manganese complex, or of one of the fungicides used against black spot in roses, such as captan, carbendazim, chlorothalonil or triforine (Ivens, 1990). Several applications at intervals of two to four weeks may be needed.

### Leaf spot caused by *Septoria hippocastani* B. & Br.

A common minor leaf spot of horse chestnut is caused by the imperfect fungus *Septoria hippocastani*. The numerous small brown spots are round or rather angular. The pycnidia, which are mainly on the upper side, are lens-shaped and up to 180 μm across. They contain long, hair-like spores that are divided into four cells by three faint septa, and measure 30–57 × 2.5–3.5 μm (Grove, 1935).

## DISEASES OF THE BARK AND TRUNK

**Bark necrosis or bleeding canker associated with *Phytophthora* spp.**
Brasier and Strouts (1976) have examined horse chestnut trees with often extensive necrotic areas of bark on the trunks (commonly on the south-facing sides) and on large branches (figure 91). In many cases sap oozed from the lesions and flowed down the trunk to form a gummy slime flux which dried to become hard and brittle. In some cases, *Phytophthora cactorum* (described in chapter 11 as the cause of seedling blight of beech) or sometimes *P. citricola* or *P. syringae* were isolated from the diseased bark. *P. cactorum* has caused similar symptoms on horse chestnut, *Acer* and many other broadleaved trees in North America, where because of the associated slime flux the disease has been aptly named bleeding canker (Pirone *et al.*, 1960).

**Figure 91** Horse chestnut with bleeding canker caused by *Phytophthora citricola* (Forestry Commission)

**Wetwood**

As noted above, slime flux is one of the symptoms of bleeding canker. The bacterial wetwood that affects many broadleaved trees, which may then show an associated fluxing, is also fairly often found in the horse chestnut. This disorder is described in chapter 3.

## WILT DISEASES

**Verticillium wilt**

*Verticillium dahliae* has been found killing boughs of a 30-year-old tree of *Aesculus hippocastanum* (Young and Strouts, 1976), and large horse chestnut trees killed by *V. albo-atrum* have also been found (Strouts *et al.*, 1982).

## ROOT DISEASES

**Phytophthora root disease**

Several cases of death of large horse chestnut trees following death of their roots have been investigated in England (Brasier and Strouts, 1976). Various species of *Phytophthora*, including *P. megasperma*, have been isolated from dying roots.

## DECAY FUNGI

The butt and stem of the horse chestnut may be decayed by *Ganoderma adspersum* and *G. applanatum*, while *Pleurotus ostreatus* may rot the stem and branches. These fungi are described in chapter 20.

## VIRUS DISEASES

**Horse chestnut mosaic**

A yellow mosaic of horse chestnut leaves has been found in Czechoslovakia, eastern Germany and Romania as well as in the United Kingdom. The leaves of affected plants become bright yellow. The symptoms are very striking, and as the virus involved can be transmitted by budding and grafting, affected trees have been propagated by nurserymen and distributed as an ornamental form (Cooper, 1979).

# Box (*Buxus sempervirens*)

Box, a small evergreen shrub or tree, is native in a few places on chalk and limestone in southern England, and in many other European countries. It is also widely planted, especially to form a hedge.

## LEAF DISEASES

**Powdery mildew: *Phyllactinia guttata***
According to Ing (1990b), box is one of the hosts of the powdery mildew *Phyllactinia guttata*, which is described below under 'hazel', on which it is more commonly found.

**Leaf fall caused by *Hyponectria buxi* (DC) Sacc. (*Trochila buxi* Capron ex Cooke)**
According to Moore (1959) and Foister (1961), the fungus *Hyponectria buxi*, which usually grows only on dead leaves, has been associated with a leaf fall of box in many parts of Scotland. The fruit bodies of *H. buxi* are olive-brown, flattened perithecia. They are at first embedded in the lower surfaces of the leaves but later exposed by the splitting of the epidermis. The colourless, elliptical ascospores measure 13–16 × 4–5 μm (Ellis and Ellis, 1985).

**The leaf rust *Puccinia buxi* DC**
The box rust, *Puccinia buxi*, is widespread and frequent throughout Europe (including Britain) and into Asia. As a result of infection in spring and early summer, the young leaves become thickened. Telia begin to form in late summer, and can be found in September and October as brown pustules on both sides of the swollen leaves. The brown, smooth-walled teliospores are two-celled, and measure 55–90 × 20–35 μm. Their very long, persistent, colourless pedicels are up to 160 μm in length (Wilson and Henderson, 1966). The teliospores germinate in the following spring, to produce the basidiospores which infect the young, tender, pale-green leaves (Gäumann, 1959). If control measures are needed, the plants can be sprayed with a fungicide such as zineb, benomyl, oxycarboxin, penconazole or propiconazole (Brooks *et al.*, 1989).

***Paecilomyces buxi* (Link ex Fr. ) Bezerra (*Verticillium buxi* Link ex Fr.)**
The fungus long known as *Verticillium buxi* but renamed *Paecilomyces buxi* by Bezerra (1963) at most seems to be only a very weak parasite (Dodge, 1944) on moribund leaves. Its erect conidiophores are up to 110 μm long, branching above to end in tapered phialides. These produce chains of colourless, elliptical or sickle-shaped spores measuring 5–8 × 2–2.5 μm (Bezerra, 1963).

**Leaf spots caused by *Phyllosticta buxina* Sacc. and *P. limbalis* Pers.**
Leaf spots of box caused by two species of *Phyllosticta* have been described by Grove (1935).

*P. buxina* has been found on Box Hill, Surrey, on pale leaf spots with a narrow purple border. The small black pycnidia are densely scattered on the upper side, and are up to 100 μm across. They contain many colourless, oblong-elliptical spores measuring 4–5 × 1.5–2 μm

*P. limbalis*, which has been found occasionally in Surrey and Hampshire

(Moore, 1959), causes white spots with a narrow brown border. Small numbers of dark-brown pycnidia arise on the spots on both sides of the leaf. The spores measure about 5–7 × 3–4 μm. *P. limbalis* appears to be the anamorph of *Mycosphaerella buxicola* (DC) Tomalin (*M. limbalis* Pers. ex Wallr.) v. Arx), the brown pseudothecia of which are embedded in the leaf tissues. Its colourless, two-celled ascospores measure 22–30 × 3–5 μm (Ellis and Ellis, 1985).

Control measures are not usually required against these leaf spots, but, if necessary, Brooks *et al.* (1989) suggest spraying with materials such as Bordeaux mixture, captan, benomyl, fenarimol, a prochloraz–manganese complex or propiconazole.


## DISEASES OF LEAVES AND TWIGS

**Leaf and twig blight caused by *Pseudonectria rousseliana* (Mont.) Wollenw. (anamorph *Volutella buxi* (Corda) Berk., syn. *Chaetodochium buxi* (DC ex Fr.) Höhn.)**

*Volutella buxi*, the anamorph of *Pseudonectria rousseliana* (Bezerra, 1963), causes a blight of leaves, twigs and branches of box in Europe and North America. Dieback and death of the leaves of individual branches occurs, so that large, conspicuous, brown patches of foliage appear on the plants (plate 51). The wood of affected twigs and branches shows a characteristic blackening. The pathogenicity of the fungus was demonstrated by Dodge (1944). *V. buxi* produces small pink pustules (*sporodochia*) up to 110 μm across on the surfaces of the leaves, mainly on the undersides. These pustules are made up of long, bristle-like hairs up to 190 μm long, and many shorter, branched conidiophores that produce colourless conidia at their ends. The conidia form a slimy mass, and individually are spindle-shaped, smooth, one-celled, and measure 8–12 × 2.5–3 μm (Bezerra, 1963). The perithecia of *Pseudonectria rousseliana* occur on the fallen leaves, and in Britain have been found at sites ·scattered over England as far north as Yorkshire. They arise in groups, and vary in colour from straw-coloured or brick-red to green. They are subglobose, and sparsely covered with stiff bristles that measure up to 100 μm in length. Their club-shaped asci contain colourless, single-celled, oval ascospores measuring 9–16 × 3–5 μm (Petch, 1938). The fungus seems to be a wound parasite, which affects stressed or otherwise damaged plants. If control is needed, fallen leaves should be collected and destroyed, and diseased shoots and branches cut out. Pirone (1970) recommends spraying with a copper-based material four times, from just before growth starts in spring until just after leaf fall.

STEM DISEASES

**Stem canker associated with *Nectria desmazierii* Beccari & de Notaris (anamorph *Fusarium buxi* Sacc.)**
*Nectria desmazierii* (anamorph *Fusarium buxi*) has on rare occasions been found associated with a stem canker of box in Britain as well as in North America. Its small conidial pustules form on leaf scars and buds. The sickle-shaped conidia measure 30–55 × 4–5 μm, and are finally four- to six-septate. Later, yellow or orange, ovate or globose perithecia up to about 175 μm across form in groups of up to 30 over the conidial pustules. They contain cylindrical, eight-spored asci measuring 65–80 × 7–9 μm. The colourless, ellipsoid, two-celled ascospores measure 12–16 × 5–7 μm (Booth, 1959). The fungus was found in Surrey in 1932, and again, in Oxfordshire, in 1968 (Batko and Strouts, unpublished).

Control measures similar to those for leaf and twig blight caused by *Pseudonectria rousseliana* can be applied if necessary.

# Hornbeam (*Carpinus betulus*)

Hornbeam is found throughout much of Europe, and is native in southeast England, where it was once managed as coppice. It has also been widely planted throughout the rest of England and Wales, and into Scotland.

LEAF DISEASES

**Powdery mildew: *Phyllactinia guttata***
*Carpinus* is one of the hosts of the powdery mildew *Phyllactinia guttata* (Ing, 1990c), which is described below under 'hazel', on which it is more commonly found.

**Leaf spot caused by *Asteroma carpini* (Lib.) Sutton (*Gloeosporium carpini* (Lib.) Desm.)**
*Asteroma carpini* (*Gloeosporium carpini*) may cause brownish, irregular, unbordered spots on hornbeam leaves. It produces many small, brownish pustules on the undersides of the spots. The spores that form in the pustules are long and cylindrical, measuring 8–10 × 1 μm (Grove, 1937). *A. carpini* is considered to be the pycnidial stage of *Mamiania fimbriata* (Pers. ex Fr.) Ces. and de Not. (*Gnomonia fimbriata* Auersw., syn. *Gnomoniella fimbriata* (Pers. ex Fr.) Sacc.). This ascomycete produces small black stromata on the fallen leaves. Long-necked perithecia embedded in the stromata mature in the spring. Their colourless, broadly elliptical ascospores measure 8–11 × 3–4 μm, and a single septum cuts off a very small cell at the base. The perfect stage is rare (Dennis, 1968).

## DISEASES OF SHOOTS

### Witches' brooms caused by *Taphrina carpini* (Rostr.) Johanss.

The fungus *Taphrina carpini* causes witches' brooms on hornbeam (figures 92, 93). From May to July its asci form a whitish layer covering the surface of pale-greenish leaves on the brooms. The latter form pendant branches of much-branched twigs, which are sometimes long, lax, open and fan-like, with their branches in one plane, or sometimes short and stubby (Reid, 1969). The eight-spored asci which cover the leaves are cylindrical, and measure 20–30 × 7–14 μm. Their one-celled elliptical spores measure 5–7 × 3–4 μm. *T. carpini* appears to be uncommon, though it has been found at sites widely scattered over England and Scotland (Dennis, 1968; Reid, 1969), and in Germany, Sweden and Russia (Mix,1949).

**Figure 92** Witches' brooms on *Carpinus betulus* caused by *Taphrina carpini* (D.H. Phillips)

# Indian bean tree (*Catalpa* spp.)

*Catalpa bignonioides*, the Indian bean-tree, from the southeastern USA, is common in European parks and gardens, and a few other *Catalpa* spp. from North America and Asia are less often grown.

   *C. bignonioides* is one of the trees known to be especially susceptible to attack by Verticillium wilt caused by *Verticillium dahliae* (Piearce and Gibbs, 1981), which is described in chapter 3.

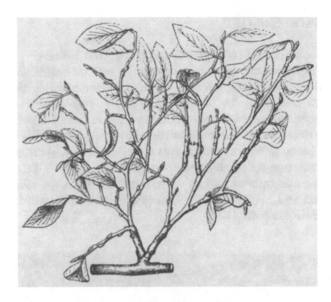

**Figure 93**  Witches' broom on *Carpinus* caused by *Taphrina carpini* (R. Hartig)

# Dogwood (*Cornus* spp.)

The dogwood, *Cornus sanguinea*, grows throughout Europe, and is common in England and Wales, especially in the south. The dwarf cornel, *C. suecica*, is much less common, but occurs locally in northern Europe. In Britain it is found in mountain moorlands in parts of Scotland and northern England. The Cornelian cherry, *C. mas*, is native in central and southeastern Europe, and is naturalised in places in Britain, where it is grown in parks and gardens, in which some other dogwoods, from North America and Asia, are also to be found.

## LEAF DISEASES

**The powdery mildew *Erysiphe tortilis* (Wallr.) Fr.**
The powdery mildew *Erysiphe tortilis* is common on the leaves of *Cornus sanguinea*, which it covers with a white layer of mycelium and conidia. Its round, brown cleistothecia are found in autumn and winter, and have very long, fasciculate appendages ten or more times the diameter of the fruit body. They usually contain three to five asci, each with from three to five spores (Salmon, 1900; Ellis and Ellis, 1985).

**Leaf spots caused by *Septoria cornicola* Desm.**
*Septoria cornicola* sometimes causes a leaf spot of *Cornus sanguinea* (Tubeuf, 1895). The greyish spots are edged with purple, and may be up to 10 mm

across. The long, narrow, colourless conidia of the fungus are three- to five-celled, and measure 30–40 × 2–3 μm (Ellis and Ellis, 1985).

# Hazel (*Corylus avellana*)

Hazel is a shrub or small tree, native and common throughout most of Europe, including Britain. In Britain it is much used in hedging, and small plantings, mainly in southeast England, are managed as coppice for the production of poles and fencing materials. *Corylus maxima*, from south and southeast Europe, is sometimes grown for its nuts.

## LEAF DISEASES

### Powdery mildew *Phyllactinia guttata* (Wallr.) Lév. (syn. *P. suffulta* (Rabenh.) Sacc., *P. corylea* (Pers.) Karst.)

The powdery mildew *Phyllactinia guttata* (CMI Descr. F. 157) forms a sparse mycelium over the leaves of hazel, and sometimes of ash, birch, oak (Batko, 1962), and other trees and shrubs. On hazel it occurs especially on young coppice growth.

Erect conidiophores on the mycelium produce usually single oidia measuring 50–120 × 8–15 μm (Brooks, 1953). Later, mainly on the under-sides of the leaves, scattered, rather flattened-globose, closed ascocarps (cleistocarps) arise. These are black when mature, and up to about ⅓ mm across. Round the equator each bears up to 15 stiff, colourless, pointed appendages with bulbous bases (figure 94). In length the appendages are up to three times the diameter of the fruit body. The top of the ascocarp is covered with a mass of short, branched outgrowths. Within each ascocarp are up to 45 ovate asci measuring 70–100 × 25–40 μm, and each ascus usually contains two ascospores that measure 30–40 × 16–25 μm (Brooks, 1953; Dennis, 1968).

By hygroscopic movements, the appendages dislodge the ascocarps so that they fall from the leaves and mature on the ground (Brooks, 1953).

Though the mildew may cause some defoliation, it is of only minor importance.

### Leaf spots caused by *Mamianiella coryli* (Batsch) Höhnel (*Gnomoniella coryli* (Batsch) Sacc.), *Piggotia coryli* (Desm.) Sutton (*Labrella coryli* (Desm.) Sacc.) and *Asteroma coryli* (Fuckel) Sutton (*Septoria avellanae* Berk. and Br.)

Several usually unimportant fungal leaf spots occur on hazel. They are caused by the ascomycete *Mamianiella coryli* and the deuteromycetes (Fungi Imperfecti) *Piggotia coryli* and *Asteroma coryli*.

*Mamianiella coryli*, which is found more often on moribund than on active living leaves, produces one to several black, long-beaked perithecia in small

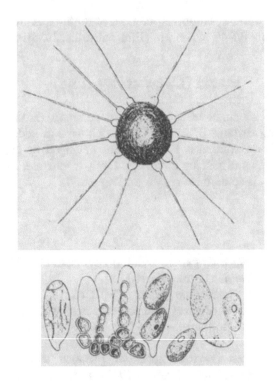

**Figure 94**  Cleistocarps, asci and ascospores of *Phyllactinia guttata*. The cleistocarp
shows the appendages with bulbous bases (K.F. Tubeuf)

stromata embedded in the leaf tissues. The eight-spored asci measure 40–50
× 8–9 μm, and contain one-celled, colourless, elliptical ascospores that
measure 8–9 × 3 μm (Winter, 1897).

*Piggotia coryli* causes rather large, irregular brown spots on the leaves.
Embedded in the spots are flattened, lens-shaped, brownish acervuli 100–150
μm across, that contain pale-brown, one-celled spores measuring 8.5–11 ×
4–5 μm (Sutton, 1980).

*Asteroma coryli* causes large brown spots on which groups of pycnidia form
on the underside of the leaf. The pycnidia measure 80–100 μm across, and
contain curved, colourless, hair-like spores measuring 12–19 × 1–1.5 μm
(Ellis and Ellis, 1985). This leaf spot is known in Britain, Austria and
Germany (Sutton, 1980), as well as in Italy and Azerbajdzhan, where severe
attacks have been recorded (Constantinescu, 1984).


## LEAF SPOTS, BUD BLIGHTS AND DIEBACKS

**Bud rot caused by *Cryptosporiopsis tarraconensis* Gené *et al.***
Gené *et al.* (1990) in Spain have described a bud rot of cultivated hazels
caused by *Cryptosporiopsis tarraconensis*. The buds die and dry out over the

winter from December to February. The dark-brown acervuli of the fungus produce white conidial masses, and the individual conidia are colourless and aseptate, measuring 7.5–16 × 4–6 μm.

## Leaf spot and dieback caused by *Xanthomonas campestris*
A pathovar of the bacterium *Xanthomonas campestris* may cause leaf-spotting, dieback of twigs and branches, and even death of hazel nut trees in nurseries and young plantations (Moore *et al.*, 1974; Prunier *et al.*, 1976).

## NUT ROTS

Nut rots may be caused by the fungi *Monilinia fructigena*, *Botrytis cinerea* and *Fusarium lateritium* Nees, the anamorph of *Gibberella baccata*. *M. fructigena* is described under 'cherry' in chapter 18, and *B. cinerea* is described in chapter 3. The sickle-shaped spores of *F. lateritium* measure 25–35 × 3.5–4.5 μm (Wormald, 1955).

# Eucalyptus

A few *Eucalyptus* spp. have been introduced into Europe from Australasia as ornamental trees. The commonest is *E. gunnii*.

## Nursery diseases

### Oedema
Small oedematous, blister-like swellings (intumescences) on the undersides of the leaves, caused by over-humid growing conditions and lack of ventilation, have been found on nursery plants of *Eucalyptus* in frames in west Scotland. Like those described by Pirone *et al.* (1960), they finally cracked open and became rust coloured.

### Frost damage and grey mould (*Botrytis cinerea*)
Nursery plants of *Eucalyptus* are very susceptible to frost (chapter 2), and may also be attacked by grey mould, *Botrytis cinerea* (chapter 3).

### Powdery mildew (*Oidium* spp.)
Glasscock and Rosser (1958) recorded a powdery mildew (an *Oidium* sp.) on seedlings of *E. perriniana* and *E. gunnii*. The conidia of the fungus measured about 24–30 × 15–18 μm, and control was achieved by spraying with capryl. Brooks *et al.* (1989) also suggest the use of benomyl, carbendazim, bupirimate or penconazole.

## Post-nursery diseases

### Winter cold
Mature trees may be killed by winter cold, and many in the eastern, more continental parts of Britain were destroyed in the hard winter of 1962–63. According to Wood (in Macdonald *et al.*, 1957), the most hardy species in Britain are *E. vernicosa, E. parvifolia, E. niphophila, E. gunnii* and *E. pauciflora*. More recent observations and experiments by Evans (1986) and Sheppard and Cannell (1987) indicate that some provenances of *E. gunnii, E. niphophila* and *E. debeuzevillei* are relatively hardy and worth planting in most of lowland Britain. If larger-scale planting is contemplated for forest use, selections should be made and propagated from hardy individuals from within known hardy provenances.

# Euonymus

The spindle tree, *Euonymus europaeus*, occurs throughout most of Europe, including Britain. *E. latifolius* is native in more southerly and southeasterly parts of Europe. *E. japonicus*, from Japan, is a much-used hedge plant, and some other *Euonymus* spp. from China and Japan are sometimes grown as ornamental shrubs.

LEAF DISEASES

### The powdery mildews *Microsphaera euonymi* (DC) Sacc. and *Oidium euonymi-japonicae* (Arcang.) Sacc.
The powdery mildew *Microsphaera euonymi* produces a rather sparse mycelium with few conidia mostly on the undersides of the leaves of *Euonymus europaeus*; it has also been found on *E. fortunei* var. *radicans* and *E. verrucosus* (Ing, 1990a,c). It also produces brown cleistothecia with colourless, dichotomously branched appendages and each containing five to ten asci (Ellis and Ellis, 1985).

*Oidium euonymi-japonicae* occurs widely on *Euonymus japonicus*, and has also been found on *E. nanus* (Ing, 1990c). In Britain, in the drier southeast it forms rather small white patches of mycelium and conidia on both sides of the leaves. In the damper western parts (for example, in the Channel Islands) it covers the leaves with a thick white felt, and the leaves are much stunted (figure 95). The conidia measure 21–36 × 7–13 μm (Ellis and Ellis, 1985). This fungus is said to be the conidial stage of *Microsphaera euonymi-japonicae* Viennot-Bourgin or sometimes of *Uncinula euonymi-japonicae* Hara (Holliday, 1989), but cleistothecia are at most very rarely seen.

Brooks *et al.* (1989) suggest that these mildews can be controlled by spraying with benomyl, carbendazim, bupirimate or penconazole.

**Figure 95** White patches of *Oidium euonymi-japonicae* on leaves of *Euonymus japonicus* (D.H. Phillips)

**The leaf rust *Melampsora euonymi-caprearum* Kleb.**
**(*M. epitea* Thüm. var. *epitea*)**
The caeomata of *M. euonymi-caprearum*, one of the forms of *M. epitea*, may occasionally be found on the leaves of *Euonymus europaeus*, mainly on the undersides. They form bright-orange spots, and their aeciospores measure 18–23 × 14–19 μm. The other stages of this rust occur on willows, under which they are described in chapter 17.

# The honey locust (*Gleditsia triacanthos*)

Cultivars of the honey locust, *Gleditsia triacanthos*, a native of North America, are grown in European parks and gardens.

Young and Strouts (1975) found a *Gleditsia* tree (probably on a *Robinia* rootstock) showing severe root decay by *Meripilus giganteus* (*Polyporus giganteus*), which is described in chapter 20.

# Holly (*Ilex aquifolium*)

The common holly, *Ilex aquifolium*, is a small evergreen, prickly tree or shrub that grows over most of Europe, including Britain. This and its many cultivars are grown also as ornamentals, often as hedging.

## LEAF DISEASES

### Leaf spots caused by *Phyllosticta aquifolina* Grove and *Coniothyrium ilicis* Smith and Ramsbottom

Grove (1935, 1937) described two pycnidial fungi causing leaf spots on holly.

The spots produced by the first of these, *Phyllosticta aquifolina*, which has been found in Kew Gardens, are round or irregular, greyish with a brown border, and up to 3 mm across. Black pycnidia up to 400 μm in diameter arise on the spots, and contain colourless, one-celled, oblong-ellipsoid spores measuring 6–7 × 2–2.5 μm (Grove, 1935).

A second fungus, *Coniothyrium ilicis*, produces white spots in the upper side of which are embedded black, thin-walled pycnidia up to 200 μm across. The pale-brown pycnospores are globose to ellipsoid, and measure 3–5 × 2–3 μm. In 1922 this fungus caused an epidemic on holly trees at Sutton Coldfield, Warwickshire (Grove, 1937).

## TWIG DISEASES

### Twig blight caused by *Vialaea insculpta* (Fr.) Sacc.
### (*Boydia remuliformis* A.L. Sm.)

From time to time the ascomycete *Vialaea insculpta* has been found apparently killing the twigs of holly in Scotland and southern England (Grove, 1921; Moore, 1959). The dull-black, globose or lens-shaped perithecia are up to 500 μm across. The asci within them soon deliquesce. The colourless ascospores are two-celled, and shaped like two clubs placed end to end. They measure 80–100 × 8–9 μm. The fungus seems to be no more than weakly parasitic. Buddenhagen and Young (1957) found that it often invaded tissues already damaged by *Phytophthora ilicis* Buddenhagen and Young.

## ROOT DISEASES

### Violet root rot

The violet root rot fungus, *Helicobasidium brebissonii* (described in chapter 3), which attacks a very wide range of herbaceous and woody hosts, has been found in Britain on holly (Moore, 1959).

# Walnut (*Juglans* spp.)

The walnut, *Juglans regia*, from southern Europe, and the black walnut, *J. nigra*, from North America, are both much grown as garden trees in the rest of Europe, including Britain. Both species produce a valuable timber, but *J. regia* is grown especially for its nuts.

## Nursery diseases

### LEAF AND SHOOT DISEASES

**Graft disease caused by *Chalaropsis thielavioides* Peyronel**
Walnut varieties grafted under glass have occasionally been affected by graft disease caused by *Chalaropsis thielavioides*, a fungus that also gives rise to a blackening of stored carrots. Affected grafts fail, and both stock and scion become covered by a dark-brown mass of powdery spores. The spores are thick-walled, round or egg-shaped, and measure about 14–19 μm across. On incubation under moist conditions the fungus produces white masses of cylindrical endospores that measure 8–15 × 2.5–4.5 μm (Wormald, 1955; Ellis and Ellis, 1985) (figure 96).

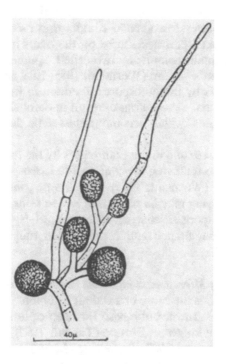

**Figure 96** *Chalaropsis thielavioides*: large, dark ascospores and cylindrical endospores (H. Wormald)

The disease is easily controlled by using formaldehyde solution to clean the propagating house before use. If necessary, material for grafting can be painted with the same chemical, which can also be used to sterilise the grafting knife (Hamond, 1931, 1935; Wormald, 1955).

## Post-nursery diseases

### LEAF DISEASES

**Leaf blotch caused by *Gnomonia leptostyla* (Fr.) Ces. and de Not. (anamorph *Marssoniella juglandis* (Lib.) Höhnel, syn. *Marssonina juglandis* (Lib.) Magn.)**

Of the fungus diseases that affect walnut leaves the best-known is the leaf blotch caused by *Gnomonia leptostyla* (anamorph *Marssoniella juglandis*). This disease occurs throughout much of Europe, North America, parts of South America, Israel, Iran, parts of the USSR (CMI Map No. 384) and India (Kaul, 1962). It is common and widespread in Britain, particularly in England and Wales.

The fungus causes brown blotches on the leaves and the young fruits. The small, black conidial pustules (*Marssoniella juglandis*) form on the dead patches, mainly on the undersides. They give rise to rather crescent or comma-shaped, colourless, two-celled conidia that measure 15–25 × 3–4.5 μm. Black, long-necked perithecia arise on the overwintering leaves. Their eight-spored asci contain colourless, two-celled, spindle-shaped ascospores that measure 17–22 × 3–4.5 μm (Wormald, 1955; Ellis and Ellis, 1985).

Primary infection is by the ascospores liberated in the spring. Secondary spread is by the conidia. Severe attacks result in defoliation, and blackening of the young green nuts, which become useless. The disease is favoured by wet weather.

Control on single trees and in nut plantings is by the raking up and burning of the fallen leaves, so removing the source of the ascospores that give rise to the spring infections (Wormald, 1955). In parts of continental Europe and North America, spraying may be needed in wet seasons, beginning in spring at the time of ascospore discharge. Many materials, including benomyl, chlorothalonil, dodine, maneb and copper-based fungicides give effective control (Berry, 1977).

**Leaf blotch caused by *Microstroma juglandis* (Bereng.) Sacc.**

*Microstroma juglandis* is the cause of a walnut leaf blotch which occurs widely wherever walnuts are grown, but has so far been of little economic importance. It has long been known in Europe (Tubeuf, 1895), though in Britain it remained unrecorded until 1982 (Strouts *et al.*, 1982). The affected leaves show yellow blotches on the upper sides, and the corresponding areas on the undersides show white mildew-like patches which bear bundles of conidiophores. These end in club-shaped cells which resemble basidia, and bear colourless, ellipsoid or spindle-shaped conidia on sterigmata (von Arx *et al.*, 1982). The conidia measure 5.5–7.5 × 3.5–5 μm (Ellis and Ellis, 1985). In North America, where the fungus also attacks *Carya* spp., it also gives rise to witches' brooms (Sinclair *et al.*, 1987).

### Bacterial blight caused by *Xanthomonas campestris* pv. *juglandis* (Pierce) Dye (*Xanthomonas juglandis* (Pierce) Dowson)

A common and damaging disease wherever walnuts are grown on a large scale is the blight caused by *Xanthomonas campestris* pv. *juglandis* (CMI Descr. B. 130). The bacterium causes small black, angular spots, particularly towards the tips of the leaflets. Large, withered patches may arise as they spread over the leaf surface. Elongated black patches on the shoots may lead to girdling, and so to death of the tissues beyond the diseased zone. Black blotches arise on the fruits, and the male catkins may also be destroyed (Wormald, 1955).

If the diseased tissues are teased out in water, a gelatinous mass of bacteria oozes out. The bacteria are small, Gram-negative, non-sporing, motile rods, each with one polar flagellum. They measure about 1.5–3 × 0.3–0.5 μm (Dowson, 1957; Stapp, 1961). Methods used in the identification of this bacterium are described by Lelliot and Stead (1987).

This disease is most damaging when cool wet weather occurs at flowering time (Zbinden, 1986).

In Britain, bacterial blight occurs mainly on young trees in the nursery, and is of relatively minor importance, but it causes serious damage in parts of Eastern Europe and in North America. It is best controlled by cutting out any affected shoots well beyond the diseased tissues, frequently sterilising the pruning tools with formaldehyde. If necessary the young foliage may be sprayed with Bordeaux mixture or copper oxychloride (Wormald, 1955; Zbinden, 1986).

## SHOOT DISEASES

### Dieback associated with *Cytospora juglandina* Sacc.

The various minor fungi described on walnut include *Cytospora juglandina*, which has been found associated with a dieback. Its pycnidia are grouped in small stromata, and contain colourless, one-celled, cylindrical spores measuring 6–7 × 1 μm (Grove, 1935).

## DECAY FUNGI

*Inonotus hispidus* and *Polyporus squamosus* both decay the stems and branches of walnut, the former causing a yellow rot and the latter a white fibrous decay. *Laetiporus sulphureus* sometimes attacks walnut, causing a brown cubical rot. These fungi are described in chapter 20.

## VIRUS DISEASES

**Walnut chlorosis, ring spot, line pattern and blackline disease**
A number of symptoms on walnut leaves (and sometimes on fruits), including
the production of yellow necrotic spots, rings and line patterns, as well as a
'blackline disease' of grafted trees, in which black lines are formed at the
point of the graft union, have been found to be due to infection by a strain or
strains of the cherry leaf roll virus known as CLRV-W. The disease is
transmitted by infected pollen, and can also be spread by grafting and
budding (Mircetich *et al.*, 1980) and by mechanical means and by seed
(Németh *et al.*, 1982). It has been found in Britain (Cooper, 1979), Hunga
(where it is widespread) (Németh *et al.*, 1982), Czechoslovakia (Novak anu
Langova, 1981), Italy (Russo *et al.*, 1978) and Bulgaria (Lazarova-Topchyiska
(1987) as well as in North America (Mircetich *et al.*, 1980; Mircetich and
Rowhani, 1984). It may cause a decline of the trees. Németh *et al.* (1982)
recommend that to reduce spread of the disease, seed should be collected
only from virus-free stock raised in isolation, and virus-free plantings for
scion production should be set up on rootstocks of *Juglans regia*.

## NUT DISEASES

Walnuts may sometimes be decayed by the common grey mould *Botrytis
cinerea* (described in chapter 3) and by the leaf blotch fungus *Gnomonia
leptostyla* and the bacterial blight organism *Xanthomonas campestris* pv.
*juglandis*, which are both described above under 'leaf diseases'.

## MISCELLANEOUS DISORDERS

**Frost damage**
Young walnut trees are very susceptible to frost damage, and in severe
winters trunks of older trees may crack, and then often bleed copiously for
long periods.

# Golden rain tree (*Koelreuteria paniculata*)

This tree, which comes from the Far East, is sometimes grown in parks and
gardens, but in Britain it flowers only in hot summers. It is very susceptible to
Verticillium wilt, which is described in chapter 3.

# Laburnum (*Laburnum anagyroides*)

The laburnum, *Laburnum anagyroides*, is native in much of Europe, and in
Britain it is a common introduced garden tree.

# Nursery diseases

### Seedling blight caused by *Peronospora cytisi* Rostrup

The downy mildew *Peronospora cytisi* (CMI Descr. F. 762) was reported in Europe about 100 years ago, when it killed many seedlings of the common laburnum and *Laburnum alpinum* in Denmark and elsewhere (Tubeuf, 1895). It does not appear to be common, and in Britain it has been found with certainty only once, in 1982 (Francis and Waterhouse, 1988). It causes brown leaf spots on the undersides of which it forms dichotomously branched conidiophores. These bear colourless conidia which measure 21–35 × 16–23 μm. Smooth-walled oospores measuring 28–35 μm across are formed in the diseased tissues.

### Brown spot caused by *Pleiochaeta setosa* (Kirchn.) Hughes
### (*Ceratophorum setosum* Kirchn.)

The imperfect fungus *Pleiochaeta setosa* (CMI Descr. F. 495) has occasionally damaged cuttings and seedlings of laburnum as well as of other leguminous plants (Green and Hewlett, 1949; Moore, 1959). Brown or black spots form on the leaves, and small groups of conidiophores on the spots produce very characteristic spores. These have a fusoid body measuring 63–98 × 13–19 μm. The body of the spore is five- to nine-celled, and the central cells are brown, while the basal and apical cells are almost colourless. The more or less conical apical cell bears a long, apical, hair-like appendage, and up to four similar but shorter lateral appendages (Hughes, 1951). The fungus may defoliate affected shoots, but Green and Hewlett (1949) found that it could be controlled by spraying once or twice with colloidal copper or Bordeaux mixture. Dichlofluanid is among the materials recommended by Brooks *et al.* (1989).

# Post-nursery diseases

## LEAF DISEASES

### The powdery mildew *Microsphaera guarinonii* Briosi and Cavara

An undescribed powdery mildew (*Oidium* sp.) has been recorded occasionally on laburnum in Britain (Moore, 1959). It appears to be the anamorphic stage of *Microsphaera guarinonii* (Ing, 1990a,d), which occurs elsewhere in Europe (Tubeuf, 1895) and has dichotomously branched appendages eight to twelve times the diameter of the cleistocarps.

### The rust *Uromyces laburni* (DC) Fuckel

The rust *Uromyces laburni* is grouped by Gäumann (1959) and Wilson and Henderson (1966) with a number of others within *U. pisi* (DC) Otth. (*U. pisi-sativi* (Pers.) Liro) (CMI Descr. F.58). It produces its uredinia and telia on the common laburnum and *Laburnum alpinum* (and also on species of *Cytisus* and *Genista*). The minute yellow uredinia occur on the undersides of the leaves, and the rounded urediniospores measure about 15–32 × 14–23 μm. The brownish telia, which occur on both sides of the leaves, bear brown, elliptical, one-celled teliospores measuring about 20–28 × 14–24 μm (Wilson and Henderson, 1966). The spermagonia and aecia are formed on *Euphorbia cyparisias* and some other *Euphorbia* spp. The fungus occurs widely across Europe, though in Britain the form on laburnum is found only occasionally.

**Leaf spots caused by *Ascochyta kabatiana* Trott. and *Septoria cytisi* Desm.**
*Ascochyta kabatiana* (of which *Phyllosticta cytisi* Desm. is said to be an early stage) may cause a leaf spot on laburnum (Moore, 1959). The spots are yellow-brown with a darker margin, and up to 2 cm across. They bear subglobose, brownish pycnidia up to 150 μm in diameter on their upper sides. The colourless, eventually two-celled, ellipsoid pycnospores measure 7–15 × 3–4 μm (Grove, 1935).

Grove (1935) and Petrescu (n.d.) also describe *Septoria cytisi* Desm. as a cause of whitish spots on laburnum leaves. The brownish-black pycnidia contain long, hair-like, multiseptate spores measuring 90–100 × 2.5–3.5 μm.

## BRANCH DISEASES

**Branch canker associated with *Cucurbitaria laburni* (Pers.) de Notaris**
The black perithecia of the ascomycete *Cucurbitaria laburni* emerge in crust-like groups through the bark on dead and dying twigs of *Laburnum*. The eight-spored asci measure up to 200 × 15 μm, and contain yellowish-brown, elliptical, muriform ascospores measuring 24–36 × 9–16 μm (Dennis, 1968). The spores are constricted at the middle transverse septum. *C. laburni* also has a *Camarosporium* anamorphic state with golden-brown muriform conidia measuring 15–25 × 8–9 μm (Ellis and Ellis, 1985), and other anamorphic states are also found (Tubeuf, 1895), and have been assigned to *Coniothyrium*, *Diplodia*, *Hendersonia* and *Phoma* (Cannon *et al.*, 1985).

The fungus is at most a very weak parasite, and Green (1932), who made a careful and detailed study of it, concluded that on normal, vigorous trees it was not even a wound parasite.

**Branch dieback caused by *Chondrostereum purpureum* (*Stereum purpureum*)**
*Chondrostereum purpureum* (*Stereum purpureum*), the cause of silver leaf in many broadleaved trees, has been found on laburnum, in which it causes a branch dieback, but no silvering of the leaves (Peace, 1962).

## VIRUS AND VIRUS-LIKE DISEASES

**Laburnum mosaic**
A sometimes brightly mottled mosaic of the leaves, which sometimes also show a yellowing along the veins, has been known for over 100 years in

various species of *Laburnum* in Europe and North America. It is spread by various aphids, and can be transmitted by grafting (Smith, 1972), and appears to be associated with the arabis mosaic virus (Cooper, 1979).

### Laburnum yellow vein
When Schultz and Harrap (1975) examined common laburnum leaves which showed an inconspicuous vein-yellowing, they found bacilliform, rhabdovirus-like particles, which they were unable to find in other leaves, which showed only mosaic-like symptoms.

## DECAY FUNGI

*Perenniporia fraxinea* (*Fomitopsis cytisina*), which attacks various broad-leaved trees including ash, elm and beech, may cause a white rot in standing laburnum trees (Cartwright and Findlay, 1946). This fungus is described in chapter 20.

# Magnolia

A good many species of *Magnolia*, all of which are introduced from North America and southeast Asia, are grown in Europe as garden trees.

## LEAF DISEASES

### Leaf spot caused by *Phyllosticta magnoliae* Sacc.
*Phyllosticta magnoliae* has been known in Europe and North America for a long time (Tubeuf, 1895; Sinclair *et al.*, 1987), and is found occasionally on magnolia leaves in Britain. It causes irregular bleached spots with an in-distinct brownish border. The spots bear a few lens-shaped pycnidia contain-ing colourless, oblong-elliptical, one-celled pycnospores measuring 8–12 × 3–4.5 μm (Grove, 1935).

# Southern beech (*Nothofagus* spp.)

Several species of *Nothofagus* have been introduced from South America and Australasia, and planted in parks and gardens. Two Chilean species, *N. procera* and *N. obliqua*, have shown some promise as forest trees. However, in hard winters (particularly in the very severe winter of 1962–63) both species, especially *N. procera*, have been badly affected by cold. This has caused dead patches and splits in the bark at the base of the trunks, and dieback in the tops of the trees (figure 2) (Nimmo, 1964).

So far *N. procera* and *N. obliqua* have been little affected by fungus diseases in Britain, though plots planted on land infested by *Heterobasidion annosum* (*Fomes annosus*) have been severely attacked and killed by that fungus. The needle rust *Micronegeria fagi* (the alternate host of which is *Araucaria*) has been found once on *Nothofagus* in Britain (Butin, private communication). The *Nothofagus* spp. have also been found to be susceptible to root damage by *Phytophthora* spp. (Strouts, 1981a).

# Foxglove-tree (*Paulownia tomentosa*)

The foxglove-tree, *Paulownia tomentosa*, introduced from China, is grown in Europe as an ornamental in parks and gardens in Britain and elsewhere in Europe.

## LEAF DISEASES

### Leaf spots caused by *Phyllosticta paulowniae* Sacc. and *Ascochyta paulowniae* Sacc. and Brun.

The two imperfect fungi *Phyllosticta paulowniae* and *Ascochyta paulowniae* sometimes cause spots on *Paulownia* leaves. Both have brownish, lens-shaped pycnidia. The colourless, oblong, one-celled pycnospores of *P. paulowniae* measure $3–7 \times 1.5–3$ μm (Grove, 1935). Those of *A. paulowniae* are pale olive-coloured, two-celled, and spindle-shaped. They measure $14–18 \times 3$ μm (Allescher, 1901).

# Plane (*Platanus* spp.)

The oriental plane *Platanus orientalis* (occasionally grown in parks and gardens) is native in the Balkans and in Crete, and into Asia. The London plane, *P. × hispanica* (*P. × acerifolia*), much grown in streets, parks and gardens, because of its resistance to air pollution and tolerance of poor rooting conditions, is usually considered to be a natural hybrid between *P. orientalis* and the North American *P. occidentalis*.

## LEAF DISEASES

### The powdery mildew *Microsphaera platani* Howe

A powdery mildew which attacks the leaves and young shoots of plane trees and is usually called *Microsphaera platani* reached Italy, apparently from the USA, in the late 1970s. It has spread rapidly into the surrounding countries, through France and into Portugal and Britain (Sequeira, 1981; Strouts *et al.*, 1984; Eliade, 1985; Vigouroux, 1986). The fungus has been studied by

Marziano *et al.* (1986), who concluded that it is in fact *M. penicillata* (Wallr.) Lév. (CMI Descr. F. 183), of which *M. platani* and *M. alni* are to be considered synonyms. In late spring it produces white floury patches on the leaves and causes a distortion of the young shoots. As it spreads, the foliage becomes a dirty grey colour, and the appearance of the trees is spoilt and their vigour may be affected, especially in southern Europe (Vigouroux, 1986). The brown cleistothecia of *M. platani* have colourless appendages which are several times dichotomously branched at the tips. On nursery plants and small trees, spraying with materials effective against other powdery mildews, such as benomyl, carbendazim, penconazole and triadimefon, may be effective.

## DISEASES OF LEAVES AND SHOOTS

**Anthracnose caused by *Apiognomonia veneta* (Sacc. and Speg.) Höhnel (*Gnomonia platani* Kleb.; anamorph *Discula platani* (Peck) Sacc., syn. *Gloeosporium nervisequium* (Fuckel) Sacc.)**

Anthracnose caused by *Apiognomonia veneta* (*Gnomonia platani*) is known throughout Europe and North America, and in Argentina, Australia and New Zealand (Milne and Hudson, 1987). It is fairly common on London plane trees in the southern half of England, but causes serious damage only in occasional years. Thus severe outbreaks, mainly in the south, but occasionally also in the northern half of the country, were experienced in the cool wet seasons of 1972, 1979, 1985 and 1986 (Redfern *et al.*, 1973; Young and Strouts, 1973; Gibbs *et al.*, 1980; Strouts *et al.*, 1986, 1987). It is more severe on the occidental plane in North America, and it also attacks this species in southern Europe (Luisi and Cirulli, 1983), where *P. orientalis*, which is generally very resistant, is also occasionally affected.

*A. veneta* is one of the species in the *A. errabunda* complex. The acervuli of the very variable anamorphic stage *Discula platani* form colourless, one-celled, oblong-ovoid conidia which generally measure 10–15 × 4–6 μm. The perithecia of the teleomorphic stage arise in winter on the fallen leaves. They are black, with long necks, and contain club-shaped, eight-spored asci. The unequally two-celled ascospores measure 14–19 × 4–5 μm (Brooks, 1953).

Neely (1976) described the symptoms of anthracnose in four separate phases which can be distinguished partly by the stage of growth of the host, and showed that the first three phases, bud, twig and shoot blight, develop in the spring following infections which occurred in the previous year.

The first phase is bud blight. During the winter and early spring a brown area develops around the base of the bud. As a result the bud dies and fails to flush and the brownish discoloration extends up and down the neighbouring stem. When growth starts in the spring a discrete canker develops on the twig around the dead bud.

Sometimes the fungus infects the whole circumference of the twig near the bud. When this occurs the disease is in the twig blight phase. The twig is girdled and the distal parts of the twig die before the buds open. This phase develops in the early spring or perhaps during the previous winter.

The third phase of the disease, called shoot blight, occurs after the bud has opened and flushed but before the new shoot is fully extended. It appears that the host tissues neighbouring the bud have been infected but the bud has not been killed. However, as the bud opens and fresh growth starts, the fungus can invade the new tissues and shoot blight then develops. Shoot blight is the most dramatic phase of symptom expression and whole trees or boughs can be affected in certain seasons.

The final phase, leaf blight, develops on the new or expanded leaves at any time during the summer. The fungus infects the leaves and gives rise to striking tongues of dead brown tissue mainly bordering the veins (plate 43).

Several workers have found that inoculation of leaves succeeds only after wounding of some kind (Schuldt, 1955; Neely and Himelick, 1967; Santamour, 1976). Milne and Hudson (1987), in inoculation experiments with London plane, using both conidia and ascospores, found that they obtained leaf blight symptoms only when they wounded the leaf veins, either by piercing them with a needle or by rubbing off the leaf hairs. They failed to infect the leaf blade itself, even after wounding. They concluded that infection took place only after wounding, probably by wind and by insects, and their findings support the suggestion by Neely (1976) that in America damage by lacewings is implicated in infection. They also found that young leaves were more susceptible than older ones until the end of the season, when old and moribund leaves were again more liable to attack. The fungus overwintered on these leaves, which gave rise to infection in the following spring by airborne ascospores and splash-dispersed conidia.

Climatic conditions in the spring appear to be critical for the development of serious disease outbreaks. As noted above, severe attacks in Britain seem to be especially associated with cool wet seasons. The same applies in New Zealand, where Hitchcock and Cole (1980) found an association between such attacks and heavy rain and low temperatures in the spring. In Italy, however, Marchetti and D'Aulerio (1983a) found that in culture the fungus grew best at 25 °C.

*Control*
Control measures for anthracnose are rarely needed as the disease tends to be so sporadic, with severe outbreaks only in occasional years. Though symptoms may then be quite spectacular, the affected trees usually make a good and rapid recovery, new shoots developing from dormant buds. If necessary, the disease can be controlled, at least on nursery plants and young trees, by spraying from the time when the buds begin to swell in the spring. Benomyl, benzimidazole, captafol and mancozeb with copper have been found effective (Marchetti *et al.*, 1984; Spies *et al.*, 1985). Neely and Himelick (1989) obtained good control of the disease on *Platanus occidentalis* by injecting trees with thiazolyl benzimidazole hypophosphite. In Italy, where attempts have been made to develop a combined tree injection technique to control both

anthracnose and the tingid *Corythucha ciliata* (Say), Tiberi *et al.* (1988) obtained the best results with a mixture of carbendazim and acephate, applied at the 'open bud' stage. In later work to the same end, D'Aulerio *et al.* (1990) reported the best results were obtained by injecting a mixture of thiabenda- zole, ipophosphate and acephate.

Some research has been undertaken into the selection and breeding of planes resistant to anthracnose. The occidental plane or American sycamore (*Platanus occidentalis*) is relatively susceptible to the disease, whereas the oriental plane (*Platanus orientalis*) is resistant. The clones of the hybrid between these two species, the London plane (*Platanus* × *hispanica*), show variable resistance (Burdekin, 1980). Santamour (1976) crossed the two parent species and made preliminary selections among the progeny for resist- ance to anthracnose.

## BRANCH AND STEM CANKER

### Canker stain caused by *Ceratocystis fimbriata* Ellis and Halstead f. sp. *platani* Walter

*Ceratocystis fimbriata* and the cankers it causes on poplars are described in chapter 16. *C. fimbriata* f.sp. *platani* is indistinguishable morphologically from the type, but it is restricted to species of *Platanus* affecting especially the London plane, *P.* × *hispanica* (*P.* × *acerifolia*), but also *P. occidentalis* and *P. orientalis*, in which it causes the disease known as canker stain, or in French-speaking countries *le chancre coloré*.

The disease has long been known in eastern North America, and its effects there have been described by Sinclair *et al.* (1987). In Europe it was first reported in Italy, in 1972, and it has since been found in France (around Marseille and in adjacent areas in the southern parts of the Rhone valley) (Vigouroux, 1986), in Armenia, in the USSR (Simonyan and Mamikonyan, 1982), and more recently in Switzerland (Accordi, 1989). It may also be present in Spain, but the record there needs confirmation (Cadahia, 1986). In Italy and France its origins have been traced back to the 1940s, when it was probably introduced during the Second World War with stores shipped in for the American forces.

The symptoms of the disease have been likened in some respects to those of Dutch elm disease in elms (Simonyan and Mamikonyan, 1982). The disease usually becomes obvious when leaves in the crown become small and yellow, and often wilt and fall. The first signs of infection, however, appear on branches, trunks and roots, around the sites of wounds, showing as blueish strips of bark surrounded by an orange or pale-brown zone. The lesions extend more rapidly above than below the infection site, and the affected bark becomes light brown, breaking up into small, jig-saw-like pieces which remain attached for two or three years, in contrast to the brownish healthy bark, which scales off in large patches in summer, to reveal the newer, paler- yellow bark beneath. If the diseased tissues are cut across, the tissues within will be found to be stained a blueish black or brownish black colour. The stained areas tend to show the form of radially arranged spindles, though eventually they may become more or less confluent (Vigouroux, 1986).

Diagnosis of the disease must be confirmed by isolation and identification of the fungus. A trap method for isolating from wood, soil and water is described by Grosclaude *et al.* (1988), who placed infected material in a damp chamber adjacent to healthy debarked plane twigs, on which perithecia of the fungus soon developed. Wood samples from diseased trees can be taken with a borer, and if the samples are incubated they soon show perithecia and also characteristic endoconidia which develop in chains and are 6–35 μm long (Vigouroux, 1986). The optimum temperature for growth and the production of conidia and ascospores is 25 °C (Mancini and Scapini, 1981). Growth of *C. fimbriata* f.sp. *platani* is impeded at temperatures above 30 °C, and the fungus is destroyed at temperatures of 35–40 °C. It may persist in wood for almost two years (Grosclaude *et al.*, 1987). In soil it can remain alive in winter for more than 105 days, but it dies out more rapidly in spring and summer (Accordi, 1989).

Though in North America nitidulid beetles may play some part in the spread of canker, and in Europe it has been shown that the fungus can spread from tree to tree through root grafts, in almost all cases in Europe the fungus is transmitted through the activities of man, on pruning tools, with wound paints, by digging machines which damage tree roots, and by vehicles which wound tree bases in car parks (Vigouroux, 1986; Accordi, 1986).

In many areas, canker stain affects isolated trees or small groups, but in some places groups of up to 500 trees have been killed, and in Marseille thousands of trees have been destroyed; severe damage has been done to London planes in the USA (Vigouroux, 1986).

### Control

As the fungus is a wound parasite, wounding should be avoided as far as possible. Pruning should be kept to a minimum, and should be done at the time of most vigorous growth, when healing is most rapid. Diseased trees should be removed and destroyed as soon as they are found. Pruning tools should be disinfected between trees, and pruning wounds treated with a protectant (Panconesi, 1981).

In the south of France, because of the importance of the disease there, the French government has intervened to encourage local authorities to take control measures against canker stain and to advise the general public (Vigouroux, 1986).

It has been found difficult to develop effective chemical control of canker stain. The fungus is known to be sensitive to carbendazim (Tawil *et al.*, 1982), while Simonyan and Mamikonyan (1982) in Armenia have recommended spraying in May with benomyl.

Attention has been drawn by Turchetti and Panconesi (1982) to the possibilities of biological control by strains of the bacterium *Bacillus subtilis*, which they found inhibitory to the growth of *C. fimbriata* f. sp. *platani*.

In the longer term it may be possible to select or breed resistant clones, and work to that end has begun in Europe.

## DECAY FUNGI

*Inonotus hispidus* may decay the stems and branches of plane, and the bases of plane trees are sometimes attacked by *Rigidoporus ulmarius*.

## MISCELLANEOUS DISORDERS

### Salt Damage
Because it is so much grown by roadsides in towns and cities, London plane has often been damaged by de-icing salt used to clear roads in cold winters. Affected trees show crown dieback, young leaves dying soon after flushing, and others later in the season showing scorching round the edges and between the veins. Patches of dead bark sometimes occur at the bases of the trunks, but there is no obvious damage to the roots (Gibbs and Burdekin, 1983). This topic is discussed further in chapter 2.

# Buckthorn (*Rhamnus catharticus*) and alder buckthorn (*Frangula alnus*)

The buckthorn and alder buckthorn are both shrubs or small trees native throughout most of Europe, including Britain. Some other *Rhamnus* spp. grow in central and southern Europe.

## LEAF DISEASES

### The powdery mildew *Microsphaera divaricata* (Wallr.) Lév.
The powdery mildew *Microsphaera divaricata*, which has sometimes been regarded as a variety of *M. alni*, produces a rather conspicuous covering of mycelium and conidia on both sides of the leaves and on the young shoots of the alder buckthorn. The brown cleistothecia contain from three to seven asci, and bear up to ten dichotomously branched appendages from 3.5 to 5.5 times the diameter of the fruit body. The ends of these appendages are recurved (Salmon, 1900; Ellis and Ellis, 1985).

### Crown rust (*Puccinia coronata* Corda)
Buckthorn and alder buckthorn are the primary hosts on which are found the spermagonia and aecia of the crown rust, *Puccinia coronata*, which is an important parasite of oats and various other grasses, on which it produces its uredinia and telia. Its attacks on the buckthorns are of minor importance.

The spermagonia occur mostly on the upper sides of the buckthorn leaves and on their petioles. Later, the aecia arise mainly on the undersides of the leaves, and between the spermagonia on the petioles. The affected leaves and leaf stalks become swollen. The aecia are cylindrical or cup-shaped, with a white margin. The orange, rounded aeciospores are 16–25 × 13–20 μm (Wilson and Henderson, 1966).

## STEM DISEASES

### Cankers caused by *Nectria coccinea*
*Nectria coccinea* may sometimes cause cankers on *Rhamnus* spp. This fungus is much more important as a cause of beech bark disease, under which it is described in chapter 11.

# False acacia or locust tree (*Robinia pseudoacacia*)

The false acacia is an ornamental tree introduced from eastern North America. The rose acacia, *Robinia hispida*, is also sometimes grown in parks and gardens.

## LEAF DISEASES

### Leaf spot caused by *Phloeospora robiniae* Höhnel
Yellowish leaf spots that later become brown, and are caused by the imperfect fungus *Phloeospora robiniae* are sometimes found on *Robinia*. The small pycnidia have thin, ill-defined walls, and from their openings emerge whitish blobs of elongated, faintly one-septate, colourless pycnospores that measure 25–28 × 2–5 μm (Grove, 1935).

## STEM SCABS AND CANKERS

### Anthracnose associated with *Phoma macrostoma*
The deuteromycete *Phoma macrostoma* is widely known as a usually weak pathogen on woody hosts. Strouts (1981b) recorded an anthracnose of two- and three-year-old nursery plants of *Robinia pseudoacacia*, the leaves and shoots of which were killed or severely damaged. Recovery shoots in the following year were also killed. *Phloeospora robiniae* was found in the leaf lesions, and *Phoma macrostoma* on both the leaves and the shoots. The

colourless, one-celled conidia of *P. macrostoma* measure 5.5–10.5 × 2.5–3.5 μm (Holliday, 1989).

### Stem canker associated with *Fusarium avenaceum* (Corda) Sacc. and *F. lateritium* Nees

Talboys and Davies (1984) isolated *Fusarium avenaceum* (the anamorph of *Gibberella avenacea* Cook)(CMI Descr. F. 25), and less often *F. lateritium* (the anamorph of *G. baccata* (Wallr.) Sacc.) from cankers on the stems of two-year-old trees of *Robinia pseudoacacia*.

## WILT DISEASES

### Verticillium wilt

Strouts (1981b) isolated *Verticillium dahliae* from stained xylem of one-year-old nursery plants of *R. pseudoacacia* which showed symptoms of Verticillium wilt (described in chapter 3). About a quarter of the plants were expected to die, and the disease regularly affected trees planted at the nursery concerned on land formerly under potatoes or strawberries. The variety 'Frisia' was badly affected, wilting plants never recovering, whereas the disease caused little damage to the variety 'Bessoniana'.

## ROOT DISEASES

### Phytophthora root disease

*Robinia* is not regarded as among the trees especially susceptible to Phytophthora root disease, though Strouts *et al.* (1985) in 1983 found *Phytophthora citricola* killing the roots of *R.* × *hillieri*. The *Phytophthora* spp. are described in chapter 3.

## DECAY FUNGI

A white rot in the butt of false acacia trees is sometimes caused by *Perreniporia fraxinea*, which also affects various other broadleaved trees. The roots of *Robinia* may also be decayed by *Meripilus giganteus*. These fungi are described in chapter 20.

# Elder (*Sambucus nigra*)

The elder is a shrub or rarely a small tree occurring as a native throughout Europe, including Britain.

LEAF DISEASES

**Leaf spots caused by *Cercospora depazeoides* (Desm.) Sacc., *Phyllosticta sambuci* Desm., *P. sambucicola* Kalchbr. and *Ramularia sambucina* Sacc.**
Several imperfect fungi have been described as the cause of leaf spots on elder. The best-known is *Cercospora depazeoides*, which gives rise to greyish spots that turn whitish or pale brown with age, and measure up to 5 mm across. The fungus produces long, colourless spores, particularly on the undersides of the spots. The spores, which have 1–6 transverse walls, measure 50–147 × 4–6 μm (Moore, 1946).

Grove (1935) also described *Phyllosticta sambuci* and *P. sambucicola* from leaf spots on elder. These both produce brownish-black pycnidia on the spots. The colourless, one-celled, oblong-ovoid spores of the former are 5–8 × 2.5–3 μm. Those of the latter are round, one-celled and 3–4 μm across, or oval, measuring 4–5 × 3 μm.

*Ramularia sambucina* has also been recorded several times on elder leaves (Wakefield and Bisby, 1941). It causes small, brownish spots that later turn white with a brown border. Its two-celled, colourless, spindle-shaped conidia are formed in chains, and measure 25–35 × 4–4.5 μm (Lindau, 1907).

DECAY FUNGI

The Jews' ear fungus, *Auricularia auricula-judae*, is common as a slow, weak parasite on old elders. The pale-brown, gelatinous, ear-like fruit bodies occur on dead branches.

# Pagoda tree and other *Sophora* spp.

The pagoda tree, *Sophora japonica*, from China and Korea, and some other *Sophora* spp. from the Far East, North and South America and New Zealand, are sometimes grown in European parks and gardens.

LEAF DISEASES

**Powdery mildew: *Microsphaera diffusa* Cooke and Petch**
*Microsphaera diffusa* attacks various legumes, and Ing (1990a, 1991) records it on *Sophora*. Its colourless conidia measure 28–51 × 17–21 μm. The colourless, dichotomously branched appendages on the brown cleistothecia are lax and irregular, with forked ends. The cleistocarps contain several asci, each with several colourless ascospores measuring about 18 × 9 μm (Salmon, 1900; Holliday, 1989).

# Lilac (*Syringa vulgaris*)

Many varieties of the common lilac, *Syringa vulgaris*, a shrub or small tree native in southern Europe, have long been grown in gardens in the rest of the continent including Britain. A few other *Syringa* spp. are also sometimes grown.

## LEAF DISEASES

### Powdery mildew: *Oidium* sp.
An *Oidium* sp. was found on lilac leaves in southern England and South Wales in 1948 and the succeeding few years (Moore, 1952), and caused a severe outbreak also in other parts of Europe. It appears to be the conidial stage of *Microsphaera penicillata sensu lato*, which is described under '*Alnus*' in chapter 13.

### Leaf spots caused by *Venturia syringae* (Syd.) Barr (*Pseudosphaerella syringae* Syd.)
The anamorphic stage of *Venturia syringae*, a *Fusarium*, causes large irregular brown leaf spots on leaves of the common lilac in Germany. The brown, one- or two-celled spores measure 10–18 × 4–6 μm and the greenish or brownish, more or less elliptical, unequally two-celled ascospores of the pseudothecial stage measure 9–11 × 4–5 μm (Sivanesan, 1977).

### Leaf blotch caused by *Heterosporium syringae* Oud.
A leaf blotch of lilac caused by *Heterosporium syringae* has occasionally been found on lilac in southern England (Moore, 1959). The fungus produces a black turf of erect conidiophores on the brown leaf blotches. The brown conidia are spiny, cylindrical with rounded ends and 1 to 3 cross-walls, and measure 25–30 × 7–9 μm (Lindau, 1910).

### Leaf spot caused by *Phyllosticta syringae* Westend.
Grove (1935) also described *Phyllosticta syringae* (which he considered to be a young state of *Ascochyta syringae* Bres.) as the cause of a lilac leaf spot. The spots are grey, with a darker border. The pycnidia, which grow on the upper sides of the leaves, are lens-shaped, and the colourless, one-celled spores measure 5–8 × 2–3 μm. According to Allescher (1901), the pycnidia of *Ascochyta syringae* are up to 130 μm across, and its colourless, two-celled pycnospores measure 8–10 × 3–3.5 μm.

## DISEASES OF LEAVES AND SHOOTS

### Leaf, twig and blossom blight caused by *Pseudomonas syringae* pv. *syringae* van Hall
The Gram-negative bacterium *Pseudomonas syringae* pv. *syringae*, which attacks many hosts, is well known as a cause of leaf and twig and blossom

blight of lilac. Buds may be blackened, and shoots girdled and killed, and the blossoms spotted and destroyed. On the older shoots, elongated blackened areas may be found. Leaves may become distorted, with crinkled edges, and angular brown spots on their surfaces. The causal bacterium appears to enter through the stomata, and also through hail wounds on the young shoots (Dowson, 1957). It is rod-shaped, and motile by several polar flagella, and measures 1.6–3.2 × 0.2–0.4 µm. It produces a green fluorescent pigment in culture (Stapp, 1961). Further information on identification is given by Lelliott *et al.* (1966) and Lelliott and Stead (1987).

Control may be achieved by cutting out the diseased shoots, and by spraying with a copper spray such as Bordeaux mixture or cupric ammonium carbonate (Brooks *et al.*, 1989).

**Leaf, shoot and stem blight caused by *Phytophthora* spp.**
*Phytophthora cactorum* and *P. syringae* (which are described in chapter 3) may both cause blights of the leaves, shoots and stems of lilac.

ROOT DISEASES

Lilacs are among the trees most susceptible to Phytophthora root disease (Strouts, 1981a). The *Phytophthora* spp. are described in chapter 3.

VIRUS DISEASES

Viruses known to affect lilacs in Europe include the elm mottle virus (CMI/AAB Descr. V 139), which causes white mosaic and ring-spot symptoms, the lilac ring mottle virus (CMI/AAB Descr. V 201), which causes chlorotic ring and line patterns, and seems to be confined to the Netherlands, and the lilac chlorotic leaf spot virus (CMI/AAB Descr. V 202), which may cause yellow line patterns or spots or a slight chlorosis.

# Lime (*Tilia* spp.)

Of the limes, the small-leaved lime, *Tilia cordata*, is found over most of Europe. It is widely scattered throughout England and Wales, and has been planted in Scotland. The large-leaved lime, *T. platyphyllos*, is also native in most of Europe. In Britain it may be native in a few limestone areas, and has been much planted elsewhere. The common lime, *T.* × *vulgaris*, a hybrid between the other two species, is also widespread in Europe. In Britain, where it is probably an introduction, it is much planted all over the country. The silver lime, *T. tomentosa*, a native of southeast Europe, is also much planted in the rest of the continent, and in Britain is found in many parks and gardens.

## LEAF DISEASES

**Leaf spots caused by *Phyllosticta tiliae* Sacc. and Speg., *Ascochyta tiliae* Kab. and Bub., *Gloeosporidium tiliae* (Oud.) Petr. and *Cercospora microsora* Sacc.**

Four imperfect fungi recorded as the cause of usually minor leaf spots on lime are *Phyllosticta tiliae*, *Ascochyta tiliae*, *Gloeosporidium tiliae* and *Cercospora microsora*.

*Phyllosticta tiliae*, which has been fairly widely noted, causes pale yellowish-brown spots with a darker margin. Its brown, lens-shaped pycnidia are up to 130 μm across, and contain colourless, ellipsoid or oblong, one-celled spores measuring 5–7 × 3 μm (Grove, 1935).

*Ascochyta tiliae*, so far observed in Britain only in Scotland, produces greyish spots up to 15 mm across, which finally drop out to leave 'shot-holes'. The rounded, brownish pycnidia are up to 150 μm in diameter, and contain colourless, two-celled, oblong spores measuring 8–10 × 2.5–3 μm (Grove, 1935).

In Britain, *Gloeosporidium tiliae* is said to be common in Scotland, and less so in England. It causes yellowish-brown, rounded leaf spots up to 2 cm across, sometimes with a darker border. Minute pustules on the undersides of the leaves contain colourless, one-celled, oblong-ovoid spores measuring 10–18 × 4–7 μm (Grove, 1937). It may cause defoliation in young trees. It is the anamorph of *Apiognomonia tiliae* (Rehm) Höhnel (*Gnomonia tiliae* Rehm), which is one of the groups of fungi centred on *A. errabunda* (Roberge) Höhnel.

*Cercospora microsora*, the anamorph of the ascomycete *Mycosphaerella microsora* Syd. ( Petrescu, n.d.) causes rounded black leaf spots about 3 or 4 mm across, with a greyish centre. On this central area the fungus produces its long narrow, colourless or slightly yellowish, usually one- to three-septate conidia which measure 35–90 × 3–4 μm (Ellis and Ellis, 1985). *C. microsora* also causes small cankers on young shoots. The perfect stage, which develops on the dead leaves, does not appear to have been found in Britain. Its colourless, two-celled spores measure 16–20 × 3.5–4 μm (Petrescu, n.d.).

The fungus was long ago described by Hartig (1900) as a cause of severe defoliation in the European mainland. More recently it has been known to cause appreciable damage to young *Tilia* plants in nursery beds in Romania, where good control has been obtained by spraying with copper oxychloride or Bordeaux mixture (Poleac *et al.*, 1948).

## DISEASES OF TWIGS AND BRANCHES

**Twig and branch blights caused by *Discella desmazierii* B. & Br. and *Pyrenochaeta pubescens* Rostr.**

Two weakly parasitic imperfect fungi have been found attacking lime twigs and branches in Britain.

The first of these, *Discella desmazierii* (*Lamproconium desmazierii* Grove), has been found from time to time over a fairly large part of mid and southern England. Its pustules on the dead bark contain narrowly oval, grey to deep-blue spores that eventually become two-celled, and measure 30–36 × 6–10 μm (Grove, 1937). The fungus also occurs elsewhere in Europe (Petrescu, n.d.).

The second, *Pyrenochaeta pubescens*, has so far been found in Britain once only, on young stock of *Tilia* × *vulgaris* 'pallida' imported from Holland (Burdekin, 1968). It forms rounded or oval, purple (finally grey) spots on the bark. The small black pycnidia that arise on the lesions are clothed with colourless, septate hairs up to 50 μm long. They contain colourless, one-celled, elliptical spores measuring 6–8 × 3–4 μm (Allescher, 1903). The disease is of minor importance, and can be controlled by pruning out the affected parts in the summer (Prihoda, 1950).

**Shoot canker caused by *Plectophomella concentrica***
Shoot canker caused by *Plectophomella concentrica* has been found in Edinburgh causing a canker on young shoots of lime. The canker is described under '*Ulmus*', on which it seems to be much more widespread.

**Mistletoe: *Viscum album***
Limes are among the trees most often colonised by the common mistletoe, which is briefly described in chapter 3.

## DISEASES OF THE TRUNK

**Bleeding canker caused by *Phytophthora* spp.**
Sometimes lime trees are affected by the bleeding canker described under 'horse chestnut'. A gummy liquid which dries to form a brittle crust exudes from broad strips of dead bark extending upwards from dead roots. *Phytophthora citricola* has been isolated from the lesions.

## WILT DISEASES

**Verticillium wilt**
Young plants of the small-leaved lime, *Tilia cordata*, and the Caucasian lime, *T. euchlora*, are among the trees most susceptible to Verticillium wilt (Piearce and Gibbs, 1981), which is described in chapter 3.

## ROOT DISEASES

Limes are among the trees most often affected by root damage by *Phytophthora* spp. (Strouts, 1981a). Affected trees often show dieback symptoms. The *Phytophthora* spp. are considered in chapter 3.

## DECAY FUNGI

The commonest decay fungus on lime is *Ustulina deusta*, which rots the butt and roots. Limes may also be attacked by *Rigidoporus ulmarius*. These fungi are described in chapter 20.

## MISCELLANEOUS DISORDERS

### Sooty moulds

Lime trees (especially the common lime, *T.* × *vulgaris*) are much infested by aphids, which cover the leaves with a sticky 'honey dew'. This sugary secretion forms a substrate for the growth of various sooty moulds, particularly *Aureobasidium pullulans* (de Bary) Arn. (*Pullularia pullulans* (de Bary) Berkh.) and *Cladosporium herbarum* (Pers.) Link ex Fr. (Friend, 1965), which form a black, saprophytic coating on the leaves.

# *Zelkova*

A few species of *Zelkova*, from the eastern Mediterranean, the Caucausus and the Far East, are occasionally found in parks and gardens.

The *Zelkova* spp. are susceptible to Dutch elm disease (described under 'elm' in chapter 14), though less so than the European and American elms.

# 20 Decay fungi of broadleaved and coniferous trees

The fungi described in this chapter are arranged in alphabetical order of their scientific names.

A great many decay fungi attack trees of both broadleaved and coniferous species and a number of those most commonly found are described in this chapter. The majority of these fungi do not seriously affect the general health of the host trees and the presence of decay is often evident only from the presence of fungal fruit bodies. However, the decay fungi described here do cause a serious reduction in the mechanical strength of the woody tissues. Decayed branches, stems and roots are all liable to break and severe damage can occur when broken limbs or uprooted trees fall on neighbouring buildings, vehicles and people. It is the duty of tree owners to inspect their trees regularly for any external signs of decay. This is an important and specialist subject and readers are referred to the publication by Young (1984) for further information. In contrast, there are two root- and butt-rot fungi, *Heterobasidion annosum* (*Fomes annosus*) and *Armillaria mellea* (*sensu lato*), which can cause widespread death of a range of tree species, and full descriptions of these fungi have been included in chapter 3.

There has been relatively little research undertaken into the biology of most of the decay fungi mentioned in this chapter. It is generally considered that these fungi enter their hosts either through wounds in above-ground parts of the tree or through the roots. Precise evidence on this and on the factors controlling invasion is generally lacking.

For many years it was thought that trees showed only passive resistance to decay. However, Shigo (1977) has put forward a concept of active resistance whereby barriers to decay development are formed within a tree in response to wounding and fungal invasion. The coordinated formation of these barriers in various parts of the wood is termed compartmentalisation. The full development of this concept, and others, can be found in a book by Shigo (1986), which readers are commended to study.

The treatment of pruning wounds to prevent the entry of decay organisms has been practised for many years. Mercer (1979) reviewed the subject and concluded that many treatments, although they may encourage callus production around the wound, do little to prevent infection by decay fungi.

Hibberd (1989) reinforced this view and considered that there are four possible approaches to the control of tree wound infections:

(1) The surface of the wound can be treated promptly with a durable sealant in the hope that infection will be excluded.

(2) The wound tissues can be treated with a chemical fungicide or a biological control agent.

(3) A treatment can be applied which makes the microenvironment of the wound tissues unsuitable for the growth of damaging fungi.

(4) Callus growth across the surface of the wound can be encouraged in the expectation that decay development will cease once occlusion is complete.

Lonsdale, in Hibberd (1989), considers that currently available wound dressings are unlikely to provide protection against decay organisms for more than 1 year. These treatments may be of assistance in excluding fresh wound parasites such as *Chondrostereum purpureum* but are not effective with respect to most decay fungi. Longer-term protection may be provided by biological control agents such as *Trichoderma viride* and at least one product incorporating this agent is commercially available. The control of the micro-environment at the surface of the wound by application of an airproof seal is still under study and much depends on the long-lasting qualities of novel sealant products.

Callus formation can be enhanced by the use of chemicals such as thiophan-ate methyl. There seems little doubt that complete callus formation over a wound is the most effective way of preventing entry or further development of decay fungi. Current practice in Britain tends to favour no treatment of wounds, except where a cosmetic effect is required. Reliance is therefore placed on natural callus development to combat decay organisms. Various measures can be taken to enhance natural growth of callus, including early pruning, to promote rapid healing, and correct pruning position, just outside the branch ridge and collar (see Lonsdale, 1983c).

In contrast to the paucity of information on the biology of many decay fungi, there is a wealth of authoritative texts on the identification of fungal fruit bodies. A list of some of the more useful references, particularly in relation to the Basidiomycetes, is included in the References section. The nomenclature of these fungi is a complex and often confusing subject. As far as possible the naming of polypores, which are well represented among decay fungi, has followed that of Pegler (1973). For descriptions of the types of decay produced by wood-rotting fungi, Cartwright and Findlay (1958), Burdekin (1979) and Greig (1981) should be consulted for further infor-mation. Some information on the cultural characters of decay fungi is given by Cartwright and Findlay (1958) and a comprehensive description of those occurring in North America is given by Nobles (1948).

Two tables of host/pathogen relationships (one for conifer and one for broadleaved species) are included at the end of the chapter in order to indicate, by tree species, the hosts on which the fungi are most commonly found.

**Armillaria spp.: see chapter 3**

**Bjerkandera adusta (Willd. ex Fr.) Karst. (*Polyporus adustus* Fr.)**
*Bjerkandera adusta* is commonly found in Europe and North America growing on dead wood of a number of broadleaved species.

The annual fruit bodies are thinnish brackets up to 4 cm across, often
imbricate, soft and pliable when fresh, later turning hard and brittle. The
upper surface is pubescent, at first white to cream, later grey to blackish
especially along the margin. The flesh is white and distinctly thicker than the
short tubes which are grey to black and separated from the flesh by a thin
black zone. The lower surface is grey to black and the pores are round to
angular, 0.2 mm in diameter. The hyaline spores are ellipsoid to cylindrical,
thin-walled and smooth, measuring $4.5 \times 2.5-3$ μm (Rea, 1922), and are pale
straw-coloured in the mass. In culture *B. adusta* grows as a smooth white felt;
it rarely fruits but the hyphae may break up into oidia.

*B. adusta* is commonly found on beech, often on trees that have been
attacked by beech bark disease. When parts of a beech stem are killed by the
disease, ambrosia beetles may carry spores of this and other fungi into the dead
wood, where a white decay subsequently develops. The stem may eventually
break, giving rise to a condition referred to as beech snap (Parker, 1974).

**Chondrostereum purpureum (Pers. ex Fr.) Pouz. (*Stereum purpureum*
Pers. ex Fr.)**
*Chondrostereum purpureum* has a world-wide distribution and is found on
wounds, dead branches, stumps and felled logs of many broadleaved species,
especially birch, beech, elm, poplar and *Sorbus* spp. It is an important wound
pathogen on trees and shrubs in the family Rosaceae, causing the disease
known as 'silver leaf' and is of major importance especially to orchard
growers of plums and other top fruit, and to a lesser extent on other members
of this family. It has also apparently caused death of coppice shoots of
*Eucalyptus archerii* following entry through an earlier cut (Strouts *et al.*,
1986). The fruit bodies may be found at any time of the year and are thin,
leathery brackets 2 cm or more across, sometimes entirely resupinate, often
imbricate and confluent. The upper surface is whitish and greying, hairy, and
the margin is sometimes lobed. The hymenium is smooth, lilac and purplish at
first, soon becoming purplish brown and fading to pale fawn. The hyaline
spores, flattened on one side, measure $6-8 \times 3-4$ μm (Rea, 1922) and are
white in the mass.

*C. purpureum* commonly causes a discoloration in beech logs. Yatsenko-
Khmelevky (1938) describes the formation of tyloses and dark-coloured
materials as a consequence of the reaction of living cells to fungal activity.
Gummy materials form in the parenchyma cells of beech wood infected by *C.
purpureum* and later small whitish areas develop giving the wood a mottled
appearance. *C. purpureum* does not appear to cause extensive decay and
Cartwright and Findlay (1958) report only slight losses in weight in laboratory
tests. Guinier (1933) suggested that *C. purpureum* can only invade wood in

which the parenchyma and medullary rays are still intact and contain reserve materials. It appears to be localised in the younger, outer layers of wood and ceases growth after about a year when the reserves have been exhausted. Rayner (1977) also indicates that *C. purpureum* is a pioneer colonist of wounds or other freshly exposed surfaces. He considers that it is poor as an agent of decay and is highly susceptible to competition from other organisms.

In culture on malt agar *C.* *purpureum* produces a colourless, cottony growth which later becomes a smooth, felted mat with yellowish or lilac tinges (Cartwright and Findlay, 1958). Small hymenial surfaces may occasionally be fcrmed. The colourless hyphae bear large clamp connections, often in whorls.

Although *C. purpureum* is an important pathogen of fruit trees, it does not cause serious decay at wound sites or on felled logs. It is often rather conspicuous, as fruit bodies may rapidly appear on freshly exposed surfaces and infected wood may become markedly stained. As a result it is easily credited with causing more damage to timber than is actually the case.

Woodgate-Jones and Hunter (1983) report excellent protection of pruning wounds on plum, following artificial inoculation with *C. purpureum*, using the fungicides triadimefon and benodanil and also with applications of *Trichoderma viride* (alone and in combination with triadimefon).

### Coriolus versicolor (L. ex Fr.) Quel. (*Polystictus versicolor* L. ex Fr., *Trametes versicolor* (Fr.) Pil.)

*Coriolus versicolor* is commonly found in all temperate countries on dead wood of many different species of broadleaved trees and occasionally of conifers. It is normally saprophytic and attacks sapwood on felled timber or on parts of trees killed by other agents, but it is occasionally found invading heartwood or live sapwood.

The annual fruit bodies of *C. versicolor* are thin, leathery, imbricate brackets, 3–8 mm across; the upper surface is velvety, with concentric satiny zones which are greyish or brownish in colour (plate 63). The white tubes are very short and the creamy-coloured pores are at first small and round, becoming irregular as they develop. The hyaline, oblong spores measure 6–8 × 3 μm (Rea, 1922).

Wood infected by *C. versicolor* turns a paler colour as decay develops and at a more advanced stage becomes white and very light in weight (figure 97). There are no markedly characteristic features of this rot but *C. versicolor* is a particularly active fungus and is able to decompose heavily lignified tropical timbers (Cartwright and Findlay, 1958).

In culture *C. versicolor* grows on malt agar as a fine colourless mycelium closely appressed to the medium. A slightly raised margin of fluffy aerial growth develops around the culture and at a later stage a tough, flat, closely felted mat is formed, white or slightly tinted in colour. Malt agar on which *C. versicolor* has been growing always becomes bleached (Cartwright and Findlay, 1958). Some chlamydospores may be formed and crystals occur in the medium.

*C. versicolor* is one of the commonest causes of decay in felled wood in

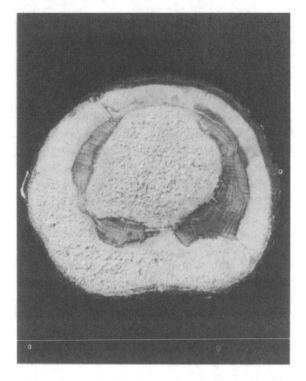

**Figure 97**    Decay in oak caused by *Coriolus versicolor* (Forestry Commission)

broadleaved species. It is frequently found growing on fence posts where they are in contact with the soil, in poorly ventilated timber stacks and in damp mining timber (Cartwright and Findlay, 1958).

### *Daedaleopsis confragosa* (Fr.) Schroet. (*Trametes rubescens* (A. & S.) Fr.)
*Daedaleopsis confragosa* is not uncommonly found in Europe, particularly on willow, birch and alder.

The annual fruit bodies are attached to the stems and are flattened, semicircular brackets measuring 5–12 cm across. The upper surface is more or less smooth and ochraceous to red-brown in colour. The flesh is corky, zoned, white at first then turning reddish or pale brown. The tubes are up to 1 cm long, pale then reddish-brown; the pores are round then elongate, white becoming greyish, turning blood-red when bruised. The hyaline spores are oblong, slightly curved and measure 10 × 2 μm (Rea, 1922).

*D. confragosa* causes an active white rot and usually occurs on severely wounded or suppressed trees (Cartwright and Findlay, 1958).

### *Daldinia concentrica* (Bolt. ex Fr.) Ces. and de Not.: King Alfred's cakes
*Daldinia concentrica* has a world-wide distribution; in Britain it is very commonly found on ash and also on birch, particularly after heath fires.

Whalley and Watling (1982) found that *D. concentrica* occurred mainly on ash in England and on birch in Scotland. In France it is most frequent on alder but also occurs on walnut and birch. It is usually considered to be a saprophyte on dead wood but may spread back into living tissues.

The fruit bodies of this ascomycete, which can be found at all times of the year on dead branches, are almost smooth, dotted with fine pores, and hemispherical in shape, measuring 3–6 cm or more in diameter (figure 98). They are at first reddish brown but soon become black and when mature have the appearance and consistency of charcoal. The young fruit body is covered with conidia which are borne on conidiophores resembling those of *Botrytis* (Cartwright and Findlay, 1958). Similar conidiophores arise from moistened, decayed wood and the hyaline conidia which they bear measure $6.5–8 \times 5–6$ μm (Panisset, 1929). The flesh is dark purplish-brown, fibrous with darker concentric zones. Perithecia form in a single layer just below the surface and narrow canals lead to tiny ostioles on the outside of the fruit body. The eight-spored asci measure $200 \times 12$ μm and the black ascospores are $12–17 \times 6–9$ μm (Dennis, 1968).

**Figure 98**  Fruit bodies of *Daldinia concentrica* (D.H. Phillips)

Ash wood decayed by *D. concentrica* (figure 99) was at one time termed 'calico wood' by timber merchants. The decay is a mottled rot with white patches among apparently healthy wood. The affected wood often contains scattered black flecks and irregular, waxy, purplish-brown or black stripes. The black flecking is due to the presence of dark fungal mycelium in the spring wood vessels. These stripes are 1–5 mm in width but in some sections they appear as circular patches or rings (Panisset, 1929).

**Figure 99**   Decay in ash caused by *Daldinia concentrica*, a fruit body of which can be seen on the right (Forestry Commission)

In culture on malt agar the fungus produces a soft, powdery mat. After the mycelium is well established, the agar and hyphae change colour to dark brown and greenish black, respectively. After repeated subculturing the culture may appear mainly a dirty fawn. The dark-coloured hyphae sometimes become aggregated into spherical bodies about 300 μm in diameter (Panisset, 1929). Typical conidia are borne in groups on short conidiophores.

*D. concentrica* probably causes decay in wood which has been killed by other agents and furthermore does not seem to spread very rapidly but it is able to resist desiccation and has been found, for example, causing decay in seasoned maple floor blocks (Cartwright and Findlay, 1958).

### *Fistulina hepatica* Schaeff. ex Fr.: beef steak fungus
*Fistulina hepatica* is commonly found in Europe and also occurs in America and Australia. It is frequently reported on oak, also on sweet chestnut and sometimes on a number of other hardwoods, including ash, walnut, willow, beech, hornbeam and elm (Rea, 1922). It is rare in Switzerland and Breitenbach and Kranzlin (1986) consider that it should be protected there.

The sporophores vary in size from 5 to 30 cm across and they can be found between August and November on branches, stems or stumps. They appear at first as creamy, soft, juicy lumps which soon become tongue-shaped or semicircular brackets. They may be sessile or stipitate, the stipe when present measuring 3–7 × 2–4 cm. The mature fruit body is purplish red or brown and the flesh is marbled (like raw steak) and reddish; it exudes a blood-red juice when broken. The tubes are separate, pale yellow and very long. The individual spores are pink, ovoid, measure 4.5–5 × 4 μm and are pinkish brown in mass (Rea, 1922). Secondary spores (conidia) are produced on the surface of the sporophore; these are ovoid and measure 5–10 × 4–6 μm (Bourdot and Galzin, 1927; Stalpers and Vlug, 1983). The fruit body is edible.

Cartwright (1937) made a detailed study of *F. hepatica* in oak wood and showed that it was responsible for the condition known as 'brown oak'. In the early stages a streaky brown discoloration of the heartwood is produced and this gradually spreads to give the whole infected area a rich brown colour. In the advanced stage the wood exhibits cubical cracking but it never becomes soft and crumbly as in the case of decay caused by *Laetiporus sulphureus*.

In the wood the mycelium is rather sparse and mostly present in medullary rays though it may be difficult to observe in heavily stained cells. In culture the fungus grows best in the dark and on 5 per cent rather than 2 per cent malt agar (Cartwright, 1937). It forms a soft, woolly mat at first creamy white but later developing tints which range from pale pinkish-cinnamon to straw-yellow and later russet. Abnormal phalloid fruit bodies are often formed with rudimentary tubes bearing spores present at their apex.

Cartwright and Findlay (1958) note that *F. hepatica* is frequently found in pollarded oaks and suggest that it enters its host through wounds which extend into the heartwood.

The production of 'brown oak' by *F. hepatica* enhances the value of the timber because of its attractive appearance. Results of mechanical tests have indicated that there is no appreciable loss in strength during the early stages of infection but at a more advanced stage considerable softening occurs and the timber is rather brittle (Latham and Armstrong, 1934; Cerny, 1982). The fungus does not continue its activity in converted timber as in the case of *Laetiporus sulphureus*.

*Fomitopsis cytisina*: see *Perenniporia fraxinea*
*Fomes annosus*: see *Heterobasidion annosum* (chapter 3)
*Fomes applanatus*: see *Ganoderma applanatum*

**Fomes fomentarius (Fries) Kickx.: tinder fungus**
*Fomes fomentarius* occurs widely in Europe and America on beech and birch and has been reported from India and Australia. It is also common in parts of the highlands of Scotland on birch but has rarely been found in England and then only on beech. Many of the early references to this fungus on beech in England may have referred to *Ganoderma applanatum* though it could be more common than the records indicate.

The perennial fruit bodies of *F. fomentarius* are found on dead trees and stumps (plate 58). They are hoof-shaped, broadly attached, up to 40 cm across, 50 cm long and 15 cm thick at the base. The upper surface is smooth, hard and concentrically zoned; its colour varies from grey or greyish brown to almost black. The flesh is light brown, firm to spongy and up to 3 cm thick. The tubes are up to 12 cm long, distinctly stratified and cinnamon-coloured. The pores measure 0.3 mm in diameter and the pore surface is beige in active growth, darkening to amber-brown. The hyaline spores are cylindrical to ellipsoid, thin-walled and measure 15–18 × 5 μm (Rea, 1922). Initial infection by *F. fomentarius* usually occurs through broken branches or other wounds on the stem. Decay in birch develops as a creamy white rot containing small black flecks. At an early stage the decayed area is surrounded by a narrow brown invasion zone. Sheets of pale creamy-coloured mycelium may be found in the cracks in the decayed wood.

In culture on malt agar a white, closely woven, tough, smooth, even mat is soon formed (Cartwright and Findlay, 1958). In the light the culture develops a light brownish coloration which darkens with age.

*F. fomentarius* is the commonest cause of decay in birch in parts of the highlands of Scotland. The flesh of the fruit body was once used as the basis of the tinder used in flint boxes and, hence, the common name 'tinder fungus' is sometimes applied to this species.

*Fomes fraxineus*: see *Perenniporia fraxinea*
*Fomes igniarius*: see *Phellinus igniarius*
*Fomes pini*: see *Phellinus pini*
*Fomes pomaceus*: see *Phellinus pomaceus*
*Fomes ulmarius*: see *Rigidoporus ulmarius*
*Fomitopsis cytisina*: see *Perenniporia fraxinea*

**Ganoderma applanatum (Pers. ex Wallr.) Pat. (*Fomes applanatus* Pers. ex Wallr., *Polyporus applanatus* Pers. ex Fr.) and *G. adspersum* (Schulz.) Donk (*G. australe* Fr. Pat.): designer's mushroom**
*Ganoderma applanatum* is a cosmopolitan polypore that causes heart rot in many broadleaved species. Its fructifications may appear throughout the year

as large, often imbricate brackets on the trunks of the host trees. They are up to 30 cm across with a reddish-brown, lumpy upper surface covered by a crust which is soft when young but hard and laccate when old. The pores are small, at first white, but becoming brownish with age. The tubes are up to 10 mm long, reddish brown or cinnamon, and may be broken away from the flesh, which is brownish, thick and hard but felt-like. The basidiospores are cocoa-brown in the mass, measuring 6.5–8.5 × 5–6.5 μm (Ryvarden, 1976). The spores are copiously produced, and often coat the tops of lower fruit bodies as a brown dust. In Switzerland the undersurface of the fruit body is often covered with a honeycomb of nipple-shaped galls of the fungus fly *Agathomyia wankowiczi*.

In Britain, *G. applanatum* causes root and butt rot of many broadleaved trees. It is especially damaging to old, over-mature beeches, but is also the commonest cause of rot in standing poplars, and is frequent also in elms. It also attacks oak, sycamore, horse chestnut, willow and walnut (Cartwright and Findlay, 1958).

Affected wood at first shows a white mottle, and tends to break up into rectangular pieces; eventually the wood becomes white, soft and spongy. In living wood the rotted zone is surrounded by a black line (Cartwright and Findlay, 1958).

The fungus is mainly a wound parasite that enters particularly through wounds in stems and roots.

In the United States, this fungus is sometimes referred to as the designer's mushroom because beautiful drawings can be inscribed on the fresh pore surface with a pin or nail.

*G. adspersum* (*G. australe*) is very closely related to *G. applanatum* but has slightly larger spores measuring 9–11.5 × 6–8 μm (Ryvarden, 1976).

**Ganoderma australe: see Ganoderma adspersum**

**Ganoderma lucidum (Leyss.) Karst. (*Polyporus lucidum* Fr.)**
*Ganoderma lucidum* causes a root and butt rot in oak and other broadleaved species in Britain and has been found world-wide on a range of broadleaved hosts. Descriptions of the sporophores of *G. lucidum* are very similar to those for *G. resinaceum* and it is possible that some records of *G. resinaceum* in Europe may be comparable to those of *G. lucidum* in Asia and America. Cartwright and Findlay (1958) report that a closely related species or variety, *G. tsugae*, grows on coniferous stumps in America. Adaskaveg and Gilbertson (1986) discuss differences between *G. lucidum* and *G. tsugae* from N. America and also between *G. lucidum* and the European *G. resinaceum*.

The annual sporophores, which measure 5–28 cm across and up to 3 cm thick, are kidney-shaped, usually with a lateral stalk, and can be found at the base of trunks or on stumps. The upper surface is at first light yellow, becoming blood-red chestnut and varnished. The margin is white or yellow and the flesh at first white, becoming reddish, spongy, then corky or woody. The tubes are white, then cinnamon and measure 0.4–1.2 cm in length; the minute pores are white, discolouring with age. The brown spores are elliptical, truncate at the base and measure 10–12 × 6–8 μm with a hyaline germination pore (Rea, 1922).

The fruit bodies are similar in many respects to those of *G. resinaceum* but *G. lucidum* is generally smaller in size and has a stalk.

In India, Bakshi *et al.* (1976) describe a root rot of *Acaciacatechu* Willd. caused by *G. lucidum*. Trees, both young and old, were killed as a result of attack by *G. lucidum* and infection could be traced back to root contacts with stumps and roots of the previous forest cover. Toole (1966) observed root infections by *G. lucidum* on several hardwood species in the Mississippi Delta and by inoculation experiments proved it pathogenic on *Albizia julibrissin* and *Liquidambar styraciflua*.

*G. lucidum* causes a white rot which does not extend far up the trunk and appears less active as a decay agent than other *Ganoderma* spp.

### Ganoderma pfeifferi Bres.

*Ganoderma pfeifferi* occurs throughout Europe; in Britain it is found mainly on beech but elsewhere in Europe it has been reported on oak.

The perennial fruit bodies are found on trunks and stumps. They are flattened or hoof-shaped brackets 30 cm or more across, 25 cm wide and 7–8 cm thick, sometimes imbricate. The upper surface is sulcate, chestnut-brown and coated with a varnished resinous crust which is sometimes covered with a brown deposit of spores. The resinous surface can easily be melted in a match flame and will on cooling become smooth and glossy (Ryvarden, 1976). The flesh is corky to firm in texture, up to 4 cm thick, dark brown in colour and zoned towards the margin. The tubes are also dark brown, 1 cm or more in length and usually stratified. The margin and lower surface are white when in active growth, becoming resinous and yellow, later turning brown but retaining yellow tints in places. The similar colour of the flesh and tubes is a feature which can be used to distinguish *G. pfeifferi* from *G. resinaceum*. The pores are small and round, 5–6 per mm. The light-brown spores are ellipsoid, verrucose and measure 9–11 × 6–9 μm (Ryvarden, 1976).

Decayed beech wood has a mottled yellow-brown appearance caused by the presence of small straw-coloured pockets of decay among darker brown areas of relatively sound wood. In Britain *G. pfeifferi* can cause severe root, stem and branch decay in beech. The fungus can extend out from the central wood of live trees into the cambial region, killing extensive areas of bark and ultimately killing the whole tree. Young and Strouts (1977) report that 15 beech trees in a park were seriously attacked by *G. pfeifferi*, while 59 neighbouring oaks were apparently unaffected.

### Ganoderma resinaceum Boud. ex Pat.

*Ganoderma resinaceum* is considered to be uncommon in Britain (Rea, 1922) but a number of cases of severe decay in oak caused by this species have been reported (figure 100) (Young and Strouts, personal communication). It has also been recorded elsewhere in Europe on oak, beech, willow and plane.

The sporophores (plate 61), which measure 10–45 cm across and 10 cm

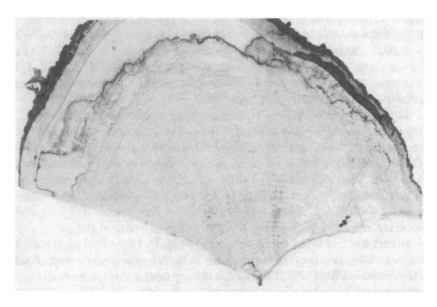

**Figure 100**   Decay in oak caused by *Ganoderma resinaceum* (Forestry Commission)

thick, are sessile (sometimes with a rudimentary stipe), semicircular, concentrically grooved with wide primary furrows and occur at the base of trunks
and on stumps. The upper surface is at first yellow and viscid, then turning
blood-red to chestnut-brown and varnished, although at this stage it may be
covered with a brownish deposit of spores. The margin is white and the flesh
pale cinnamon, thick, soft and corky. The tubes are dark cinnamon, 0.5–3 cm
long and the pores at first white, becoming dark cinnamon, 0.4–0.5 mm in
diameter. The spores are fuscous, ovate, oblong or obovate truncate at the
base, 10–12 × 6–8 μm. The fruit body of *G. resinaceum* is similar to that of *G.
lucidum* (Rea, 1922), though the latter generally has a stalk and is smaller.

Bourdot and Galzin (1927) describe *G. resinaceum* as a 'dangerous parasite
whose decay affects the heartwood; under the action of the mycelium the
wood melts away, so to speak, and is replaced by a mycelial felt'. It causes a
more serious decay on oak than *G. lucidum* (Cartwright and Findlay, 1958).

### *Grifola frondosa* (Dicks. ex Fr.) S. F. Gray (*Polyporus frondosus* Dicks. ex Fr.): Japanese 'maitake'

*Grifola frondosa* is not uncommon in Britain and the rest of Europe on oak
and hornbeam and is also found in North America on a number of hardwoods
including oak and elm and occasionally on coniferous species. Fruit bodies
occur on stumps and roots and at the base of the trunk.

The sporophore is stipitate, the stem branching many times and giving rise
to numerous overlapping pilei (Overholts, 1953). It occurs at the bases of
trees, on roots and on stumps. The overall size may vary from 15 to 30 cm; the
individual pileus measures 2–6 cm across and the stipe 10–30 × 5–10 cm. The

upper surface is rough and greyish and the margin is white, spatulate, lobed and intricately relobed. The flesh is at first white and fleshy, not more than 5 mm thick and becomes discoloured and firm with age. The tubes are a similar colour, decurrent and 0.5 mm long. The pores are also white, 0.3–1.0 mm diameter, round at first and later becoming angular. The hyaline spores are subglobose and measure 6 × 5 μm (Rea, 1922).

Cartwright (1940) described various stages of butt rot in oak caused by *G. frondosa*. The first signs of decay are streaks or pockets of white rot surrounded by a water-soaked, reddish-brown invasion zone. Greig and Gulliver (1976) noted the presence of a characteristic bright-orange zonation in association with the white pocket rot. At a very advanced stage the wood is severely decayed and only a soft, spongy pulp remains interspersed with white mycelium. Young and Strouts (1979) describe a root system of *Quercus robur* severely infected by *G. frondosa* but showing no decay at the butt.

In culture the fungus grows moderately rapidly (8 cm in 3 to 4 weeks) as a dense, white, raised mycelial mat with irregular yellow and chamois patches (Greig and Gulliver, 1976). Numerous clamp connections, sometimes arched, and globose to ovoid chlamydospores are present in culture. The cultures often have a characteristic 'carbide' smell.

*G. frondosa* can cause a very severe butt rot in oaks in Britain making them unsafe and also markedly reducing the value of any timber which might be harvested. However, the fungus has so far not been found widely there.

During the 1980s, the Japanese discovered that maitake (*G. frondosa*) was a source of grifolan, an antitumour chemical (e.g. Hishida *et al.*, 1988) and of a substance with blood pressure-lowering activity (Adachi *et al.*, 1988).

### *Heterobasidion annosum*: see chapter 3

### *Hypholoma fasciculare* (Huds. ex Fr.) Kummer: sulphur tuft
*Hypholoma fasciculare* is commonly found fruiting in clusters on broadleaved and coniferous stumps and less commonly as a cause of decay, e.g. in larch, Norway and Sitka spruce and western red cedar (Greig, 1981).

The cap of the sporophore is 2–7 cm across, convex with remains of the pale-yellow veil often adhering to the edge and is bright yellow with orange tints towards the centre. The stem is 40–100 × 5–10 mm, often curved, often with a faint ring. The flesh and gills are both sulphur-yellow, the gills later becoming greenish, and the oval spores measure 6–7 × 4–4.5 μm. The fruit bodies occur in dense clusters throughout the year. They have a very bitter taste and are not edible.

The decay is a soft, fibrous, yellow-brown rot, somewhat similar to that caused by *H. annosum*.

### *Inonotus dryadeus* (Pers. ex Fr.) Murr. (*Polyporus dryadeus* Pers. ex Fr.)
*Inonotus dryadeus* is widespread and fairly common on oak in Britain and Europe and has occasionally been found in North America. Burdekin (1977) reported that it was observed in the trunks of oaks broken following a severe gale. It has once been recorded on beech in Britain (Strouts *et al.*, 1984).

The annual sporophore (plate 59) can be seen from May to December usually at the base of oak trees or on their stumps. It is a thick, sessile bracket, semicircular in shape, measuring from 7 to 30 cm across and up to 8 cm thick, and is sometimes found in overlapping groups. The upper surface is thin, soft and yellowish at first, then turning light brown. Small, round depressions at the margin of an actively growing fruit body are often filled with a very astringent watery exudate. The flesh is ferruginous in colour, soft at first and later becoming corky. The tubes are similarly coloured, 10–30 mm in length and the pores are whitish, 0.2–0.4 mm across. The spores are hyaline or pale yellow, subglobose, 1-guttulate and measure 6–8 × 6–7 μm (Rea, 1922). The setae are brown, thick-walled, irregular, sickle- or hook-shaped.

Fruit bodies are not regular in their occurrence and several years may elapse between the appearance of consecutive sporophores on an infected tree. They often appear in early autumn and decompose during the winter.

At an early stage, infection by *I. dryadeus* causes a dark-brown, water-soaked area in which appear irregular stripes of a yellowish colour (Cartwright and Findlay, 1958). These gradually expand and the decayed wood becomes lighter in colour, finally becoming a soft, white filamentous mass. The severely decayed wood is broken up into sections by sheets of white mycelium. Decay is largely confined to the heartwood.

In culture growth is very slow; at first it appears as a whitish, furry mycelium tinged with yellow, later becoming yellow and darkening to brown.

Decay is confined to the roots and the butt region and rarely spreads more than 2 metres up the stem. Although it may not cause serious loss of timber it is important because the decay at the base of the stem and in the roots renders oaks susceptible to windthrow.

### *Inonotus hispidus* (Bull. ex Fr.) Karst. (*Polyporus hispidus* Bull. ex Fr.)

*Inonotus hispidus* is the most frequent cause of decay in ash in Britain; it is also found on apple and plane and occasionally on sycamore, walnut and elm. It has been reported from Europe, America, Asia and Australia. In North America it has been found on a range of hosts including ash, oak, willow and walnut.

The annual sporophore (plate 60) can be found in Britain from May to February attached to the main stem or branches of the host tree. In the United States, in Mississippi, the fruit bodies appear from July to October, the majority in September and October. They reach full size in 1 to 2 weeks and discharge their spores for 2 to 4 weeks (McCracken and Toole, 1969). They are semicircular in shape, 6–35 cm across and 2–10 cm deep. The upper surface is at first orange-rust in colour and tomentose, darkening to black and becoming hispid with age. The tubes are similarly coloured, 0.5–4.0 cm long and they open into fine pores, 0.25–0.40 mm in diameter. The yellowish, subglobose spores measure 8.3–11 × 7–9 μm (CMI Descr. F 193); brown setae are sometimes present. The black fruit bodies may remain attached to the host during the winter months although they often fall and gradually disintegrate on the ground at the base of the tree.

Infection usually occurs through a branch wound or stub. Nutman (1929) observed that although *I. hispidus* was usually saprophytic in the heartwood, in inoculation experiments it could penetrate and grow slowly in the living wood. Toole (1955) found elongated cankers associated with infection by *I. hispidus* on oak and reported that once well-established in the heartwood it grows out through the sapwood to the cambium. Fruit bodies appeared on the surface of the cankers. Similar observations have been made on ash in Britain although the narrow, elongated cankers associated with the branch wounds are relatively inconspicuous.

The decay (figure 101) shows at first as white or yellow tongues of discoloration, limited by a brown zone resulting from the formation of a gummy material (Cartwright and Findlay, 1958). In ash and walnut the more advanced decay is yellowish brown and spongy.

Cartwright *et al.* (1936) tested the strength of small sample blocks of ash and found that resistance to a sudden blow was reduced by 27 per cent two weeks after inoculation with *I. hispidus* and by 90 per cent after twelve weeks.

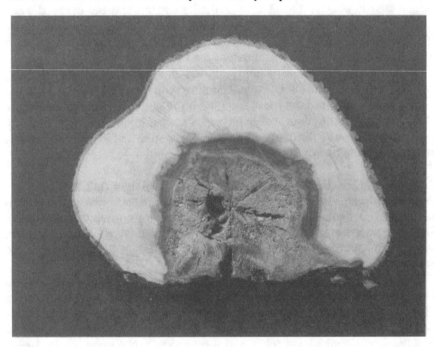

**Figure 101**   Decay in *Acer* caused by *Inonotus hispidus* (Forestry Commission)

In culture *I. hispidus* grows readily but rather slowly. On malt agar it forms a thick mat, broadly zoned and coloured yellow turning to brown. No clamp connections and no secondary spores have been observed in culture.

Burdekin (1977) reported many cases of branch and stem breakage in hedgerow ash in Britain where decay from *I. hispidus* was evident. As ash timber is used for wooden handles for a wide variety of tools and sports equipment, even slight decay could render the implement dangerous and infected wood must be rejected.

### *Inonotus obliquus* (Pers.) Pilat (*Poria obliqua* (Pers.) Bres.): Russian 'Tchaga'

*Inonotus obliquus* is common on birch in certain countries in Europe, including parts of Britain, Russia, Sweden and Yugoslavia, and in North America. In Britain its distribution appears to be limited largely to parts of northeast Scotland although there are occasional records from other parts of northern England and Scotland (Batko, 1950).

*I. obliquus* frequently produces a sterile sporophore and its true identity was first discovered by Campbell and Davidson (1938) in North America. They demonstrated that the sterile form was associated with a fertile fruit body and that cultures derived from these two sources were identical. Reid (1976) found the fertile form of *I. obliquus* for the first time in Britain in 1975.

The sterile sporophore or conk is produced on trunks of live birch trees and it appears as a black outgrowth varying in size from 5 to 20 cm diameter with a very irregular and fissured surface. It closely resembles a partly burnt piece of coal or wood. The texture of the sterile sporophore is hard and woody and when broken the internal surfaces are rusty brown, somewhat granular and often mottled with white or cream streaks leading towards the point of attachment. No spores, either conidia or basidiospores, are produced on these structures and their function, if any, remains unknown.

In contrast to the sterile sporophores, the fertile fructification has only rarely been observed and then only beneath the bark of dead birch trees. The fruit bodies are resupinate, poria-like fructifications. They may extend over several metres and at intervals horizontal outgrowths of the fungus, measuring up to 2 cm across, develop beneath the bark and produce a pore surface on their underside. These structures may be up to 10 cm in length and they resemble pilei but there is no differentiated surface layer. The upper surface and context are golden brown, the tubes are up to 6 mm long and the pores, 2–4 per mm, are covered with a creamy to buff bloom. The basidiospores are broadly elliptic, hyaline becoming faintly coloured, and measure $8.0–9.2 \times 5.2–5.75$ μm (Reid, 1976).

Cultures of the fungus on malt agar develop a pore surface which produces spores of a similar size.

*I. obliquus* causes a white rot in which dark brownish-black zones may be found. The annual rings of decayed wood may separate from one another and in the later stages a weft of mycelium replaces the medullary rays (Cartwright and Findlay, 1958).

*I. obliquus* may cause a serious heart rot in birch leading to breakage of stems, often where the sterile fruit body is found.

In Russia, it is called 'Tchaga' and an infusion from the sterile conk has been used for medicinal purposes since the sixteenth and seventeenth centuries (Reid, 1976). It is said to relieve pain and has been used in the treatment of the early stages of cancer.

***Inonotus radiatus* (Sow. ex Fr.) Karst. (*Polyporus radiatus* Sow. ex Fr.)**
*Inonotus radiatus* is commonly found in Europe and America and occurs frequently on alder, occasionally on birch, willow and poplar.

The fruit body is often found on the trunks of dead trees, and is annual, developing between September and April. It is a thick short bracket, 2–6 cm across, and often imbricate. The upper surface is radially grooved, velvety becoming smooth and tawny or rust-brown in colour; the margin is golden yellow when young. The flesh is hard, fibrous and yellow to rust-brown; the tubes are a similar colour and measure 5–10 mm in length. The minute pores are at first silvery and glistening, finally turning rust-brown. The hyaline spores are subglobose, measuring 4.5–7 × 3–4.5 μm (Pegler, 1973), and the dark-brown setae are both fusiform and hooked, measuring 20–30 × 5–8 μm.

Decay in alder caused by *I. radiatus* is whitish or pale biscuit-coloured and breaks with a flaky fracture (Cartwright and Findlay, 1958).

In culture on malt agar, the fungus forms a velvety, slightly tufted, compact mat, which eventually has a somewhat powdery appearance. The culture develops various yellow-brown tints, particularly when grown in the dark. The submerged mycelium and surrounding medium are rich dark brown (Cartwright and Findlay, 1958).

*I. radiatus* is the most important cause of decay in alder in Britain.

***Laetiporus sulphureus* (Bull. ex Fr.) Murr. (*Polyporus sulphureus* Bull. ex Fr.): sulphur polypore, chicken of the woods**
*Laetiporus sulphureus* occurs commonly in both Europe and America, attacking both hardwood and coniferous species. In Britain it is probably the most important decay fungus found in old standing oaks; it has also been found on sweet chestnut, yew, cherry and sometimes on pine, larch and beech.

The annual sporophores (plate 64) appear from May to November on the main stem and often at the site of a wound. They occur in overlapping groups which measure 10–40 cm across. The individual parts are thin and resupinate and have a wavy margin. The upper surface is a brightly coloured orange and the flesh is soft and cheesy. The lower surface is a pale sulphur-yellow at first but colours on both surfaces fade with time. Young fructifications often exude a pale yellow juice if bruised. The pores are small, 0.3–0.8 mm, sometimes absent and the tubes short, 0.5 mm. The spores are white, elliptical and measure 7–8 × 5 μm (Rea, 1922). Reid (1985) refers to an imperfect stage, *Ptychogaster* (*Ceriomyces*) *aurantiacus* (Pat.) Sacc. found on oak stumps.

The first visible signs of infection are a yellow or reddish discoloration in the wood (Cartwright and Findlay, 1958). At a later stage the wood becomes a dark reddish brown and breaks up into cubical blocks (figures 102, 103). The radial cracks between the blocks may be filled with sheets of a pale yellowish mycelium, 2–4 mm thick, which resembles chamois leather in colour and texture.

In culture, *L. sulphureus* grows moderately rapidly; it is white to pale buff in colour, sparse and powdery. Numerous spores are produced in culture; they are thin-walled, oval to round and measure 5–9 × 5–7 μm (Greig and

**Figure 102**   Decay in oak caused by *Laetiporus sulphureus*, showing the wood breaking up into brown, cubical blocks (D.H. Phillips)

Gulliver, 1976). Terminal and intercalary thick-walled chlamydospores are
often present.

   *L. sulphureus* is commonly found causing serious decay in open-grown oaks
and also in larger structural timbers. Even small traces of infection can
continue their development in sawn timber, particularly if the wood remains
moist. Cartwright and Findlay (1958) consider that *L. sulphureus* can only
enter when the heartwood is exposed, as, for example, when a large branch is
broken. Greig and Gulliver (1976) frequently observed sporophores of *L.
sulphureus* on branch stubs of oak but report that in some cases the decay was
most extensive at ground level and likely to have entered through the roots.

### *Lentinus lepideus* Fr.

*Lentinus lepideus* is mainly important as a cause of decay in felled timber such
as telegraph poles but it has also been found causing a brown cubical rot in
Scots pine (Greig, 1981).

   The cap is up to 10 cm in diameter, pale yellow with reddish-brown scales
and having a serrate edge to the whitish gills. The stipe is 2–7 cm long, 2.5 cm
thick and white.

**Figure 103**  Decay in oak caused by *Laetiporus sulphureus*: final break-up of the wood into separate blocks. The white patches and streaks on the wood blocks are the yellow mycelial sheets of the fungus (D.H. Phillips)

### *Meripilus giganteus* (Pers.) P. Karst. (*Polyporus giganteus* (Pers.) Fr.)

*Meripilus giganteus* is widely distributed throughout Europe and has been reported from North America (Cartwright and Findlay, 1958). In Britain it is found commonly on beech and also on oak and robinia.

The fruit body is annual (from July to January) and consists of a rounded mass of imbricated pilei which may be 30 cm or more across and can weigh up to 12 kg. It develops either on stumps or on decayed roots near the base of the trunk. The pileus when fresh is brownish yellow to chestnut brown on top with a yellowish margin. The upper surface is somewhat granular and fibrillose, the flesh is white, and the lower surface, with its angular or irregular pores, is creamy white. The lower surface darkens when touched, as does the whole fruit body with age. The fruit body usually deteriorates into a black, slimy mass following the first severe autumn frost. The tubes are 4–6 mm long and the spores are hyaline, subglobose or broadly elliptical and measure 6 × 5 μm (Cartwright and Findlay, 1958).

*M. giganteus* causes a decay in the roots and stumps particularly of beech (figure 104). Decayed roots are at first generally brown with small areas apparently filled with white mycelium; these gradually enlarge and the roots may subsequently become hollow.

In culture the fungus produces a soft, white, cotton-woolly growth; mycelial hummocks may develop and also tubes bearing spores. The mycelium becomes more leathery and finally tawny or cinnamon-brown patches appear.

Cartwright and Findlay (1958) suggest that it is usually a pathogen of secondary importance but more recently Young and Strouts (1974) have found it commonly associated with windblown beech and sometimes with

**Figure 104** Decay in beech root caused by *Meripilus giganteus* (Forestry Commission)

large trees showing thin and sparse foliage. Such trees may have severely decayed roots though death and decay of roots may be limited to the smaller roots at a depth of 45 cm or more.

### *Perenniporia fraxinea* (Fr.) Ryv. (*Fomitopsis cytisina* (Berk.) Bond. and Sing., *Fomes fraxineus* (Bull.) Fr.)

*Perenniporia fraxinea* is comparatively rare in Britain and in the rest of Europe and North America. The most important hosts in Britain are ash, robinia and laburnum but it also occurs on elms, poplar and beech. It is found on these host species throughout Europe; in the United States it has been found on other hardwoods, including oak and maple.

The bracket-shaped fruit body measures 5–40 cm across, is sometimes imbricate and is found at the base of infected trees or on stumps. The upper surface is initially whitish but darkens with age to become fuscous and then dark brown or black, sometimes greenish with algae. The flesh is soft and yellowish at first but soon turns hard and woody. The tubes, 5–25 mm in length, are similar in colour to the flesh, in contrast to the otherwise rather similar fruit body of *Rigidoporus ulmarius*, where the tubes and flesh differ in colour. The pores are small, 0.25 mm in diameter, and pinkish brown. The hyaline spores are subglobose and measure 6–7 μm (Rea, 1922). Decay is usually restricted to the basal part of the main trunk. At an early stage, infected wood tends to break readily across the grain; later the wood decomposes into a felt-like mass of white mycelium.

Montgomery (1936) and Campbell (1938) have described the fungus in culture. Growth starts with fine radiating hyphae appressed to the surface of the culture medium. A white felted mat develops on 2 per cent malt agar and pore surfaces may form over an extensive area. Normal basidiospores are produced on these surfaces. A pale buff coloration may be seen in the centre of the culture but this tends to lighten and become creamy-white in older cultures. *P. fraxinea* may cause a severe butt rot in ash and other species but it is not of great significance because of its limited occurrence.

### *Phaeolus schweinitzii* (Fr.) Pat. (*Polyporus schweinitzii* Fr.): velvet top

*Phaeolus schweinitzii* is widely distributed, and is known to occur on many coniferous species throughout Europe and North America, parts of Central America and the West Indies, Japan and a few other areas in East Asia, North and South Africa, New Zealand, and perhaps Australia and South America (CMI Map No. 182). As a cause of loss it is far less important than *Heterobasidion annosum* (*Fomes annosus*), though it may cause considerable local damage, and was one of the more important fungi found by Peace (1938) in his butt-rot survey. In Britain it does not kill trees, but gives rise to an extensive rot in the tree stem.

It is found mainly in mature trees fifty or more years old (Peace, 1938; Greig, 1962), though this is often because the infection is not discovered until the tree is felled or windblown. In fact it may be already present in the stem when trees are no more than 12 to 15 years of age (Greig, unpublished). It occurs chiefly in conifer plantations following hardwoods or in second rotation conifers after pines (Greig, 1962, 1981).

The sporophores of the fungus arise annually in summer or early autumn (plate 62). They are most often found on old thinning stumps or inside the remains of wind-shattered trees. They also occur on standing infected trees (sometimes on the sites of fire scars or other wounds), or on the ground around them, often several feet from their bases. Those on the ground (apparently on the litter surface, but in fact attached below to the root system by thick, well-marked mycelial cords) are toadstool-like, with a thick, brown, velvety stalk attenuated below, and a cap about 30 cm across. Fruit bodies growing on tree trunks are bracket-like and unstalked. The upper surface of the fructification is bright brown, becoming a deep rusty-brown, with a coarsely velvety texture that gives the fungus its popular name of 'velvet top'. The rim is thick and bright yellow, and the underside is greyish green, with irregular angular pores. The pore surface is water-soaked when fresh, and turns brown when bruised. The basidiospores are white, elliptical to oval, 7–8 × 3–4 μm. The flesh is spongy-fibrous, and at first yellowish brown, but later a deep rusty brown. The sporophores remain as described for from 2 to 3 months, but then die, and become dark brown, corky and light in weight. It is in this condition that they are most often seen, remaining many months before decomposing, so that they often occur alongside those of the current year (Gill, 1925; Wakefield and Dennis, 1981; Gladman and Greig, 1965; Barrett, 1968).

*P. schweinitzii* occurs as many strains, which vary in their cultural characters and growth rate (Childs, 1937; Barrett, 1968). Typically in culture it produces velvety cushions, at first greenish yellow, later bright yellow, becoming reddish brown, and smelling of aniseed (Cartwright and Findlay, 1958).

## Mode of entry

It is usually considered that the fungus enters the tree through damaged roots and wounds on the stem. York *et al.* (1936) showed experimentally that it could enter the roots of seedlings of *Pinus strobus*. Gremmen (1961b) and Greig (1962) both found the fungus in tree root systems, and the former traced it into the stem from main roots which it appeared to have killed. Most tree infections in Britain originate from the roots, particularly the tap roots. Barrett (1968) did pot experiments which showed that *P. schweinitzii* could invade damaged roots of young conifers, though in his trials it did not spread far beyond the damaged area. In the field he studied infested root systems, and concluded that infection took place only after injury by other agents, such as adverse physical conditions or *Armillaria mellea*. Barrett (1970) also found in laboratory experiments that growth of *P. schweinitzii* through sterilised wood colonised by *A. mellea* was not inhibited and indeed might have been stimulated. These studies were reinforced by Barrett and Greig (1984), who provided further evidence of the predisposition to attack as a result of *Armillaria* infection.

Studies by Hepting and Chapman (1938) in shortleaf and loblolly pines in the USA indicated that the fungus sometimes invaded trees through the roots, but it usually gained entry through basal wounds, particularly those caused by fires. They noted that in Arkansas, where there was little fire damage, few trees were infested by *P. schweinitzii*, but in Texas fire damage was more extensive, and also more trees were attacked by the fungus. Acording to Harvey (1962), in young Douglas fir in the USA infection takes place mainly through the roots, but most damage is found in older trees, in which entry is through fire scars. In Britain, Greig (1962) obtained some evidence that infection of 60-year-old pines had taken place through fire scars.

Greig made other observations that suggested occasional entry through wounds made by snow break, singling of double leaders, and in extraction, though this fungus was not among those found by Pawsey and Gladman (1965) in their survey of extraction damage. Boyce (1924) recorded entry through an old knot.

Barrett (1985) showed that basidiospores of *P. schweinitzii* are effective sources of a persistent saprophytic infestation in the soil. Further studies by Barrett and Greig (1985) indicated that *P. schweinitzii* was present most frequently in soils from ex-broadleaved woodland sites which were subsequently planted with Sitka spruce. However, it was also present in soils of

Sitka spruce plantations without a previous broadleaved history and where no disease had been observed. Hence, the potential for infection exists on non-woodland as well as previously wooded sites.

### Host range
*P. schweinitzii* occurs on most of the commoner conifer species, and was found by Rhoads (1921) on charred stumps of *Eucalyptus globosus*. In Britain it has been noted especially on Sitka spruce and Douglas fir, but it also attacks Scots and Corsican pine and larches; it has also been recorded there on lodgepole pine, *Pinus radiata*, *Picea omorika* and *Cedrus* spp. Elsewhere it has also been noted on southern pines (Hepting and Chapman, 1938), *Pinus ponderosa* (Gill, 1925), *P. strobus* (York, *et al.*, 1936) and *Chamaecyparis* spp. (Yde-Andersen, 1961).

### Symptoms of decay
Incipient decay may be present in trees that appear quite healthy, and the first sign of attack may be the presence of fructifications on or around the bases of the affected stems (Gremmen, 1961b). Even at an early stage, the wood becomes weak and useless (Kimmey and Bynum, 1961). At first the fungus stains the wood a uniform pale yellow or yellowish brown. Later the wood becomes dry, soft and crumbly. It soon begins to show radial cracks, and breaks up into irregularly cubical chunks, sometimes with thin sheets of white or yellowish mycelium between (figure 105). The rotten chunks finally become dark brown in colour and very light in weight, and can be readily picked out if the tree is felled, or shattered by wind (Greig, 1981). The infected wood smells strongly of aniseed, and in the later stages of rot can be crushed to a smooth powder between the fingers (Harvey, 1962).

### Damage caused
The damage caused to individual trees is often severe, as the rot may progress for several metres up the stem, and affect most of its diameter. In Britain decay may reach a height of 2 to 3 metres or more, particularly in Sitka spruce, and because of weakening of the wood in the earlier stages, and later its complete breakdown into separate pieces, windsnap often follows infection (Gladman and Low, 1963). Losses overall are not great, however, because in spite of the high loss of timber in each affected tree, the fungus is only sporadic in its distribution (Gladman and Greig, 1965).

Measures for its control are not yet available; infested sites should not be replanted with susceptible species.

### Phellinus igniarius (L. ex Fr.) Quel. (*Fomes igniarius* (Linn.) Fr.)
*Phellinus igniarius* is commonly found in Europe and America and has been reported from India and Australia. It generally occurs in Britain on willow but has occasionally been found on ash. Rea (1922) describes a form of this fungus, *F. igniarius* var. *nigricans*, occurring on birch. In America, Verrall (1937) found three different types of sporophore of *P. igniarius* occurring on

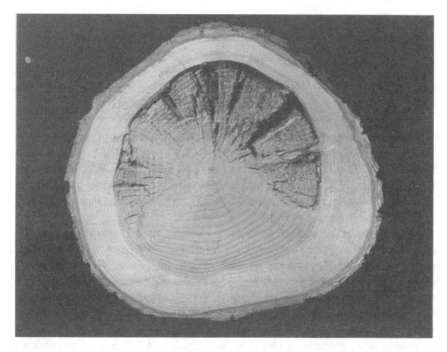

**Figure 105** Cubical rot in Douglas fir caused by *Phaeolus schweinitzii* (Forestry Commission)

aspen, birch and other hosts, respectively, which Ryvarden (1976) distinguishes as three closely related but separate species, *P. populicola*, *P. nigricans* and *P. igniarius*. Breitbach and Kranzlin (1986) report it commonly found on apples and also on willow, *Sorbus* and *Alnus*.

The perennial fruit body of *P. igniarius* is hoof-shaped or a thick, broadly attached bracket measuring up to 20 cm across. The upper surface is greyish black, rough, often cracked, and the margin is fawn when in active growth. The flesh is dark brown and hard, the stratified tubes are cinnamon, 2–8 mm long and the pores are 0.25 mm in diameter, yellowish or greyish brown. The spores are hyaline, subglobose and measure 5–6 × 4–5 μm (Bourdot and Galzin, 1927). Dark-coloured pointed setae measuring about 20 μm in length are present but sometimes rather scarce.

The fungus invades the tree through wounds, particularly dead branch stubs. A central heart rot develops which is soft and yellowish white with a distinct dark zone separating decayed from healthy wood. At a later stage fine black zone lines are formed in the spongy decayed wood (Ohman and Kessler, 1964).

Cartwright and Findlay (1958) describe the growth of *P. igniarius* in culture as thin, sparse, with appressed hyphae later developing into a smooth, brown, felted mat with only traces of zonation. Verrall (1937), as noted above, found considerable variation in isolates of *P. igniarius* in the USA and distinguished three main groups which more or less corresponded with three different types of sporophore.

*P. igniarius* is a fairly common cause of decay in willow in Britain but it assumes far greater significance on poplars and birches elsewhere in Europe and in the USA. Ohman and Kessler (1964) claim that *P. igniarius* is the most serious cause of decay in broadleaved species in the USA.

### Phellinus pini (Thore) Pilat (*Fomes pini* (Thore) P. Karst., *Trametes pini* (Thore) Fr.)

*Phellinus pini* is one of the most serious heart-rotting fungi in conifers in North America; it is commonly found in Northern Europe and also reported from Asia. It is very rare in England (Strouts, 1981) and occasionally found in Scotland, always on old pine trees.

Sporophores often appear on branch stubs or on the site of old injuries. They are often perennial and are variable in shape, from a thin bracket to a more ungulate form, ranging in size from 5 to 15 cm across. The upper surface is rough, becoming encrusted with age and is at first rusty brown, darkening to almost black. The flesh is tawny-ferruginous, tough and woody. The tubes are often stratified and are a similar colour to the flesh; the pores vary in size from 0.16 to 0.4 mm and may be round or oblong in cross-section. The spores are pale yellow, maturing to light brown, 5.5–6 × 4.5–5.5 μm (Bourdot and Galzin, 1927). Pointed, conical, dark-brown setae are present, measuring 40–65 × 6–10 μm.

Initial infection takes place through branch stubs where heartwood is present and therefore is largely restricted to older trees. In the early stage of fungal invasion, sometimes known as red-heart, a pinkish or purplish-red discoloration occurs in ring- or crescent-shaped areas. There is often an abundance of resin present. At a later stage a white pocket rot develops, often in rings around the stem separated by apparently healthy wood. This type of decay is often described as a ring scale (figure 106) and it is usually found in the upper parts of the tree where the original infection occurred (Cartwright and Findlay, 1958). Blanchette (1980) has studied this white pocket rot with a scanning electron microscope and reviews its use for biological pulping and for livestock feed production.

### Phellinus pomaceus (Pers.) Maire (*Fomes pomaceus* (Pers.) Lloyd)

*Phellinus pomaceus* occurs throughout Europe and America and has been recorded in Australia and Japan. In Britain it is found on plum, cherry, hawthorn and other Rosaceae and on poplar. It has been found on a range of ornamental species in Italy (Bisiac and Minervini, 1980), including *Prunus* spp. *Acacia umbrellifera*, *Fagus sylvatica purpurea* and *Platanus* spp. It is closely related to *Phellinus igniarius*, of which some authorities (for example, Bourdot and Galzin, 1927) classify *P. pomaceus* as a subspecies. The fruit bodies are found on live or dead stems and branches. They are hard and woody, either hoof-shaped or resupinate or semiresupinate and measure about 5 cm across. The upper surface, where present, has concentric furrows

**Figure 106**   Ring scale in pine caused by *Phellinus pini* (B.J.W. Greig)

and is ash or brownish grey. In active growth the rounded margin is grey and velvety, later becoming cinnamon. The flesh is light brown, the tubes are cinnamon, 4–6 mm long and stratified in older specimens. The pores are brown, with a greyish bloom at first, and measure about 0.2 mm in diameter. The pale-brown or hyaline spores are almost spherical and measure 6 × 5–6 μm (Rea, 1922). Flask-shaped cystidia are abundant, dark brown at the base and hyaline at the apex, and they measure 15–20 × 7–8 μm (Rea, 1922).

Decay caused by *P. pomaceus* is commonly found on old plum trees. At an advanced stage of decay the wood in the centre of the stem is white and crumbling. A dark purplish zone, 1–2 cm of apparently sound wood, surrounds the decayed region.

In culture on malt agar the mycelium is at first white but in the light it soon develops colours ranging from buff to tawny. No clamp connections have been observed in culture but crystals are numerous in older cultures (Cartwright and Findlay, 1958). *P. pomaceus* may cause serious decay in old orchards. It appears to infect trees through natural or pruning wounds.

### Pholiota squarrosa (Müller) Fr.

*Pholiota squarrosa* is commonly found in Europe and America on a range of broadleaved species including aspen, poplar, apple, cherry and occasionally on conifers (Peace, 1938).

The fruit bodies develop from July to December and are often found in clusters at the base of the trunk. The cap is 3–12 cm across, convex then flattened and ochre-yellow to yellowish rust in colour (Wakefield and Dennis, 1981). The upper surface is crowded with reddish-brown recurved shaggy scales. The flesh is light yellow and the gills are yellowish when young, then pale rust-coloured, broadly adnate with a decurrent tooth. The stipe measures 6–20 × 1–2.5 cm, concolorous with the cap and similarly squarrose up to the ring. The ring is small, dark brown, often torn and attached to the upper part of the stipe. The brown spores are smooth, elliptical and measure 6–8 × 3.5–4 μm (Wakefield and Dennis, 1981). It is not edible.

In culture on Noble's medium after one month a golden-brown mat of mycelium has developed with a white advancing zone. Yellow strands of mycelium, covered with crystals, may be present near the inoculum plug and occasional conidia may be seen.

There is little information on the nature of the decay caused by *P. squarrosa*, though Peace (1938) mentions that it causes a soft brown rot in Norway spruce. Davidson *et al.* (1959) suggest that it may infect aspen poplars through the roots.

### Piptoporus betulinus (Bull. ex Fr.) P. Karst (*Polyporus betulinus* Fr.): birch polypore

*Piptoporus betulinus* occurs widely on birch in Europe, Asia and America.

The annual fruit bodies are frequently found on dead trees or logs but also develop on live stems and branches. They measure 5 to 30 cm across, are hoof- or kidney-shaped and are attached by a narrow base which sometimes forms a thick short stalk. The upper surface is rounded, pale brown in colour, smooth when young and cracking with age. The flesh is white, soft at first, then corky. The tubes are white, 2–8 mm in length and the pores, also white, are 0.3 mm in diameter. The hyaline cylindrical spores measure 4.4–5.5 × 1.4–2 μm (Macdonald, 1937).

Infection appears to occur largely through stem or branch wounds. At an advanced stage of decay the wood becomes reddish brown in colour, is brittle in texture and breaks up into rectangular blocks often separated by thin sheets of whitish mycelium. Macdonald (1937) observed the appearance of dark zone lines in wood blocks artificially inoculated with the fungus.

In culture, on potato dextrose agar, growth is at first thin but soon develops a cotton wool appearance (Macdonald, 1937). The mycelium is at first white, later becoming pale brown or faintly pink, and often exudes drops of clear or amber-coloured liquid. Numerous crystals are formed in the medium. Flat pore surfaces and roughly rounded sporophores may develop on drier parts of the culture medium.

*P. betulinus* is the commonest cause of decay in birch in England, though its place is taken by *Fomes fomentarius* in parts of the highlands of Scotland. Although *P. betulinus* is frequently found on dead birch it also occurs on live trees which have been damaged by branch or stem breakage or by fire.

## *Pleurotus ostreatus* (Jacq.) Kummer: the oyster fungus or oyster mushroom

*Pleurotus ostreatus* is commonly found in Europe and America and has also been reported from Africa, Asia and Australia. It occurs on a wide range of broadleaved species in Britain, particularly on beech and also horse chestnut and poplar. It has been reported on oak in the United States.

The annual sporophores often occur in clusters and usually appear on a wound on the trunk or at a branch stub in autumn, though they can develop at any time of year. They are thick, fleshy, fan- or shell-shaped and measure 7–13 cm across. The fruit bodies are often sessile but the stipe when present is 2–4 cm long and 2 cm thick. The upper surface is smooth and moist, at first almost black, soon becoming pale grey or fawn. The white flesh is soft when young but becomes firm at a later stage. The under surface consists of gills which are white, becoming yellow with age, decurrent and said by some authorities to anastomose at the base. This fungus is very variable and this is exemplified by the range of spore sizes quoted by various authors. Rea (1922) states that the spores measure 9–11 × 4.5–6 µm whereas Romagnesi (1962) records measurements of 7.2–10 × 3.2–4.2 µm. The fruit bodies are edible.

*P. ostreatus* causes a white flakey rot, decayed areas being separated from healthy wood by a narrow, dark invasion zone. Marked cracking across the grain has been observed in decayed poplar wood.

In culture on malt agar *P. ostreatus* grows rapidly as a white, woolly, felted mat which becomes rough and almost leathery and develops occasional creamy patches (Cartwright and Findlay, 1958). Small fruit bodies with pilei and long stipes usually appear though the pileus may be absent. If the fungus is cultured in the dark on bread the mycelium is at first white but later a central flesh-ochre-coloured patch develops. In the light, drops of a yellowish-red liquid are exuded when grown on this medium. These colours help to distinguish *P. ostreatus* from other common species of *Pleurotus* which remain white when cultured on bread.

This fungus can cause extensive decay in beech and other hardwood species and its development can be particularly rapid once it has gained entry.

## *Pleurotus ulmarius* (Bull.) Fr.

*Pleurotus ulmarius* is fairly common in Britain and Europe and is usually found on elm.

The annual fruit bodies are found from June to December, often on wounds on trunks of elm. The pileus is 6–20 cm across, more or less circular, fleshy, convex then plane, ochraceous and often cracked (Wakefield and Dennis, 1981). The near-central stipe is stout, measuring 5–11 cm long and 1.5–4 cm thick, white becoming tinged with yellow, firm, often curved upward, the base often downy. The flesh is white and tough. The gills are broad, pale ochraceous, adnate and somewhat crowded. The spores are globose, 5–6 μm in diameter (Rea, 1922). The fruit bodies are edible.

Cartwright and Findlay (1958) report that *P. ulmarius* causes a brown rot in the heartwood and that in such wood the annual rings of decayed portions become separated.

*Polyporus adustus*: see *Bjerkandera adusta*
*Polyporus albidus*: see *Tyromyces stipticus*
*Polyporus betulinus*: see *Piptoporus betulinus*
*Polyporus dryadeus*: see *Inonotus dryadeus*
*Polyporus frondosus*: see *Grifola frondosa*
*Polyporus giganteus*: see *Meripilus giganteus*
*Polyporus hispidus*: see *Inonotus hispidus*
*Polyporus lucidus*: see *Ganoderma lucidum*
*Polyporus radiatus*: see *Inonotus radiatus*
*Polyporus schweinitzii*: see *Phaeolus schweinitzii*

### *Polyporus squamosus* Huds.: saddle-back fungus, dryad's saddle

*Polyporus squamosus* is commonly found in Britain and other parts of Europe and has been reported from India, Australia and America, where it is described as rare. In Britain it occurs on a number of hardwoods, particularly elm and sycamore, but also on other *Acer* species, *Pyrus* spp. and walnut. It is probably the principal cause of stem rot in elm.

The annual, fan-shaped sporophores can be found from April to December usually on pruning wounds or branch scars fairly high in the tree. They are often imbricate and when fully grown vary in size from 10 to 60 cm across. The short, thick, lateral stalk is black at its base. The upper surface is pale fawn and has numerous brown appressed scales. The tubes are white, then yellowish, 5–10 mm long. The lower surface is pale cream and the angular pores are relatively large, measuring 1–2 mm across. The spores, which are produced in vast numbers, are hyaline, oblong and measure $10–12 \times 4–5$ μm (Rea, 1922).

Infection usually occurs through a branch wound and extensive decay may develop both up and down the heartwood. At an advanced stage the decay is spongy or stringy, and includes a mass of white mycelium. The stem may eventually become hollow and a tough brown stromatic sheet may line the cavity. Campbell and Munsun (1936) report the presence of black zone lines, particularly in decayed wood close to the sporophore. These authors describe the development of the dark, bladder-shaped cells which form the pseudo-sclerotial plates or zone lines.

In culture, *P. squamosus* grows relatively slowly and the mycelium tends to aggregate at the sides of agar slopes into groups of short papillae (Campbell and Munsun, 1936). A brownish skin forms over the hyaline mycelium and in some cases it has a dusty yellow appearance due to the presence of large numbers of ovoid oidia with average measurements of $18 \times 8$ μm (Cartwright and Findlay, 1958). Numerous rod-shaped crystals are formed in the medium, sometimes in clusters. Occasionally small fertile sporophores are formed but more frequently only the stipes develop; pore surfaces have also been observed.

Elm boughs are notoriously liable to breakage and although in some cases the wood may be free from decay, in others, heartrot, particularly that caused by *P. squamosus*, is often present. The young fruit body is said to be edible and when mature it has been used as a razor strop.

*Polyporus stipticus*: see *Tyromyces stipticus*
*Polyporus sulphureus*: see *Laetiporus sulphureus*
*Polyporus ulmarius*: see *Rigidoporus ulmarius*
*Polystictus versicolor*: see *Coriolus versicolor*

### *Rigidoporus ulmarius* (Sow. ex Fr.) Imazeki *(Fomes ulmarius* Sow., *Polyporus ulmarius* Sow.)

*Rigidoporus ulmarius* is commonly found in Britain and is rather less common elsewhere in Europe and absent from Scandinavia. It has also been reported from Asia and North Central America. Cartwright and Findlay (1958) considered that *Fomes geotropus* Cooke, found in southeastern USA, was closely related to and perhaps a distinct variety of *R. ulmarius*. Lowe (1957) critically studied these species and concluded that the American species was specifically identical with *R. ulmarius*. Lombard *et al.* (1960) point out that cultures of *R. ulmarius* previously used to describe the species in the United States were incorrectly identified and this had led to some confusion in nomenclature.

The most important host of *R. ulmarius* in Britain is elm but it has also been found on lime, plane, poplar and sycamore. In the USA it occurs in large, old hardwoods, especially *Magnolia grandifolia* L.

The perennial fruit bodies are bracket-shaped, measure from 7 to 30 cm across and are usually found at the base of trees or on stumps. The upper surface is white or yellowish with age, usually lumpy and sometimes the sporophore consists of irregular rounded masses. Green algae often grow on the surface of older fruit bodies. The flesh is at first white, tough and fibrous, becoming yellowish, hard and woody with age. The stratified tubes are cinnamon, up to 15 mm long, and the pores are a similar colour, measuring 0.25–0.4 mm across The spores are hyaline, smooth, more or less globose and measure 6–7 × 6 μm (Rea, 1922). The contrasting colours of the flesh and tubes in *R. ulmarius* afford a ready means of distinguishing this fungus from the rather similar *Perenniporia fraxinea* (*Fomitopsis cytisina*).

Decay is usually found at the base of the tree and at an advanced stage the wood is dark brown, friable and tends to break up into rectangular blocks. No thick mycelial sheets develop in the cracks between blocks.

The fungus makes moderate growth in culture and is difficult to isolate from the sporophores, even when active, owing to heavy contamination with bacteria. Growth on malt agar is loose and spidery; in places the hyphae become aggregated into small lumps which are white or fawn in colour.

*R. ulmarius* is one of the most important causes of butt rot in elm in Britain.

It is particularly common in parkland and hedgerows where trees have been damaged, for example following bark stripping by deer.

### *Sparassis crispa* Wulf. ex Fr. (*Sparassis radicata* Weir): cauliflower or brain fungus

*Sparassis crispa* is commonly found causing a root and butt rot in old pine, Sitka spruce, Douglas fir, larches and cedar in Europe and has also been reported from Japan. The same fungus is known as *S. radicata* in North America (Martin and Gilbertson, 1976), where it is found on Douglas fir, spruces, pines and larches. A closely related but distinct species, *S. laminosa* Fr., is found on oak (Delatour, 1975, Van Zanen, 1988) in Europe and North America.

The fleshy sporophores of *S. crispa* are annual, up to 30 cm across, cream or white in colour and grow annually in late summer at the base of living trees and on stumps. They form a dense mass, shaped like a cauliflower or sponge, with many flattened branches or lobes, the edges of which turn brown with age (figure 107). Spores are pale ochraceous in mass, individually hyaline, subglobose, 6–7 μm in diameter or elliptical, 6 × 4 μm (Rea, 1922).

**Figure 107**   Fruit body of *Sparassis crispa* growing on pine (D.H. Phillips)

In Britain *S. crispa* causes a brown cubical rot in conifers over 60 years of age, especially Scots pine (Greig, 1981). The decayed wood is dry, crumbly and light brown in colour and closely resembles that caused by *P. schweinitzii*. A possible distinction can be observed in the colour of the mycelium which develops in the radial cracks in advanced decay; that of *S. crispa* is always white, whereas the mycelium of *P. schweinitzii* is often yellow (Pawsey, 1971; Greig, 1981). Trees which are severely decayed frequently break during gale force winds, usually close to ground level.

Infection occurs through the root system. Control measures are rarely needed but where infection is widespread replanting with Sitka spruce should be avoided.

*Sparassis radicata*: see *Sparassis crispa*

### Stereum gausapatum Fr. (*Stereum spadiceum* Fr.)
*Stereum gausapatum* is commonly found in Europe and America on a range of broadleaved trees, especially on oak.

The fruit bodies may be found at any time of year on branches or stems, stumps and logs as a thin, leathery skin 5 cm or more across, often confluent, imbricate and either resupinate or reflexed. The upper surface is hirsute, greyish or brownish with a lobed margin which is white when in active growth. The hymenium is smooth, brown, darkening with age and when fresh bleeds on cutting or bruising. The hyaline spores are oblong, incurved and measure 7–8 × 4–5 μm (Rea, 1922). *S. gausapatum* is somewhat similar to *S. rugosum* (see below).

Cartwright and Findlay (1958) describe the decay in small branches as a soft, yellowish-white rot. In large branches, dark-brown streaks occur, in the middle of which are yellowish bands where the wood is completely decayed. The wood between decayed areas appears to be perfectly sound, though fungal hyphae may be present. Long pipes of rot may thus develop in stems or branches and the term 'pipe rot' is often used for this type of decay (figure 108). At an advanced stage of decay the whole branch may become light-coloured and rotten following coalescence of the decayed areas.

In culture on malt agar the mycelium is at first white but later a wood-brown mat develops around the inoculum plug. Considerable variation in the general appearance of the cultures has been noted (Cartwright and Findlay, 1958).

*S. gausapatum* is an important cause of decay in the branches and stems of standing trees though it can also cause rot in the sapwood of felled trees. It has also been found causing a butt rot in oak but this is not a frequent occurrence. Davidson *et al.* (1942) found *S. gausapatum* causing decay in oak sprouts (coppice shoots) in the USA, where parent stumps were the main source of infection.

### Stereum hirsutum Fr.
*Stereum hirsutum* is commonly found in Europe and America on wounds, dead branches and timber of a number of broadleaved trees, including oak, birch and beech.

The fruit bodies occur from January to December and consist of thin leathery brackets, 2–10 cm across, sometimes entirely resupinate, often imbricate and confluent. The upper surface is hairy, yellowish or greyish in

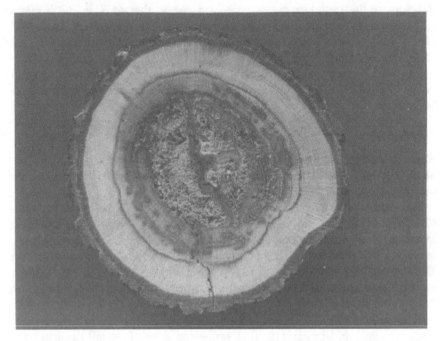

**Figure 108**   Pipe rot in oak branch caused by *Stereum gausapatum* (Forestry Commission)

colour and somewhat zoned. The hymenium is smooth, bright yellow when young, later turning greyish. The hyaline spores are elliptical, flattened on one side and measure 6–8 × 3–4 μm (Rea, 1922).

*S. hirsutum* is often found decaying the sapwood of oak, the wood becoming soft and lighter in colour and weight. Cartwright and Findlay (1958) report that on occasion the fungus is able to invade the trunk from dead branches and cause a pipe rot in the heartwood. However, this type of rot is more commonly caused by *S. gausapatum*, *S. hirsutum* more commonly being confined to the sapwood. In contrast to *Chondrostereum purpureum*, *S. hirsutum* can attack the wood in the absence of living tissues (Guinier, 1933). In support of this view, the same author observed the presence of *S. hirsutum* on 2-year-old fire scars on beech in France.

In culture on malt agar, *S. hirsutum* grows first as a loose aerial mycelium soon developing into a thicker mat and finally forming a tough skin over most of the surface (Cartwright and Findlay, 1958). The mycelium is at first white, later becoming yellow.

The broad young hyphae bear whorled clamp connections which may be very large, sausage-shaped and loosely attached (Cartwright and Findlay, 1958). A few crystals may be produced in the mycelium.

*S. hirsutum* is a very common cause of decay in the sapwood of logs of oak and other broadleaved species. Decayed wood is paler than normal and at an advanced stage in oak it is yellowish white and very light in weight. It does not normally cause serious decay in live standing trees.

**Stereum purpureum: see Chondrostereum purpureum**

**Stereum rugosum Pers.**
*Stereum rugosum* is said to occur commonly in Europe on broadleaved trees, including oak and beech, but it can be confused with *S. gausapatum*.

The fruit bodies of *S. rugosum* can be found at any time of year on stems or branches, stumps and logs. They are 2 cm or more across, widely effused or shortly reflexed, silky, then glabrous. Where visible, the upper surface is pinkish-buff, later becoming grey, and the hymenium is similarly coloured and bleeds when bruised. The flesh is whitish, becoming discoloured and relatively rigid (in contrast to *S. gausapatum*, which is thinner and softer). The spores are hyaline, oblong, incurved and measure $10-12 \times 4-5$ μm (Rea, 1922) or $6.5-9 \times 3.5-4.5$ μm (Breitenbach and Kranzlin, 1986).

*S. rugosum* can cause a pipe rot in oak and both Potter (1901–2) and Liese (1930) observed a canker on oak associated with decay. Strouts *et al.* (1982) reported large perennating cankers on red oak apparently related to infection by this fungus.

**Stereum sanguinolentum Fr.: red rot**
*Stereum sanguinolentum* is a common saprophyte on conifers, appearing on the cut surfaces of stumps and felled trees left in the forest, and sometimes causing a trunk rot in standing trees. It is known in Europe, temperate North America, East Africa and New Zealand, and is variously called red rot, red heart, *sapin rouge*, or (in Douglas fir) mottled bark disease (Basham *et al.*, 1953; Hubert, 1935).

The fruit bodies are thin, resupinate crusts or small brackets (figure 109), up to about 8 cm across (Cartwright and Findlay, 1958), and may form in large, imbricate groups. The upper surfaces of the brackets are greyish or fawn, with silky radiating fibrils, and sometimes zoned. The hymenium is smooth, greyish or buff, and bleeds when scratched or bruised. The basidiospores are colourless (white in mass), elliptical or cylindrical, and measure $8-9 \times 2-2.5$ μm (Wakefield and Dennis, 1981). Cultures of the fungus on malt agar stain the medium red, and have a characteristic sweet smell (Cartwright and Findlay, 1958). As a saprophyte the fungus has a wide host range, occurring on stumps and brash and fallen trees of many conifers (Gladman and Low, 1963). On living trees in Britain it has been found causing damage mainly on Norway and Sitka spruce (particularly the former), though it has also been found on larches and sometimes on other conifers. In North America, however, it is known mainly as a cause of trunk rot of silver firs (Basham *et al.*, 1953; Risley and Silverborg, 1958). In Kenya it has rotted standing trees of *Pinus patula* and *P. radiata* (Gibson, 1966).

*Mode of entry*
*S. sanguinolentum* is a wound parasite, and in Britain enters the standing tree chiefly through carelessly made brashing and pruning wounds (Banerjee,

**Figure 109**  Fruit bodies of *Stereum sanguinolentum* on cut end of Scots pine log
(Forestry Commission)

1955), branches broken by snow and ice, and wounds made during extraction
(Gladman and Low, 1963; Pawsey and Gladman, 1965). Atta and Hayes
(1987) also found *S. sanguinolentum* as the most common fungus isolated
from extraction wounds on Norway spruce. Decay columns were more extensive with larger wounds. Roll-Hansen and Roll-Hansen (1980a,c) also found
that *S. sanguinolentum* was the most common and the fastest-growing of
natural colonisers of artificial wounds in Norway spruce. Elsewhere it has
entered through blazes cut to mark trees (Spaulding *et al.*, 1935) and through
wounds made by elephants and other game (Gibson, 1966). There is evidence
that dead branches do not provide places of entry (Etheridge, 1963).

*Symptoms*
Growth of the fungus in the wood may be relatively fast, but rotting is fairly
slow. Hence, in Britain staining and incipient decay are much more often seen
than are the later stages of rot (Cartwright and Findlay, 1958). On wound
surfaces and on transverse sections of affected stems, infection shows in the
earlier stages as an irregular yellowish-brownish stain. Wood with incipient
decay is firm and brownish yellow, and at a more advanced stage the rotten
wood becomes dry, fibrous and orange-brown with longitudinal lemon-yellow
streaks (Greig, 1981). Evidence of rot in standing trees sometimes shows

externally as white streaks and resin flow on the lower part of the trunk (Hubert, 1935; Banarjee, 1955).

Pawsey and Gladman (1965) calculated that the average rate of growth of the fungus in the tree was 6.7 in (17 cm) per year, but in some Scottish stands it reached 16 in (39.6 cm) per year in Norway spruce, 10.5 in (26.7 cm) per year in Sitka spruce, and 9.6 in (24.4 cm) per year in Japanese larch. Rate of decay increases with the size of the wounds that permit the fungus to enter (Hunt and Krueger, 1962; Davidson and Etheridge, 1963).

Pawsey and Gladman (1965) considered that of all the fungi they found causing butt rot following extraction damage in conifers, *S. sanguinolentum* was potentially the most important. So far in Britain losses caused by this fungus have been relatively small. This may be due partly to the age of many British forests, as losses have been greater in North America, where they have occurred mainly in old plantations. Thus Davidson (1951) in New Brunswick recorded no loss in Balsam fir under 40 years old, but in trees over 80 years old decay gave rise to a loss of a quarter of the marketable volume, and in trees over 170 years old losses were three times as great as this.

In order to avoid infection and rot by *S. sanguinolentum*, live branches should be brashed or pruned carefully by sawing close to the stem. Wounding when extracting produce should be kept to the minimum. Steps to deal with snow- and ice-damaged trees can be taken only in gardens and arboreta, but in such places broken branches should be trimmed.

**Trametes pini: see Phellinus pini**
**Trametes rubescens: see Daedaleopsis confragosa**
**Trametes versicolor: see Coriolus versicolor**

**Tyromyces stipticus (Pers. ex Fr.) Kotl. and Pouz. (*Polyporus stipticus***
**Pers. ex Fr., *Polyporus albidus* Secr.)**
*Tyromyces stipticus* is fairly commonly found associated with extraction damage to coniferous trees or on logs in Europe and has occasionally been reported from North America. The most common hosts are Norway and Sitka spruce and larch. The sporophores occur as small brackets (5–8 cm across and 3–15 mm thick at the base) on stem wounds, or on the cut ends of logs or sometimes as irregular pads on the underside of logs (Greig, 1981). The fruit bodies are white, fleshy, easily broken, and the pores are clearly visible on the under surface. Older specimens are cream or grey in colour and become hard when dry. The spores are oblong-ellipsoid to subcylindrical, smooth, thin-walled and hyaline, measuring $3.5–4.5 \times 1.5–2.0$ μm (Ryvarden, 1976).

At an early stage *T. stipticus* produces a light-yellow stain in the wood and as incipient decay develops the wood becomes light brown, dry and softish. Advanced decay takes the form of a brown cubical rot, with broad radial cracks filled with white mycelium.

This decay is very similar to that caused by *Phaeolus schweinitzii* and *Sparassis crispa*, particularly in Sitka spruce (Pawsey, 1971). However, it can usually be distinguished in the forest, since only decay caused by *T. stipiticus* is normally associated with damage to the stem of the host.

**Figure 110**   Decay in beech caused by *Ustulina deusta* (Forestry Commission)

## *Ustulina deusta* (Hoffm.) Lind. (*Ustulina vulgaris* Tul., *Ustulina zonata* Lev.)

*Ustulina deusta* causes a serious root and butt rot in lime, beech and elm (Wilkins, 1934, 1936, 1939b, 1943) and is also commonly found on old hardwood stumps. Burdekin (1977) found it occurring commonly in wind-blown beech where severe decay had rendered the base of the main trunk or the major roots liable to breakage. Young and Strouts (1977) report that they had observed many cases of severe rot and stem rot of lime and beech caused by *U. deusta*, invading usually via the roots but sometimes via aerial wounds. In the USA Blaisdell (1939) found it frequently in hardwoods such as maple and oak. It has also been reported in the tropics on a wide range of hosts such as cocoa, rubber and tea in Asia, Australia, Africa and America (Cartwright and Findlay, 1958). *U. zonata* is the synonym which has been generally used to describe the tropical form of the fungus and *U. vulgaris* is its temperate counterpart.

*Ustulina* is one of the few ascomycete genera that can cause decay in standing trees; it is a pyrenomycete and a member of the family Xylariaceae. The conidia are produced on a thin disc that measures up to 5 cm across, at first bluish grey with a white margin, later becoming yellowish grey and powdery. The conidia are oval, thin-walled and hyaline and measure $7 \times 3$ μm (Wilkins, 1939a). The mature sexual fructification consists of a black stromatic crust, peppered with the ostioles of embedded perithecia, about 1 mm across, and having the consistency of charcoal. The fructifications usually

**Figure 111** Fractured end of a wind-snapped lime tree decayed by *Ustulina deusta*. Black stromatic sheets of the fungus can be seen (Forestry Commission)

merge and form a mass that may be up to 0.5 metre long. They are often found in the clefts between buttress roots or on the surface of decayed wood and they may persist for several years. The asci are cylindrical, about 300 × 15 μm and eight-spored; the ascospores are uniseriate, black, fusiform, with one side somewhat flattened and measure 28–34 × 7–10 μm (Dennis, 1968).

Wilkins (1936) describes *U. deusta* in lime, noting that it may have entered the tree through a basal wound, and the same author (Wilkins, 1939b, 1943) studied decay in elm and beech in which it may have started in the tap root area. Campbell and Davidson (1940) observed that *U. deusta* commonly infected sugar maple stems through the parent stump and through neighbouring dead stubs. They also found infection present in roadside maples with large wounds; trees with basal infections were frequently killed by girdling. Wilkins (1939a) suggests that conidia may be important in the infection process. In culture the fungus grows as a white, smooth and flat mat; it turns grey at the centre and then gradually black (Cartwright and Findlay, 1958).

Wilkins (1936,1939b,1943) gives a detailed description of the decayed wood (figures 110, 111), which is paler than healthy tissues and contains many fine, black zone lines. There may be a discoloured reaction zone marking the radial limit of decay. The decay column may extend for 5 metres up the tree and above this there may be a reddish-brown zone caused by infiltration of the cell walls with colouring matter.

*Ustulina vulgaris*: see *Ustulina deusta*
*Ustulina zonata*: see *Ustulina deusta*

**Table 20.1**    Summary of common broadleaved trees and their associated decay fungi

| Tree species | Position of rot | Appearance of rot | Decay fungus |
| --- | --- | --- | --- |
| Alder | Stem | White rot | *Daedaleopsis confragosa* |
| | Stem | White flaky rot | *Inonotus radiatus* |
| Ash | Root and butt | White, stringy to spongy with dark zone lines | *Armillaria mellea sensu lato* |
| | Stem and branch | White with black specks and streaks | *Daldinia concentrica* |
| | Butt | White, felt-like mass | *Perenniporia fraxinea* |
| | Stem and branch | Yellow with dark gummy reaction zone | *Inonotus hispidus* |
| Beech | Root and butt | White, stringy to spongy with dark zone lines | *Armillaria mellea sensu lato* |
| | Stem and branch | White flaky rot | *Bjerkandera adusta* |
| | Stem and branch | Brown stain later becoming mottled decay | *Chondrostereum purpureum* |
| | Root and butt | White pocket rot | *Heterobasidion annosum* |
| | Butt, stem and branch | White mottled decay with dark reaction zone | *Ganoderma adspersum/applanatum* |
| | Butt, stem and branch | Yellow mottled decay | *Ganoderma pfeifferi* |
| | Root | Soft brown decay with white pockets | *Meripilus giganteus* |
| | Stem and branch | White flaky rot | *Pleurotus ostreatus* |
| | Root and butt | Light coloured decay with many fine zone lines | *Ustulina deusta* |
| Birch | Stem and branch | Brown stain, later mottled decay | *Chondrostereum purpureum* |
| | Stem and branch | White, flaky rot | *Coriolus versicolor* |
| | Root and butt | White pocket rot | *Heterobasidion annosum* |
| | Stem | Creamy white rot with black flecks | *Fomes fomentarius* |
| | Stem | White decay with dark reaction zone | *Inonotus obliquus* |
| | Stem | Reddish-brown, brittle, decay | *Piptoporus betulinus* |
| Cherry | Stem | Brown cubical rot | *Laetiporus sulphureus* |
| | Stem and branch | White crumbly decay | *Phellinus pomaceus* |
| Elm | Stem | Brown decay | *Pleurotus ulmarius* |
| | Stem | White spongy decay | *Polyporus squamosus* |
| | Butt | Brown cubical rot | *Rigidoporus ulmarius* |
| | Root and butt | Light-coloured decay with many fine zone lines | *Ustulina deusta* |

*Table 20.1–continued*

| Tree species | Position of rot | Appearance of rot | Decay fungus |
|---|---|---|---|
| Horse Chestnut | Butt and stem | Mottled decay with dark reaction zone | *Ganoderma adspersum/applanatum* |
| | Stem and branch | White flaky rot | *Pleurotus ostreatus* |
| Lime | Root and butt | Light-coloured decay with many fine zone lines | *Ustulina deusta* |
| Oak | Stem and branch | White flaky rot | *Coriolus versicolor* |
| | Stem and branch | 'Brown oak' | *Fistulina hepatica* |
| | Butt | White decay | *Ganoderma lucidum* |
| | Butt | Leathery white rot | *Ganoderma resinaceum* |
| | Butt | White decay with orange zones | *Grifola frondosa* |
| | Butt | Soft white rot | *Inonotus dryadeus* |
| | Stem and branch | Brown cubical rot | *Laetiporus sulphureus* |
| | Stem | Pipe rot | *Stereum gausapatum* |
| | Stem | Soft white decay | *Stereum hirsutum* |
| | Stem | Pipe rot | *Stereum rugosum* |
| Plane | Stem and branch | Yellow with dark reaction zone | *Inonotus hispidus* |
| Poplar | Stem and branch | Brown stain, later mottled decay | *Chondrostereum purpureum* |
| | Stem and branch | White flaky decay | *Pleurotus ostreatus* |
| | Butt | Soft brown decay | *Pholiota squarrosa* |
| Sweet chestnut | Stem and branch | Rich brown stain | *Fistulina hepatica* |
| | Stem and branch | Brown cubical rot | *Laetiporus sulphureus* |
| Sycamore | Stem and branch | Yellow with dark reaction zone | *Inonotus hispidus* |
| | | Yellow spongy rot | *Polyporus squamosus* |
| Walnut | Stem and branch | Yellow with dark reaction zone | *Inonotus hispidus* |
| | Stem and branch | White spongy rot | *Polyporus squamosus* |
| Yew | Stem and branch | Brown cubical rot | *Laetiporus sulphureus* |

**Table 20.2** Susceptibility of major conifer species to the six most common decay fungi. HS: highly susceptible, MS: moderately susceptible, SS: slightly susceptible, R: resistant (after B.J.W. Greig, 1981)

| Conifer hosts | Armillaria mellea s.l. | Heterobasidion annosum | Phaeolus schweinitzii | Sparassis crispa | Stereum sanguinolentum | Tyromyces stipticus |
|---|---|---|---|---|---|---|
| Scots pine* (*Pinus sylvestris*) | R | R | MS | MS | SS | SS |
| Sitka spruce (*Picea sitchensis*) | HS | HS | HS | HS | HS | MS |
| Norway spruce (*P. abies*) | HS | HS | R | R | HS | HS |
| Omorika spruce (*P. omorika*) | HS | MS | MS | MS | HS | MS |
| Larches† (*Larix* spp.) | SS | MS | MS | MS | MS | MS |
| Douglas fir (*Pseudotsuga menziesii*) | SS | SS | MS | MS | MS☆ | |
| Western hemlock (*Tsuga heterophylla*) | HS | HS | R | R | MS☆ | |
| Western red cedar (*Thuja plicata*) | HS | HS | R | R | MS☆ | |
| Lawson cypress‡ (*Chamaecyparis lawsoniana*) | HS | MS | R | R | MS | |
| Grand fir§ (*Abies grandis*) | SS | R or SS | R | R | MS☆ | |

* Similar ratings for Corsican pine (*Pinus nigra* var. *maritima*) and lodgepole pine (*P. contorta*).
† Includes European (*Larix decidua*), Japanese (*L. kaempferi*) and hybrid larch (*L. eurolepis*).
‡ Similar ratings for Leyland cypress (*Cupressocyparis leylandii*).
§ Similar ratings for noble fir (*Abies procera*) and other silver firs (*Abies* spp.).
☆ The susceptibility rating is taken from studies outside Britain.

# References

**Note**: The references CMI Map . . ., CMI . . ., CMI Descr. B . . ., CMI Descr. F . . . and CMI/AAB Descr. V . . . given in the text are explained in the Preface and on p. 4 (Chapter 1)

Aalto-Kallonen, T. and Kurkela, T. (1985). Gremmeniella disease and site factors affecting the condition and growth of Scots pine. *Comm. Inst. For. Fenn.*, No. 126, 28pp.

Accordi, S.M. (1989). Sopravvivenza ñel terreno di *Ceratocystis fimbriata* f. sp. *platani*. *Inftore. fitopatol.*, **39**, 57-62 (abstract in *Rev. Pl. Path.*, **69**, No. 1262, 1990)

Adachi, K., Namba, H., Otsaka, M. and Kuroda, H. (1988). Blood pressure lowering activity present in the fruit body of *Grifola frondosa* maitaki. *Chem. Pharm. Bull. (Tokyo)*, **36** (3)

Adams, G.C. and Roberts, D.L. (1988). Epidemic of Lophodermium needle cast of Scots pine in Michigan. *Pl. Dis.*, **72**, 801

Adaskaveg, J.E. and Gilbertson, R.L. (1986). Cultural studies and genetics of sexuality of *Ganoderma lucidum* and *Ganoderma tsugae* in relation to the taxonomy of the *Ganoderma lucidum* complex. *Mycologia*, **78** (5), 694-705

Aerts, R., Soenen, A. and Beauduin, E. (1959). Essais de lutte contre le Dothichiza du Peuplier (*Dothichiza populea* Sacc. et Br.). *Agricultura Louvain*, Ser. 2, **7**, 269-87

Albertson, F.W. and Weaver, J.E. (1945). Injury and death of trees in prairie climate. *Ecol. Monogr.*, **15**, 393-433

Alcock, N.L. (1924). A dieback and bark disease of willows, attacking the young twigs. *Trans. R. Scott. arboric. Soc.*, **38**, 128-30

Alcock, N.L. (1928). *Keithia thujina*, Durand: a disease of nursery seedlings of *Thuja plicata*. *Scott. For. J.*, **42**, 77-9

Alcock, N.L. and Maxwell, I. (1925). Successional diseases on willow. *Trans. R. Scott. arboric. Soc.*, **39**, 34-7

Alcock, S. and Wheeler, B.E.J. (1983). Variability of *Cryptostroma corticale*, the causal fungus of sooty bark disease of sycamore. *Pl. Path.*, **32**, 173-6

Aldhous, J.R. (1972). *Nursery Practice*, Bull.For.Commn, No. 43

Alexopoulos, C.J. and Mims, C.W. (1979). *Introductory Mycology*, 3rd edn, Wiley, New York

Alford, D.V. and Locke,T.(1989). Pests and diseases of fruit and hops. In N.Scopes and L.Stables (editors), *Pest and Disease Handbook*, 3rd edn, pp. 323-403. BCPC Publications, Bracknell

Allescher, A. (1901). *Die Pilze Deutschlands, Oesterreichs, und der Schweiz*, Abt. 6, Kummer, Leipzig

Allescher, A. (1903). *Die Pilze Deutschlands, Oesterreichs, und der Schweiz*, Abt. 7, Kummer, Leipzig

Alonso, O., Peredo, H., Barria, R. and Chacon, S. (1985). Metodo de laboratorio para evaluar la efectividad de fungicidas contra hongos patogenos forestales. *Boletin micologico*, **2**, 119-24

Anagnostakis, S.L. (1987). Chestnut blight; the classsical problem of an introduced pathogen. *Mycologia*, **79**, 23-7

Anderson, J.B., Korhonen, K. and Ullrich, R.C. (1980). Relationships between European and north American biological species of *Armillaria mellea*. *Experl. Mycol.*, **4**, 87-95

Anderson, N.A. and Anderson, R.L. (1980). Leaf and shoot blight of aspen caused by *Venturia macularis* in northern Minnesota. *Pl. Dis.*, **64**, 558-9

Anderson, R.L. (1986). *Checklist of micro-organisms associated with tree seeds in the world, 1985*. General Technical Report,Southeastern Forest Experiment Station, USDA Forest Service No. SE-39, 34pp.

Anon (1927). Abbanerscheinungen bei der amerikanischen Weide. *Dt. Korbweidenzüchten, Juliheft*, 75-6

Anon (1930). Chronique forestière. La maladie du rond. *Bull. Soc. Centr. For. Belg.*, **36**, 522-6

Anon (1938). Elm disease eradication in the United States. *Chron. Bot.*, **4**, 429-32

Anon (1939). *Effect of Sulphur Dioxide on Vegetation*, National Research Council, Canada, Ottawa, 447pp.

Anon (1951a). Hedges defoliated at Huntley. *Phytopath. Intell. Rep. (Scot. )*, No. 13

Anon (1951b). *Phomopsis Disease of Conifers*, Leafl.For.Commn, No. 14

Anon (1952). Editorial notes. Diseases of the cricket bat willow. *Q. Jl. For.*, **46**, 1

Anon (1954). *International Convention for the Protection of Plants and Plant Products*. Treaty Series 16 (1954). HMSO, London

Anon (1956a). *Two Leaf-cast Diseases of Douglas Fir*. Leafl. For. Commn, No. 18, 5pp.

Anon (1956b). Oak mildew, *Leafl. For. Commn*, No. 38, 6pp.

Anon (1961). *Armillaria Root Rot*, Advis.Leafl.Minist.Agric.Fish.Fd. No. 500

Anon (1962). Field excursions. *Conference and Study Tour on Fomes annosus Scotland, 1960*, I.U.F.R.O., Firenze

Anon (1963). *Forest Disease Research. 1962 at the Southern Forest Experiment Station*, 43-44. U.S. Dept.Agric. Forest Service

Anon (1964). *Fomes annosus root rot. Ann. Rep. N. E. For. Exp. Sta., 1963*, pp. 58-9

Anon (1967a). *Honey Fungus*, Leafl.For.Commn, No. 6

Anon (1967b). *Keithia disease of Western red cedar*, Leafl. For. Commn, No. 43, 7pp.

Anon (1967c). *Maple Diseases and Their Control*. U.S. Dept. Agric. Home and Gardens Bull. No. 81, 8pp.

Anon (1968). *Cultivation of the Cricket Bat Willow*, Bull.For.Commn, No. 17, 37pp.

Anon (1969). *Coral Spot*, Advis.Leafl.Minist.Agric.Fish.Fd., No. 23.

Anon (1971). *Apple and Pear Canker*, Advis Leafl.Minist. Agric.Fish.Fd., No. 100

Anon (1975). Plant pathology, in *Report of Forest Research Institute for 1974*, New Zealand Forest Service, Wellington, New Zealand, pp. 44-8

Anon (1979). *Les Maladies des Peupliers*. (FAO/CIP/79). FAO

Anon (1989). *OECD Environmental Data-Compendium 1989*, HMSO

Anselmi, N. (1977). Gleosporiosi del salice. *Cellulosa e Carta*, **28**, 19-39

Anselmi, N. (1979). La 'bronzatura' del salice causata da *Marssonina salicicola* (Bres.). Magn. *Cellulosa e Carta*, **30**, 3-19 (abstract in *Rev. Pl. Path.*, **60**, No. 4686, 1981)

Anselmi, N. (1980). Studies on *Drepanopeziza salicis* (Tul.). v. Höhnel, perfect state of *Gloeosporium salicis* West. *Eur. J. For. Path.*, **11**, 438-47

Anselmi, N. (1986). Resurgence of *Cryptodiaporthe populea* in Italy. *EPPO Bull.*, **16**, 571-83

Appen, A. von (1927). Wedenschorf (*Fusicladium saliciperdum*). *Illus. Landw. Zeit.*, **47**, 67-8

Arthaud, J., David, A., Faye, M., Lung-Escarment, B. and Taris, B. (1980). Processus d'infection par *Armillariella ostoyae* Romagn. de racines de *Pinus pinaster* Sol. isolées et cultiveés sur un milieu synthétique. *Bull. Trimestrial Soc. mycol. Fr.*, **96**, 261-9

Arvidson, B. (1954). En studie av granrotrotans (*Polyporus annosus* Fr.). ekonomiska Konsekvenser. *Svenska SkogFörs. Tidskr.*, **52**, 381-412

Arx, J.A. von (1957a). Ueber *Fusicladium saliciperdum* (All. et Tub.). Lind. *Tijdschr. PlZiekt.*, **63**, 232-6

Arx, J.A. von (1957b). Die Arten der Gattung Colletotrichum Cda. *Phytopath. Z.*, **29**, 413-68

Arx, J.A. von, Walt, J.P.van der and Liebenberg, N.V.D.M. (1982). The classification of *Taphrina* and other fungi with yeast-like cultural states. *Mycologia*, **74**, 285-6

Ashcroft, J.M. (1934). *European Canker of Black Walnut and Other Trees*, Bull.W.Va.agric.Exp.Sta., No. 261, 52pp.

Ashworth, L.J., Jr., Gaona, S.A. and Surber, E. (1985). Verticillium wilt of pistachio: the influence of potassium nutrition on susceptibility to infection by *Verticillium dahliae*. *Phytopathology*, **75**, 1091-3

Atkinson, M.A. and Cooper, J.I. (1976). Ultrastructural changes in leaf cells of poplar naturally infected with poplar mosaic virus. *Ann. appl. Biol.*, **83**, 395-8

Atta, H.Ali el and Hayes, A.J. (1987). Decay in Norway spruce caused by *Stereum sanguinolentum* Alb. & Schw.ex Fr. developing from extraction wounds.*Forestry*, **60**, 101-12

Baker, K.F. (1946). Observations on some Botrytis diseases in California. *Pl. Dis. Reptr.*, **30**, 145-55

Bakshi, B.K. (1950). Fungi associated with ambrosia beetles in Great Britain. *Trans. Br. mycol. Soc.*, **33**, 111-20

Bakshi, B.K. (1952). *Oedocephalum lineatum* is a conidial stage of Fomes annosus. *Trans. Br. mycol. Soc.*, **35**, 195

Bakshi, B.K., Reddy, M.A.R. and Sujan Singh (1976). Ganoderma root rot mortality in Khair (*Acacia catechu* Wild.). in reforested stands. *Eur. J. For. Path.*, **6**, 20-38

Banerjee, S. (1955). A disease of Norway spruce (*Picea excelsa*(Lam.). Link). associated with *Stereum sanguinolentum* (A. and S.)Fr. and *Pleurotus mitis* (Pers.). Berk. *Indian J. mycol. Res.*, **1**, 1-30

Banerjee, S. (1956). An oak (*Quercus robur* L.) canker caused by *Stereum rugosum* (Pers.)Fr. *Trans Br. mycol. Soc.*, **39**, 267-77

Banfield, W.M. (1941). Distribution by the sap stream of spores of three fungi that induce vascular wilt diseases of elm. *J. agric. Res.*, **62**, 637-81

Banfield, W.M. (1968). *Dutch Elm Disease; Recurrence and Recovery in American Elm.* Bull. agric. Exp. Stn. Massachussets, No. 568, 60pp.

Barklund, P. (1987). Occurrence and pathogenicity of *Lophodermium piceae* appearing as an endophyte in needles of *Picea abies*. *Trans. Br. mycol. Soc.*, **89**, 307-13

Barnard, E.L. and Schroeder, R.A. (1984/85). Anthracnose of Acacia in Florida; occurrences and fungicidal control. *Proc. Florida State Hort. Soc.*, **97**, 244-47 (abstract in *Rev. Pl. Path.*, **65**. No. 2989, 1986)

Barr, Margaret E. (1978). *The Diaporthales in North America. Mycologia* Memoir No. 7. J.Cramer, Lehre

Barrett, D.K. (1968). In Tinsley, T.N., Barrett, D.K. and Biddle, P.G., Forest pathology. *Ann. Rep. Commonw. For. Inst. Oxford*, 1966-67, pp. 23-6

Barrett, D.K. (1970). *Armillaria mellea* as a possible factor predisposing roots to infection by *Polyporus schweinitzii*. *Trans. Br. mycol Soc.*, **55**, 459-62

Barrett, D.K. (1985). Basidiospores of *Phaeolus schweinitzii*, a source of soil infestation. *Eur. J. For. Path.*, **15**, 417-25

Barrett, D.K. and Greig, B.J.W. (1984). Investigation into the infection biology of *Phaeolus schweinitzii. Proceedings of the 6th International Conference on Root and Butt rots of forest trees (IUFRO Working Party S2-06-01), CSIRO, Melbourne*, 95-103

Barrett, D.K. and Greig, B.J.W. (1985). The occurrence of *Phaeolus schweinitzii* in the soils of Sitka spruce plantations with broadleaved or non-woodland histories. *Eur. J. For. Path.*, **15**, 412-417

Barrett, D.K. and Pearce, R.B. (1981). Giant leaf blotch disease of sycamore (*Acer pseudoplatanus*) in Britain. *Trans. Br. mycol. Soc.*, **76**, 317-45

Barthelet, J. and Vinot, M. (1944). Notes sur les maladies des cultures meriodionales. *Ann. Epiphyt., N. S.*, **10**, 11-23

Basham, J.T., Moole, P.V. and Davidson, A.G. (1953). New information concerning balsam fir decays in Western North America. *Can. J. Bot.*, **31**, 334-62

Basova, S.V. (1987). [Seasonal dynamics of powdery mildew of pedunculate oak under conditions in a seed grafting plantation]. *Mikol. Fitopatol.*, **21**, 269-273 (Abstract in *Rev. Pl. Path*.**67**, No. 4126, 1988).

Bassett, C. and Zondag, R. (1968). Protection against fungal and insect attack in New Zealand forests. *Pap. 9th Commonw. For. Conf., India, 1968*, No. 357

Batko, S. (1950). A note on the distribution of *Poria obliqua* (Pers.)Bres. *Trans. Br. mycol. Soc.*, **33**, 105-6

Batko, S. (1956). *Meria laricis* on Japanese and hybrid larch in Britain. *Trans. Br. mycol. Soc.*, **39**, 13-16

Batko, S. (1962). Perithecia of oak mildew in Britain, in 'New or uncommon plant diseases and pests'. *Pl. Path.*, **11**, 184

Batko, S. (1974). Notes on new and rare fungi on forest trees in Britain. *Bull. Br. mycol. Soc.*, **8**, 19-21

Batko, S., in G.D.Holmes and G.Buszewicz (1954, 1955, 1956, 1957a,b). Forest tree seed investigations. *Rep. For. Res., Lond., 1953*, pp. 14-7; 1954, pp. 1-4; 1955, pp. 13-14; 1956, pp. 15-17; 1957, pp. 17-19

Batko, S., Murray, J.S. and Peace, T.R. (1958). *Sclerophoma pithyophila* associated with needle cast of pines and its connection with *Pullularia pullulans*. *Trans. Br. mycol. Soc.*,**41**, 126-8

Batko, S. and Pawsey, R.G. (1964). Stem canker of pine caused by *Crumenula sororia*. *Trans. Br. mycol. Soc.***47**, 257-61

Baule, H. and Fricker, C. (1970). *The Fertilizer Treatment of Forest Trees*, Trans.C.L.Whittles, BLV, Munich

Bavendamm, W. (1936). Der Rindenbrand der Pappeln. *Tharandt. forstl. Jb.*, **87**, 177-9

Baxter, D.V. (1953). *Observations on Forest Pathology as Part of Forestry in Europe*, Univ.Mich.Sch.For.& Cons.Bull. No. 2, 39pp.

Bazzigher, G. (1956a). *Pilzkrankheiten in Auf-forstungen. Kurzmitt.* Schweiz Anst. forstl.Versuchsw, No. 12, 3pp.

Bazzigher, G. (1956b). Pilzschaden an Kastanien nordlich der Alpen. *Schweiz. Z. Forstwes.*, **107**, 694-5

Bazzigher, G. (1981). Selection of blight-resistant chestnut trees in Switzerland. *Eur. J. For.*

*Path.*, **11**, 199-207

Beattie, R.K. (1937). The Dutch elm disease in Europe. *Am. Forests*, **43**, 159-62

Bedker, P.J. and Blanchette, R.A. (1983). Development of cankers caused by *Nectria cinnabarina* on honey locusts after root pruning. *Pl. Dis.*, **67**, 1010-3

Bedker, P.J. and Blanchette, R.A. (1984a). Identification and control of cankers caused by *Nectria cinnabarina* on honey locust. *J. Arb.*, **10**, 33-9

Bedker, P.J. and Blanchette, R.A. (1984b). Control of *Nectria cinnabarina* cankers on honey locust. *Pl. Dis.*, **68**, 227-30

Bélanger, R.R., Manion, P.D. and Griffin, D.H. (1989). *Hypoxylon mammatum* ascospore infection of *Populus tremuloides* clones; effects of moisture stress in tissue culture. *Phytopathology*, **79**, 315-7

Benetti, M.P. and Motta, E. (1979). Il 'cancro' della tuia da *Seiridium* (*Coryneum*) *cardinale*. *Annali dell'Istituto Sperimentale per la Patalogia Vegetale Roma*, **5**, 39-48

Benzian, B. (1965). *Experiments in Nutrition Problems in Forest Nurseries*, Bull.For.Commn, No. 37, Vol. 1

Benzie, J.W. (1958). A red pine plantation problem in Upper Michigan. *Lake St. Forest Exp. Stn Tech. Note*, No. 524

Beradze, I.A. and Dzimistarishvili, N.B. (1985). [Results of testing some fungicides against grey mould on hazelnuts] *Subtropicheskie Kul'tury* No. 2, 156-7 (abstract in *Rev. Pl. Path.*, **65**, No. 2459, 1986).

Berg, T.M. (1964). Studies on poplar mosaic virus and its relation to the host. *Meded. LandbHoogesch. Wageningen*, **64** (11), 59pp. and appendix

Bergstrom, L. and Gustafson, A. (1985). Hydrogen ion budgets for four small runoff basins in Sweden. *Ambio*, **14**, 346-348

Berkeley, G.H., Madden, G.O. and Willison, R.S. (1931). Verticillium wilts in Ontario. *Scient. Agric.*, **11**, 739-59

Bernaux, P. (1956). Contribution a l'étude de la biologie des *Gymnosporangium*. *Ann. Inst. Rech. Agron.*, Ser.C (Ann. Epiphyt.), **7**, 1-210

Berry, F.H. (1965). *Treat stumps to prevent Fomes annosus in shortleaf pine plantations*. U.S. For.Serv.Res.Not., CS-34

Berry, F.H. (1977). Control of walnut anthracnose with fungicides in a black walnut plantation. *Pl. Dis. Reptr.*, **61**, 378-9

Bevan, R.J. and Greenhalgh, G.N. (1976). *Rhytisma acerinum* as a biological indicator of pollution. *Environmental Pollution*, **10**, 271-85

Bevercombe, G.P. and Rayner, A.D.M. (1978). *Dichomera saubinetii* and bark diamond canker formation in sycamore. *Trans. Br. mycol. Soc.*, **71**, 505-7

Bevercombe, G.P. and Rayner, A.D.M. (1980). Diamond-bark diseases of sycamore in Britain. *New Phytol.*, **86**, 379-92

Bevercombe, G.P. and Rayner, A.D.M. (1984). Population structure of *Cryptostroma corticale*, the causal fungus of sooty bark disease of sycamore. *Pl. Path.*, **33**, 211-7

Bezerra, J.L. (1963). Studies on *Pseudonectria rousseliana*. *Acta Bot. Neerl.*, **12**, 58-63

Biddle, P.G. and Tinsley, T.W. (1968). Virus diseases of forest trees. *Rep. Forest Res., Lond., 1968*, pp. 149-51

Biddle, P.G. and Tinsley, T.W. (1971a). Poplar mosaic virus in Great Britain. *New Phytol.*, **70**, 61-6

Biddle, P.G. and Tinsley, T.W. (1971b). Some effects of poplar mosaic virus on the growth of poplar trees. *New Phytol.*, **70**, 67-75

Bier, J.E. (1959). The relation of bark moisture to the development of canker diseases caused by native, facultative parasites. I. Cryptodiaporthe canker on willow. *Can. J. Bot.*, **37**, 229-38

Biggs, A.R., Davis, D.D. and Merrill, W. (1983). Histopathology of cankers on *Populus* caused by *Cytospora chrysosperma*. *Can. J. Bot.*, **61**, 563-74

Biggs, P. (1957). Studies on *Meria laricis*, needle cast disease of larch. *Rep. Forest Res. Lond., 1957*, pp. 102-4

Biggs, P. (1959). Studies on *Meria laricis*, needle cast disease of larch. *Rep. Forest Res. Lond., 1958*, pp. 112-13

Billing, E.B. (1984). Principles and application of fireblight assessment systems. *Acta Horticulturae*, **151**, 15-22

Binns, W.O. and Fourt, D.F. (1980). Surface workings and trees, in *Research for Practical Arboriculture*, Occasional Pap.For.Commn, No. 10.

Binns, W.O., Mayhead, G.J. and Mackenzie, J.M. (1980). *Nutrient Deficiencies of Conifers in British Forests; an Illustrated Guide*, Leafl.For.Commn, No. 76.

Biraghi, A. (1949). Il disseccamento fogliare del castagno causato da *Sphaerella maculiformis*. *Italia for. Mont.*, **4**, 21-4

Biraghi, A. (1951). *Endothia parasitica* e gen. *Quercus. Ital. for. mont.*, **6**, 15-6

Bisby, G.R. and Mason, E.W. (1940). List of pyrenomycetes recorded for Britain. *Trans. Br. mycol. Soc.*, **24**, 127-243

Bisiach, M. and Minervini, J. (1980). A resupinate form of *Phellinus pomaceus* agent of dieback of ornamental trees. *Riv. Patol. Veg.*, **16** (3-4), 101-8

Blaisdell, D.J. (1939). Decay of hardwoods by *Ustulina vulgaris* and other Ascomycetes. *Phytopathology*, **29**, 2

Blamey, M. and Grey-Wilson, C. (1989). *The Illustrated Flora of Britain and Northern Europe*. Hodder and Stoughton, London

Blanchette, R.A. (1980). Wood decomposition by *Phellinus pini* a scanning electron microscope study. *Can. J. Bot.*, **58** (13), 1496-1503

Blenis, P.V., Patton, R.F. and Spear, R.N. (1984). Effect of temperature on the ability of *Gremmeniella abietina* to survive and to colonise host tissue. *Eur. J. For. Path.*, **14**, 153-64

Bliss, D.E. (1941). Relation of soil temperature to Armillaria root rot in California. *Phytopathology*, **31**, 3 (abstract)

Bliss, D.E. (1944). *Controlling Armillaria Root Rot in Citrus*, California Agric.Exp.Sta., 7pp.

Bliss, D.E. (1946). The relation of soil temperature to the development of Armillaria root rot. *Phytopathology*, **36**, 302-18

Bliss, D.E. (1951). The destruction of *Armillaria mellea* in citrus soils. *Phytopathology*, **41**, 665-83

Blomfield, J.E. (1924). Witches-brooms. *J. R. micr. Soc., Lond.*, 190-4

Bloomberg, W.J. (1962a). Cytospora canker of poplars: factors influencing the development of the disease. *Can. J. Bot.*, **40**, 1271-80

Bloomberg, W.J. (1962b). Cytospora canker of poplars: the moisture relations and anatomy of the host. *Can. J. Bot.*, **40**, 1281-92

Bloomberg, W.J. and Funk, A. (1960). Willow blight in British Columbia. *Bi-m. Progr. Rep. Div. For. Biol., Dep., Agric. Can.*, No. 16, pp. 3-4

Boa, E.A. and Preece, T.F. (1981). *Pseudomonas savastanoi* var.*fraxini*: symptoms of ash canker disease in the field in Scotland and England. *Proc. 4th Int. Conf. Pl. Path. Bacteria (Angers)*, **2**, 917-28

Boer, S.H.de (1980). Leaf spot of cherry laurel caused by *Pseudomonas syringae*. *Can. J. Pl. Path.*, **2**, 235-8

Boesewinkel, H.J. (1983). New records of the three fungi causing cypress canker in New Zealand, *Seiridium cupressi* (Guba). comb. nov. and *S. cardinale* on *Cupressocyparis* and *S. unicorne* on *Cryptomeria* and *Cupressus*. *Trans. Br. mycol. Soc.*, **80**, 544-7

Bolli, M. (1943). Influenza dei raggi ultra-violetti sull'oidio della quercia. *Ann. Fac. Agr. Perugia*, **2**, 65-8

Bond, T.E.T. (1956). Notes on *Taphrina*. *Trans. Br. mycol. Soc.*, **39**, 60-6

Bond, T.E.T. (1961). In British Records. *Trans. Br. mycol. Soc.*, **44**, 608-14

Booth, C. (1959). *Studies of pyrenomycetes: IV. Nectria (Part 1)*. Mycol.Pap., No. 73, 115pp.

Borg, P. (1935). Report of the plant pathologist. *Rep. Insp. Agric. Malta*, 1933-34 pp. 43-6

Bormann, F. H. (1985). Air pollution and forests: an ecosystem perpspective. *Bioscience*, **35**, 434-41

Borthwick, A.W. (1909). A new disease of *Picea*. *Notes R. Bot. Gdn Edinb.*, **20**, 259-61

Boudier, B. (1978). Essais de lutte contre l'oidium du laurier-cerise. *Phytiatrie-Phytopharmacie*, **27**, 215-20

Boudier, B. (1983a). *Didymascella thujina*, principal ennemi du *Thuja* dans l'ouest de la France. *Phytoma Nov. 1983*, **56**, 51-3

Boudier, B. (1983b). Les ennemis du peuplier d'Italie. Lutte contre la brunissure des feuilles due à *Marssonina populi*. *Revue. Hort.*, **238**, 35-42

Boudier, B. (1985). Le marronier en France. Comment lutter contre sa principale maladie: le black-rot. *P. H. M. -Revue hort.* No. 259, 11-17

Boudier, B. (1986). Essai de mise au point de methode de lutte contre *Didymascella thujina*. Ministère de l'Agriculture, Service de la Protection des Végétaux, Hérouville Saint-Clair, France.

Boudier, B. (1987a). Le flutriafol: un progrès important dans la lutte contre les oidiums du chêne et de l'érable. *Revue hort.* No. 273, 54-6

Boudier, B. (1987b). Lutte contre 'le rouge cryptogamique du pin'. *Revue for. fr.*, **39**, 107-12

Bouhot, D. and Perrin, R. (1980). Mise en évidence de résistances biologiques aux *Pythium* en sols forestiers.*Eur. J. For. Path.*, **10**, 77-89

Boullard, B. (1961). Etude d'une attaque de 'l'*Armillaria mellea*' (Vahl). Quél. sur l'Epicea de Sitka. *Rev. for. fr.*, **1**, 16-24

Bourdot, H. and Galzin, A. (1927). *Hymenomycetes de France*, Société Mycologique de France

Bowen, P.R. (1930). A maple leaf disease caused by *Cristulariella depraedans*. *Bull. Conn. agric. Exp. Stn.*, **316**, 625-47

Boyce, J.S. (1924). An unusual infection of *Polyporus schweinitzii* Fr. *Phytopathology*, **14**, 588

Boyce, J.S. (1927). Observations on forest pathology in Great Britain and Denmark. *Phytopathology*, **17**, 1-18

Boyce, J.S. (1961). *Forest Pathology*, 3rd edn, McGraw-Hill, New York

Bracher, R. (1924). Notes on *Rhytisma acerinum* and *Rhytisma pseudoplatani*. *Trans Br. mycol. Soc.*, **9**, 183-6

Bradbury, J.F. (1986). *Guide to Plant Pathogenic Bacteria*. Commonwealth Agricultural Bureaux, Slough

Brandt, C.S. (1962). Effects of air pollution on plants, in A.C.Stern (Ed.), *Air Pollution*, 1st edn, Vol. 1, Academic Press, New York, pp. 255-81

Brandt, C.S. and Heck, W. (1967). Effects of air pollutants on vegetation, in A.C.Stern (Ed.). *Air Pollution*, 2nd edn, Vol. 1, Academic Press, New York, pp. 401-43

Brandt, R.W. (1960). The rhabdocline needle cast of Douglas fir. *Tech. Publ. N. Y. St. Coll. For.*, No. 84.

Brandt, R.W. (1964). *Nectria Canker of Hardwoods*, Forest Pest Leafl. No. 84

Brasier, C.M. (1971). In Forest Pathology. *Rep. Forest Res., Lond., 1971*, pp. 77-84

Brasier, C.M. (1979). Dual origin of recent Dutch elm disease outbreaks in Europe. *Nature, Lond.*, **281**, 78-80

Brasier, C.M. (1981). The pathogen, in J.Stipes and R.J.Campana (Eds). *A Compendium of Elm Diseases*, American Phytopathological Society

Brasier, C.M. (1983). Problems and prospects in Phytophthora research. In *Phytophthora: its Biology, Ecology, Taxonomy and Pathology* (Ed.D.C.Erwin, S.Bartnich-Garcia and P.H.Tsao), pp. 351-364. St Paul, Minnesota, USA: American Phytopathological Society

Brasier, C.M. (1986). The population biology of Dutch elm disease: Its principal features and some implications for other host-pathogen systems. In Ingram, D.S., Williams, P.H. (Eds) *Advances in Plant Pathology, Vol. 5*, 53-118, Academic Press Inc. London

Brasier, C.M. (1989). *Phytophthora megasperma*. In Smith *et al.* (1989), q.v.

Brasier, C.M. (1990). China and the origins of Dutch elm disease: an appraisal. *Plant Path.*, **39**, 5-16

Brasier, C.M. (1991). *Ophiostoma novo-ulmi* sp.nov., causative agent of current Dutch elm disease pandemics. *Mycopathologia*, in press

Brasier, C.M. and Gibbs, J.N. (1973). Origin of the Dutch elm disease epidemic in Britain. *Nature, Lond.*, **242**, 607-9

Brasier, C.M. and Gibbs, J.N. (1978). Origin and Development of the current Dutch elm disease epidemic, in P.R. Scott and A. Bainbridge (Eds), *Plant Disease Epidemiology*, Blackwell, Oxford.

Brasier, C.M. and Strouts, R.G. (1976). New records of Phytophthora on trees in Britain. 1. Phytophthora root rot and bleeding canker of horse chestnut (*Aesculus hippocastanum* L.). *Eur. J. For. Path.*, **6**, 129-36

Braun, G., Schonborn, A.von and Weber, E. (1978). Untersuchungen zur resistenz von geholzen gegen auftausalz (Natriumchlorid). *Allg. Forstz.*, **149**, 216-35

Braun, H. and Hubbes, M. (1957). Sporeninfektion und Antagonismus bei *Dothichiza populea*. *Naturwissenschaften*, **44**, 333

Braun, H.J. (1956). Zur frage der Infektion von Schal- und Schurfwunden durch den Wurzelschwamm *Fomes annosus*. *Forst- u. Holzwirt*, **11**, 430-1

Braun, H.J. (1958). Untersuchungen über den Wuezelschwamm *Fomes annosus* (Fr.). Cooke. *Forstwiss. Centralb.*, **77**, 65-88

Braun, H.J. (1976). Das Rindensterben der Buch, *Fagus sylvatica* L., verursacht durch die Buchenschildlaus *Cryptococcus fagi* Bär. I. Die anatomie der Buchenrinde als BasisUrsache. *Eur. J. For. Path.*, **6**, 136-46

Braun, H.J. (1977). Das Rindensterben der Buche, *Fagus sylvatica* L., verursacht durch die Buchenschildlaus *Cryptococcus fagi* Bär. II. Ablauf der Krankheit. *Eur. J. For. Path.*, **77**, 76-93

Breitenbach, J. and Kranzlin, F. (1986). *Fungi of Switzerland. Vol. 2. Non-gilled fungi*. Verlag Mykologia, Switzerland

Brendel, G. (1965). Untersuchungen über Hybriden der Gattung *Populus*, Sektion 'Aegeiros' und ihren Einfluss auf die Biologie von *Dothichiza populea* Saccardo et Briard. *Phytopath. Z.*, **53**, 1-34

Breuel, K. and Börtitz, S. (1966). Beiträge zur Ätiologie des Braunfleckengrindes, eine Rindennekrose der Gattung *Populus*. 2. Mitteilung: Untersuchungen über das pathogene Agens. *Phytopath. Z.*, **57**, 59-78

Broekhuijsen, M.J. (1934). Wilgenkanker, veroorzaakt door *Discella carbonacea* (Fries). Berk. et Br. *Tijdschr. PlZiekt.*, **40**, 62-3

Bronchi, P. (1956). Origine dell'abetina pura artificiale nella Foresta Demaniale di Badia Prataglia in relazione ai recente danni da *Fomes annosus* su '*Abies alba*'. *Monti Boschi*, **7**, 368-73

Brooks, A.V., Halstead, A.J., Smith, P.M. and Evans, E.J. (1989). Pests and diseases of outdoor ornamentals, including hardy nursery stock, bedding plants and turf. In Scopes and Stables, 1989, q.v., pp. 513-601

Brooks, F.T. (1910). *Rhizina undulata* attacking newly planted conifers. *Q. Jl. For.*, **4**, 308-9

Brooks, F.T. (1953). *Plant Diseases*, 2nd edn, Oxford University Press. London

Brooks, F.T. and Walker, M.M. (1935). Observations on *Fusicladium salicipe**r**dum*. *New Phytol.*, **34**, 64-7

Brown, J.M.B. (1953). *Studies in British Beechwoods*, Bull.For.Commn, Lond., No. 20, iv, 100pp.

Bryce, J. (1950). Watermark disease of willows. *Essex Farmers' J.*, **28**, 24

Bubak, F. (1914). Eine neue *Rhizosphaera*.*Ber. dt. bot. Ges.*, **32**, 188-90

Buchwald, N.F. (1939). Douglasiens Sodskimmel (*Phaeocryptopus gaumannii*). En ny svamp paa Douglasgran i Danmark.*Dansk. Skovforen. Tidsskr.*, 1939, 357-82

Buckland, D.C. (1953). Observations on *Armillaria mellea* in immature Douglas fir. *For. Chron.*, **229**, 344-7

Buczacki, S.T. (1973a). Some aspects of the relationship between growth vigour, canker and dieback of European larch. *Forestry*, **46**, 71-9

Buczacki, S.T. (1973b). Factors governing mycelial establishment in the larch canker disease. *Eur. J. For. Path.*, **3**, 39-49

Buczacki, S.T. (1973c). Observations on the infection biology of larch canker. *Eur. J. For. Path.*, **3**, 228-32

Buczacki, S.T. and Harris, K. (1981). *Collins Guide to the Pests, Diseases and Disorders of Garden Plants*. Collins, London

Buddenhagen, I.W. and Young, R.A. (1957). A leaf and twig disease of English holly caused by *Phytophthora ilicis* n.sp. *Phytopathology*, **47**, 95-101

Buijsman, E., Maas, J.F.M. and Asman, W.A.H. (1985). *Ammonia emission in Europe*. Report R-85-1, Institute for Meteorology and Oceanography, State University of Utrecht, The Netherlands.

Buisman, C. (1932). *Ceratocystis ulmi*, de geslachlitijke vorm van *Graphium ulmi* Schwarz. *Tijdschr. PlZiekt*, **38**, 1-8

Bulit, J. (1957). Contribution à l'étude biologique du *Nectria galligena* Bres., agent du chancre du pommier. *Ann. Inst. Rech. agron., Sér. C. (Ann. Epiphyt.)*, **8**, 67-89

Bull, R.A. (1951). A new gall disease of *Sequoia sempervirens*. *Gdnrs' Chron.*, Series 3, **130**, 10-1

Burchill, R.T. (1978). In D.M.Spencer (Ed.), *The Powdery Mildews*. Academic Press, London, pp. 473-93

Burdekin, D.A. (1966). The role of Agromyzid flies in relation to bacterial canker. *Publ. FAO Int. Poplar Commn Res. Dis. Group*, 1pp

Burdekin, D.A. (1968). Forest Pathology. *Rep. Forest Res., Lond.*, 1968, pp. 108-11

Burdekin, D.A. (1969). Forest Pathology. *Rep. Forest Res., Lond.*, 1969, pp. 106-10

Burdekin, D.A. (1970). Forest Pathology. *Rep. Forest Res., Lond.*, 1970, pp. 114-19

Burdekin, D.A. (1972). Bacterial canker of poplar. *Ann. appl. Biol.*, **72**, 295-9

Burdekin, D.A. (1977). Gale damage to amenity trees. *Arb. J.*, **3**, 181-9

Burdekin, D.A. (1979). *Common decay fungi in broadleaved trees*. Arboric Leafl. No. 5, HMSO, London

Burdekin, D.A. (1980). *Susceptibility of London plane clones to anthracnose*. Proceedings of the seminar research for Practical Arboriculture, Preston. Forestry Commission Occasional Paper 10

Burdekin, D.A. (1986). European plant health requirements for forest pests. *EPPO Bull.*, **16**, 505-12

Burdekin, D.A. and Phillips, D.H. (1970). Chemical control of *Didymascella thujina* on western red cedar in forest nurseries. *Ann. appl. Biol.*, **67**, 131-6

Burdekin, D.A. and Phillips, D.H. (1977). *Some important foreign diseases of broadleaved trees.* Forest Rec., Lond., No. 111

Burden, R.D. and Bannister, M.H. (1985). Growth and morphology of seedlings and juvenile cuttings in six populations of *Pinus radiata. N. Z. Jl. For. Sci.*, **15**, 123-34

Burger, F.W. (1932). Bacterieziekte van der wilg. *Ned. BoschbTijdschr.*, **5**, 75-84

Burke, M.J., Gusta, L.V., Quarume, H.A., Weiser, C.J. and Li, P.H. (1976). Freezing and injury in plants. *A. Rev. Pl. Physiol.*, **27**, 507-28

Burmeister, P. (1966). Beobachtungen uber einige wichtige Pilzkrankeiten an Zierkoniferen im oldenburgischen Baumschulgebiet. *Gartenbauwissenschaft*, **31**, 469-506

Burnett, J.H. (1977). *Fundamentals of Mycology*, 2nd edn, Crane-Russak Co., New York

Burnham, C.R. (1988). The restoration of the American Chestnut. *Am. Scient.*, **76**, 478-87

Burnham, C.R., Rutter, P.A. and French, D.W. (1986). Breeding blight-resistant chestnuts. *Pl. Breed. Rev.*, **4**, 347-97

Butin, H. (1956a). Beobachtungen über das vorjährige Auftreten der Dothichiza-krankheit der Pappel. *NachBl. dt. PflSchutzdienst (Braunschweig), Stuttgart*, **8**, 55-8

Butin, H. (1956b). Untersuchungen über Resistenz und Krankheitsanfälligkeit der Pappel gegen über *Dothichiza populea* Sacc. et Br. *Phytopath. Z.*, **28**, 353-74

Butin, H. (1958a). Über die auf *Salix* und *Populus* vorkommenden Arten der Gattung *Cryptodiaporthe* Petr. *Phytopath. Z.*, **32**, 399-415

Butin, H. (1958b). Untersuchungen über ein Toxin in Kulturfiltraten von *Dothichiza populea* Sacc. et Br. *Phytopath. Z.*, **33**, 135-46

Butin, H. (1960). *Die Krankheiten der Weide und deren Erreger*, Berlin

Butin, H. (1962). (Germination and viability of spores of *Dothichiza populea* under various atmospheric humidities). *Ber. dt. bot. Ges.*, **75**, 221-32; abstract in *Rev. appl. Mycol.*, **42**, 222, 1963

Butin, H. (1965). Untersuchungen zur ökologie einiger Blauepilze an verarbeiten Kiefernolz. *Flora, Jena*, **155**, 400-40

Butin, H. (1973). Morphologie und taxonomische Untersuchungen an *Naemacyclus niveus* (Pers. ex Fr.). Fuck. ex Sacc. und verwandten Arten. *Eur. J. For. Path.*, **3**, 146-63

Butin, H. (1981). Über den Rindenbranderreger *Fusicoccum quercus* und andere Rindenpilze der Eich. *Eur. J. For. Path.*, **11**, 33-44

Butin, H. (1983). *Krankheiten der wald und Parkbäume*. Georg Thieme Verlag, Stuttgart

Butin, H. and Hackelberg, L. (1978). Uber den Verlauf einer *Scleroderris lagerbergii* Epidemie in einem Schwarzkiefernbestand. *Eur. J. For. Path.*, **8**, 369-79

Butin, H. and Paetzholdt, M. (1974). Schäden an *Juniperus virginiana* L. durch *Phomopsis juniperovora. NachrBl. dt PflSchutzdienst Berl.*, **26**, 36-9

Butin, H. and Richter, J. (1983). Dothistroma needle blight: a new pine disease in the German Federal Republic. *Nach. Dt. Pfldienst.*, **35**, 129-31

Butin, H. and Wagner, C. (1985). Mykologische Untersuchungen zur 'Nadelrote' der Fichte. *Forstwiss. ZentBl.*, **104**, 178-86

Buxton, E.W., Sinha, I. and Ward, V. (1962). Soil-borne diseases of Sitka spruce seedlings in a forest nursery. *Trans. Br. mycol. Soc.*, **45**, 433-48

Byrom, N.A. and Burdekin, D.A. (1970). *Drepanopeziza punctiformis*. British records, No. 101. *Trans. Br. mycol. Soc.*, **54**, 139-41

Cadahia, D. (1986). Problèmes phytosanitaires forestiers dus aux organismes de quarantaine OEPP introduits récement en Espagne. *EPPO Bull.*, **16**, 537-41

Cadman, W.A. (1953). *Shelterbelts for Welsh Hill Farms.* For. Rec., Lond., No. 22

Caetano, M.F.F. (1980). Uma grave doença das Cupressaceas em Portugal. *Agros* **63**, 5-9

Caetano, M.F.F. (1985). *Guignardia aesculi* (Peck). Setw. Uma nova dença do 'Castanheiro da India' em Portugal. *Publicacao do Laboratorio de Patologia Verissimo de Almeida, Lab. Pl. Path., Lisbon, Portugal* (abstract in *Rev. Pl. Path.*, **65**, No. 2460. 1986)

Callan, E. McC. (1939). *Cryptorrhynchus lapathi* L. in relation to the watermark disease of the cricket-bat willow. *Ann. appl. Biol.*, **26**, 135-7

Callen, E.O. (1938). Some fungi on yew. *Trans. Br. mycol. Soc.*, **22**, 94-106

Cambonie, L. (1932). Nos châtaigniers sont malades; la jaunisse ou maladie des taches de feuille. *La vie Agric. et Rurale* **21**, 336

Cameron, C.A. (1874). *Garden Chronicle*, **1**, 274-5

Campbell, A.H. (1934). Zone lines in plant tissues. II. The black lines formed by *Armillaria mellea* (Vahl). Quél. *Ann. appl. Biol.*, **21**, 1-22

Campbell, A.H. and Munsun, R.R. (1936). Zone lines in plant tissues. III. The black zone lines produced by *Polyporus squamosus*. *Ann. appl. Biol.*, **23**, 453-64

Campbell, A.H. and Vines, A.E. (1938). The effect of *Lophodermellina macrospora* (Hartig). Tehon on leaf abscission in *Picea excelsa* Link. *New Phytol.*, **37**, 358-69

Campbell, W.A. (1938). The Cultural Characteristics of the Species of *Fomes*. *Bull. Torrey bot. Cl.*, 31-69

Campbell, W.A. and Davidson, R.W. (1938). A Poria as the fruiting stage of the fungus causing the sterile conks on birch. *Mycologia*, **30**, 553-60

Campbell, W.A. and Davidson, R.W. (1940). *Ustulina vulgaris* decay in sugar maple and other hardwoods. *J. For.*, **38**, 474-7

Cannell, M.G.R. (1985). Analysis of risks of frost damage to forest trees in Britain. *Crop Physiology of Trees*. July 1984.

Cannon, P.E., Hawksworth, D.L. and Sherwood-Pike, M.A. (1985). *The British Ascomycotina. An annotated checklist*. Commonwealth Agricultural Bureaux, Slough

Cantiani, M. (1960). Note sulla diffusione del *Marciunne radicale* nelle abetine di Vallombrosa. *Italia for. mont.*, **15**, 122-4

Cao, C.L. (1987). [Black spot of poplar seedlings] *Journal of Northeast Forestry University, China*, **15**, 100-4 (abstract in *Rev. Pl. Path.*, **67**, No. 2625, 1988)

Capretti, P. (1983). Damni da *Hypoxylon mammatum* (Wahl.). Mill. sul pioppi tremolo. *Inftore fitopatol.*, **33**, 47-9

Capretti, P. and Moriondo, F. (1983). Danni in alcuni impianti di conifere associati alla presenza di *Heterobasidion annosum* (*Fomes annosus*).*Phytopathologia Mediterranea*, **22**, 157-67

Capretti, P. and Mugnai, L. (1987). Disseccamenti del faggio da *Nectria ditissima* Tul. *Inftore fitopatol.*, **37**, 49-53

Capretti, P. and Mugnai, L. (1989). Saprophytic growth of *Heterobasidion annosum* in silver-fir logs interred in different types of forest soils. *Eur. J. For. Path.*, **19**, 257-62

Caroselli, N.E. (1954). Verticillium wilt of maple. *Diss. Abstr.*, **14**, 2186-7

Caroselli, N.E. (1955a). Investigations of toxins produced *in vitro* by the maple wilt fungus *Verticillium* sp. *Phytopathology*, **45**, 183 (abstract)

Caroselli, N.E. (1955b). The relation of soil-water content and that of sapwood-water to the incidence of maple wilt caused by *Verticillium* sp. *Phytopathology*, **45**, 184 (abstract)

Caroselli, N.E. (1957a). Juniper blight and progress on its control. *Pl. Dis. Reptr.*, **41**, 216-8

Caroselli, N.E. (1957b). Verticillium wilt of maples. *Rhode Island Exp. Sta. Bull.*, **335**, 1-84

Caroselli, N.E. (1959). The relation of sapwood moisture content to the incidence of maple wilt caused by *Verticillium albo-atrum*. *Phytopathology*, **49**, 496-8

Carter, C.I. and Gibbs, J.N. (1989). Pests and diseases of forest crops. In Scopes, N. and Stables, L. (Eds), *Pest and Disease Control Handbook*, 3rd edn, BCPC, Thornton Heath, pp. 619-34

Carter, J.C. (1945). Wetwood of elms. *Bull. Ill. Nat. Hist. Survey*, **23**, 401-48

Cartwright, K.St.G. (1937). A reinvestigation into the cause of 'Brown oak', *Fistulina hepatica* (Huds.) Fr. *Trans. Br. mycol. Soc.*, **21**, 68-83

Cartwright, K.St.G. (1940). Note on a heart rot of oak trees caused by *Polyporus frondosus* Fr. *Forestry*, **14**, 38-41

Cartwright, K.St.G., Campbell, W.G. and Armstrong, F.H. (1936). The influences of fungal decay on the properties of timber. I. Effect of progressive decay by *Polyporus hispidus* Fr. on the strength of English ash (*Fraxinus excelsior* L.). *Proc. R. Soc. B.*, **120**, 76-95

Cartwright, K.St.G. and Findlay, W.P.K. (1958). *Decay of timber and its prevention*. HMSO London

Casaliccho, G. (1965). La punteggiatura cloro-necrotica del Fragsino maggiore. *Monti Boschi*, **16**, 39-46

Castellani, E. (1967). A che punto siamo nella lotta contro *Marssonina* del pioppo. *Bull. Agricoltura 1967*, **11**, 1-11

Castellani, E. and Cellerino, G.P. (1964). Una pericolosa malattia dei pioppi euramericani determinata da *Marssonina brunnea* (Ell.& Ev.). P.Magn. *Cellulosa Carta*, **15**, 3-16

Castellani, E. and Cellerino, G.P. (1967). Resultati ti tre anni di lotta contro la *Marssonina brunnea* del pioppo. Ex *Giornate fitopatologiche 1967*, *Bologne*, 213-219; abstract in *Rev. appl. Mycol.*, **47**, No. 2853, 1968

Cayley, D.M. (1921). Some observations on the life history of *Nectria galligena* Bres. *Ann Bot.*, **35**, 75-92

CEC (1977). Council directive of 1976 on protective measures against the introduction into Member States of harmful organisms of plants or plant products (77/93/EEC). *Official Journal of the European Communities*, **1.26**, 20-50 (with later amendments).

Cech, M., Kralik, O. and Blattny, C. (1961). Rod-shaped particles associated with virosis of spruce. *Phytopathology*, **51**, 183-5

Cellerino, G.P. (1966). Prove di lotta contro la *Marssonina brunnea* del pioppo. *Cellulosa e Carta*, **12**, 13-28

Cerny, A. (1982). Biology and economic importance of the beafstéak fungus *Fistulina hepatica* in Czeckoslovakia. *Lesnictvi (Prague).*, **28**, 399-412

Chase, T.E. and Ullrich, R.C. (1983). Sexuality, distribution, and dispersal of *Heterobasidion annosum* in pine plantations of Vermont. *Mycologia*, **75**, 825-31

Chastagner, G.A. and Byther, R.S. (1983a). Control of Swiss needle cast of Douglas fir Christmas trees with aerial applications of chlorothalonil. *Pl. Dis.*, **67**, 790-2

Chastagner, G.A. and Byther, R.S. (1983b). Infection period of *Phaeocryptopus gaumannii* on Douglas fir needles in Western Washington. *Pl. Dis.*, **67**, 811-3

Chastagner, G.A., Byther, R.S., MacDonald, J.D. and Michaels, E. (1984). Impact of Swiss needle cast on postharvest hydration and needle retention of Douglas fir Christmas trees. *Pl. Dis.*, **68**, 192-5

Chet, I.(editor)(1987). *Innovative approaches to Plant Disease Control*. Wiley, New York

Chiba, O. (1964). Studies on the variation in susceptibility and the nature of resistance of poplars to the leaf rust caused by *Melampsora larici-populina* Klebahn. *Bull. For. Exp. Sta. Meguro*, **166**, 86-157

Chiba, O. and Zinno, Y. (1960). Uredospores of the poplar leaf rust *Melampsora larici-populina* Kleb. as a source of primary infection. *J. Jap. For. Soc.*, **42**, 406-10

Childs, T.W. (1937). Variability of *Polyporus schweinitzii* in culture.*Phytopathology*, **27**, 29-50

Chock, A.K. (1979). The International Plant Protection Convention. In D.L.Ebbels and J.E.King (Eds), *Plant Health. The Scientific Basis for Administrative Control of Plant Diseases and Pests*, pp. 1-11. Blackwell, Oxford

Choi, D. and Simpson, J. A. (1991). Ascocarp development and cytology of *Cyclaneusma minus*. *Mycol. Res.*, **95**, 795-806

Christensen, C.M. (1940). Studies on the biology of *Valsa sordida* and *Cytospora chrysosperma*. *Phytopathology*, **30**, 459-75

Christensen, C.M. and Hodson, A.L. (1954). Artificially induced senescence of forest trees. *J. For.*, **52**, 126-9

Ciccarone, C. (1988). Defogliazione di *Populus nigra* da *Venturia populina* in Emilia. *Micologia Italiana*, **17**, 39-41

Clancy, K.J. and Lacey, B. (1986). Studies on the *Armillaria mellea* complex in Ireland. In *Research Report 1984-85. Faculty of General Agriculture, University College Dublin.*

Clapham, A.R., Tutin, T.G. and Moore, D.M. (1987). *Flora of the British Isles*.Ed.3. Cambridge University Press, Cambridge

Clapham, D.H. and Ekberg, I. (1986). Induction of tumours by various strains of *Agrobacterium tumefasciens* on *Abies nordmanniana* and *Picea abies*. *Scand. J. For. Res.*, **1**, 435-7

Clapham.D.H., Ekberg, I., Erikson, G., Hood, E.E. and Worrell, L. (1990). Within-population variation in susceptibility to *Agrobacterium tumefasciens* A 281 in *Picea abies* (L.). Karst. *Theoretical and Applied Genetics* **79**, 654-6

Claydon, N., Grove, J.F. and Pople, M. (1985). Elm bark beetle boring and feeding deterrents from *Phomopsis oblonga*.*Phytochemistry*, **24**, 937-43

Clinton, G.P. (1929). Willow scab-blight. *Proc. Am. Meeting Fifth Nat. Shade Tree Conf., 1929*, pp. 61-3

Clinton, G.P. and McCormick, F.A. (1929). The willow scab fungus *Fusicladium saliciperdum*. *Connecticut Agric. Exp. Sta. Bull.* No. 302, pp. 443-69

Cobb, F.W.Jr, and Schmidt, R.A. (1964). Duration of susceptiblity of Eastern White pine stumps to *Fomes annosus*. *Phytopathology*, **54**, 1216-8

Coleman, J.S., Murdoch, C.W., Campana, R.J. and Smith, W.H. (1985). Decay resistance of elm wetwood.*Can. J. Pl. Path.* **7**, 151-4

Coley-Smith, J.R., Verhoef, K. and Jarvis, W.R. (editors) (1980). *The Biology of Botrytis*. Academic Press, London

Collis, D.G. (1971). *Rhabdocline Needle Cast of Douglas Fir in British Columbia*. For. Insect and Dis. Survey Pest Leafl., No. 32

Conners, I.L., McCallum, A.W. and Bier, J.E. (1941). Willow blight in British Columbia. *Phytopathology*, **31**, 1056-8

Constantinescu, O. (1984). Taxonomic revision of *Septoria*-like fungi parasitic in Betulaceae. *Trans. Br. mycol. Soc.*, **83**, 383-98

Cooke, M.C. (1885). Some remarkable moulds. *J. Queckett microsc. Club*, *S II*, No. 12, 138-43

Cooley, S.J. (1984). *Meria laricis* on nursery seedlings of western larch in Washington. *Pl. Dis.*, **68**, 826

Cooper, J.I (1979). *Virus Diseases of Trees and Shrubs*, Institute of Terrestrial Ecology, c/o Unit of Invertebrate Virology, Oxford.

Cooper, J.I. (1988). In Smith *et al.* , 1988, q.v.

Cooper, J.I. and Atkinson, M.A. (1975). Cherry leaf roll virus causing a disease of *Betula* spp. in the United Kingdom. *Forestry*, **48**, 193-203

Cooper, J.I. and Edwards, M.-L. (1981). The distribution of poplar mosaic virus in hybrid poplars and virus detection by ELISA. *Ann. appl. Biol.*, **99**, 53-61

Cooper, J.I., Edwards, M.-L., Arnold, M.K. and Massalski, P.R. (1983). A tabovirus that invades *Fraxinus mariesii* in the United Kingdom. *Pl. Path.*, **32**, 469-72

Cooper, J.I., Edwards, M.-L. and Siwecki, R. (1986). The detection of poplar mosaic virus and its occurrence in a range of clones in England and Poland. *Eur. J. For. Path.*, **16**, 116-25

Cooper, J.I. Massalski, P.R. and Edwards, M.-L. (1984). Cherry leaf roll virus in the female gametophyte and seed of birch and its relevence to vertical virus transmission. *Ann. appl. Biol.*, **105**, 55-64

Corke, A.T.K. (1980). Biological control of tree diseases.*Ann. Rep. Long Ashton Res. Stn.* 1979, 190-8

Costa, M.E.A.P. da (1982). Uma doença do *Salix babylonica* em Portugal. *Garcia de Orta, Estudos Agronomicos*, **9**, 155-158 (abstract in *Rev. Pl. Path.*, **65**, No. 397, 1986)

Costonis, A.C. and Sinclair, W.A. (1967). Seasonal development of symptoms of ozone injury on eastern white pine. *Phytopathology*, **57**, 339

Cotter, H.van T. (1977). *Beech Bark Disease: Fungi and Other Associated Organisms*. M.S.Thesis, University of New Hampshire, 138 pp.

Coutts, M.P. and Rishbeth, J. (1977). The formation of wetwood in Grand fir. *Eur. J. For. Path.*, **7**, 13-22

Crandall, B.S (1950). The distribution and significance of the chestnut root rot Phytophthoras, *P. cinnamomi* and *P. cambivora*. *Pl. Dis. Reptr.* , **34**, 194-6

Crandall, B.S., Gravatt, G.F. and Ryan, M.M. (1945). Root disease of *Castanea* species and some coniferous and broadleaf nursery stocks, caused by *Phytophthora cinnamomi*. *Phytopathology*, **35**, 162-180

Crane, A.J. and Cocks, A.T. (1989). The transport, transformation and deposition of airborne emissions from power stations. In *Acid Deposition* (edited by Longhurst, J.W.S.). British Library Technical Publications

Cristinzio, M. (1938). Il male dell'inchiostro del castagno nella Campania. *Ric. Ossvz. Divulg. fitopat. Campania ed Mezzogiorne (Portici)*, **7**, 64-71

Cunningham, G.H. (1957). Thelephoraceae of New Zealand. Parts XII & XIII. *Trans. R. Soc. N. Z.*, **84**, 479-96

Curl, E.A. and Arnold, M.M. (1964). Influence of substrate and microbiological interactions on growth of *Fomes annosus*. *Phytopathology*, **54**, 1486-7

Curschmann, C.H. (1960). Über die sogenannte 'Ackersterbe' der Kiefe und ihre waldbaulichen Auswirkungen. *Forst Jagd*, **10**, 464-5

Dahte, A. (1960). Vorbeugung gegen Dothichiza-befall der Baumschulpflanzen durch Chemikalienspritzung. *Holzzentralblatt*, **86**, 934

Dam, B.C. van (1985). Niewe problemen met roest bij populier? *Populier*, **22**, 28-31 (abstract in *Rev. Pl. Path.*, **65**, No. 5741, 1986)

Dam, B.C. van and Kam, M. de (1984). *Sphaeropsis sapinea* (= *Diplodia pinea*) oorzaak van het afsterven van eindscheuten bij *Pinus* in Nederland. *Ned. BoschbTijdschr.*, **56**, 173-7

Dance, B.W. (1961a). Leaf and shoot blight of poplars (section Tacamahaca Spach). caused by *Venturia populina* (Vuill.). Fabric. *Can J. Bot.*, **39**, 875-90

Dance, B.W. (1961b). Spore dispersal in *Pollaccia radiosa* (Lib.). Bald. & Cif. *Can. J. Bot.*, **39**, 1429-35

Darker, G.D. (1932). The Hypodermataceae of conifers. *Contr. Arnold Arbor.*, **1**, 1-13

Darker, G.D. (1967). A revision of the genera of the Hypodermataceae. *Can. J. Bot.*, **45**, 1399-444

Darley, E.F. and Middleton, J.T. (1966). Problems of air pollution in plant pathology. *A. Rev. Pl. Path.*, **4**, 103-18

Darley, E.F. and Wilbur, W.D. (1954). Some relationships of carbon disulfide and *Trichoderma viride* in the control of *Armillaria mellea*. *Phytopathology*, **44**, 485 (abstract)

D'Aulerio, A.Z., Marchetti, L., Valle, E.Dalla, Giovanni, G.de, Badiali, G., Boselli, M. and Lodi, M. (1990). Controllo simultaneo su platano di *Gnomonia platani* (Kleb.). e *Corythucha ciliata* (Say). con iniezione di fitofarmaci al tronco. *Inftore fitopatol.*, **40**, 59-63. (abstract in *Rev. Pl. Path.*, **69**, No. 6658, 1990)

D'Aulerio, A.Z., Zambonelli, A. and Biavati, B. (1987). Ceppi di *Endothia parasitica* provenienti da varie regioni italiane in relazione alla virulenza ed alla lotta biologica. *Inf. fitopat.*, **37**, 43-8

Davidson, A.G. (1951). Decay of balsam fir in New Brunswick. *Bi-m. Prog. Rep. Div. For. Biol. Dep. Can.*, **7** (5), 1

Davidson, A.G. and Etheridge, D.E. (1963). Infection of balsam fir, *Abies balsamea* (L.). Mill., by *Stereum sanguinolentum. Can. J. Bot.*, **41**, 759-65

Davidson, A.G. and Fowler, M.E. (1967). Scab and black canker of willow, in Important Forest Insects and Diseases of Mutual Concern to Canada, the United States and Mexico, *Dep. For. rur. Dev. Can. Publ.*, No. 1180, pp. 201-3

Davidson, R.W. (1935a). Forest pathology notes. *Pl. Dis. Reptr.*, **19**, 94-7

Davidson, R.W. (1935b). Fungi causing stain in logs and lumber in the Southern states including five new species. *J. agric. Res.*, **50**, 789-807

Davidson, R.W., Campbell, W.A. and Vaughan, D.B. (1942). *Fungi causing decay in living oaks in the Eastern United States and their cultural identification.* U.S. Dept.Agric.Tech.Bull., No. 785

Davidson, R.W., Hinds, T.E. and Hawkesworth, F.G. (1959). *Decay of aspen in Colorado.* U.S. Dept. Agric.Rocky Mount.Forest Range Exp.Stn Pap., No. 45

Davis, S.H. (1948). Organic fungicides in the control of certain shade and ornamental tree diseases; abstract in *Phytopathology*, **38**, 575

Davis, W.C. and Latham, D.H. (1939). Cedar blight on wilding and forest tree nursery stock. *Phytopathology*, **29**, 991-2

Davison, A. W. (1971). The effects of de-icing salt on roadside verges. I: Soil and plant analysis. *J. Appl. Ecol.*, **8**, 555-61

Day, W.R. (1924). *The Watermark Disease of the Cricket Bat Willow*, Oxf.For.Mem., No. 3, 30pp.

Day, W.R. (1927a). The parasitism of *Armillaria mellea* in relation to conifers. *Ql. Jl. For.*, **21**, 9-21

Day, W.R. (1927b). The oak mildew *Microsphaera quercina* (Schw.). Burrill and *Armillaria mellea* (Vahl). Quel. in relation to the dying back of oak. *Forestry*, **1**, 108-12

Day, W.R. (1929). Environment and disease. A discussion on the parasitism of *Armillaria mellea* (Vahl)Fr.*Forestry*, **9**, 60-1

Day, W.R. (1930a). Mycological investigations. *J. For. Comm.* **9**, 63

Day, W.R. (1930b). The 'javart' disease of sweet chestnut, *Cytodiplospora castanea* Oudemans (=*Diplodina castaneae* Prilleux et Delacroix). *Q. Jl For.*, **24**, 114-7

Day, W.R. (1931). The relationship between frost and larch canker. *Forestry*, **5**, 41-56

Day, W.R. (1938). Root-rot of sweet chestnut and beech caused by species of *Phytophthora*. I.Cause and symptoms of disease: its relation to soil conditions. *Forestry*, **12**, 101-16

Day, W.R. (1939). Root-rot of sweet chestnut and beech caused by species of *Phytophthora*. II. Inoculation experiments and methods of control. *Forestry*, **13**, 46-58

Day, W.R. (1946). The pathology of beech on chalk soils. *Q. Jl. For.*, **40**, 72-86

Day, W.R. (1950). *Cambial Injuries In a Pruned Stand of Norway Spruce*, Forest Rec., Lond., No. 4, 11pp.

Day, W.R. and Peace, T.R. (1934). *The experimental production and the diagnosis of frost injury on forest trees*, Oxf. For. Mem. No. 16

Défago, G. (1937). *Cryptodiaporthe castanea* (Tul.). Wehmeyer, parasite du châtaignier. *Phytopath. Z.*, **10**, 168-77

Defraiteur, M. P. and Schumacher, R. (1988). Plateau des Hautes-Fagnes ou plateau des fontaines-salees? Une novelle ateinte a la reserve naturelle: les sels de deneigement. *Hautes Fagnes*, **1**, 9-13

Dehorter, B. and Perrin, R. (1983a). Production in vitro de périthèces du *Nectria ditissima*, agent du chancre du hêtre (*Fagus sylvatica*). I. Influence du milieu de culture et de la température. Application à la réalisation d'infections artificielles du hêtre. *Can. J. Bot.*, **61**, 1941-6

Dehorter, B. and Perrin, R. (1983b). Production in vitro de périthèces du *Nectria ditissima*, agent du chancre du hêtre (*Fagus sylvatica*). II. Effets de la composition carbon/azote du milieu nutritif et influence de la lumière. *Can. J. Bot.*, **61**, 1947-54

Del Cañizo, J. (1942). A new method of controlling ink disease of the European chestnut. *Int. Bull. Pl. Prot.*, **16**, 2M-3M

Delabraze, P. and Lanier, L. (1972). Contribution à la lutte chimique contre le gui (*Viscum album* L.). *Eur. J. For. Path.*, **2**, 95-103

Delatour, C. (1975). Comportement *in vitro* du *Sparassis laminosa* Fr. *Eur. J. For. Path.*, **5**, 240-7

Delatour, C. (1986a). Le problème de *Phytophthora cinnamomi* sur le chêne rouge (*Quercus rubra*). *EPPO Buli.*, **16**, 499-504

Delatour, C. (1986b). Le problème des *Ceratocystis* europeéns des chênes. *EPPO Bull.*, **16**, 521-5

Delatour, C. and Morelet, M. (1979). La pourriture noire des glands. *Rev. for. fr.*, **31**, 101-15

Delbos, R.P. and Dunez, J. (1988). *Apple chlorotic leaf spot virus (ACLSV)* In Smith *et al.*, 1988, q.v., pp. 5-7

Delen, N. (1980). Studies on the control possibilities of Chestnut blight (*Endothia parasitica* (Murr.). A. and A.). in Turkey. II. Appearance possibility of resistance after continuous application of effective systemic fungicides against the pathogen in vitro. *J. Turkish Phytopathol.*, **9**, 27-47 (abstract in *Rev. Pl. Path.*, **60**, No. 2792, 1981).

Delforge, P. (1930). Le *Chrysomyxa abietis* (rouille des aiguilles de l'Epicéa). *Bull. Soc. Centr. For. Belg.*, **37**, 419-23

Dennis, R.W.G. (1931). The black canker of willows. *Trans. Br. mycol. Soc.*, **16**, 76-84

Dennis, R.W.G. (1968). *British Ascomycetes*, J. Cramer, Lehre

Dennis, R.W.G. and Wakefield, E.M. (1946). New or interesting British fungi. *Trans. Br. mycol. Soc.*, **29**, 141-66

Derbyshire, D.M. and Crisp, A.F. (1978). Studies on treatments to prolong the storage life of carrots. *Expl. Hort.*, **30**, 23-8

Desvignes, J.C. (1988). *Cherry green ring mottle disease.* In Smith *et al.* , 1988, q.v., pp. 103-4

Dewey, F.M. and Brasier, C.M. (1988). Development of ELISA for *Ophiostoma ulmi* using antigen-coated wells. *Pl. Path.*, **37**, 28-35

Dewey, F.M., Munday, C.J. and Brasier, C.M. (1989). Monoclonal antibodies to specific components of the Dutch elm disease pathogen *Ophiostoma ulmi*. *Pl. Path.*, **38**, 9-20

Diamandis, S. (1978a). 'Top-dying' of Norway spruce, *Picea abies* (L.) Karst., with special reference to *Rhizosphaera kalkhoffii* Bubak. I.Development of foliar symptoms. *Eur. J. For. Path.*, **8**, 337-45

Diamandis, S. (1978b). 'Top-dying' of Norway spruce, *Picea abies* (L.) Karst., with special reference to *Rhizosphaera kalkhoffii* Bubak. II. Status of *R. kalkhoffii* in 'top-dying' of Norway spruce. *Eur. J. For. Path*, **8**, 345-56

Diamandis, S. (1979). 'Top-dying' of Norway spruce, *Picea abies* (L.) Karst., with special reference to *Rhizosphaera kalkhoffii* Bubak. VI. Evidence related to the primary cause of 'top-dying'. *Eur. J. For. Path.*, **9**, 183-91

Diamandis, S. and Kam, M de (1986). A severe attack on Scots pine by resin top disease in N. Greece. *Eur. J. For. Path.*, **16**, 247-9

Dick, M. (1985). Stem canker of one-year-old *Pinus radiata*. *N. Z. Jl. For.*, **30**, 87-93

Dickenson, Susan and Wheeler, B.E.J. (1981). Effects of temperature, and water stress in sycamore, on growth of *Cryptostroma corticale*. *Trans Br. mycol. Soc.*, **76**, 181-5

Dierks, R. (1959). Zur chemischen Bekämpfung der Korbweiden-Parasiten *Fusicladium saliciperdum* (All. et Tub.). Lind. und *Glomerella miyabeana* (Fuk.). v. Arx. *Pflanzenschutz*, **11**, 125-30

Dimitri, L. (1967). Untersuchen über die ätiologie des 'rindensterbens' der Buche. *Forstwiss. ZentBl.*, **86**, 257-76

Dirr, M.A. (1976). Selection of trees for tolerance to salt injury. *J. Arboric.*, **2**, 209-16

Diwani, S.A. and Millar, C.S. (1986). Needle-cast disease in nurseries in north-east Scotland. *Scott. For.*, **40**, 185-9

Diwani, S.A. and Millar, C.S. (1987). Pathogenicity of three *Lophodermium* species on *Pinus sylvestris* L. *Eur. J. For. Path.*, **17**, 53-8

Doane, C.C. (1962). Evaluation of insecticide for control of the smaller European elm bark beetle, *J. econ. Ent.*, **55**, 414-5

Dobbs, C.G. (1951). Fruiting of the oak mildew (*Microsphaera alphitoides*). *Nature, Lond.*, **167**, 357

Dobbs, R.C. and McMinn, R.G. (1973). Hail damage to a new White spruce and Lodgepole pine plantation in central British Columbia. *For. Chron.*, **49**, 174-5

Dobson, M. (1990). De-icing salt damage to trees and shrubs and its amelioriation. *Proceedings of Arboriculture Conference, York, 1990.* Forestry Commission Bulletin, No. 97.

Dochinger, L.S. (1956a). Maple wilt progress report. *Shade Tree*, **29**, 1-3

Dochinger, L.S. (1956b). New concepts of the Verticillium wilt disease of maple.

*Phytopathology*, **46**, 467 (abstract)

Dodge, B.O. (1944). *Volutella buxi* and *Verticillium buxi*. *Mycologia*, **36**, 416-25

D'Oliveira, A.L.B. and Pimento, A.AL. (1953). Infecçoes latentes de *Melampsoridium betulinum* (Pers.) Kleb. em gomos de *Betula celtiberica* Rothm. et Vasc. *Estud. Inform serv. flor. aqüic. Portugal*, No. 9-C3, 3pp., abstract in *For. Abstr.*, **15**, No. 2646

D'Oliveira, M de L. (1939). Inoculacaoes experimentais com o *Bacterium savastanoi* E.F.Smith e o *Bacterium savastanoi* var. *fraxini* N.A.Brown. *Agron. lusit.*, **1**, 88-102

Donandt, S. (1932). Untersuchungen über die Parthogenität des Wirtelpilzes *Verticillium alboatrum* R.u.B. *Z. Parasitenkunde*, **4**, 653-711

Donaubauer, E. (1964). Untersuchungen über die Variation der Krankheitsanfalligkeit verschiedener Pappeln. *Mitt. Forstl. Bundes-Versuchsamst. Mariabrunn*, **63**, pp. 121 (Vienna); abstract in *For. Abs.*, **26**, 1965, No. 817

Donaubauer, E. (1965). [The Marssonina disease of poplar in Austria]. *Allg. Forstz.*, **76**, suppl. (Informationsdienst No. 96), 2pp.; abstract in *For. Abs.*, **27**, 1966, No. 2424

Donaubauer, E. (1967). Über die Verbreitung von *Marssonina brunnea* (Ell. & Ev.). Magn. und *M. populi-nigrae* Kleb. *Pap. IUFRO Congr. Munich 1967*, Vol. 5, pp. 279-84

Donnelly, D.M.X. and Sheridan, M.H. (1986). Anthraquinones from *Trichoderma polysporum*. *Phytochemistry*, **25**, 2303-94

Dooley, H.L. (1984). Temperature effects on germination of uredospores of *Melampsoridium betulinum* and on rust development. *Pl. Dis.*, **68**, 686-8

Dorworth, C.E. and Krywienczyk, J. (1975). Comparisons among isolates of *Gremmeniella abietina* by means of growth rate, conidia measurement and immunogenic reaction. *Can J. Bot.*, **53**, 2506-25

Dowding, P. (1970). Colonisation of freshly bared pine sapwood surfaces by staining fungi. *Trans. Br. mycol. Soc.*, **55**, 399-412

Dowding, P. (1973). Effects of felling time and insecticide treatment on the interrelationships of fungi and arthropods in pine logs. *Oikos*, **24**, 422-9

Dowson, W.J. (1937). *Bacterium salicis* Day, the cause of the watermark disease of the cricket-bat willow. *Ann. appl. Biol.*, **24**, 528-44

Dowson, W.J. (1957). *Plant diseases due to bacteria*, 2nd edn, Cambridge University Press, Cambridge

Dowson, W.J. and Callan, E. McC. (1937). The watermark disease of the white willow. *Forestry*, **11**, 104-8

Dowson, W.J. and Dillon-Weston, W.A.R. (1937). Brown rot of hawthorn. *Gdnrs' Chron.*, **101**, No. 2624, 426

Driver, C.H. (1963a). Effect of certain chemical treatments on colonisation of slash pine stumps by *Fomes annosus*. *Pl. Dis. Reptr.*, **47**, 569-71

Driver, C.H. (1963b). Further data on borax as a control of surface infection of slash pine stumps by *Fomes annosus Pl. Dis. Reptr.*, **47**, 1006-9

Driver, C.H. and Dell, T.R. (1961). *Fomes annosus* root-rot in slash pine plantations of the Eastern Gulf Coast States *Pl. Dis. Reptr.*, **45**, 38-40

Driver, C.H. and Ginns, J.H., Jr. (1966). A method of mass screening of Southern pines for resistance to a root-rot induced by *Fomes annosus* (Fr.). Cke, in H.D.Gerhold, E.J.Shreiner, R.E. McDermott and J.A. Winieski (Eds), *Breeding pest-resistant trees*, Pergamon Press, London, pp. 421-2

Duchesne, L.C., Ellis, B.E. and Peterson, R.L. (1989). Disease suppression by the ectomycorrhizal fungus Paxillus involutus: contribution of oxalic acid. *Can. J. Bot.*, **67**, 2726-30

Duda, B. and Sierota, Z.H. (1987). Survival of Scots pine seedlings after biological and chemical control of damping-off in plastic greenhouses. *Eur. J. For. Path.*, **17**, 110-7

Dufrénoy, J. (1922a). Les maladies du châtaignier. *C. r. Congres Regional a Brive*, 1922, 45-63

Dufrénoy, J. (1922b). Tumeurs de *Sequoia sempervirens*. *Bull. Soc. Path. veg. France*, **9**, 277-81

Dufrénoy, J. and Dufrénoy, M.L. (1927). Hadromycoses. *Annls Epiphyt.*, **13**, 195-212

Durand, E.J. (1913). The genus *Keithia*. *Mycologia*, **5**, 6-11

Durrieu, G., Beneteau, A. and Niocel, S. (1985). *Armillaria obscura* dans l'écosystème forestier de Cerdagne. *Eur. J. For. Path.*, **15**, 350-5

Ebrahim-Nesbat, F. and Heitefuss, R. (1985). Rickettsien-ahnliche Bakterien (RLO) in Feinwurzeln erkrankter Fichten unterschiedlichen Alters. *Eur. J. For. Path*, **15**, 182-7

Ebrahim-Nesbat, F. and Heitefuss, R. (1989). Isolierung eines Potyvirus aus erkranker Fichte im Bayerischen Wald. *Eur. J. For. Path.*, **19**, 222-30

Edmonds, R.L., Hinshaw, R.W. and Leslie, K.B. (1984a). A 24 hour deposition sampler for spores of *Heterobasidion annosum*. *Phytopathology*, **74**, 1032-4

Edmonds, R.L., Leslie, K.B. and Driver.C.H. (1984b). Spore deposition of *Heterobasidion annosum* in thinned coastal western hemlock stands in Oregon and Washington. *Pl. Dis.*, **68**, 713-5

Edson, H.A. (1938). United States of America: bark canker of Monterey cypress. *Int. Bull. Pl. Prot.*, **12**, 98

Edwards, M.C. and Ayres, P.G. (1981). Cell death and cell wall papillae in the resistance of oak species to powdery mildew disease. *New Phytol.*, **89**, 411-8

Edwards, M.C. and Ayres, P.G. (1982). Seasonal changes in resistance of *Quercus petraea* (sessile oak) leaves to *Microsphaera alphitoides*. *Trans. Br. mycol. Soc.*, **78**, 569-71

Edwards, M.V. (1952). Provenance studies. *Rep. Forest Res. Lond.*, 1951, pp. 38-43

Ehrlich, J. (1934). The beech bark disease, a *Nectria* disease of a *Fagus* following *Cryptococcus fagi* (Baer). *Can. J. Res.*, **10**, 593-692

Elgersma, D.M. (1967). Factors determining resistance of elms to *Ceratocystis ulmi*. *Phytopathology*, **57**, 641-2

Eliade, E. (1985). Fainarea platanului produsa de *Microsphaera platani* Howe in Romania. *Analele Universitatii Bucarasti, Biologie*, **34**, 55-7 (abstract in *Rev. Pl. Path.*, **67**, No. 977, 1988)

Ellis, M.B. and Ellis, J.P. (1985). *Microfungi on Land Plants: An Identification Handbook*. Croom Helm, London

Ende, G. van den (1958). Untersuchungen über den Pflanzenparasiten *Verticillium albo-atrum* Reinke et Berthe. *Acta bot. neerl.*, **7**, 665-740

Engelhard, A.W. (1957). Host index of *Verticillium albo-atrum* Reinke & Berthe. (including *Verticillium dahliae* Kleb.). *Suppl. Pl. Dis. Reptr.*, **244**, 24-49

Englerth, G.H. (1942). *Decay of Western Hemlock in Western Oregon and Washington*. Bull.Sch.For.Yale, No. 50, pp. i and 53

Epners, Z. (1964). A new psychrophilic fungus causing germination failure of conifer seeds. *Can. J. Bot.*, **42**, 1589-604

EPPO(1978). Data sheets on quarantine organisms, set 1. *EPPO Bull.* **8** (2). and as separate sheets

EPPO(1979a). Data sheets on quarantine organisms, set 2. *EPPO Bull.* **9** (2). and as separate sheets

EPPO (1979b). *Report of the Joint EPPO/IUFRO Working Party Meeting on Phytosanitary regulations in Forestry*. EPPO, Paris

EPPO (1980). Data sheets on quarantine organisms, set 3. *EPPOBull.*, **10** (1) and as separate sheets

EPPO (1990). Oak decline and the status of *Ophiostoma* spp. on oak in Europe. *EPPO Bull.*, **20**, 405-22

Epstein, A.H. (1978). Control tactics in research and practice.Preventing root graft transmission. In W.A.Sinclair and R.J.Campana (Eds), *Dutch Elm Disease, Perspectives After 60 Years, Search, Agriculture*, **8**, 32-3

Escarment, B.L., Taris, B., Ducom, P., Lanusse, G.P. and Malato, G. (1985). Action et comportement in situ du bromure de méthyle sur le développement d'*Armillaria obscura*, parasite du pin maritime dans le Sud-Ouest de la France. *EPPO Bull.*, **15**, 97-104

Etheridge, D.E. (1955). Comparative studies of North American and European cultures of the root-rot fungus *Fomes annosus* (Fr.). Cooke. *Can. J. Bot.*, **33**, 416-28

Etheridge, D.E. (1963). Infection of balsam fir, *Abies balsamea* (L.). Mill. by *Stereum sanguinolentum* (Alb. and Schw. ex Fr.)Fr. *Trans. R. Soc. Can.* **1**, Ser.4, Sect.3, pp 357-60

Etheridge, D.E. and Morin, L.A. (1967). The microbiological condition of wood of living balsam fir and black spruce in Quebec. *Can. J. Bot.*, **45**, 1003-10

Evans, J. (1984). *Silviculture of broadleaved woodland*. Bull.For.Commn. Edin., 232pp.

Evans, J. (1986). A re-assessment of cold-hardy Eucalypts in Great Britain. *Forestry*, **59**, 223-42

Everard, J.E. (1974). *Fertilizers in the Establishment of Conifers in Wales and Southern England*, Bookl.For.Commn, No. 41

Fahy, P.C. and Persley, G.J. (1983). *Plant Bacterial Diseases: a Diagnostic Guide*. Academic Press, London

Falk, S.P., Griffin, D.H. and Manion, P.D. (1989). Hypoxylon canker incidence and mortality in naturally occurring aspen clones. *Pl. Dis.*, **73**, 394-7

Farsky, O. and Fekete, D. (1965). *Glomerella miyabeana* (Fukushi). Arx (*Physalospora miyabeana* Fukushi), pricinou odumirani rizku stromovykh Vrb na jiznum Slovensku. *Lesn. Cas Ust. ved. Inform. MZLVH*, **10**, 745-58

Faulkner, R. and Holmes, G.D. (1954). Experimental work in nurseries. *Rep. Forest Res., Lond.*, 1953. pp. 17-31

Federov, N.I. and Bobko, I.N. (1988a). [Protection of pine plantations from root rot induced by honey fungus. I. Chemical control]. *Mikol. Fitopatol.*, **22**, 255-61 (abstract in *Rev. Pl. Path.*, **68**, No. 3961, 1989)

Federov, N.I. and Bobko, I.N. (1988b). [Protection of pine stands from root rot caused by honey fungus. II. Biological measures]. *Mikol. Fitopatol.*, **22**, 456-460. (abstract in *Rev. Pl. Path.*, **68**, No. 3959, 1989).

Federov, N.I. and Poleshchuk, Yu.M. (1982). [Using a preparation of *Peniophora gigantea* in Scots pine stands ]. *Les. Khoz.*, No. 6, 53-5 (abstract in *Rev. Pl. Path.*, **63**, No. 5158, 1984).

Ferdinandsen, C. and Jørgensen, C.A. (1938-39). *Skovtraernes Sygdomme*, Gyldendalske Boghandel, Copenhagen

Fergus, C.L. (1953). Relation of weather to the severity of white oak anthracnose. *Phytopathology*, **43**, 103-5

Fernandes, C.T. (1955). A luta contra a 'doenca da tinta' nos santos do norte de Portugal e ensoios diversos para a sua major eficiencia e economica. *Publ. Serv. flor. Portugal*, **22**, 53-9

Findlay, W.P.K. (1959a). Sap stain of timber. *For. Abstr.*, **20**, 1-7Findlay, W.P.K. (1959b). Sap stain of timber. *For. Abstr.*, **20**, 167-74

Fischer, E. (1933). Die Rostepidemie der Rottane in den Alpen in herbst 1932.*Mitt. Naturforsch. Ges. Bern*, 1932, 20-1,

Fischer, E. (1938). Die Douglasienschütte. *Blumen-u PflBau ver. Gartenwelt*, **62**, 331-2

Fitzpatrick, H.M. (1930). *The Lower Fungi*, McGraw-Hill, New York and London

Flack, N.J. and Swinburne, T.R. (1977). Host range of *Nectria galligena* Bres. and the pathogenicity of some Northern Ireland isolates. *Trans. Br. mycol. Soc.*, **68**, 185-92

Flemming, S., Cwielong, P. and Hüttermann, A. (1982). The cell wall of *Fomes annosus* (*Heterobasidion annosum*) as a target for biochemical control. 2. The influence of inhibitors of cell wall synthesis. *Eur. J. For. Path.*, **12**, 273-80

Fluckiger, W. and Braun, S. (1981). Perspectives of reducing the deleterious effect of de-icing salt upon vegetation. *Plant Sci.*, **63**, 527-9

Foex, E. (1941). L'invasion des chênes de l'Europe par le blanc ou oidium (*Microsphaera alphitoides* Griffon et Maublanc). *Rev. Eaux For.*, **79**, 338-49

Foister, C.E. (1948). Report of the annual conference of the cryptogamic section, 1946. *Trans. bot. Soc. Edinb.*, **34**, 392-6

Foister, C.E. (1961). *The Economic Plant Diseases of Scotland*. Dep.Agric.Fish.Scotl., Tech. Bull., No. 1, HMSO, Edinburgh

Forbes, R.S., Stone, G.L. and Moran, G.V. (1965). The beech disease in the Maritime Provinces. *Bi-mon. Prog. Rep. Div. For. Biol., Ottawa*, **21**, 1

Foster, R.E. (1962). *Fomes annosus* in British Columbia.*Conference and Study Tour on Fomes annosus, Scotland, 1960*, IUFRO, Firenze, pp. 19-20

Fowler, D. and Cape, J.N. (1982). Air pollutants in agriculture and horticulture. In *Effects of gaseous pollution in agriculture and horticulture* (edited by Unsworth, M.H.; Ormrod, D.P.). London, UK, Butterworths, 3-26

Fowler, M.E. (1938). Twig cankers of asiatic chestnuts in the eastern United States. *Phytopathology*, **28**, 693-704

Fowler, M.E. (1962). *Fomes annosus* in North Eastern United States, *Conference and Study Tour on Fomes annosus, Scotland, 1960, IUFRO, Firenze*, pp. 20-22

Francis, S.M. (1986). Needle blight of conifers. *Trans. Br. mycol. Soc.*, **87**, 397-400

Francis, S.M. and Waterhouse, G.M. (1988). List of Peronosporaceae reported from the British Isles. *Trans. Br. mycol. Soc.***91**, 1-62

Francke-Grosman, H. (1962). Under what conditions can *Fomes annosus* grow in non-sterilised soil? *Conference and Study Tour on Fomes annosus, Scotland, 1960, IUFRO, Firenze*, pp. 22-8

Franken, E. (1956). Witterungsverlauf im Frühjahr 1955 und Pilzkrankheiten der Pappeln. *Holzzucht*, **9**, 30

Frankie, G.W. and Parmeter, J.R. Jr. (1972). A preliminary study of the relationship between *Coryneum cardinale* (Fungi Imperfecti). and *Laspeyresia cupressana* (Lepidoptera: Tortricidae). *Pl. Dis Reptr.*, **56**, 992-5

Friend, R.J. (1965). A study of sooty moulds on lime trees (*Tilia* × *vulgaris*). *Trans. Br. mycol. Soc.*, **48**, 367-70

Fukushi, T. (1921). A willow canker disease caused by*Physalospora miyabeana* and its conidial form *Gloeosporium*. *Ann. phytopath. Soc. Japan*, **1**, 1-12

Fulbright, D.W., Paul, C.P. and Garrod, S.W. (1988). Hypovirulence: a natural control of chestnut blight. In *Biocontrol of Plant Diseases*. Vol. II. (Edited by Mukerji, K.G. and Garg, K.L). CRC Press, USA

Funk, A (1972). Sirococcus shoot-blight of western hemlock in British Columbia and Alaska. *Pl. Dis. Reptr.*, **56**, 645-7

Gadgil, P.D. (1967). Infection of *Pinus radiata* needles by *Dothistroma pini. N. Z. Jl. Bot.*, **5**, 499-503

Gadgil, P.D. (1984). *Cyclaneusma* (*Naemacyclus*) needle cast of *Pinus radiata* in New Zealand. 1: Biology of *Cyclaneusma minus. N. Z. Jl. For. Sci.*, **14**, 179-96

Gaggini, J.B. (1970). Anthracnose of weeping willows. *Gdnrs' Chron.*, **13**, 168

Gaisberg, E. von (1928). Studien uber'den Larchenkrebspilz *Dasyscypha willkommii*, insbesondere uber die Keimung seiner sporen. *Centralb. fur Bakt.*, *Ab.* **2**, 73, pp. 206-33

Gaisberg, E. von (1937). Uber die Adelopus-Nadelschutte in Wurttembergischen Douglasienbestanden mit Hinweis auf die bisher hier bekanntgewordene Verbreitung von *Rhabdocline. Silva*, **25**, 37-42, 45-8

Garbowski, L. (1936). Przyczynek do znajomosci mikroflory grzybrej nasion drzew lesnych. *Prace Wydz. Chor. Rosl panstw. Inst. Nauk Gosp. wiejsk Bydgoszczy*, **15**, 5-30; abstract in *Rev. appl. Mycol.*, **16**, 147, 1937

Gardiner, A.S. (1965). 'Flamy' birch and its frequency in some highland populations. *Scott. For.*, **19**, 180-4

Gardner, M.W., Yarwood, C.E. and Duafala, T. (1972). Oak mildews. *Pl. Dis. Reptr.*, **56**, 313-7

Gareth Jones, D. (1987). *Plant Pathology: Principles and Practice.* Open University Press, Milton Keynes

Garrett, C.M.E., Panagopolos, C.G. and Crosse, J.E. (1966). Comparison of plant pathogenic pseudomonads from fruit trees. *J. appl. Bact.*, **29**, 342-56

Garrett, S.D. (1956a). *Biology of Root-Infecting Fungi*, Cambridge University Press, Cambridge

Garrett, S.D. (1956b). Rhizomorph behaviour in *Armillaria mellea* (Vahl). Quél. II. Logistics of infection. *Ann. Bot.*, **20**, 193-209

Garrett, S.D. (1958). Inoculum potential as a factor limiting lethal action by *Trichoderma viride* Fr. on *Armillaria mellea* (Fr.). Quel. *Trans. Br. mycol. Soc.*, **41**, 157-64

Gäumann, E. (1959). *Die Rostpilze Mitteleuropas, Beitrage zur Kryptogamenflora der Schweiz, 12*. Buchdruckerei Buchler, Bern

Gäumann, E., Roth, C. and Anliker, J. (1934). Uber die Biologie der *Herpotrichia nigra* Z. *PflKrankh., PflPath., Pflschutz*, **44**, 99-116

Gene, J., Guarro, J. and Figueras, M.J. (1990). A new species of *Cryptosporiopsis* causing rot of *Corylus avellana. Mycol. Res.*, **94**, 309-12

Georgescu, C. and Gasmet, V. (1954). Un atac de *Rosellinia byssiseda* (Tode). Schroet., la pinettii de Molid. *Revista padurilor*, **68**, 31-34.

Georgevitch, P. (1926). Susenje hrastovich suma u Slavoniji. Abstract in *Rev. appl. Mycol.*, **6**, 5, 1927

Gergacz, J. (1967). *Marssonina* Karositasa Nyarakon. *Erdo*, **16**,304-8; abstract in *Rev. appl. Mycol.*, **47**, 1968, No. 3585

Gerlings, J.H.J. (1939). Herkomstonderzoek van den Douglaspar aan de afdeeling houtteelt van het Instituut voor Boschbouwkundig Onderzoek. *Ned. BoschbTijdschr.*, **12**, 405-32

Gerwen, C.P.van(1980). Kanker bij benk, een gevaar voor verjonging onder scherm. *Ned. Boschb. Tijdschr.*, **52**, 1-5

Gibbs, J.N. (1967a). A study of the epiphytic growth habit of *Fomes annosus.Ann. Bot.*, **31**, 755-74

Gibbs, J.N. (1967b). The role of host vigour in the susceptibility of pines to *Fomes annosus. Ann. Bot.*, **31**, 803-15

Gibbs, J.N. (1978a). Development of the Dutch elm disease epidemic in southern England, 1971-1976. *Ann. appl. Biol.*, **88**, 219-28

Gibbs, J.N. (1978b). Oak wilt.*Arboric. J.*, **3**, 351-6

Gibbs, J.N. (1979). Measures to prevent oak wilt from reaching Europe, in D.L.Ebbels and J.E.King (Eds), *Plant Health: The Scientific Basis For the Administrative Control of Plant Diseases and Pests*, Blackwell, Oxford

Gibbs, J.N. (1981). European forestry and *Ceratocystis* species. *EPPO Bull.*, **11**, 193-7

Gibbs, J.N. (1984). *Oak Wilt.* Forest Rec., Edin., No. 111

Gibbs, J.N. and Brasier, C.M. (1973). Correlation between cultural characters and pathogenicity in *Ceratocystis ulmi* isolates from Britain, Europe and North America. *Nature, London.*, **241**, 381-3

Gibbs, J.N. and Burdekin, D.A. (1983). De-icing salt and crown damage to London plane. *Arb. J.*, **7**, 227-37

Gibbs, J.N., Burdekin, D.A. and Brasier, C.M. (1977). *Dutch Elm Disease*, Forest Rec., Lond.,

No. 115

Gibbs, J.N., England, N. and Wolstenholme, R. (1988). Variation in the pine stem rust fungus *Peridermium pini* in the United Kingdom. *Pl. Path.*, **37**, 45-53

Gibbs, J.N., Greig, B.J.W. and Hickman, I.T. (1987). An analysis of Peridermium stem rust of Scots pine in Thetford Forest in 1984 and 1985. *Forestry*, **60**, 203-18

Gibbs, J.N. and Wainhouse, D. (1986). Spread of forest pests and pathogens in the Northern Hemisphere. *Forestry*, **59**, 141-53

Gibbs, J.N., Young, C.W.T. and Strouts, R.G. (1980). In Forest Pathology. *Rep. Forest Res. Edinb.* 1980, p.34

Gibson, I.A.S. (1957). Saprophytic fungi as destroyers of germinating pine seeds. *E. Afr. agric. J.*, **22**, 203-6

Gibson, I.A.S. (1966). A note on *Stereum sanguinolentum*, a new record for Kenya forests. *E. Afr. agric. For. J.*, **32**, 38-40

Gibson, I.A.S. (1972). Dothistroma blight of *Pinus radiata A. Rev. Phytopathol.*, **10**, 51-72

Gibson, I.A.S., Christensen, P.S. and Munga, F.N. (1964). First observations in Kenya on a foliage disease of pines caused by *Dothistroma pini* Hulbary. *Commonw. For. Rev.*, **43**, 31-48

Gibson, I.A.S. and Goodchild, N.A. (1961). *Armillaria mellea* in Kenya tea plantations. *Rep. 6th Commonw. mycol. Conf.*, pp. 39-48

Gibson, I.A.S., Kennedy, P. and Dedan, J.K. (1966). Further observations in Kenya on a foliage disease of pines caused by *Dothistroma pini* Hulbary. *Commonw. For. Rev.*, **45**, 67-76

Gill, L.S. (1925). Notes on sporophores of *Polyporus schweinitzii* Fr. on yellow pine in California. *Phytopathology*, **15**, 492-3

Ginns, J.H.Jr. (1968). *Rhizina Root Rot of Douglas Fir in British Columbia*, For. Pest Leafl.Victoria, 5pp.

Ginzberg, M. (1961). Biologia i szkodliwosc grzyba *Venturia tremulae* Aderh. w Polsce. *Prace Inst. badaw. Lesn.*, **211**, 3-37

Gladman, R.J. and Greig, B.J.W. (1965). *Principal Butt Rots of Conifers*, Bookl.For.Commn, No. 13

Gladman, R.J. and Low, J.D. (1963). Conifer heart rots in Scotland. *Forestry*, **36**, 227-44

Glasscock, H.H. and Rosser, W.R. (1958). Powdery mildew on *Eucalyptus*, in 'New or uncommon plant diseases and pests'. *Pl. Path.*, **7**, 152

Glerum, C. and Farrar, J.L. (1966). Frost ring formation in the stems of some coniferous species. *Can. J. Bot.*, **44**, 879-86

Goidanich, G. (1932). La verticilliosi dell' '*Acer campestre*' L. e alcuni altri casi di tracheomicosi in Italia. *Boll. R. Staz. Pat. Veg., N. S.*, **12**, 285-97

Goidanich, G. (1934). La verticilliosi dell' *Acer platanoides* L., dell'*Acer pseudoplatanus* L. e della *Maclura aurantiaca* L. *Boll. R. Staz. Pat. Veg., N. S.*, **14**, 268-72

Goidanich, G. (1940). La 'necrosi corticale' del Pioppo causata da *Chondroplea populea* (Sacc. & Br.). Kleb. Repr. from *Riv. cellulosa*, **18**, 5, 29pp.

Gojkovic, G. (1970). Chemical protection of poplars against *M. brunnea* in Jugoslavia. *Topola*, **14**, 3-68

Goode, M.J. (1953). Control of oak leaf blister in Mississippi. *Phytopathology*, **43**, 472(abstract)

Goode, M.J. and Parris, G.K. (1954). Additions to our knowledge of the life cycle of *Taphrina caerulescens*. *Phytopathology*, **44**, 332 (abstract)

Gooding, G.V. (1964). Effect of temperature on growth and survival of *Fomes annosus* in freshly cut pine bolts. *Phytopathology*, **54**, 893-4 (abstract)

Gordon, A.G., Salt, G.A. and Brown, R.M. (1976). Effect of pre-sowing moist-chilling treatments on seedbed emergence of Sitka spruce seed infected by *Geniculodendron pyriforme* Salt. *Forestry*, **49**, 143-51

Gorter, G.J.M.A. (1984). Identity of oak powdery mildew in South Africa. *South African Forestry Journal*, **129**, No. 81-2

Gothe, H. (1957). Beobachtungen uber stockfaule in Schlitzer Larchen-bestanden. 2 Mitteilung. *Forst- u. Holzwirt*, **12**, 70-4

Gourbiere, F., Pepin, R. and Bernillon, D. (1986). Microscopie de la mycoflore des aiguilles de sapin (*Abies alba*). II. *Lophodermium piceae. Can J. Bot.*, **64**, 102-7

Govi, G. and D'Aulerio, A.Z. (1982). Resistenza varietale del castagno all'*Endothia parasitica. Inftore fitopatol.*, **32**, 59-63

Govi, G. and Deserti, F. (1980). Interventi chimico contro il *Coryneum cardinale* del cipresso. *Inftore fitopatol.*, **30**, 19-21

Graniti, A. (1986). *Seiridium cardinale* and other cypress cankers. *EPPO Bull.***16**, 479-86

Graniti, A. (1988). In *European Handbook of Plant Diseases* (Ed.Smith, I.M., Dunez, J.,

Lelliott, R.A., Phillips, D.H. and Archer, S.A.). Blackwell, Oxford

Grant, T.J. (1937). *Reduction of Nectria Canker in Hardwood Forests of the Northeast*, U.S. Dept.Agric.N.East Forest Exp.Stn.Occasional Pap., No. 6

Grasso, V. (1952). Conifere suscettibili ed immuni al *Coryneum cardinale* Wag. *Italia for. mont.* **7**, 148-9

Grasso, V. (1969). Attachi di *Coryneum cardinale* Wag. su galbule di cupressi. *Italia for. mont.*, **24**, 181-3

Grasso, V., Panconesi, A. and Raddi, P. (1979). Testing for resistance to cypress canker disease in Italy. *Phytopathologia Mediterranea*, **18**, 166-71

Gravatt, G.F. (1926). *Maple Wilt*, U.S. Dept.Agric.Circ., No. 382

Green, D.E. and Hewlett, M.A. (1949). Die-back of *Cytisus* cuttings. *Jl. R. hort. Soc.*, **74**, 310-2

Green, F.M. (1932). Observations on *Cucurbitaria laburni* (Pers.). de Not. *Trans. Br. mycol. Soc.*, **16**, 289-303

Greenhalgh, G.N. (1976). Aberdeen foray. *Bull. Br. mycol. Soc.*, **10**, 55-63

Greenhalgh, G.N. and Bevan, R.J. (1978). Response of *Rhytisma acerinum* to air pollution. *Trans Br. mycol. Soc.*, **71**, 491-523

Gregory, P.H., Peace, T.R. and Waller, S. (1949). Death of sycamore trees associated with an unidentified fungus. *Nature, Lond.*, **164**, 275

Gregory, P.H. and Waller, S. (1951). *Cryptostroma corticale* and sooty bark disease of sycamore (*Acer pseudoplatanus*). *Trans. Br. mycol. Soc.*, **34**, 579-97

Gregory, S.C. (1982). Bark necrosis of *Acer pseudoplatanus* L. in northern Britain. *Eur. J. For. Path.*, **12**, 157-67

Gregory, S.C. (1989). *Armillaria* species in northern Britain. *Pl. Path.*, **38**, 93-7

Gregory, S.C., Redfern, D.B., MacAskill, G.A. and Pratt, J.E. (1988). In Pathology. *Rep. Forest Res. Edinb.*1988, p.39

Gregory, S.C. and Watling, R. (1985). Occurrence of *Armillaria borealis* in Britain. *Trans Br. mycol. Soc.*, **84**, 47-55

Greig, B.J.W. (1962). *Fomes annosus* (Fr.)Cke. and other root-rotting fungi in conifers on ex-hardwood sites. *Forestry*, **35**, 164-82

Greig, B.J.W. (1971). Death and decay caused by *Fomes annosus*. *Rep. Forest Res., Lond.*, pp. 77-8, 1971

Greig, B.J.W. (1979). Species susceptibility to Fomes butt rot. *Q. Jl. For.*, **73**, 21-5

Greig, B.J.W. (1981). *Decay in conifers*. Leafl.For.Commn, No. 79

Greig, B.J.W. (1984). Management of East England pine plantations affected by *Heterobasidion annosum* root rot. *Eur. J. For. Path.*, **14**, 392-7

Greig, B.J.W. (1987a). History of Peridermium stem rust of Scots pine (*Pinus sylvestris* L.). in Thetford Forest, East Anglia. *Forestry*, **60**, 193-202

Greig, B.J.W. (1987b). In Pathology. *Rep. Forest Res. Edin.*, 1987, pp. 46-7

Greig, B.J.W. and Burdekin, D.A. (1970). Control and eradication of *Fomes annosus* in Great Britain. *Proc. 3rd Int. Conf. on Fomes annosus*, U.S. Dept.Agric.For.Serv., pp. 21-32

Greig, B.J.W., Gregory, S.C. and Strouts, R.G. (1991). *Honey fungus*. Bull. For. Commn., Edinb., **100**

Greig, B.J.W. and Gulliver, C.C. (1976). Decay in oaks in the Forest of Dean. *Q. Jl. For.*, **70**, 157-9

Greig, B.J.W. and Redfern, D.B. (1974). *Fomes annosus*. Leafl. For.Commn, No. 5

Greig, B.J.W. and Redfern, D.B. (1987). *Plectophomella concentrica* on *Ulmus procera*. *Trans. Br. mycol. Soc.*, **89**, 399

Greig, B.J.W. and Strouts, R.G. (1983). *Honey fungus*. Arboric. Leafl. 2, HMSO, London.

Gremmen, J. (1954). Op *Populus* en *Salix* voorkomende *Melampsora*-soorten in Nederland. *Tijdschr. PlZiekt.*, **60**, 243-50

Gremmen, J. (1956). Een blad en twijgziekte van populieren veroorzaakt door *Venturia tremulae* en *Venturia populina*. *Tijdschr. PlZiekt.*, **62**, 236-42

Gremmen, J. (1958). Bijdrage tot de biologie van *Cryptodiaporthe populea* (Sacc.) Butin (*Dothichiza populea* Sacc. & Bri.). *Ned. BoschbTijdschr.*, **30**, 251-60

Gremmen, J. (1959). Uber zwei Phacidiaceae von *Pinus silvestris* L. *Phytopath. Z.*, **35**, 27-30

Gremmen, J. (1961a). Naaldhout aftsterving door *Rhizina undulata* in het bijzonder na takkenbranden. *Ned. BoschbTijdschr.*, **33**, 5-10

Gremmen, J. (1961b). *Polyporus schweinitzii* Fr., de orzaak van stamrot in naaldhout. *Ned. Bosb. Tijdsschr.*, **33**, 354-8

Gremmen, J. (1962). De Marssonina-ziekte van de populier. 1. Het voorkomen van apothecien en de functie van de ascosporen voor de verspreiding van de ziekte. *Ned. Bosb. Tijdschr.*, **34**,

428-32

Gremmen, J. (1963). De biologiske bestrijding van de Wortzelzwam *Fomes annosus* (Fr.). Cke middel van *Peniophora gigantea* (Fr.). Massee. *Ned. Boschb. Tijdschr.*, **35**, 356-67

Gremmen, J. (1964). Three poplar-inhabiting *Drepenopeziza* species and their life history. *Nova Hedwigia*, **9**, 170-6

Gremmen, J. (1968a). Stem cankers van groveden en Corscaanse den veroorzaakt door *Crumenula sororia* Karst. *Ned. BoschbTijdschr.*, **40**, 176-82

Gremmen, J. (1968b). Bijdrage tot de biologie van *Brunchorstia pinea* (Karst.) Hohn., de oorzak van het taksterven bij Oostenrijkse en Corsicaanse. *Ned. BoschbTijdschr.*, **40**, 221-31

Gremmen, J. (1978). Research on Dothichiza-bark necrosis (*Cryptodiaporthe populea*) in poplar. *Eur. J. For. Path.*, **8**, 362-8

Gremmen, J. (1980). Problems and prospects in breeding Melampsora rust-resistant poplars. *Proceedings of the IUFRO Working Party S2-05, resistance in pines to Melampsora pinitorqua, June 1979, Suonenjoki, Finland. Folia For.* No. 422, 5-9

Gremmen, J. and Kam, M. de (1970). *Erwinia salicis* as the cause of dieback in *Salix alba* in the Netherlands and its identity with *Pseudomonas saliciperda*. *Neth. J. Pl. Path.*, **76**, 249-52

Gremmen, J. and Kam, M. de (1977). *Ceratocystis fimbriata*, a fungus associated with poplar canker in Poland. *Eur. J. For. Path.*, **7**, 44-7

Gremmen, J. and Kam, M. de (1981). New developments in research into the watermark disease of white willow (*Salix alba*) in the Netherlands. *Eur. J. For. Path.*, **11**, 334-339.

Grenfell, A.L. (1985). Mixed bag. *BSBI News*, No. 39, p.9

Grente, J. (1965). Les formes Hypovirulentes d'*Endothia parasitica* et les espoirs de lutte contre le chancre du chataignier. *C. R. Hebd. Seances Acad. Agr. France*, **51**, 1033-7

Grente, J. and Solignat, G. (1952). La maladie de l'encre du chataignier et son evolution. *C. r. Acad. Agric. Fr.*, **38**, 126-9

Griffin, D.H., Manion, P.D., Valentine, F.A. and Gustavson, L. (1984). Canker elongation, branch death, and callus formation as resistance or susceptibility responses in *Populus tremuloides* and virulence or avirulence characteristics of *Hypoxylon mammatum*. *Phytopathology*, **74**, 683-7

Griffin, D.M. (1955). Fungal damage to roots of Sitka spruce seedlings in forest nurseries. *Rep. Forest Res., Lond.*, 1954 pp. 52-3

Griffin, D.M. (1956). Fungal damage to roots of Sitka spruce seedlings in forest nurseries. *Rep. Forest Res., Lond.*, 1955 pp. 75-6

Griffin, D.M. (1957). Fungal damage to roots of Sitka spruce seedlings in forest nurseries. *Rep. Forest Res., Lond.*, 1956, pp. 86-7

Griffin, D.M. (1965). A study of damping-off, root damage and related phenomena in coniferous seedlings in British forest nurseries, in B.Benzian (Ed.), *Experiments on Nutrition Problems in Forest Nurseries, 1*, Bull.For.Commn, Lond., No. 37, pp. 212-227

Griffin, G.J. (1986). Chestnut blight and its control. *Hort. Rev.*, **8**, 291-336

Griffin, G.J., Hebard, F.V., Wendt, R.W. and Elkins, J.R. (1983). Survival of American chestnut trees: evaluation of blight resistance and virulence in *Endothia parasitica*. *Phytopathology*, **73**, 1084-92

Griffon, E. and Maublanc, A. (1912). Les *Microsphaera* des chênes. *Bull. Soc. mycol. Fr.*, **28**, 88-103

Grigor'ev, A.G. (1984). [Bioecological features of some Mediterranean oaks introduced into the northern Crimea] *Byulletin Gosudarstvennogo Nikitiskogo Botanicheskogo Sada* No. 53, 10-14 (abstract in *Rev. Pl. Path.*, **64**, No. 1773, 1985)

Grill, D., Fachbach, G. and Brunegger, A. (1983). REM-Betrag zum Befall *Chrysomyxa abietis* Teleutosporenlagern durch Muckenlarven.*Eur. J. For. Path.*, **13**, 409-14

Grill, D., Hafellner, J. and Waltinger, H. (1980). Rasterelektronenmikroskopische Untersuchungen an *Chrysomyxa abietis* befallenen Fichtennadeln. *Phyton, Austria*, **20**, 279-84

Grill, D., Pfeifhofer, Q. and Esterbauer, H. (1984). Carotenoids in *Chrysomyxa abietis* infected spruce needles. *Eur. J. For. Path.*, **14**, 296-301

Grosclaude, C., Olivier, R., Pizzuto, J.C., Romiti, C. and Madec, S. (1988). Détection par piégage du *Ceratocystis fimbriata* f.*platani*. Application à l'étude de la persistence dans du bois infecté. *Eur. J. For. Path.*, **18**, 385-90

Gross, H.L. and Weidensaul, T.C. (1967). *Phacidiopycnis pseudotsugae* associated with bleeding cankers on hemlock. *Pl. Dis. Reptr.*, **51**, 807-8

Grove, W.B. (1921). Mycological notes. *J. Bot., Lond.*, **59**, 13-7

Grove, W.B. (1935). *British stem- and leaf-fungi (Coelomycetes), Vol. 1*, Cambridge University Press, Cambridge

Grove, W.B. (1937). *British Stem and Leaf Fungi (Coelomycetes), Vol. 2*, Cambridge University Press, Cambridge.

Groves, J.W. (1969). *Crumenulopsis*, a new name to replace *Crumenula* Rehm. *Can. J. Bot.*, **47**, 47-51

Guba, E.F. (1961). *Monograph of Monochaetia and Pestalotia*. Harvard University Press, Harvard

Guerreiro, G. (1949). Estudos realizados no Castanheiro em 1948. *Bol. Jun. Frut., Lisb.*, **9**, 13-40

Guilivo, C. (1973). Hagelschutz im obst-und weinbau mit kunststoffnetzen. *Erwerbsobstbau*, **15**, 187-90

Guillaumin, J.J. (1988). In Smith *et al.*, 1988, q.v.

Guillaumin, J.J. and Berthelay, S. (1981). Determination specifique des armillaires par la methode des groupes de compatibilite sexuelle. Specialisation ecologique des especes francaises. *Eur. J. For. Path.*, **18**, 401-8

Guillaumin, J.J. and Lung, B. (1985). Etudes de la specialisation d'*Armillaria mellea* (Vahl). Kumm. et *Armillaria obscura* (Secr.). Herink en phase saprophytique et en phase parasitaire. *Eur. J. For. Path.*, **15**, 342-9

Guillaumin, J.J., Lung, B., Romagnesi, H., Marxmüller, H., Lamoure, D., Durrieu, G., Berthelay, S. and Mohammed, C. (1985). Systmatique des Armillaires du groupe mellea. Consequences phytopathologiques. *Eur. J. For. Path.*, **15**, 268-77

Guinier, P. (1933). Sur la biologie de deux champignons lignicoles. *C. R. Soc. Biol. Nancy*, **112**, 1363-6

Guldemond, J.L. and Kolster, H.W. (1966). De bestrijding van *Marssonina* bij populieren. *Ber. BosbProefstn Dorschk.*, **51**, 6pp.

Gundersen, K. (1962). The physiology of *Fomes annosus*. *Conference and Study Tour on Fomes annosus, Scotland, 1960., IUFRO, Firenze*, pp. 31-7

Gundersen, K. (1963). Nytt kemisk medel mot rotrotra. *Skogen*, **50**, 288-9

Gundersen, K. (1967). Nitrite as a nutrient for microfungi of the outer stem cortex of pine and spruce and its toxicity to *Fomes annosus*. *Studia forestal suec.*, No. 43, 22pp.

Guseinov, E.S. (1976). [Benomyl against powdery mildew of oak]. *Zashchita Rastenii*, **5**, 32; abstract in *Rev. Pl. Path.*, **50**, 1976, No. 5363

Guyot, R. (1933). De la maladie du rond; de l'influence des foyers ou des foyers d'incendie dans sa propagation. *Rev. gen. Sci.*, **44**, 239-47

Gyorfi, J. (1957). Az erdei fak racos niegbetegedesei. *Erdesz. Kutat.*, **1-2**, 83-94, abstract in *Rev. appl. Mycol.*, **37**, 559, 1958

Haas, P.G. and Wennemuth, G. (1962). [Refrigated storage of woody plants from nurseries. III. Botrytis and Fusarium infection of woody plants in refrigerated storage]. *Gartenbauwissenschaft*, **27**, 231-42; abstract in *Rev. appl. Mycol.*, **42**, 1963, p. 323.

Hagner, M. (1962). Nagra faktorer av betydelse for rotmurklans skadegorelse. *Norrlands Skogsv. Tidskr.*, **2**, 245-70

Hahn, G.G. (1920). *Phomopsis juniperovora* a new species causing blight of nursery cedars. *Phytopathology*, **10**, 249-53

Hahn, G.G. (1926). *Phomopsis juniperovora* and closely related strains on conifers. *Phytopathology*, **16**, 899-914

Hahn, G.G. (1930). Life history studies of the species of *Phomopsis* occurring on conifers.*Trans. Br. mycol. Soc.*, **15**, 32-93

Hahn, G.G. (1931). Life history studies of the species of *Phomopsis* occurring on conifers. Part 1. *Trans. Br. mycol. Soc.*, **15**, 32-93

Hahn, G.G. (1943). Taxonomy, distribution and pathology of *Phomopsis occulta* and *P. juniperovora*. *Mycologia*, **35**, 112-29

Hahn, G.G. (1957). *Phacidiopycnis* (*Phomopsis*) canker and dieback of conifers. *Pl. Dis. Reptr.*, **41**, 623-33

Hahn, G.G. and Ayers, T.T. (1934). Dasyscyphae on conifers in North America. I. The large-spored, white excipled species. *Mycologia*, **26**, 73-101

Hahn, G.G. and Ayers, T.T. (1943). Role of *Dasyscypha willkommii* and related fungi in the production of canker and dieback of larches. *J. For.*, **41**, 483-95

Halber, M. (1963). *Botrytis* sp. on Douglas fir seedlings. *Pl. Dis. Reptr.*, **47**, 556.

Hallaksela, A-M. and Nevalainen, S. (1981). [Control of root rot fungus (*Heterobasidion annosum*) by treating Norway spruce stumps with urea]. *Folia Forestalia*, No. 470, 10pp.

Hamacher, J. (1987). Infektionsverlauf und Haustorienultrastruktur des Echten Mehltaus *Phyllactinia guttata* (Wallr.ex Fr.). Lev.auf Birken. *Mededelingen van de Faculteit*

*Landbouwwetenschappen Rijksuniversiteit Gent*, **52**, 831-40

Hamilton, W.D. (1980). Wetwood and slime flux in landscape trees. *Journal of Arboriculture*, **6**, 247-9

Hamm, P.B. and Hansen, E.M. (1987). Identification of *Phytophthora* spp. known to attack conifers in the Pacific Northwest. *Northwest Science* **61**, 103-9

Hamond, J.B. (1931). Some diseases of walnuts. *Ann. Rep. East Malling Res. Sta.*, *1928, 1929 and 1930, II suppl.*, pp. 142-9

Hamond, J.B. (1935). A graft disease of walnut caused by a species of *Chalaropsis*. *Trans. Br. mycol. Soc.*, **19**, 158-9

Hangyal-Balul, W. (1983). Akacmayes csiracsemeteh gombas betegsegei es az ellenuk valo vedekezes lehetosegei. *Erdesz. Kutat. (1981, publ. 1983)*, **74**, 343-349 (abstract in *Rev. Pl. Path.*, **65**, 920, 1986

Hangyal-Balul, W. (1986). [Fungi causing deterioration of acorns and germinating oak seedlings and methods of protection against them]. A tolgymakk es-csiacsemetek pusztulasat okozo gombak es az elleniik valo vedekezes lehetosegei.*Erdesz. Kutat.*, **76-77**, 293-304 (abstract in *Rev. Pl. Path.*, **68**, No. 1464, 1989).

Hangyalne,B.W. and Toth,J.(1987). [Chemical protection of Acacia [*Robinia pseudoacacia*] propagation material]. *Erdeszeti Kutatasok* (1986, publ.1987), **78**, 373-6 (abstract in *Rev. Pl. Path.*, **69**, No. 5220, 1990)

Hansen, E.M.(1985). *Forest Pathogens of N. W. North America and their potential for damage in Britain*. Forest Rec.Lond., No. 129

Hansen, E.M., Brasier, C.M., Shaw, D.S. and Hamm, P.B. (1986). The taxonomic stucture of *Phytophthora megasperma*: evidence for emerging biological species groups. *Trans. Br. mycol. Soc.*, **87**, 557-73

Hansen, H.N. (1956). The perfect stage of *Coryneum cardinale*. *Phytopathology*, **46**, 636-9 (abstract).

Hanso, M. and Torva, A. (1975). Black snow mould in Estonia. 1. On ecology and morphology of *Herpotrichia juniperi*. *Metsanduslikud Uuremused*, **12**, 262-79

Haraldstad, A.R. (1962). Investigations on *Fomes annosus* in Hoylandskomplekset, South Western Norway. *Nyt. Mag. Bot.*, **9** (1961), 175-98

Harrison, K.A. (1965). Willow blight and the survivial of some *Salix* species in Nova Scotia. *Can. Pl. Dis. Surv.*, **45**, 94-5

Hart, J.H. (1965). Economic impact of Dutch elm disease in Michigan. *Pl. Dis. Reptr.*, **49**, 830-2

Hartig, R. (1878). Die krebsartigen Krankheiten der Rotbuche. *Z. Forst-und Jagdwesen*, **9**, 377-83

Hartig, R. (1880). Der Ahornkeimlingspilz *Cercospora acerina*. *Untersuchungen aus dem forstbotanischen Institut zu Munchen*, **1**, 58-61

Hartig, R. (1894). *Text-book of the Diseases of Trees*, English ed. trans L. W. Somerville, revised and edited H. Marshall Ward, Macmillan, London

Hartig, R. (1900). *Lehrbuch der Pflanzenkrankheiten*, 3rd edn, Springer, Berlin

Hartley, C. (1921). *Damping-off in Forest Nurseries*. Bull. U.S. Dept Agric., No. 934, pp. 1-99

Hartley, C., Davidson, R.W. and Crandall, B.S. (1961). *Wetwood, bacteria, and increased pH in trees*. U.S. Dept.Agric.Forest Service, No. 2215, 34pp.

Hartley, C., Pierce, R.G. and Hahn, G.G. (1919). Moulding of snow smothered nursery stock. *Phytopathology*, **9**, 521-31

Hartline, B.K. (1980). Fighting the spreading Chestnut blight. *Science, USA*, **209**, 892-3

Harvey, G.M. (1962). *Heart rots of Douglas fir*, U.S. Dept.Agric.Pest Leafl., No. 73

Hawksworth, D.L., Sutton, B.C. and Ainsworth, G.C. (1983). *Ainsworth and Bisby's Dictionary of the Fungi*. Commonwealth Agricultural Bureaux, Slough

Hawksworth, F.G. and Staley, J.M. (1968). *Rhizosphaera kalkhoffii* on spruce in Arizona. *Pl. Dis. Reptr.*, **52**, 804-5

Hawksworth, F.G. and Wiens, D. (1972). *Biology and classification of the Dwarf Mistletoes*. USDA Forest Service Handbook, No. 40

Haworth, R.H. and Spiers, A.G. (1988). Characterization of bacteria from poplars and willows exhibiting leaf spotting and stem cankering in New Zealand. *Eur. J. For. Path.*, **18**, 426-36

Hayes, A.J. (1973). The occurrence of *Crumenula sororia* Karst. on Lodgepole pine in the United Kingdom. *Forestry*, **46**, 125-38

Hayes, A.J. (1975). The mode of infection of Lodgepole pine by *Crumenula sororia* Karst. and the susceptibility of different provenances to attack. *Forestry*, **48**, 99-113

He, B.Z., Yang, D.Q., Deng, X.L., Liu, G.Q., Yue, Y.P. and Liu, C.Y. (1985). [Study on the pathogen and control of needle-cast of *Pinus sylvestris* var. *mongolica* caused by

*Lophodermium seditiosum*] *Journal of North-eastern Forestry College, China*, **13**, 77-81. (abstract in *Rev. Pl. Path.*, **66**, No. 1167, 1987)

Heck, O. (1927). Muss man die Hexenbesen der Weisstanne·verfolgen? *Fortwiss. ZentBl.*, **49**, 132-40

Heggestad, H.H. (1968). Diseases of crops and ornamental plants incited by air pollutants. *Phytopathology*, **58**, 1089-97

Henderson, D.M. (1954). The genus *Taphrina* in Scotland. *Notes R. bot. Soc. Edin.*, **21**, 165-80

Henderson, D.M. and Watling, R. (1978). *Fungi in The Island of Mull. A survey of its flora and environment (Jermy, A. C. and Crabb, J. A., editors)*. London, British Museum (Natural History)

Henman, G.S. and Denne, M.P. (1984). Control of wood quality in the British oaks. *Rep. Forest Res. Edin.*, 1984, p.59

Henman, G.S. and Denne, M.P. (1986). Control of wood quality in the British oaks. *Rep. Forest Res., Edin.*, 1986, p.57

Henman, G.S. and Denne, M.P. (1988). Control of shake in oak. *Rep. Forest Res., Edin.*, 1988, p.70

Hennebert, G.L. (1964). L'identification des rouilles du peuplier. *Agricultura*, **12**, 661-71

Henriksen, H.A. and Jorgensen, E. (1954). Rodfoerderverangreb i relation til udhugninsgrad. *Forst. Fors Vaes. Danm.*, **21**, 215-57

Hepting, G.H. (1971). *Diseases of Forest and Shade Trees of the United States*. U.S. Dept.Agric. Handbook 386, 658pp.

Hepting, G.H. and Chapman, A.D. (1938). Losses from heart rot in two shortleaf and loblolly pine stands. *J. For.*, **36**, 1193-201

Hepting.G.H. and Downs, A.A. (1944). Root and butt rot in planted white pine at Biltmore, North Carolina. *J. For.*, **42**, 119-21

Hewitt, H.G. (1974). Conidial germination in *Microsphaera alphitoides*. *Trans. Br. mycol. Soc.*, **63**, 587-9

Heybroek,H.M.(1966). Dutch elm disease abroad. Reprinted (unpaginated) from *Am. Forests*, June 1966

Heybroek, H.M. (1967). The Dutch elm disease in the old world. *Pap. 14th I. U. F. R. O. Conf. Sec. 24, Munich, 1967*, pp. 447-54

Hibberd, B.G. (Ed.) (1989). *Urban Forestry Practice*, For.Commn Handbook 5, pp. 94-100

Hiley, W.E. (1919). *The Fungal diseases of the Common Larch*, Clarendon Press, Oxford

Hiley, W.E. (1921). The larch needle cast fungus, *M. laricis*. *Q. Jl. For.*, **15**, 57-62

Hill, S.A. (1984). *Methods in Plant Virology*. Blackwell, Oxford

Hintikka, V. (1973). A note on the polarity of *Armillaria mellea*. *Karstenia*, **12**, 32-9

Hiruki, C.(editor)(1988). *Tree Mycoplasmas and Mycoplasma Diseases*. University of Alberta Press, Edmonton

Hishida, I., Namba, H. and Kuroda, H. (1988). Antitumor activity inhibited by orally administered extract from fruit body of *Grifola frondosa* maitake. *Chem. Pharm. Bull. (Tokyo).*, **36** (5).

Hitchcock, L.A. and Cole, A.L.J. (1980). Plane-tree anthracnose disease and climate. *N. Z. J. Sci.*, **23**, 69-72

Hodges, C.S. (1964). The effect of competition by *Peniophora gigantea* on the growth of *Fomes annosus* in stumps and roots. *Phytopathology*, **54**, 623 (abstract)

Hodges, C.S. and Green, H.J. (1961). Survival in the plantation of eastern red cedar seedlings infected with Phomopsis blight in the nursery. *Pl. Dis. Reptr.*, **45**, 134-6

Hoff, R., Bingham, R.T. and MacDonald, G.I. (1980). Relative blister rust resistance of white pines. *Eur. J. For. Path.*, **10**, 307-16

Hoffmann, G.M. and Fliege, F. (1967). *Kabatina juniperi* als Ursache eines Zweigsterbens an verschiedenen juniperusarten. *Z. Pfl Krankh. Pflpath. Pflschutz.*, **74**, 587-93

Hoitink, H.A., Sydnor, T.D. and Wilson, C.L. (1979). Resistance of maple cultivars and species to Verticillium wilt—a preliminary report. *Research Circular, Ohio. Agricultural Research and Development Center* No. 246, 46-47 (abstract in *Rev. Pl. Path.*, **60**, No. 2204, 1981).

Holdenrieder, O. (1984). Untersuchungen zur biologischen Bekampfung von *Heterobasidion annosum* an Fichte (*Picea abies*). mit antagonischen Pilzen II. Interaktionstests auf Holz. *Eur. J. For. Path.*, **14**, 117-53

Holdenrieder, O. (1986). Beobachtungen zum Vorkommen von *Armillaria obscura* und *Armillaria cepistipes* an Tanne in Sudbayern. *Eur. J. For. Path.*, **16**, 375-9

Holliday, P. (1989). *A Dictionary of Plant Pathology*. Cambridge University Press, Cambridge

Holmes, F.W. (1967). Resistance of certain elm clones to *Ceratocystis ulmi* and *Verticillium albo-*

*atrum. Phytopathology*, **57**, 1247-9

Holt, C.E., Gockel, H. and Huttermann, A. (1983). The mating system of *Fomes annosus* (*Heterobasidion annosum*). *Eur. J. For. Path.*, **13**, 174-81

Hood, I.A. and Sandberg, C.J. (1985). *Some major fungi of conifers (Phomopsis pseudotsugae, P. juniperovora, and Rhizosphaera kalkhoffii)*. *Forest Pathology in New Zealand*. Forest Research Institute, New Zealand No. 13, 8pp.

Hood, I.A. and Vanner, A.L. (1984). *Cyclaneusma (Naemacyclus)*. needle-cast of *Pinus radiata* in New Zealand. 4: Chemical control research. *N. Z. Jl. For. Sci.*, **14**, 215-22

Hoog, G.S. de (1984). *Ceratocystis* versus *Ophiostoma*: a reappraisal. *Mycologia*, **76**, 292-9

Hoog, G.S. de (1974). The genera *Blastobotrys, Sporothrix, Calcarisporium* and *Calcarisporiella* gen. nov. *Stud. Mycol., Baarn*, **7**

Hoog, G.S.de, Rahman, M.A. and Boekhout, T. (1983). *Ramichloridium, Veronaea* and *Stenella*: generic delimitation, new combinations and two new species. *Trans. Br. mycol. Soc.*, **81**, 485-90

Hopp, P.J. (1957). Zur Kenntnis des Larchenkrebses (*Dasyscypha willkommii*) an *Larix decidua. Forstwiss. Zbl.*, **76**, 334-54

Houston, D.R. (1976). Protection against beech scale, *Cryptococcus fagi*, by *Dichaena rugosa*, a bark fungus of European and American trees. *Proc. Am. Phytopath. Soc.*, **3**, 306 (abstr.)

Houston, D.R. (1983a). American beech resistance to *Cryptococcus fagisuga*. In Houston, D.R. and Wainhouse, D. (1983). q.v., 38-42

Houston, D.R. (1983b). Influence of lichen species on colonization of *Fagus grandifolia* by *Cryptococcus fagisuga*: preliminary observations from certain Nova Scotia forests. In Houston, D.R. and Wainhouse, D. (1983), q.v., 105-8

Houston, D.R. (1983c). Effects of parasitism by *Nematogonum ferrugineum* (*Gonatorrhodiella highlei*) on pathogenicity of *Nectria coccinea* var.*faginata* and *Nectria galligena*. In Houston, D.R. and Wainhouse, D. (1983), q.v. 109-14

Houston, D.R., Parker, E.J., Perrin, R. and Lang, K.J. (1979). Beech bark disease: a comparison of the disease in North America, Great Britain, France and Germany.*Eur. J. For. Path.*, **9**, 199-211

Houston, D.R. and Wainhouse, D. (1983). *Proceedings, IUFRO beech bark disease Working Party Conference, 1982*.USDA Forest Service General Technical Report WO-37

Hsiang, T. and Edmonds, R.L. (1989). Physiological specialization of *Heterobasidion annosum* on conifer hosts. *Can J. Bot.*, **67**, 2396-400

Hsiang, T., Edmonds, R.L. and Driver, C.H. (1989). Conidia of *Heterobasidion annosum* from *Tsuga heterophylla* forests in western Washington. *Can. J. Bot.*, **67**, 1262-1266.

Hubbes, M. (1959). Untersuchungen uber *Dothichiza populea* Sacc. et Briard, den Erreger des Rindenbrandes der Pappel. *Phytopath. Z.*, **35**, 58-96

Hubert, F.E. (1935). A disease of conifers caused by *Stereum sanguinolentum*. *J. For.*, **33**, 485-9

Hudler, G.W., Banik, M.T. and Miller, S.G. (1987). Unusual epidemic of tar spot on Norway maple in upstate New York. *Pl. Dis.*, **71**, 65-8

Hudson, H.J. (1987). *Guignardia* leaf blotch of horsechestnut. *Trans. Br. mycol. Soc.*, **89**, 400-1

Hughes, S.J. (1951). *Studies in Micro-fungi. III. Mastigosporium, Camposporium, and Ceratophorum*. Mycol. Pap. No. 36

Hulbary, R.L. (1941). A needle blight of Austrian pine. *Illinois Nat. Hist. Survey Bull.*, **21**, 231-6

Hull,S.K. and Gibbs,J.N.(1991). *Ash dieback – a survey of non-woodland trees*. Bull. For. Commn, Edinburgh, 93

Humphreys-Jones, D.R. (1977). Leaf and shoot death (*Coniothyrium fuckelii* Sacc.). of *Juniperus communis* L.var.*compressa* Carr.*Pl. Path.*, **26**, 47-8

Humphreys-Jones, D.R. (1980). Leaf and Shoot death (*Coniothyrium fuckelii*) on *Thuja orientalis* cv *Aurea Nana*. *Pl. Path.*, **29**, 199-200

Humphreys-Jones, D.R. (1982). *Phoma exigua* on *Ginkgo biloba*. *Pl. Path.*, **31**, 91-3

Hunt, J. (1956). Taxonomy of the genus *Ceratocystis*. *Lloydia*, 1-56

Hunt, J. and Krueger, K.W. (1962). Decay associated with thinning wounds in young-growth Western Hemlock and Douglas fir. *J. For.*, **60**, 336-40

Hunter, T. and Stott, K.G. (1978). A disease of *Salix alba* var. *coerulea* caused by a fluorescent *Pseudomonas* sp. *Pl. Path.*, **27**, 144-5

Huss, E. (1952). Skogsforskningsinstitutets metodik vid frounderskoningar. *Medd. Skogsforsken Inst. Stockh.*, **40** (1951), 1-82

Ibragimov, I.A. (1957). (On the question of poplar withering in the Bashkir, A.S.S.R.); abstract in *Rev. appl. Mycol.*, **38**, 1959, p.427

Ibragimov, I.A. (1964). Tsitosporoz Topolei i voprosy ikh ot zabolevaniya. Abstract in *Rev.*

*appl. Mycol.*, **44**, No. 2630, 1965

Illingworth, K. (1973). Variation in the susceptibility of lodgepole pine provenances to Sirococcus blight. *Can. J. For. Res.*, **3**, 585-589.

Ing, B. (1990a). An introduction to British powdery mildews 1. *The Mycologist*, **4**, 46-8

Ing, B. (1990b). An introduction to British powdery mildews 2. *The Mycologist*, **4**, 88-90

Ing, B. (1990c). An introduction to British powdery mildews 3. *The Mycologist*, **4**, 125-8

Ing, B. (1990d). An introduction to British powdery mildews 4. *The Mycologist*, **4**, 172-7

Ing, B. (1991). An introduction to British powdery mildews 5. *The Mycologist*, **5**, 24-7

Innes, J.L. (1987). Air pollution and forestry. *For. Comm. Bull.* , No. 70, HMSO

Innes, J.L. (1989). Acid rain and trees. In *Acid deposition* (edited by Longhurst, J.W.S.). British Library Technical Publications

Innes, J.L. and Boswell, R.C. (1987). *Forest health surveys (1987). Part 1. Results.* Bull.For.Commn, Edin. 74.

Innes, J.L. and Boswell, R.C. (1989). *Forest Health Surveys (1988).* Bull.For.Commn, Edin. 88.

Innes, J.L., Boswell, R.C., Binns, W.O. and Redfern, D.B. (1986). *Forest health and air pollution: 1986 survey.* Res.Dev. Pap.For.Commn, Edin. 150

Intini, M.G. (1988). Un caso di moria dell'abete bianco associato ad *Armillaria ostoyae* (Romagnesi). Herink. *Inftore fitopatol.*, **38**, 67-70

Isaac, I. (1949). A comparative study of pathogenic isolates of *Verticillium*. *Trans. Br. mycol. Soc.*, **32**, 137-57

Isaac, I. (1953). The spread of diseases caused by species of *Verticillium*. *Ann. appl. Biol.*, **40**, 630-8

Ito, K. and Hosaka, Y. (1951). Grey mold and sclerotial disease of 'sugi' (*Cryptomeria japonica* D.Don). seedlings, the causes of the so-called 'snow mould'. *Bull. For. Exp. Sta. Tokyo*, **51**, 1-27

Ito, K., Zinno, Y. and Kobayashi, T. (1963). Larch canker in Japan. *Bull. Govt. Forest Exp. Sta. Tokyo*, **155**, 23-47

Ivens, G.W.(editor)(1990). *The UK Pesticide Guide 1990.*CAB International, Wallingford and BCPC, Bracknell

Ivory, M.H. (1967). Spore germination and growth in culture of *Dothistroma pini* var. *keniensis*. *Trans. Br. mycol. Soc.*, **50**, 563-72

Ivory, M.H. (1968). Reaction of pines in Kenya to attack by *Dothistroma pini* var. *keniensis*. *E. Afr. agric. For. J.*, **33**, 236-44

Ivory, M.H. and Patterson, D.N. (1969). Progress in breeding *Pinus radiata* resistant to Dothistroma needle blight in East Africa.*Silvae Genet.*, **19**, 38-42

Jacobson, J.S., Weinstein, L.H., McCune, D.C. and Hitchcock, A.E. (1966). The accumulation of fluorine in plants.*J. Air Pollution Ass.*, **16**, 412-7

Jagielski, A. (1969). [Investigations on the effectiveness of some fungicides against *Microsphaera alphitoides* on oak]. *Biul. Inst. Ochr. Rosl., Poznan*, **45**, 45-50; abstract in *Rev. Pl. Path.*, **50**, No. 1403, 1971

Jahn, H. (1979). *Pilze die an Holz wachsen.* Herford: Bussescha Verlagshandlung. 268pp.

Jahn, H. and Jahn, M.A. (1980). Wo kommt der Honiggelbe Hallimasch, *Armillaria mellea* (Vahl ex Fr.). Karst. sensu stricto vor? *Westf. Pilzor.*, **11**, 154-9

Jahnel, H. and Junghans, B. (1957). Uber eine wenig bekannte KiefenKrankheit (*Sclerophoma pithyophila*). *Forstwiss. Zbl.*, **76**, 129-32

Jalaluddin, M. (1967a). Studies on *Rhizina undulata* I. Mycelial growth and ascospore germination. *Trans. Br. mycol. Soc.*, **50**, 449-59

Jalaluddin, M. (1967b). Studies on *Rhizina undulata*. II. Observations and experiments in East Anglian plantations. *Trans. Br. mycol. Soc.*, **50**, 461-72

Jalkanen, R. (1985). The occurrence and importance of *Lophodermella sulcigena* and *Hendersonia acicola* on Scots pine in Finland. *Karstenia*, **25**, 53-61

Jalkanen, R. and Laakso, R. (1986). *Hendersonia acicola* in an epidemic caused by *Lophodermella sulcigena* with special reference to biological control. *Karstenia*, **26**, 49-56

Jamalainen, E.A. (1961). Havupuiden taimistojen talvituhosienivauriot ja niiden kemiallinen torjunta. Reprinted from *Silva Fenn.*, **108**, 15pp.

James, R.L. and Cobb, F.W. (1984). Spore deposition by *Heterobasidion annosum* in forests in California. *Pl. Dis.*, **68**, 246-8

James, R.L., Cobb, F.W., Jr., Miller, P.R. and Parmeter, J.R.Jr. (1980a). Effects of oxidant air pollution on susceptibility of pine roots to *Fomes annosus*. *Phytopathology*, **70**, 560-3

James, R.L., Cobb, F.W., Jr., Wilcox, W.W. and Rowney, D.L. (1980b). Effects of photochemical oxidant injury of ponderosa and Jeffrey pines on susceptibility of sapwood and freshly cut stumps to *Fomes annosus*. *Phytopathology*, **70**, 704-8

James, R.L. and Woo, J.Y. (1984). Fungicide trials to control Botrytis blight at nurseries in Idaho and Montana. *Tree Planters' Notes*, **35**, 16-19

Jamison, V.C. (1956). Pertinent factors in governing availability of soil moisture to plants. *Soil Sci.*, **81**, 459-71

Jancarik, V. (1964). Skotska sypavka Douglasky. *Lesn. Pr., Czechoslovakia*, **43**, 420-3

Janse, J.D. (1981a). The bacterial disease of ash (*Fraxinus excelsior*) caused by *Pseudomonas syringae* subsp.*savastanoi* pv.*fraxini*. I. History, occurrence and symptoms. *Eur. J. For. Path.*, **11**, 306-15

Janse, J.D. (1981b). The bacterial disease of ash (*Fraxinus excelsior*) caused by *Pseudomonas syringae* subsp. *savastanoi* pv. *fraxini*. II. Etiology and taxonomic considerations. *Eur. J. For. Path.*, **11**, 425-38

Janse, J.D. (1982). The bacterial disease of ash (*Fraxinus excelsior*) caused by *Pseudomonas syringae* subsp. *savastanoi* pv. *fraxini*. III. Pathogenesis. *Eur. J. For. Path.*, **12**, 218-31

Janson, A. (1927). Ueber den Schorf und andere Korbweidenschadlinge. *Nachricht uber Schadlingsbekampf*, **2**, 161-4

Jauch, C. (1952). La 'antracnosis' de los sauces cultivados en el Delta del Parana. *Rev. Fac. Agron. B. Aires*, **13**, 285-308

Jaynes, R.A. and DePalma, N.K. (1984). Natural infection of nuts of *Castanea dentata* by *Endothia parasitica*. *Phytopathology*, **74**, 296-9

Jeng, R.S., Bernier, L. and Brasier, C.M. (1988). A comparative study of cultural and electrophoretic characteristics of Eurasian and North American races of *Ophiostoma ulmi*. *Can. J. Bot.*, **66**, 1325-33

Jenkins, C.L. and Starr, M.P. (1982). The pigment of *Xanthomonas populi* is a non-brominated aryl-heptaene belonging to xanthomonadin pigment group 11. *Current Microbiology* **7**, 195-8

Jenkins, P.T. (1952). Armillaria on fruit trees. *J. Dep. Agric. Vict.*, **50**, 88-90

Jennings, O.E. (1934). Smoke injury to shade trees. *Proc. 10th National Shade Tree Conf.*, pp. 44-8

Jindal, K.K. and Blardwag, L.N. (1986). Occurrence of crown gall on *Eucalyptus tereticornis* Sm. in India. *Indian Forester*, **112**, 1121

Jobling, J. and Young, C.W.T. (1965). Apparent variations in the resistance of poplar clones to bacterial canker. *Rep. Forest Res., Lond., 1964*, pp. 151-7

Johansson, M. and Stenlid, J. (1985). Infection of roots of Norway spruce (*Picea abies*) by *Heterobasidion annosum*. I. Initial reactions in sapwood by wounding and infection. *Eur. J. For. Path.*, **15**, 32-45

Jokinen, K. (1983). [The effect of fertilisation on the occurrence of *Heterobasidion annosum*]. *Folia Forestalia, Institutum Forestale Fenniae*, **573**, 22pp.

Jokinen, K. (1984). [The spread of *Heterobasidion annosum* and its control with *Phlebiopsis gigantea* during thinning in young stands of Scots pine]. Mamyn tyvitervastaudin leviaminen ja torjunta harmaaorvakalla (*Phlebiopsis gigantea*) mamyn taimikoiden harvennuksessa. *Folia Foerestalia, Institutum Forestale Fenniae*, No. 607, 12 pp. (abstract in *Rev. Pl. Path.*, **65**, No. 4128, 1986)

Jones, A.T. and Mayo, M.A. (1973). Purification and properties of elm mottle virus. *Ann. appl. Biol.*, **75**, 347-57

Jones, C., Griffin, G.J. and Elkins, J.R. (1980). Association of climatic stress with blight on Chinese Chestnut in the eastern United States. *Pl. Dis.*, **64**, 1001-4

Jones, E.W. (1944). Biological flora of the British Isles. *Acer* L. *J. Ecol.*, **32**, 215-52

Jones, S.G. (1923). Life-history of *Rhytisma acerinum* (preliminary account). *Ann. Bot.*, **37**, 731-2

Jones, S.G. (1925). Life-history and cytology of *Rhytisma acerinum* (Pers.). Fr. *Ann. Bot.*, **39**, 41-75

Jorgensen, C.A., Lund, A. and Treschow, C. (1939). Underogelsen over Rodfordaervesen, *Fomes annosus* (Fr.). Cke. *K. Vet. Hojeh. Aarsskr., 1939*, 71-128; abstract in *Rev. appl. Mycol.*, **18**, 772-3, 1939

Jorgensen, E. (1955). Trametes angreb i laehegn. *Dansk Skovforen. Tidsskr.*, **40**, 279-85

Jorgensen, E. (1962). Fomes root rot investigations in Ontario. *Conference and Study Tour on Fomes annosus, Scotland, 1960, IUFRO, Firenze*, pp. 44-7

Jorgensen, H.A. (1952). Studies on *Nectria cinnabarina*. Hosts and variations. *Arsskr. K. Vet. -Landbohojsk., 1952*, pp. 57-120

Ju, G.Z., He, B.Z., Yin, C.C., Zhang, Y.Q. and Li, C.L. (1984). [A study on blister rust of *Pinus sylvestris* var. *mongolica*]. *Scientia Silvae Sinicae*, **20**, 149-55 (Abstract in *Rev. Pl. Path.*,

**64**, No. 2748, 1985).

Juhasova, G. and Bencat, F. (1981). Rozsirenie a vyskyt hub *Melanconis modonia* Tul. a *Phytophthora cambivora* (Petri). Buism. na gastane jedlom na Slovensku. *Pol'nohospodarsivo,* **27**, 70-81 (abstract in *Rev. Pl. Path.,* **60**, No. 5132, 1981)

Kais, A.G., Cordell, C.E. and Affeltranger, C.E. (1986). Benomyl root treatment controls brown-spot disease on longleaf pine in the southern United States. *For. Sci.,* **32**, 506-11

Kalandra, A. (1962). Skodlivy vyskyt korni spaly na Topolech v ceskych krajich v CSSR. *Prace vyzk. Ust. lesn. CSSR, 1961* pp. 273-303

Kam, M. de (1978). *Xanthomonas populi* subsp. *salicis,* cause of bacterial canker in *Salix dasyclada. Eur. J. For. Path.* **8**, 334-7

Kam, M. de (1981). The identification of two subspecies of *Xanthomonas populi* in vitro. *Eur. J. For. Path.,* **11**, 25-9

Kam, M. de (1982a). Detection of soluble antigens of *Erwinia salicis* in leaves of *Salix alba* by enzyme-linked immunosorbent assay. *Eur. J. For. Path.,* **12**, 1-6

Kam, M. de (1982b). Detection and transport of soluble antigens of *Erwinia salicis* and their role in symptom expression of the watermark disease. *Eur. J. For. Path.,* **12**, 365-76

Kam, M. de (1982c). Damage to poplar caused by *Pseudomonas syringae* in combination with frost and fluctuating temperature. *Eur. J. For. Path.,* **12**, 203-9

Kam, M. de (1983). The watermark disease is not transmitted with one-year-old cuttings of *Salix alba. Eur. J. For. Path.,* **13**, 212-5

Kam, M. de (1984). Het vastellen van de gevoeligheid van wilgen voor de watermerkziekte: problemen en perspectiven. *Ned. BosbTijdschr.,* **56**, 22-6

Kam, M. de and Heisterkamp, S.H. (1987). Comparison of two methods to measure the susceptibility of poplar clones to *Xanthomonas populi. Eur. J. For. Path.,* **17**, 33-46

Kam, M. de and Tol, G. van (1988). Watermerkziekte en de toehomst van de wilg in Nederland. *Ned. BosbTijdschr.,* **60**, 320-7

Kamara, A.M., El-Lakang, M.H., Badran, O.A. and Attia, Y.G. (1981). Seed pathology of *Araucaria* spp. 1. A survey of seed-borne fungi associated with *Araucaria* spp. *Australian Forest Research,* **11**, 269-74

Kamiri, L.K. and Laemmlen, F.F. (1981a). Effects of drought-stress and wounding on Cytospora canker development on Colorado blue spruce. *Journal of Arboriculture,* **7**, 113-6

Kamiri, L.K. and Laemmlen, F.F. (1981b). Epidemiology of Cytospora canker caused in Colorado blue spruce by *Valsa kunzei. Phytopathology,* **71**, 941-7

Kampfer, M. (1931). Neue Seuche an Pappeln. *Gartenwelt,* **35**, 525

Karadzic, D. (1987). Efikasnost nekih fungicida u suzbijanju gljive *Dothistroma pini* Hulbari u kuturama crnoz bora. *Zastita Bilja,* **38**, 15-31 (abstract in *Rev. Pl. Path.,* **67**, No. 4142, 1988).

Karadzic, D. (1989a). *Zastita Bilja,* **40**, 35-46 (abstract in *Rev. Pl. Path.,* **69**, No. 7479, 1990)

Karadzic, D. (1989b). *Scirrhia pini* Funk et Parker. Life cycle of the fungus in plantations of *Pinus nigra* Arn. in Serbia. *Eur. J. For. Path.,* **19**, 231-6

Karadzic, D. and Zoric, R. (1981). Prilog pozavanju biologije gljive *Naemacyclus minor* Butin — prouzrokovaca osipanja cetina belog bora. *Zastita Bilja,* **32**, 79-90

Karlman, M. (1986). Damage to *Pinus contorta* in Northern Sweden with special emphasis on pathogens. *Studia for. suec.,* No. 176, 42pp. (abstract in *Rev. Pl. Path.,* **67**, No. 985, 1988)

Karoles, K. (1985). Infeksioonilise tousmepoletiken tehitajate liigiline koosseis Eesti metsataim-lates aastatel 1977-1981. *Metsanduslikud Uurimused, Estonian SSR,* **20**, 108-21

Karsten, P.A. (1871). *Mycologica Fennica. I. Discomycetes,* p.211.

Kartavenko, N.T. (1958). [The significance of shading in the control of oak mildew]. *Bot. Z.,* **43**, 399-400; abstract in *Rev. appl. Mycol.,* **38**, 341, 1959

Kaul, T.N. (1962). Occurrence of *Gnomonia leptostyla* (F.). de Not. on walnut in India. *Curr. Sc.,* **31**, 349

Kechel, H.G. and Boden, E. (1985). Resistanzprufung an Pappeln aus Gewebekultur. *Eur. J. For. Path.,* **15**, 45-51

Kerling, L.C.P. (1955). Reactions of elm wood to attacks of *Ophiostoma ulmi* (Buism.). Nannf. *Acta bot. neerl.,* **4**, 398-403

Kerling, L.C.P. (1966). The hibernation of the oak mildew. *Acta Bot. Neerl.,* **15**, 76-83

Kern, F.D. (1973). *A revised taxonomic account of* Gymnosporangium. Pennsylvania State University Press

Kerr, A. (1980). Biological control of crown gall through the production of agrocin 84. *Pl. Dis.,* **64**, 25-30

Kerr, A. and Panagopoulos, C.G. (1977). Biotypes of *Agrobacterium radiobacter* var. *tumefasciens* and their biological control. *Phytopath. Z.,* **90**, 172-9

Kershaw, D.J., Gadgil, P.D., Legatt, G.J., Ray, J.W. and Pas, J.B. van der (1982). *Assessment and control of Dothistroma needle blight*. FRI Bull. No. 18, 1982, 45pp.

Kersters, K. and De Ley, J. (1984). Genus III. *Agrobacterium* Conn 1942, 359. In Bergey's *Manual of Systematic Bacteriology. Vol. 1* (Ed. N.R.Krieg), pp. 244-54. Williams and Wilkins, Baltimore

Kessler,W.(1988). Wurzelfaule an Jungpflanzen von Eiche und Rotbuche durch *Cylindrocarpon destructans*. *Sozialistische Forstwirtschaft*, **38**, 110-11

Khabibullina, F.M. (1984). *Les. Khoz.*, **4**, 47-9 (abstract in *Rev. Pl. Path.*, **65**, No. 1548, 1986).

Khairi, S.M. and Preece, T.F. (1978a). Hawthorn powdery mildew: occurrence, survival, and ascospore production of *Podosphaera clandestina* cleistothecia in England. *Trans. Br. mycol. Soc.*, **71**, 289-93

Khairi, S.M. and Preece, T.F. (1978b). Hawthorn powdery mildew: diurnal and seasonal distribution of conidia in air near infected plants. *Trans. Br. mycol. Soc.*, **71**, 395-7

Kikishuma, M., Hiratsuka, Y., Shibata, H. and Sato, S. (1984). [Cronartium blister rust on *Pinus densiflora* having *Pedicularis resupinata* var. *caespitosa* as an alternate host]. *Trans. mycol. Soc. Japan*, **25**, 315-18 (abstract in *Rev. Pl. Path.*, **64**, No. 1783, 1985).

Kile, G.A. and Watling, R. (1983). Occurrence of *Armillaria borealis* in Britain. *Trans. Br. mycol. Soc.*, **84**, 47-55

Kimmey, J.W. and Bynum, H.H., Jr. (1961). *Heart rot of red and white firs*. U.S. Dept. Agric.For.Pest Leafl., No. 52

Kispatic, J. (1962). [Leaf blotch on horse chestnut]. *Sumarski List*, **86**, 45-6; abstract in *Rev. appl. Mycol.*, **42**, 577, 1963

Klebahn, H. (1934). Eine Blattkrankheit der Edelkastanie und einige sie begleitende Pilze. *Z. PflKrankh. u. Pflanzenschutz*, **44**, 1-23

Kleiner, B.D. (1965). Tsitosporoz Topolei v Uzbekistane. Ex Sporovye Rasteniya Srednei Azu i Kazakhstana (131-138). Tashkent, Nauka; abstract in *Rev. appl. Mycol.*, **46**, No. 449, 1967

Kliejunas, J. (1989). Fungicides for control of *Sirococcus* tip blight of pine at a northern California nursery. *Tree Planters' Notes*, **40**, 30-2

Kobayashi, T. (1967). Critical revision of the genera *Rhizosphaera* Mangin et Hariot and *Rhizophoma* Petrak et Sydow, a little known fungus group associated with needle disease of conifers. *Bull. Tokyo Govt. For. Exp. Sta.*, **204**, 91-112

Kochman, J. (1929). Studja biologiczne nad paroszytem Wierzby *Fuscladium saliciperdum* (All. et Tub.). Lind. *Mm. Inst. nat. polon. Econ. Rur. Putawy*, **10**, 555-73

Kock, G. (1935). Eichmehltau und Rauchgasschden. *Z. PflKrankh.*, **45**, 44-5

Koike, T. and Tanaka, K. (1986). Photosynthesis and respiration in leaves of *Betula maximowiczii* infected with *Taphrina betulina*. *J. Jap. For. Soc.*, **68**, 194-6

Kontzog, H.G. (1989). Untersuchungen zur ubertragung des Pappelmosaik-Virus (poplar mosaic virus). *Archiv fur Phytopathologie und Pflanzenschutz*, **25**, 185-7

Korczyk, A. (1984). Studies on the susceptibility of the progeny of plus trees of 8 Polish Scots pine provenances to attack by *Lophodermium pinastri*. *Sylwan*, **128**, 43-57

Korhonen, K. (1978). Interfertility and clonal size in the *Armillaria mellea* complex. *Karstenia*, **18**, 31-42

Kowalski, T. (1982). Fungi affecting *Pinus sylvestris* needles of various ages. *Eur. J. For. Path.*, **12**, 182-90

Kowalski, T. (1983). *Scleroderris lagerbergii* and its pathogenicity towards conifers. *Sylwan*, **127**, 33-43

Kowalski, T. (1988). *Cyclaneusma (Naemacyclus) minus* an *Pinus sylvestris* in Polen. *Eur. J. For. Path.*, **18**, 176-83

Kozlowski, T.T. (1964). *Water Metabolism in Plants*. Harper, New York.

Kozlowski, T.T. (1986). Effects on seedling development of direct contact of *Pinus resinosa* seeds on young seedlings with captan. *Eur. J. For. Path.*, **16**, 87-90

Kozlowski, T.T. and Constantinidou, Helen A. (1986). Responses of woody plants to environmental pollution. Part 1.Sources and types of pollutants and plant responses. *For. Abstr.*, **47**, 1, 5-35

Kramer, P.J. (1962). The role of water in tree growth. In T.T. Kozlowski (Ed.), *Tree Growth*. Ronald Press, New York, pp. 171-82

Krampe, O. and Rehm, H.J. (1952). Untersuchungen über den Befall von *Pseudotsuga taxifolia viridis* mit *Adelopus gaumannii* Rhode. NachrBl. dt. PflSchutzdienst, Berl., **6**, 208-12

Kristensen, H.R. (1963). Oak mosaic. *Tidsskr. Pl. Av.*, **67**, 616

Krzan, Z. (1981a). Cykl rozwojowy i morfologia grzba *Melampsora larici-populina* Kleb.— sprawcy rdzy topoli. *Arboretum Korn.*, **26**, 35-50 (Abstract in *Rev. Pl. Path.*, **62**, No. 407,

1983).

Krzan, Z. (1981b). Poplar resistance to infection by the fungus *Melampsora larici-populina* in field conditions. *Arboretum Korn.*, **26**, 123-42 (abstract in *Rev. Pl. Path.*, **62**, No. 410, 1983)

Kubler, H. (1988). Frost cracks in stems of trees. *Arboric. J.*, **12**, 163-75

Kujala, V. (1950a). Uber die Kleinpilze der Koniferen in Finland. *Metsatiet. Tutkimuslait Julk.*, **38**, 4

Kujala, V. (1950b). Uber die Kleinpilze der Koniferen in Finnland. *Commun. Instr. For. Fenn.* 1-121

Kunkel, H. (1968). Untersuchungen uber die Buchenwollschildlaus *Cryptococcus fagi* (Baer.). (Insecta, Coccina), einen Vertreter der Rindenparenchymsauger. *Z. Ang. Ent.*, **61**, 373-80

Kunze, E.H. (1952-53). Uber den antagonismus zwischen *Armillaria mellea* und *Pleurotus ostreatus in vitro. Wiss. J. Univ. Jena*, **2**, 97-9

Kupevich, V.F. and Transhel, V.G. (1957). *Cryptogamic plants of the USSR. IV: Fungi. (1). Rust fungi.* Transl. Israel Programme for Scientific Translations

Kurkela, T. (1980). Serological tests for identifying species of *Melampsora* on aspen. *Proceedings of the IUFRO Working Party S2-05, resistance in pines to Melampsora pinitorqua, June 1979, Suonenjoki, Finland. Folia For.* No. 422, 17-8

Kurkele, T. (1981). Growth reduction in Douglas fir caused by *Rhabdocline* needle cast. *Metsatet. Tutkimuslait. Julk.* No. 102, 16pp.

Lagerberg, T. (1933). *Ascochyta parasitica* (Hartig), en Skadesvamp pa Granplantor. *Svenska skogesvFr. Tidskr.*, **3**, 1-10.

Lagerburg, T., Lundberg, G. and Melin, E. (1927). Biological and practical researches into blueing in pine and spruce. *Svenska SkogsvFor. Tidskr.*, **25**, 145-272

Laine, L. (1968), Kuplamorsky (*Rhizina undulata* Fr.). uusi metsan tuhosieni maasamme. *Fol. For., Helsinki*, **44**, 11pp.

Laing, E.V. (1929). Notes from the Forestry Department, Aberdeen University. *Scott. For. J.*, **43**, 48-82

Laing, E.V. (1948). Preliminary note on a disease of Sitka spruce in Cairnhill Plantation, Durris, Kincardineshire. *Forestry*, **21**, 217-20

Lang, K.J. (1981). *Cylindrocarpon cylindroides* var. *tenue* als Pathogen an *Abies concolor* und einigen Tannenbastarden. *Eur. J. For. Path.*, **11**, 191-2

Lang, K.J. (1983). Present state of beech bark disease in Germany. In Houston and Wainhouse (1983), q.v.10-12

Lang, K.J. (1987). *Dothistroma pini* an jungen Fichten (*Picea abies*). *Eur. J. For. Path.*, **17**, 316-7

Lang, K.J. and Karadzic, D. (1987). Is *Dothistroma pini* a danger to *Pinus sylvestris*? *Forst. Cent.*, **106**, 45-50

Lange, M. and Hora, B. (1963). *Collins Guide to Mushrooms and Toadstools*, Collins, London

Langner, W. (1936). Untersuchungen uber Larchen-, Apfel- und Buchenkrebs. *Phytopath. Z.*, **9**, 111-45

Langridge, Y.N. and Dye, D.W. (1982). A bacterial disease of *Pinus radiata* seedlings caused by *Pseudomonas syringae. N. Z. Jl. agric. Res.*, **25**, 273-6

Lanier, L. (1962). *Contribution a l'étude des maladies du rond.* Note Technique Forestière, Nancy, 9, 7pp.

Lanier, P., Joly, P., Bondoux, P. and Bellemere, A. (1978). *Mycologie et Pathologie Forestières (2 volumes).* Masson, Paris

Larue, J.H., Paulus, A.O., Wilbur, W.D., O'Reilly, M.J. and Darley, E.F. (1962). Armillaria root rot fungus controlled with methyl bromide soil fumigation.*Calif. Agric.*, **16**, 8-9

Latham, J. and Armstrong, F.H. (1934). The mechanical strength properties of 'Brown oak'. *Forestry*, **8**, 131-5

Latour, J.M. (1950). Intervention de la gelée dans la formation du chancre du melèze d'Europe (*Larix decidua* Mill.). *Bull. Soc. cent. for. Belg.*, **53**, 239-41

Laubert, R. (1933). Mehltau und *Rhytisma* auf *Acer negundo. NachrBl. dt. Pflschutzdienst.*, **13**, 94

Laubert, R. (1936). Die Blattfallkrankheit der Pappeln. *Kranke Pflanze*, **15**, 196-7

Lazarev, V. (1983). Bolesti iglica bijelog bora (*Pinus sylvestris* L.). *Zastita Bilja*, **34**, 265-74

Lazarev, V. (1985). Bolesti kore bukve u izdanackim sumama. *Zastits bilja*, **35**, 197-206

Lazarova-Topchyiska, M. (1987). [Walnut black line (preliminary communication)]. (abstract in *Rev. Pl. Path.*, **69**, No. 7471, 1990)

Leach, R. (1937). Observations on the parasitism and control of *Armillaria mellea. Proc. R. Soc., Ser. B. 131*, 561-73

Leibundgut, H. (1969). Untersuchungen uber die Anfalligkeit verschiedener Eichenherkunfte fur die Erkrankung an Mehltau. *Schweiz. Z Forstwes.*, **120**, 486-93

Leibundgut, H., Dafis, S. and Bezanon, M. (1964). Etudes sur diverses provenances de mélèze européen (*Larix decidua* L.). et la variabilité de leur infection par le chancre du mélèze (*Dasyscypha willkomii* Hart.). *Schweiz. Z. Forstwes.*, **115**, 255-60

Leibundgut, H. and Frick, L. (1943). Eine Buchenkrankheit im schweizerischen Mitteland. *Schweiz. Z. Forstw.*, **94**, 297-306

Leith, I.D. and Fowler, D. (1988). Urban distribution of *Rhytisma acerinum* (Pers.). Fr. (tar spot). on sycamore. *New Phytol.*, **108**, 175-81

Lelliott, R.A. (1959). Fireblight of pears in England. *Agriculture, Lond.*, **65**, 564-9

Lelliott, R.A. and Billing, E.B. (1984). *Fireblight of Apple and Pear.* Leafl.Min.Agric.Fish.Fd., No. 571 (amended)

Lelliott, R.A., Billing, E. and Hayward, A.C. (1966). A determinative scheme for fluorescent plant pathogenic bacteria. *J. appl. Bact.*, **29**, 470-8

Lelliot, R.A. and Stead, D. (1987). *Methods for the Diagnosis of Bacterial Diseases of Plants.* Blackwell, Oxford.

Lesovskii, A.V. and Martyshechkina, A.F. (1985). [Protection of medium-aged oak stands from *Microsphaera alphitoides*]. *Lesovodstvo i Agrolesomelioratsiya*, No. 70, 63-66 (abstract in *Rev. Pl. Path.*, **65**, No. 5177, 1986)

Lester, D.T. (1978). Control tactics in research and practice.Exploiting host variation. In W.A.Sinclair and R.J.Campana (Eds), Dutch Elm Disease, Perspectives After 60 Years, *Search, Agriculture*, **8**, 39-42

Levitt, J. (1969). Dormancy and survival. *Symp. Soc. Exp. Biol.*, **23**, 395-448

Li, C. (1984). [Two specialized forms of *Marssonina populi* (Lib.). Magn.] *Journal of the Nanjing Institute of Forestry*, No. 4, 10-17 (abstract in *Rev. Pl. Path.*, **64**, No. 3210, 1985)

Lies, E. (1980). La déperissement du hêtre au Grand-Duche de Luxembourg. *Ann. Sci. forest.*, **37**, 275-7

Liese, J. (1930). Beobachtungen uber Stamm- und Stockfaulen unserer Waldbaume. *Zeitschrift fur Forst- und Jagdwesen*, **26**, 587-91

Liese, J. (1931). Starke Schade durch den Pappelkrebs. *Dt. Forstz.*, **46**, 465-6

Liese, J. (1932). Zur Biologie der Douglasiennadelschutte. *Zeitsch. fur Forst- u. Jagdwesen*, **64**, 680-93

Liese, J. (1939). The occurrence in the British Isles of the Adelopus disease of Douglas fir. *Q. Jl. For.*, **33**, 247-52

Lilja, S. (1986). Disease and pest problems on *Pinus sylvestris* nurseries in Finland. *EPPO Bull.*, **16**, 561-4

Lin, D., Dumas, M.I. and Hubbes, M. (1989). Isozyme and general protein patterns of *Armillaria* spp. collected from boreal mixed forest of Ontario. *Can. J. Bot.*, **67**, 1143-7

Lindau, G. (1907). *Die Pilze Deutschlands, Osterreichs und der Schweiz*, Abt. 8, Kummer, Leipzig

Lindau, G. (1910). *Die Pilze Deutschlands, Osterreichs und der Schweiz*, Abt. 9, Kummer, Leipzig

Lindeijer, E.J. (1931). Een bacterie-ziekte van de wilg. *Tijdschr. PlZiekt.*, **37**, 63-7

Lindgren, R.M. (1942). *Temperature, moisture and penetration studies of woodstaining Ceratostomellae in relation to their control.* U.S. Dept.Agric.Tech.Bull., No. 807

Liss, B., Blaschke, H. and Schutt, P. (1984). Vergleichende Feinwurzeluntersuchungen an gesunden und erkrankten Altfichten auf zwei Standorten in Bayern—ein Beitrag zur Waldsterbensforschung. *Eur. J. For. Path.*, **14**, 90-102

Liu, S.C. and Teng, H.M. (1986). [Aeciospore surface morphology of pine blister rusts of China observed under a scanning electron microscope]. *Acta Mycologica Sinica*, **5**, 7-9 (abstract in *Rev. Pl. Path.*, **65**, No. 4571, 1986)

Lohman, M.L. and Watson, A.J. (1943). Identity and host relations of *Nectria* species associated with diseases of hardwoods in the Eastern States. *Lloydia*, **6**, 77-108

Lohwag, K. (1962). Die Blattbraune der rosskastanie. *Pflanzenarzt*, **15**, 40

Lombard, F.F., Davidson, R.W. and Lowe, J.L. (1960). Cultural characteristics of *Fomes ulmarius* and *Poria ambigua*. *Mycologia*, **1**, 280-94

Longhurst, J.W.S. (1989). *Acid deposition. Sources, effects and controls.* British Library Technical Publications.

Longo, B.N., Longo, N., Moriondo, F. and Drovandi, F. (1985). Observations on some Italian provenances of *Melampsora populnea* I. Studies for identification of *Melampsora pinitorqua* and *M. larici-tremulae*. *Eur. J. For. Path.*, **15**, 432-44

Longo, N., Moriondo, F. and Longo, B.N. (1979). Some aspects of biology of *Melampsora pinitorqua* Rostr. in Italy also compared to other European countries. *Phytopathologia Mediterranea*, **19**, 30-4

Lonsdale, D. (1980a). Nectria infection of beech bark in relation to infestation by *Cryptococcus fagisuga* Lind. *Eur. J. For. Path.*, **10**, 161-8

Lonsdale, D. (1980b). *Nectria coccinia* infection of beech: variations in disease in relation to predisposing factors. *Ann. Sci. forest.*, **37**, 307-17

Lonsdale, D. (1983a). Wood and bark anatomy of young beech in relation to Cryptococcus attack. In Houston, D.R. and Wainhouse, D. (1983), q.v. 43-9

Lonsdale, D. (1983b). Fungal associations in the build-up and decline of *Cryptococcus fagisuga* populations. In Houston, D.R. and Wainhouse, D. (1983), q.v. 99-104

Lonsdale, D. (1983c). *A definition of best pruning position.* Arboriculture Information Note 48/83. Forestry Commission

Lonsdale, D. (1986). In Pathology. *Rep. Forest Res. Edin.*, 1986, pp. 39-40

Lonsdale, D. (1987). In Pathology. *Rep. Forest Res. 1987*, pp. 47-8

Lonsdale, D. and Pratt, J.E. (1981). Some aspects of the growth of beech trees and the incidence of beech bark disease on chalk soils. *Forestry*, **54**, 183-95

Lonsdale, D., Pratt, J.E. and Aldsworth, F.G. (1979). Beech bark disease and archeological crop marks. *Nature, Lond.*, **277**, 414

Lonsdale, D. and Sherriff, C. (1983). Some aspects of the ecology of Nectria on beech. In Houston, D.R. and Wainhouse, D. (1983), q.v. 59-68

Lonsdale, D. and Wainhouse, D. (1987). *Beech bark disease.* Bull.For.Commn, Edinb., No. 69, 13pp.

Lopez, N. (1986). Control del mal de almacigos en semilleros de *Pinus radiata*. *Revista* de la Facultad de Ciencias Agricolas, Universidad Nacional de Loja, Ecuador. (abstract in *Rev. Pl. Path.*, **67**, No. 437, 1988)

Lortie, M. (1964). Pathogenesis in cankers caused by *Nectria galligena*. In Symposium on cankers of forest trees. *Phytopathology*, **54**, 250-78

Lortie, M. (1969). Inoculations of *Nectria galligena* on northern hardwoods. *Contr. Fonds. Rech. for. Univ. Laval.*, **13**, 21pp.

Lortie, M. and Kuntz, J.E. (1962). Studies on spore discharge of *Nectria galligena*. *Phytopathology*, **52**, 740 (abstract)

Lortie, M. and Kuntz, J.E. (1963). Ascospore discharge and conidium release by *Nectria galligena* Bres. under field and laboratory conditions. *Can. J. Bot.*, **41**, 1205-10

Lousley, J.E. (1976). *Flora of Surrey.* David and Charles, Newton Abbott

Low, J.D. and Gladman, R.J. (1960). *Fomes annosus in Great Britain*. Forest Rec., Lond., No. 41

Low, J.D. and Gladman, R.J. (1962a). Butt-rot of conifers in Scotland. A brief account of its distribution and occurrence. *Conference and Study Tour on Fomes annosus, Scotland, 1960*, IUFRO, Firenze, pp. 48-50

Low, J.D. and Gladman, R.J. (1962b). Present day research on *Fomes annosus* in Britain by the Forestry Commission.*Conference and Study Tour on Fomes annosus, Scotland, 1960, IUFRO, Firenze*, pp. 56-65

Low, J.D. and Greig, B.J.W. (1973). Spring frosts affecting the establishment of second rotation crops in Thetford Chase. *Forestry*, **46**, 139-55

Lowe, J.L. (1957). *Polyporaceae of North America, the genus Fomes.* N.Y.State Univ. Coll. Forestry at Syracuse Univ., Tech. Pub. No. 80

Luisi, N. and Cirulli, M. (1983). Gravi attacchi di *Gnomonia platani* Kleb. sul platano in Puglia. *Inftore Fitopatol.*, **33**, 43-5. (abstract in *Rev. Pl. Path.*, **63**, No. 2519, 1984)

Luisi, N. and Grasso, V. (1973). Biologia delle erysiphaceae di alcune cupolifere nell'Italia meridionale. *Ann. Acc. It. Sc. For.*, **22**, 211-67

Lukomski, S. (1962). *Verticillium alboatrum* moze byc grozne. *Las pol.*, **2**, 10-1

Lundquist, J.E. (1984). The occurrence and distribution of Rhizina root rot in South Africa and Swaziland. *South African Forestry Journal*, **131**, 2-4

Lung-Escarment, B., Poincot, D., Arthaud, J. and Taris, B. (1984). Le *Phytophthora cinnamomi* Rands, menace grave pour les plantations d'eucalyptus du Sud-ouest. *C. r. Séanc. Acad. Agric. Fr.*, **70**, 1539-41

Luque, J. and Girbal, J. (1988). Dieback of cork oak (*Quercus suber*). in Catalonia (NE Spain). caused by *Botryosphaeria stevensii. Eur. J. For. Path.*, **19**, 7-13

Lyr, H. (1967). Uber die Ursachen der Buchenrindennekrose (beech bark disease). *Arch. Forstw.*, **16**, 803-7

McBeath, J.H. (1984). Symptomatology on spruce trees and spore characteristics of a bud rust pathogen. *Phytopathology*, **74**, 456-61

MacBrayne, C.G. (1981). Canker and dieback of *Sorbus* spp. *Eur. J. For. Path.*, **11**, 325-33

McCain, A.H. (1984). Cypress canker control with fungicides. *Journal of Arboriculture*, **10**, 212-4

McCracken, F.I. and Toole, E.R. (1969). Sporophore development and sporulation of *Polyporus hispidus*. *Phytopathology*, **59**, 884-5

Macdonald, J.A. (1937). A study of *Polyporus betulinus* (Bull.)Fr.*Ann. appl. Biol.*, **24**, 289-310

Macdonald, J.A. and Russell, J.R. (1937). *Phomopsis scobina* (Cke). v.Höhn. and *Phomopsis controversa* (Sacc.) Trav. on ash. *Trans. bot. Soc. Edinb.*, **32**, 341-52

Macdonald, J., Wood, R.F., Edwards, M.V. and Aldhous, J.R. (1957). *Exotic Forest Trees in Great Britain*, Bull.For.Commn, No. 30

McKay, R. and Clear, T. (1953), Association of *Rhizina inflata* with group dying of Sitka spruce. *Ir. For.*, **10**, 58-9

McKay, R. and Clear, T. (1955). A further note on group dying of Sitka spruce and *Rhizina inflata*. *Ir. For.*, **12**, 58-63

McKay, R. and Clear, T. (1958). Violet root rot (*Helicobasidium purpureum*) on Douglas fir (*Pseudotsuga taxifolia*) and *Pinus contorta*. *Ir. For.*, **14**, 90-7

McKenzie, M.A. (1954). Maple wilt. *Trees, Cleveland, Ohio*, **14**, 19-20

Magasi, L.P. and Newell, W.R. (1983). The status of beech bark disease in the Maritime Provinces of Canada in 1980. In Houston, D.R. and Wainhouse, D. (1983), q.v. 13-7

Magasi, L.P. and Pond, S.E. (1982). European larch canker: a new disease in Canada and a new North American host record. *Pl. Dis.*, **66**, 339

Magnani, G. (1960). Bolla fogliare del pioppo. *Cellulosa Carta*, **11**, 27-31

Magnani, G. (1962). Septoriosi del *Populus nigra* L. da *Septoria populi* Desm. *Pubbl. Cent. Sper. agric. for., Rome*, **6**, 7-25

Magnani, G. (1963a). Necrosi su piantine di Eucalitto causate da *Botrytis cinerea* Pers. *Publ. Cent. Sper. Agric. For. Roma*, **6**, 211-23

Magnani, G. (1963b). Prove di resistenza di alcuni Pioppi euramericani a *Dothichiza populea* Sacc. et Briard. *Pubbl. Cent. Sper. agric.*, **6**, 155-78

Magnani, G. (1964). Alterazioni su foglie di pioppi causata da specie di Marssoninae. *Publ. Cent. Sper. agric. for.*, **7**, 251-81

Magnani.G. (1965). Contributo alla conoscenza della *Marssonina brunnea* (Ell. & Ev.). P.Magn. *Pubbl. Cent. Sper. agric. for.*, **8**, 123-6

Magnani, G. (1966). [Investigations on the possibility of physiological specialisation in *Melampsora alii-populina*]. *Pubbl. Cent. Sper. agric. for.*, **8**, 127-33; abstract in *For. Abs.*, **28**, No. 2506, 1967

Magro, P., Lenna, P.D. and Marciano, P. (1984). *Trichoderma viride* on cypress shoots and antagonistic action against *Seiridium cardinale*. *Eur. J. For. Path.*, **14**, 165-70

Malvick, D.K. and Moore, L.W. (1988). Survival and dispersal of a marked strain of *Pseudomonas syringae* in a maple nursery. *Pl. Path.*, **37**, 573-80

Mamedov, K.D. (1974). [Chemical and biological preparations against powdery mildew of oak]. *Les. Khoz. (1974)*, **7**, 83-84; abstract in *Rev. Pl. Path.*, **54**, No. 1890, 1975

Manap Ahmad, A. and Hayes, A.J. (1971). *Crumenula soraria* Karst. associated with cankering and dieback of Corsican pine. *Q. Jl. For.*, **25**, 185-200

Mancini, G. and Scapini, I. (1981). Influenza di alcuni fattori nutritivi ed ambientali sullo sviluppo *in vitro* di *Ceratocystis fimbriata* f.*platani*. *Riv. Patol. veg.*, *Padova*, **17**, 163-74

Mandahar, C.L. (editor). (1989). *Plant Viruses*. CRC Press, London

Mangin, L. (1894). Sur la maladie du rouge dans les pépinières et les plantations de Paris. *C. r. hebd. Séanc. Acad. Sci., Paris*, **119**, 753-6

Manka, K. (1953). *Badania terenowe i laboratoryjne nad Opienka miodowa*. Inst.Bad.Lesnictwa (Warsaw)., Pr. No. 94, 96pp.

Manka, K. (1986). Thinning as a treatment against *Heterobasidion annosum* and *Gremmeniella abietina*. *Sylwan*, **130**, 1-8,

Manners, J.G. (1953). Studies on larch canker. I. The taxonomy and biology of *Trichoscyphella willkommii* (Hartig). Nannf. and related species. *Trans. Br. mycol. Soc.*, **36**, 362-74

Manners, J.G. (1957). Studies on larch canker. II. The incidence and anatomy of cankers produced experimentally either by inoculation or by freezing. *Trans. Br. mycol. Soc.*, **40**, 500-8

Manners, J.G. (1982). *Principles of Plant Pathology*. Cambridge University Press, Cambridge

Manshard, E. (1927). Der Buchenheimlingspilz *Phytophthora omnivora* de Bary und seine Bekämpfung. *Forstarchiv.*, **3**, 84-6

Maraite, H. and Meyer, J.A. (1966). Incidence de quelques facteurs du milieu sur la pourriture rouge de l'épicea. *Bull. Soc. Roy. For. Belg.*, **7**, 493-509

Maramorosch, K. and Raycaudhuri, S.P.(editors)(1982). *Mycoplasma Diseases of Trees and Shrubs*. Academic Press, New York

Marchetti, L. and D'Aulerio, A.Z. (1980). L'oidio della quercia. *Inftore fitopatol.*, **30**, 23-4

Marchetti, L. and D'Aulerio, A.Z. (1982). Indicazioni di lotta chimica contro l'agente del cancro del cipresso (*Coryneum cardinale* Wag.). *Inftore fitopatol.*, **32**, 55-87

Marchetti, L. and D'Aulerio, A.Z. (1983a). *Gnomonia platani*: aspetti biologici e lotta chimica. *Inf. Fitopatol.*, **33**, 30-3

Marchetti, L. and D'Aulerio, A.Z. (1983b). Interventi chirurgici contro il cancro del cipresso con applicazione di mastici protettivi. *Inftore fitopatol.*, **33**, 51-4

Marchetti, L., D'Aulerio, A.Z. and Badiali, G. (1984). Un bienno di lotta chimica contro l'antracnosi del platano. *Inf. Fitopatol.*, **34**, 35-6

Marchetti, L., D'Aulerio, A.Z. and Grassi, S. (1986). Esperienze di lotta contro il cancro del cipresso con impiego di *Trichoderma viride* Pers. *Inftore fitopatol.*, **36**, 43-5

Marcu, G. (coordinator). *et al.* (1966). *Studiul cauzelor si al metodelor de provenire si combatere a uscarii stejarului*. Centrul de Documentare Tehnica Pentu Economia Forestiera, Romania

Margus, M. (1959). Lehisekultuuridest ja nende tervislikust seisundist Eesti NSV-s. *Eesti NSV teadur, Akad. Toimet., Ser. biol.*, **8**, 204-15; abstract in *Rev. appl. Mycol.*, **39**, 197, 1960

Marinkovic, P. and Karadzic, D. (1985). *Nectria coccinea* (Pers.ex Fr.)Fries, uzrok susenja bukve u Srbiji. *Zastita Bilja*, **36**, 263-72

Marinkovic, P. and Karadzic, D. (1987). Delovanje nekik fungicida na klijanje reprodutivnik organa i porast micelije *Sphaeropsis sapinea* Dyko a. Sutton. *Zastita Bilja*, **38**, 5-13 (abstract in *Rev. Pl. Path.*, **67**, No. 4141, 1988)

Marsh, R.W. (1939). Observations on apple canker. II. Experiments on the incidence and control of shoot infection. *Ann. appl. Biol.*, **26**, 458-69.

Marsh, R.W. (1952). Field observations on the spread of *Armillaria mellea* in apple orchards and in a black currant plantation. *Trans. Br. mycol. Soc.*, **35**, 201-7

Martin, E.I. (1957). Neoplastisches Wachstum bei *Sequoiadendron giganteum* Buchholz. *Phytopath. Z.*, **30**, 342-3

Martin, K.J. and Gilbertson, R.L. (1976). Cultural and other morphological studies of *Sparassis radicata* and related species. *Mycologia*, **68**, 622-39

Martinez, A.T. and Ramirez, C. (1983). *Rhizosphaera oudemansii* (Sphaeropsidales). associated with a needle cast of Spanish *Abies pinsapo*. *Mycopathologia*, **83**, 175-82

Martinsson, P. (1985). The influence of pine twist rust (*Melampsora pinitorqua*). on growth and development of Scots pine (*Pinus sylvestris*). *Eur. J. For. Path.*, **15**, 103-11

Maruyama, P.J. (1984). A new host and distribution record of a larch needle blight, *Meria laricis* Vuill., in Alberta. *Canadian Plant Disease Survey*, **64**, 19

Marxmüller, H. (1982). Etude morphologique des *Armillaria* ss.str. a anneau. *Bull. Soc. mycol. France*. **98**, 87-124

Marxmüller, H. and Printz, P. (1982). Honningsvampe. *Svampe*, **5**, 1-10, 59-60

Marziano, F., Scalcione M. and Noviello, C. (1986). Considerazioni sull'oidio del platano. *Anali della Facolta di Sienze Agraree della Universita degli Studi di Napoli Portici*, **20**, 60-70 (abstract in *Rev. Pl. Path.*, **67**, No. 3643, 1988)

Massee, G. (1914). Black knot of birch. *Kew Bull.*, pp. 322-3

Matheron, M.E. and Mirchetich, S.M. (1985). Control of Phytophthora root and crown rot and trunk canker in walnut with metalaxyl and fosetyl Al. *Pl. Dis.*, **69**, 1042-3

Mathon, B. (1982). Les dépérissements des Cyprès et des Thujas dus à deux champignons: *Seiridium cardinale* (Wag.). Sutton et Gibson (syn.*Coryneum cardinale* (Wag.)). et *Didymascella thujina* (Durand). Maire. *Revue hort.*, **228**, 41-4

Matta, A. and Kerling, L.C.P. (1964). *Verticillium alboatrum* as a parasite of *Senecio vulgaris*. *Neth. J. Pl. Path.*, **70**, 727-32

Matthews, R.E.F. (1981). *Plant Virology*, 2nd edn, Academic Press, New York

Matthews, R.E.F. (1982). Classification and nomenclature of viruses. *Intervirology*, **17**, 1-199

Maxwell, H. (1933a). The sycamore fungus. *Nature, Lond.*, **132**, 409

Maxwell, H. (1933b). The sycamore fungus. *Nature, Lond.*, **132**, 752

May, C. and Gravatt, G.F. (1951). Vascular diseases of hardwoods (caused by *Verticillium alboatrum* and *Dothiorella ulmi*). From *Important Tree Pests of the Northeast. Committee on tree pest leaflets, New England Section, Society of American Foresters, New Hampshire*, Tree Pest Leafl. No. 19

Mayer, M. and Allen, D.C. (1983). *Chilocorus stigma* (Coleoptera: Coccinellidae). and other predators of beech scale in central New York. In Houston, D.R. and Wainhouse, D. (1983), q.v., 89-98

Mayhead, G.J. (1976). Forest fertilising in Great Britain. *Proc. Fertil. Soc.*, **158**, 2-17

Meer, F.A. van der, Maat, D.Z. and Vink, J. (1980). Poplar mosaic virus: purification, antiserum preparation, and detection in poplars with the enzyme-linked immunosorbent assay (ELISA) and with infectivity tests on *Nicotiana megalosiphon*. *Nederlands Journal of Plant Pathology*, **86**, 99-110

Meer, J.H.H.van der (1926). Verticillium wilt of maple and elm seedlings in Holland. *Phytopathology*, **16**, 611-4

Mehlisch, K. (1938). Einepilzlicher schadling an Abies. Blumen — u. PflBau ver., *Gartenwelt*, **42**, 92

Melchior, G.H. (1975). Research in resistance in *Pinus sylvestris* to *Lophodermium pinastri*, in *Lophodermium in Pines*, pp. 5-10, Reinbek, No. 103

Melchior, G.H., Stephan, B.R. and Mohrdiek, O. (1983). Growth performance and Marssonina attack of *Populus deltoides* Bartr. grown in northern Germany. *Silvae Genet.*, **32**, 65-71

Mendel, Z., Golan, Y., Madar, Z. and Solel, Z. (1983). (Insect pests and diseases of cypress in Israel). *La-Yaaran*, **77**, 37-41 (Abstr. in *Rev. Pl. Path.*, **67**, No. 450, 1988)

Mercer, P.C. (1979). Attitudes to pruning wounds. *Arboric. J.*, **3**, 457-65

Meredith, D.S. (1959). Infection of pine stumps by *Fomes annosus* and other fungi. *Ann. Bot.*, **23**, 454-76

Meredith, D.S. (1960). Further observations on fungi inhabiting pine stumps. *Ann. Bot.*, **24**, 63-78

Merkle, R. (1951). Über die Douglasien-Vorkommen und die Ausbreitung der Adelopus-Nadelschütte in Württemberg-Hohenzollern. *Forst. -u-Jagdztg*, **122**, 161-91

Mesturino, L. and Mugnai, L. (1986). Disseccamenti delle foglie di faggio da *Gloeosporium fagi* (Desm. & Rob.). Westend. *Inftore fitopatol.*, **36**, 39-41

Metcalfe, G. (1940). The watermark disease of willows. I. Host-parasite relationships. *New Phytol.*, **39**, 322-32

Metcalfe, S.E., Atkins, D.H.F. and Derwent, R.G. (1989). Acid deposition modelling and the interpretation of the United Kingdom secondary precipitation network data. *Atmospheric Environment*, **23**, No. 9, 2033-2052

Michaels, E. and Chastagner, G.A. (1984a). Distribution, severity and impact of Swiss needle cast on Douglas fir Christmas trees in western Washington and Oregon. *Pl. Dis.*, **68**, 939-42

Michaels, E. and Chastagner, G.A. (1984b). Seasonal availability of *Phaeocryptopus gaeumannii* ascospores and conditions that influence their release. *Pl. Dis.*, **68**, 942-4

Mielke, J.L. (1957). *The yellow witches' broom of subalpine fir in the Intermountain region.* Res. Note Intmtn For. Exp. Sta., No. 47, 5pp.

Mielke, M.E., Haynes, C. and Macdonald, W.L. (1982). Beech scale and *Nectria galligena* on beech in the Monongahela National Forest, West Virginia. *Pl. Dis.*, **66**, 851-2

Mijuskovic, M. and Vujanovic, V. (1989). *Gymnosporangium gaemannii* Zogg, jedna rijetka, a za Jugoslaviju nova vrsta. *Zastita Bilja*, **40**, 443-51 (abstract in *Rev. Pl. Path.*, **69**, No. 7483, 1990)

Milatovic, I. (1956). Palez lisea Divljeg Kestena. *Zasht. Bilja*, **38**, 109-11

Milburn, M. and Gravatt, G.F. (1932). Preliminary note on a Phytophthora root disease of chestnut. *Phytopathology*, **22**, 977-8

Millar, C.S. (1975). Deposition of ascospores of Lophodermium from pines, in *Lophodermium in Pines*, pp. 11-20, Reinbek, No. 103

Miller, D.J. and Goodall, B. (1981). Blue staining in ponderosa pine sapwood at moderate and low temperatures. *Dep. For. Prod., Sch. For., Oregon State Univ. Corvallis.*

Miller, D.R. (1974). *Sydowia polyspora* found on white fir twigs in California. *Pl. Dis. Reptr.*, **58**, 94-5

Miller-Jones, D.N., Houston, D.R. and Preece, T.F. (1977). The use of electrical resistance measurements to detect watermark disease of cricket bat willow. *Pl. Dis. Reptr.*, **61**, 268-72

Miller-Weeks, M. (1983). Current status of beech bark disease in New England and New York. In Houston, D.R. and Wainhouse, D. (1983), q.v., 21-3

Miller-Weeks, M. and Stark, D. (1983). European larch canker in Maine. *Pl. Dis.*, **67**, 448

Milne, K.L. and Hudson, H.J. (1987). Artificial infection of leaves of the London plane by ascospores and conidia of *Apiognomonia veneta*. *Trans. Br. mycol. Soc.*, **88**, 399-432

Minkevich, I.I. and Bazova, S.V. (1984). [The development of leaf rust of poplar in relation to weather conditions] *Mikologiya i Fitopatologiya*, **18**, 149-52 (abstract in *Rev. Pl. Path.*, **63**, No. 5147, 1984)

Minkevich, I.I. and Stoyanov, S.M. (1987). [Forecasting the development of oak powdery mildew from data on the phenology of the host plant in Bulgaria]. *Izvestiya Vysshikh*

*Uchebnykh Zavedenii, Lesnoi Zhurnall.* (Abstract in *Rev. Pl. Path.*, **68**, No. 989, 1989)

Minter, D.W. (1981). *Lophodermium on pines.* Mycol. Pap. No. 147, 54pp. CMI Kew

Minter, D.W. and Millar, C.S. (1980). Ecology and biology of three *Lophodermium* spp. on secondary needles of *Pinus sylvestris. Eur. J. For. Path.*, **10**, 169-81

Minter, D.W., Staley, J.M. and Millar, C.S. (1978). Four species of *Lophodermium* on *Pinus sylvestris. Trans. Br. mycol. Soc.*, **71**, 295-301

Mircetich, S.M. and Rowhani, A. (1984). The relationship of cherry leaf roll virus and blackline disease of English walnut trees. *Phytopathology*, **74**, 423-8

Mircetich, S.M.J., Refsguard, J. and Matheron, M.E. (1980). Blackline of English walnut trees traced to graft-transmitted virus. *California Agriculture*, **34**, 8-10 (abstract in *Rev. Pl. Path.*, **60**, No. 5140, 1981)

Mirchev, S. and Alam, M. (1985). [The resistance of some oak species to powdery mildew, and the possibility of early diagnosis]. *Gorskostopanska Nauka*, **22**, 24-31 (Abstr. in *Rev. Pl. Path.*, **65**, No. 6200, 1986)

Mitchell, C.P., Millar, C.S. and Haworth, M.N. (1976a). Effect of the needle cast fungus *Lophodermella sulcigena* on growth of Corsican pine. *Forestry*, **49**, 153-8

Mitchell, C.P., Williamson, B. and Millar, C.S. (1976b). *Hendersonia acicola* on pine needles infected by *Lophodermella sulcigena. Eur. J. For. Path.*, **2**, 92-102

Mitchell, C.P., Millar, C.S. and Williamson, B. (1978). The biology of *Lophodermella conjuncta* Darker on Corsican pine needles. *Eur. J. For. Path.*, **8**, 108-18

Mitchell, L.A. (1988). A sensitive dot immunoassay employing monoclonal antibodies for detection of *Sirococcus strobilinus* in spruce seed. *Pl. Dis.*, **72**, 664-7

Mitchell, L.A. and Sutherland, J.R. (1986). Detection of seed-borne *Sirococcus strobilinus* with monoclonal antibodies in an enzyme-linked immunosorbent assay.*Can. J. For. Res.*, **16**, 945-8

Mittal, R.K., Pritam Singh and Wang, B.S.P. (1987). Botrytis: a hazard to reforestation. A literature review. *Eur. J. For. Path.*, **17**, 369-84

Mix, A.J. (1949). A monograph of the genus *Taphrina. Kans. Univ. Sci. Bull.*, **33**, 3-167

Mohr, E. (1984). Wasserstoffsperoxid—ein Mittel zur Verhutung der Stockinfektion von Fichtenbeständen durch *Heterobasidion annosum? Eur. J. For. Path.*, **14**, 291-6

Molin, N. and Rennerfelt, E. (1959). *Honungsskivlingen, Armillaria mellea (Vahl). Quél., som parasit pa barrtrad.* Medd. Skogsforsken Inst., Stockh., No. 48, 26pp.

Monod, M. (1983). Monographie taxonomique des Gnomoniaceae. *Sydowia, Beih.*, **9**, 1-315

Montgomery, H.B.S. (1936). A study of *Fomes fraxineus* and its effects on ashwood. *Ann. appl. Biol.*, **23**, 465-86

Mooi, J.C. (1948). Kanker en Takinsterving van der Wilg veroorzaakt door *Nectria galligena* en *Cryptodiaporthe salicina*, Thesis, University of Amsterdam, 119pp.; abstract in *Rev. appl. Mycol.*, **28**, pp. 92-4, 1949

Moore, F.J. (1952). Some powdery mildews on ornamental plants. *Pl. Path.*, **1**, 53-5

Moore, L.W., Lagerstedt, H.B. and Hartmann, N. (1974). Stress predisposes young filbert trees to bacterial blight. *Phytopathology*, **64**, 1537-40

Moore, W.C. (1946). New and interesting plant diseases. *Trans. Br. mycol. Soc.*, **29**, 250-8

Moore, W.C. (1959). *British Parasitic Fungi*, Cambridge University Press, Cambridge

Moorman, G.W., Lukezic, F.L. and Levine, R. (1988). *Pseudomonas syringae*, causal agent of a leaf spot on *Ulmus* sp. *Pl. Dis.*, **72**, 801

Moreau, C. and Moreau, M. (1953). Les maladies du châtaignier en forêt de Marly. *Rev. for. fr.*, **6**, 411-4

Moreau, C. and Moreau, M. (1954). Nouvelle observations sur le dépérissement des érables. *Bull. Soc. Linn. de Normandie*, **7**, 66-7

Moreau, R. and Schaeffer, R. (1962). *Fomes annosus* in the French jura. *Conference and Study Tour on Fomes annosus, Scotland, 1960, IUFRO, Firenze*, pp. 66-69

Morehart, A.L., Donohue, F.M.III amd Melchior, G.L. (1980). Verticillium wilt of yellow poplar. *Phytopathology*, **70**, 756-60

Morehart, M. and Melchior, G.L. (1982). The influence of water stress on Verticillium wilt of yellow poplar. *Can. J. Bot.*, **60**, 201-9

Morelet, M. (1980). Brunchorstia disease. 2. Distribution in France. *Eur. J. For. Path.*, **10**, 354-9

Morelet, M. (1982). La brunissure cryptogamique des cupressacée en France. *Revue hort.*, **227**, 35-9

Morelet, M. (1985). Les Venturia des peupliers de la section Leuce. 1. Taxonomie. *Cryptogamie Mycol.* **6**, 101-17

Morelet, M. (1986). Les risques d'adaption de *Venturia tremulae* aux trembles sélectionnés. *EPPO Bull.*, **16**, 589-92

Moriondo, F. (1972). Il cancro del cipresso da *Coryneum cardinale* Wag. 1 contributo: La progressione del processo infettivo nei tessuti cauliari. *Annali Accad. ital. Sci. for.*, **21**, 399-426

Moriondo, F. (1980). Features of *Cronartium flaccidum* and its hosts in Italy. In Powers, H.R., Grasso, V. and Raddi, P. Rusts of hard pines. Proceedings of the meeting of IUFRO Working Group S2.06.10, September, 1979, Florence, Italy. *Phytopath. Mediterranea*, **19**, 35-43

Moriondo, F., Capretti, P., Liperoti, A. and Mugnai, L. (1988). Preliminary observations on intersterility groups of *Heterobasidion annosum* in Italy. *Phytopathologia Mediterranea*, **27**, 178-9

Morton, H.L. and Miller, R.E. (1982). Chemical control of Rhabdocline needle cast of Douglas fir. *Pl. Dis.*, **66**, 999-1000

Motta, E. (1979). Presenza di spermazi in *Seiridium cardinale*. *Annali dell'Istituto Sperimentali di Patologia Vegetale, Roma*, **5**, 39-47

Motta, E. (1984). *Seiridium cardinale*: insediamento del patogeno su semi di Cupressace e possibilità di lotta chimica. *Annali dell'Istituto Sperimentale per la Patologia Vegetale Roma*, **9**, 205-10

Motta, E. and Saponaro, A. (1983). Micoflora dei semi di Cupressaceae. *Annali dell'Istituto Sperimentale per la Patologia Vegetale, Roma*, **8**, 71-5

Müller, E. (1950). Eine Knospenkrankheit der Stechfichte. *Schweiz Beitr. Dendrol.*, **2**, 69-72

Müller, K. (1912). Über das biologische Verhalten von *Rhytisma acerinum* auf verschiedenen Ahornarten. *Ber. dt. bot. Ges.*, **30**, 385-91

Müller, R. (1953a). Zur Frage des Pappelrindentodes. Ergebnisse und Folgerungen aus einem versuch. *Schweiz. Z. Forstwes.*, **104**, 408-28

Müller, R. (1953b). Weiters Erfahrungen über den Pappelrindentod (*Dothichiza populea*). *Schweiz. Z. Forstwes.*, **104**, 534-5

Müller-Kögler, E. (1954). Bekampfung des Eichenmehltaus. *Forstsch. Merkbl. niedersachs. forstl. VersAnst., Abt. B*, **5**, 4pp.

Müller-Stoll, W.R. and Hartmann, U. (1950). Über den Cytospora-Krebs der Pappel (*Valsa sordida* Nitschke). und die Bedingungen für eine parasitäre Ausbreitung. *Phytopath. Z.*, **16**, 443-78

Munk, A. (1957). Danish pyrenomycetes. A preliminary flora. *Dansk. bot. Ark.*, 1-491

Munson, R.G. (1939). Observations on apple canker. I. The discharge and germination of spores of *Nectria galligena* Bres. *Ann. appl. Biol.*, **26**, 440-56

Murdoch, C.W. and Campana, R.J. (1983). Bacterial species associated with wetwood of elm. *Phytopathology*, **73**, 1270-3

Murdoch, C.W. and Campana, R.J. (1984). Stem and branch distribution of wetwood and relationship of wounding to bleeding in American elms. *Pl. Dis.*, **68**, 890-2

Murray, B.J. (1926). Three fungus diseases of *Salix* in New Zealand, and some saprophytic fungi found on the same hosts. *Trans. N. Z. Inst.*, **56**, 58-70

Murray, J.S. (1953a). A note on the outbreak of '*Chrysomyxa abietis* Unger'(spruce needle rust) in Scotland, 1951. *Scott. For.*, **7**, 52-4

Murray, J.S. (1953b). Group dying of spruce in Eire. *Ir. For.*, **10**, 55-7

Murray, J.S. (1955). *Rusts of British Forest Trees*, Bookl.For.Commn, No. 4

Murray, J.S. (1957). *Top dying of Norway spruce in Great Britain*. Reprinted from Proc.7th Br.Commnw.For.Conf., 1957

Murray, J.S. (1962a). Forest pathology. *Rep. Forest Res., Lond.*, 1961, pp. 57-9

Murray, J.S. (1962b). Fomes annosus. Its importance in Britain. *Conference and Study Tour on Fomes annosus in Scotland, 1960, IUFRO, Firenze*, pp. 69-71

Murray, J.S. (1974). The fungal pathogens of oak, in M.G.Morris and F.H.Perring (Eds), *British Oak*, Classey, Farringdon, for the Botanical Society of the British Isles

Murray, J.S. (1978). Death of bark in *Acer pseudoplatanus* associated with drought. *Eur. J. For. Path.*, **8**, 65-75

Murray, J.S. and Batko, S. (1962). *Dothistroma pini* Hulbary: a new disease of pine in Great Britain. *Forestry*, **35**, 57-65

Murray, J.S., Millar, C.S. and van der Kamp, B.J. (1969). Incidence and importance of *Peridermium pini* (Pers.). Lév. in north-east Scotland. *Forestry*, **42**, 165

Murray, J.S. and Young, C.W.T. (1956). The effect of brashing and thinning on the incidence of *Lophodermium pinastri*. *Q. Jl. For.*, **50**, 75-6

Murray, J.S. and Young, C.W.T. (1961). *Group Dying of Conifers*, Forest Rec., Lond., No. 46, 19pp.

Mutto, S. and Panconesi, A. (1987). Ultrastructural modifications in *Cupressus sempervirens* tissues invaded by *Seiridium cardinale*. *Eur. J. For. Path.*, **17**, 193-204

Myren, D.T. (1984). *Meria laricis* found in European larch in Ontario. *Pl. Dis.*, **68**, 732

Naidenov, Ya. (1985). [*Marssonina castagnei* (Desm. et Mort.). P.Magn. on poplars of Sect. Leuce in Bulgaria]. *Gorskostopanska Nauka* **22**, 51-7 (abstract in *Rev. Pl. Path.*, **65**, No. 4112, 1986)

Nattrass, R.M. (1928). The Physalospora disease of the basket willow. *Trans. Br. mycol. Soc.*, **13**, 286-304

Nattrass, R.M. (1930). *A Note on Two* Marssonina *Diseases on Willows*, Minist. Agric., Egypt (Pl. Prot. Sect. Bull. No. 99, 19pp.)

Nedkvitne, K. (1966). Dyrkning av Edelgran *Abies alba* Müll., i Vest-Norge. *Meddr. Vestland. forstl ForstStn*, **40**, 12, 8pp. unnumbered, plus pp. 135-219

Neely, D. (1967). Dutch elm disease in Illinois cities. *Pl. Dis. Reptr.*, **51**, 511-4

Neely, D. (1972). Municipal control of Dutch elm disease in Illinois cities. *Pl. Dis. Reptr.*, **56**, 460-2

Neely, D. (1976). Sycamore anthracnose. *J. Arb.*, **2**, 153-7

Neely, D. and Himelick, E.B. (1963). *Aesculus* species susceptible to leaf blotch. *Pl. Dis. Reptr.*, **47**, 170

Neely, D. and Himelick, E.B. (1967). Characteristics and nomenclature of the oak anthracnose fungi. *Phytopathology*, **57**, 1230-6

Neely, D. and Himelick, E.B. (1989). Control of sycamore anthracnose. *American Nurseryman*, **170**, 89-91

Neergaard, P. (1957). 8.areberetning vedrorende fropatologisk kontrol, 1, juni 1955-31 maj 1956. *Ann. Rep. Seed Path. Control.* June 1955-31 Maj 1956. State Plant Control, Copenhagen, 15pp.

Neergaard, P. (1979). *Seed Pathology.* Two volumes. Macmillan, London

Negrutskii, S.F. (1986). [*Heterobasidium annosum*] Kornevaya gubka, Moscow, USSR; *Agropromizdat* (1986), 2nd edn, 196pp. (abstract in *Rev. Pl. Path*, **68**, No. 1473, 1989)

Nelson, A.C. (1965). The morphology and cytology of three noteworthy ascomycetes. *Diss. Abstr.*, **25**, 3812

Nelson, D.L. and Krebill, R.G. (1982). Occurrence and effect of *Chrysomyxa pirolata* cone rust on *Picea pungens* in Utah (USA). *Great Basin Naturalist*, **42**, 262-72 (abstract in *Rev. Pl. Path.*, **63**, No. 283, 1984)

Nelson, E.E., Silen, R.R. and Mandel, N.L. (1989). Effects of Douglas fir parentage on Swiss needle cast expression. *Eur J. For. Path.*, **19**, 1-6

Nelson, R.M. (1940). Vigorous young poplar trees can recover from injury by Nectria cankers. *J. For.*, **38**, 587-8

Nembi, V. and Panconesi, A. (1982). Il risanamento chirurgico del *Cupressus sempervirens* dagli attachi di *Seiridium cardinale*. *Inftore fitopatol.*, **32**, 59-62

Németh, M.V., Szentivanyi, P. and Kolber, M. (1982). [Infection of walnut by cherry leaf roll virus. 1. Identification and distribution of the virus in Hungary]. A dio cherry leaf roll virus fertozöttsége. I. A virus azonositasa és elofordulasa magyarorszagon. *Növényvédelem*, **18**, 1-10 (abstract in *Rev. Pl. Path.*, **62**, No. 1661, 1983)

Nesme, X., Michel, M.F. and Digat, B. (1987). Population heterogeneity of *Agrobacterium tumefasciens* in galls of *Populus* L. from a single nursery. *Applied and Environmental Microbiology*, **55**, 655-9

Neves, N., Moniz, F., Azevedo, N.de, Ferreira, M.C. and Ferreira, G.W.S. (1986). Present phytosanitary situation of Portuguese forests. *EPPO Bull.*, **16**, 505-8

Newhook, F.J. (1959). The association of *Phytophthora* spp. with mortality of Pinus radiata and other conifers. *N. Z. J. agric. Res.*, **2**, 808-43

Newhook, F.J., Waterhouse, G.M. and Stamps, D.J. (1978). *Tabular key to the species of* Phytophthora *de Bary*. Mycol.Pap., No. 143

Nicholls, T.H. (1973). Fungicide control of *Lophodermium pinastri* on red pine seedlings. *Pl. Dis. Reptr.*, **57**, 263-6

Nienhaus, F. (1985). Infectious diseases in forest trees caused by viruses, mycoplasma-like organisms and primitive bacteria. *Experientia*, **41**, 597-603

Nienhaus, F., Ebrahim-Nesbat, E., Fricke, M., Buttner, C. and Welter, K. (1985). Investigations on viruses from declining beech trees (*Fagus sylvatica* L.). in Rhineland and Westfalia, Federal Republic of Germany. *Eur. J. For. Path.*, **15**, 402-11

Nifren, D.T. (1984). *Meria laricis* found in European larch in Ontario. *Pl. Dis.*, **68**, 732

Nimmo, M. (1964). Winter damage in Southern beeches, *Nothofagus* species. *Rep. Forest Res. Lond.*, 1964, p.29

Nobles, M.K. (1948). Studies in forest pathology. VI. Identification of cultures of wood-rotting fungi. *Can. J. Res.*, **C26**, 281-431

Nohara, Y., Kodama, T. and Aoyama, Y. (1961). Studies on the control of poplar rusts. Part 1. Control tests with fungicides. *Bull. For. Exp. Sta., Meguro*, **130**, 45-50

Novak, J.B. and Langova, J. (1981). [Cherry leaf roll virus from walnut (*Juglans regia*) in Czechoslovakia]. Virus svinutky tresne z oresaku vlasského (*Juglans regia*) v Czechoslovensku. *Sbornik UVTIZ, Ochrana Rostlin*, **17**, 1-8 (abstract in *Rev. Pl. Path.*, **61**, No. 1410, 1982).

Nuesch, J. (1960). Beitrag zur Kenntnis der Weidenbewohnenden Venturiaceae. *Phytopath. Z.*, **39**, 329-60

Nutman, M.J. (1929). Studies of wood-destroying fungi. I.*Polyporus hispidus* (Fries). *Ann. appl. Biol.*, **16**, 40-64

O'Brien, J.T. (1973). Sirococcus blight of red pine.*Pl. Dis. Reptr.*, **57**, 246-7

Oellette, G.B. (1961). Studies in the infection process of *Ceratocystis ulmi* (Buism.). C. Moreau in American elm trees.*Can. J. Bot.*, **40**, 1568-75

Oganova, E.A. (1957). [On bacterial canker of ash]; abstract in *Rev. appl. Mycol.*, **38**, p.281, 1959

Ogilvie, L. (1924). Observations on the 'slime-fluxes' of trees. *Trans. Br. mycol. Soc.*, **9**, 167-82

Ogilvie, L. (1932). Note on the rusts of basket willows and their control. *Rep. agric. hort. Res. Stn. Univ. Bristol, 1931*, pp. 133-8

Ogilvie, L. and Hutchinson, H.P. (1933). *Melampsora amygdalinae*, the rust of basket willows (*Salix triandra*). I.Observations and experiments in 1932. II. Spore germination experiments. *Rep. agric. hort. Res. Stn. Univ. Bristol, 1932*, pp. 125-38

Ohman, J.H. (1966). *Scleroderris lagerbergii* Gremmen: tha cause of dieback and mortality of red and jack pine in Upper Michigan plantations. *Pl. Dis. Reptr.*, **50**, 402-5

Ohman, J.H. and Kessler, K.J. (1964). *White trunk rot of hardwoods.* U.S. Dept.Agric.For.Pest Leafl., No. 88

Old, K.M. and Kobayashi, T. (1988). Eucalypts are susceptible to the chestnut blight fungus, *Endothia parasitica*. *Aust. J. Bot.*, **36**, 599-603

Oleksyn, J. and Przybyl, K. (1987). Oak decline in the Soviet Union—Scale and hypothesis. *Eur. J. For. Path.*, **17**, 321-36

O'Riordain, F. and Kavanagh, T. (1965). Marssonina leaf spot of poplar. *Ir. J. agric. Res.*, **4**, 233-5

Orlos, H. (1951). Zwalczanie maczniaka *Microsphaera alphitoides* Griff. et Maubl. w szkolkach debowych.*Inst. Bad. Lesn. Prace*, No. 67

Orlos, H. (1954). Proba zwalczania opienki w swierczynach gorskich w skali polgospodarczej. *Las pol.*, **28**, 9-11,

Orlos, H. (1957). Badania nad zwalczaniem opienki miodowej (*Armillaria mellea* Vahl). metoda bioloziczna. *Roczn. Nauk. lesn.*, **15**, 195-236

Osmaston, L.S. (1927). Mortality among oaks. *Q. Jl. For.*, **21**, 28-30

Ostaff, D.P. (1985). Age distribution of European larch canker in New Brunswick. *Pl. Dis.*, **69**, 796-8

Ostrofsky, A. and Ostrofsky, W.D. (1984). *Kabatina juniperi* associated with branch tip dieback of eastern red cedar in Maine. *Pl. Dis.*, **68**, 351

Ostrofsky, A. and Peterson, G.W. (1981). Etiologic and cultural studies of *Kabatina juniperi*. *Pl. Dis.*, **65**, 908-10

Ostry, M.E. and Anderson, N.A. (1983). Infection of trembling aspen by *Hypopxylon mammatum* through cicada oviposition wounds. *Phytopathology*, **73**, 1092-6

Ostry, M.E. and McNabb, H.S., Jr (1986). *Populus species and hybrid clones resistant to Melampsora, Marssonina and Septoria.* Research Paper, North central Forest experiment Station, USDA Forest Service, No. NC-272, 6pp. (abstract in *Rev. Pl. Path.*, **68**, No. 290, 1989)

Ostry, M.E. and Nicholls, T.H. (1989). Effects of *Lophodermium seditiosum* on growth of pine nursery seedlings in Wisconsin. *Pl. Dis.*, **73**, 798-800

Oszako, T. (1985). Grzyby jako bioindykatory zmian w ekosystemach lesnych pod wplywem przemyslowych zanieczyszczen powietrza. *Sylwan*, **129**, 33-42 (abstract in *Rev. Pl. Path.*, **67**, No. 966, 1988)

Overholts, L.O. (1953). *The Polyporaceae of the United States, Alaska, and Canada.* University of Michigan Press

Owen, D.R., Lindahl, K.Q., Wood, D.L. and Parmeter, J.R. (1987). Pathogenicity of fungi isolated from *D. brevicornis* and *D. ponderosae* to ponderosa pine seedlings. *Phytopathology*, **77**, 631-6

Paden, J.W., Sutherland, J.R. and Woods, T.A.D. (1978). *Caloscypha fulgens* (Ascomycetidae,

Pezizales): the perfect state of the conifer seed pathogen *Geniculodendron pyriforme* (Deuteromycotina, Hyphomycetes). *Can. J. Bot.*, **56**, 2375-9

Pagony, H., Koren, E.Sz. and Hezedus, P. (1983). [Decline of pine stands caused by *Fomes annosus* Cooke on sandy soils]. *Növenyvedelem*, **19**, 529-35 (abstract in *Rev. Pl. Path.*, **63**, No. 2535, 1984).

Palleroni, N.J. (1984). Genus I. *Pseudomonas*. In *Bergey's Manual of Systematic Bacteriology*, *Vol. I.* (ed. *N. R. Krieg)*, pp. 141-99. Williams and Wilkins, Baltimore

Palmer, M.A. and Nicholls, T.H. (1985). Shoot blight and collar rot of *Pinus resinosa* caused by *Sphaeropsis sapinea* in forest nurseries. *Pl. Dis.*, **69**, 739-40

Palmer, M.A., Nicholls, T.H. and Croghan, C.F. (1986). Fungicidal control of shoot blight caused by *Sphaeropsis sapinea* on red pine nursery seedlings. *Pl. Dis.*, **70**, 194-6

Pan, L.P. and Liu, C.Z. (1984). [A preliminary report of study on biological characteristics of *Lachnellula willkommii* of *Larix gmelinii*]. *Forest Science and Technology (Linye Keji Tongxun)*. No. 8, 19-23 (Abstract in *Rev. Pl. Path.*, **66**, No. 2085, 1987)

Pan, X.R. and Liu, C.Z. (1985). [Research of biological character of *Lachnellula willkommii* (Hartig). Dennis fungus]. *Journal of North-east Forestry University, China*, **13**, 55-61 (Abstract in *Rev. Pl. Path*, **66**, No. 1165, 1987)

Panagopoulos, C.G. (1988). *Agrobacterium tumefasciens* (Smith & Townsend). Conn. In Smith *et al.* 1988, q.v., pp. 176-9

Panconesi, A. (1981). *Ceratocystis fimbriata* of plane trees in Italy: Biological agents and control possibilities. *Eur. J. For. Path.*, **11**, 385-95

Panconesi, A. and Raddi, P. (1986). The influence of some chemical treatments on cypress canker disease development. *Eur. J. For. Path.*, **16**, 83-6

Panisset, T.E. (1929). *Daldinia concentrica* attacking the wood of *Fraxinus excelsior*. *Ann. appl. Biol.*, **16**, 400-21

Pankhurst, E.S. (1980). The effects of natural gas on trees and other vegetation. *Landscape Design, February* 1980, 32-3,

Pape, H. (1925). Der Pilz *Fusicladium saliciperdum* (All. et Tub.). Lind. als Korbweidenschädling. *Dt. Erwerbsgartenbau*, **24**, 326-9

Parameswaran, N. (1980). Occurrence of mycoplasma-like bodies in phloem cells of beech trees affected by 'bark necrosis'. *Ann. Sci. Forest.*, **37**, 371-2

Parker, A.K. (1970). The effect of relative humidity and temperature on needle cast disease of Douglas fir. *Phytopathology*, **60**, 1270-3

Parker, A.K. and Reid, J. (1969). The genus *Rhabdocline* Syd. *Can. J. Bot.*, **47**, 1533-45

Parker, E.J. (1974a). *Beech Bark Disease*, Forest Rec., Lond., No. 96

Parker, E.J. (1974b). Some investigations with beech bark disease Nectria in Southern England. *Eur. J. For. Path.*, **5**, 118-24

Parker, E.J. (1983). Beech bark disease in Great Britain. In Houston, D.R. and Wainhouse, D. (1983), q.v.1-6

Parker, J. (1968). Drought resistance mechanisms, in T.T.Kozlowski (Ed.), *Water Deficits and Plant Growth, Vol. 1*, Academic Press, New York, pp. 195-234

Parrini, C. and Panconesi, A. (1981). Attachi di *Seiridium* (*Coryneum*) *cardinale* (Wag.). Sutton e Gibson su alcune Cupressaceae in vivace. *Inftore fitopatol.*, **31**, 7-15

Parris, G.K. and Byrd, J. (1962). Oak anthracnose in Mississippi. *Pl. Dis. Reptr.*, **46**, 677-81

Pas, J.B.van der, Slater-Hayes, J.D., Gadgil, P.D. and Bulman, L. (1984a). *Cyclaneusma* (*Naemacyclus*) needle cast of *Pinus radiata* in New Zealand. 2: Reduction in growth of the host, and its economic importance. *N. Z. Jl. Forestry Sci.*, **14**, 197-209

Pas, J.B.van der, Bulman, L. and Slater-Hayes, J.D. (1984b). *Cyclaneusma* (*Naemacyclus*) needle cast of *Pinus radiata* in New Zealand. 3: Incidence and severity of the needle cast *N. Z. Jl. For. Sci.*, **14**, 210-4

Pas, J.B.van der, Bulman, L. and Horgan, G.P. (1984c). Disease control by aerial spraying of *Dothistroma pini* in tended stands of *Pinus radiata* in New Zealand. *N. Z. Jl. For. Sci.*, **14**, 23-40

Patton, R.F. and Riker, A.J. (1959). Artificial inoculations of pine and spruce trees with *Armillaria mellea*. *Phytopathology*, **49**, 615-22

Paucke, H. (1968). Frostungen an Buchen zur Unduktion von Rindennekrosen. *Arch. Forstw.*, **17**, 565-70

Pavari, A. (1949). Chestnut blight in Europe. *Unasylva*, **3**, 8-13

Pawsey, R.G. (1960). An investigation into·Keithia disease of *Thuja plicata*. *Forestry*, **33**, 174-86

Pawsey, R.G. (1962a). Leaf blotch of horse chestnut, in 'New or uncommon plant diseases and pests'. *Pl. Path.*, **11**, 137-8

Pawsey, R.G. (1962b). Resurgence of sooty bark disease of sycamore, in 'New or uncommon plant diseases and pests'. *Pl. Path.*, **11**, 138

Pawsey, R.G. (1963). Rotation sowing of *Thuja* in selected nurseries to avoid infection by *Keithia thujina*. *Q. Jl For.*, **56**, 206-9

Pawsey, R.G. (1964a). *Grey Mould in Forest Nurseries*, Leafl. For.Commn, No. 50, 7pp.

Pawsey, R.G. (1964b). *Resin-top Disease of Scots Pine*, Leafl.For.Commn, No. 49

Pawsey, R.G. (1965). Cycloheximide fungicide trials against *Didymascella thujina* on western red cedar, *Thuja plicata*. *Rep. Forest Res., Lond., 1964*, pp 141-150

Pawsey, R.G. (1971). Some recent observations on decay in conifers associated with extraction damage, and on butt rot caused by *Polyporus schweinitzii* and *Sparassis crispa*. *Q. Jl. For.*, **65**, 193-208

Pawsey, R.G. (1974). *Cronartium ribicola*, a potential threat to *Pinus contorta* in Britain. *Forestry*, **47**, 89-91 (and plates I and II following p.28)

Pawsey, R.G. (1984). *Ash dieback survey: summer 1983*. Arboriculture research Note 51/84/EXT. Arboricultural Advisory and Information Service, Forestry Commission, Edinburgh

Pawsey, R.G. and Gladman, R.J. (1965). *Decay in standing conifers developing from extraction damage*, Forest Rec., Lond., No. 54

Pawsey, R.G. and Rahman, M.A. (1976). Chemical control of infection by honey fungus, *Armillaria mellea*: a review. *Arboric. J.*, **2**, 468-79

Pawsey, R.G. and Young, C.W.T. (1969). A reappraisal of canker and dieback of European larch. *Forestry*, **42**, 145-64

Peace, T.R. (1936). Spraying against *Meria laricis*, the leaf cast disease of larch. *Forestry*, **10**, 79-82

Peace, T.R. (1938). Butt rot of conifers in Great Britain. *Q. Jl. For.*, **32**, 81-104

Peace, T.R. (1955a). Forest pathology. *Rep. Forest Res., Lond.*, 1954, pp. 31-3

Peace, T.R. (1955b). Sooty bark disease of sycamore a disease in eclipse. *Q. Jl. For.*, **49**, 197-204

Peace, T.R. (1960), *The Status and Development of Elm Disease in Britain*, Bull. For. Commn., Lond., No. 33, 44pp.

Peace, T.R. (1962). *Pathology of Trees and Shrubs*, Clarendon Press, Oxford

Peace, T.R. and Holmes, C.H. (1933). *Meria laricis. The Leaf Cast Disease of Larch*. Oxf. For. Mem. No. 15, 29pp.

Peace, T.R. and Murray, J.S. (1956). Forest pathology. *Rep. Forest Res., Lond.*, 1955, pp. 52-7

Peacock, J.W. (1975). Research on chemical and biological control of bark beetles, in D.A.Burdekin and H.M. Heybroek (Eds), *Dutch Elm Disease, Proc. IUFRO Conf. Sept. 1973, USDA* pp. 76-87

Pegler, D.N. (1973). The Polypores. Suppl., *Bull. Br. mycol. Soc.* **7**, 43pp.

Pei, M.H. and Shang, R.Z. (1984). [Study on the leaf rust of Cathay poplar caused by *Melampsora larici-populina* Kleb.]. *Journal of North-Eastern Forestry Institute, China*, **12**, 40-49. (Abstract in *Rev. Pl. Path.*, **65**, No. 1538, 1986).

Peno, M. (1983). Epifitna i endofitna mikroflora semena smrce (*Picea excelsa* Link). *Zastita Bilja*, **34**, 5-13 (abstract in *Rev. Pl. Path.*, **63**, No. 933, 1984)

Pentland, G.D. (1965). Stimulation of rhizomorph development of *Armillaria mellea* by *Aureobasidium pullulans* in artificial culture. *Can. J. Microbiol.*, **11**, 345-50

Perrin, R. (1976). Clef de détermination des Nectria d'Europe. *Bull. trimest. Soc. mycol. Fr.*, **92**, 335-47

Perrin, R. (1977). Le dépérissement du hêtre. *Rev. for. fr.*, **29**, 100-129

Perrin, R. (1979). Contribution à la connaissance de l'étiologie de la maladie de l'écorce du hêtre. I. Etat sanitaire des hêtraies françaises. Role de *Nectria coccinea* (Pers. ex Fries). Fries. *Eur. J. For Path.*, **9**, 148-66

Perrin, R. (1980a). Les infections chez le hêtre. *Rev. for. fr.*, **32**, 129-78

Perrin, R. (1980b). Contribution à la connaissance de l'étiologie de la maladie de l'écorce du hêtre. II. Etude experimentale de l'association *Cryptococcus fagisuga* Lind. *Nectria coccinea* (Pers. ex Fries) Fries. Role respectif des deux organismes. *Ann. Sci. forest*, **37**, 319-31

Perrin, R. (1983a). Current status of beech bark disease in France. In Houston, D.R. and Wainhouse, D. (1983), q.v., 7-9

Perrin, R. (1983b). Specificity of *Cryptococcus fagisuga* and *Nectria coccinea* association in beech bark disease in Europe. In Houston, D.R. and Wainhouse, D. (1983), q.v., 50-3

Perrin, R. (1983c). Pectinase activity of *Nectria coccinea* (Pers. ex Fries). Fries in relation to beech bark disease. In Houston, D.R. and Wainhouse, D. (1983), q.v., 54-8

Perrin, R. (1984a). Les activités pectinasiques de *Nectria ditissima* Tul. agent du chancre du hêtre. *Eur. J. For. Path.*, **14**, 219-25

Perrin, R. (1984b). La variabilité de la sensibilité du hêtre (*Fagus sylvatica* L.). à *Nectria coccinea*, un des agents de la maladie de l'écorce. *Eur. J. For. Path.*, **14**, 321-5

Perrin, R. (1985). La variabilité de la sensibilité du hêtre au chancre dû à *Nectria ditissima* Tul. *Annal. Sci. for. Nancy*, **42**, 225-37

Perrin, R. (1986). Maladies d'origine tellurique en pépinière forestière: problème du diagnostic et lutte intégre. *EPPO Bull.*, **16**, 553-60

Perrin, R. and Gerwen, C.P.van (1984). La variabilité du pouvoir pathogène de *Nectria ditissima* agent du chancre du hêtre. *Eur. J. For. Path.*, **14**, 170-6

Perrin, R. and Sampangi, R. (1986). La fonte des semis en pepinière forestière. *Eur. J. For. Path.*, **16**, 309-21

Perry, R.G. and Peterson, J.L. (1982). Susceptibility and response of juniper species to *Kabatina juniperi* infection in New Jersey. *Pl. Dis.*, **66**, 1189-91

Persson, A. (1955). Kronenmykose der Hybridaspe. 1. Untersuchungen über Auftreten selektive Wirkung und Pathogenität des Erregers. *Phytopath. Z.*, **24**, 55-72

Petaisto, R.L. (1982). Juurten leikkaamisen jälkainen sienitautiriski havupuun tainilla taimitarhalla. *Folia Forestalia* No. 505, 8pp. (abstract in *Rev. Pl. Path.*, **61**, 7190, 1982

Petch, T. (1938). British Hypocreales. *Trans. Br. mycol. Soc.*, **21**, 243-305

Petersen, G.W. (1967). Dothistroma needle blight of Austrian and Ponderosa pines: epidemiology and control. *Phytopathology*, **57**, 437-41

Petersen, R.S. and Jewell, F.F. (1968). Status of American stem rusts of pine. *A. Rev. Phytopathol.*, **6**, 13-40

Peterson, G.W. (1973). Infection of *Juniperus virginiana* and *J. scopulorum* by *Phomopsis juniperovora*. *Phytopathology*, **63**, 246-51

Peterson, R.S. (1964). *Fir broom rust*, U.S. Dept. Agric. For. Pest Leafl., No. 87, 7pp.

Peterson, R.S. (1965). Notes on western rust fungi. IV. *Mycologia*, **57**, 465-71

Petrescu, M. (n.d.). *Aspecte fitopatologice din padurile Republice Socialiste România*. Editura Agro-Silvica, Bucharest.

Petri, L. (1923). Sur la formation des chlamydospores chez l'oidium des chênes. *Congrès Path. vég. (Centenaire de Pasteur), Strasburg, 1923*, pp. 36-7

Phillips, D.H. (1963a). Forest Pathology. *Rep. Forest Res. Lond., 1962*, pp. 60-65

Phillips, D.H. (1963b). *Leaf Cast of Larch, Meria laricis*. Leafl. For.Commn, No. 21, 4pp.

Phillips, D.H. (1965). Forest pathology. *Rep. Forest Res. Lond. 1964*, pp. 57-60

Phillips, D.H. (1966). Stem crack in *Abies grandis*. *Rep. Forest Res., Lond., 1965*, pp. 61-2

Phillips, D.H. (1967a). Forest pathology. *Rep. Forest Res., Lond., 1966*, pp. 70-74

Phillips, D.H. (1967b). Forest pathology. *Rep. Forest Res., Lond., 1967*, pp. 96-102

Phillips, D.H. (1978). *The EEC Plant Health Directive and British Forestry*, Forest Rec., Lond., No. 116

Phillips, D.H. (1979). *British Forestry: the problem of alien pests and diseases*, in D.L.Ebbels and J.E.King (Eds), *Plant Health: The Scientific Basis For the Administrative Control of Plant Diseases and Pests*, Blackwell, Oxford

Phillips, D.H. (1980a). *International Plant Health Controls: Conflicts, Problems and Co-operation. A European Experience*. Res.Dev. Pap.For Commn, Lond., No. 125

Phillips, D.H. (1980b). Plant health legislation and forest trees in Britain. *Arboric. J.*, **4**, 152-7

Phillips, D.H. (1981). Legislating for introduced species: the lessons from plant health. *Discussion Paper in Conservation No. 30, BANC/University College, London*, pp. 50-9

Phillips, D.H. and Bevan, D. (1966). Forestry Quarantine and its biological background. *Proc. Wld For. Congr. Madrid, 1966.*, **2**, 1928-34 (and Forest Rec. Lond. No. 63, 1967)

Phillips, D.H. and Greig, B.J.W. (1970). Some chemicals to prevent stump colonisation by *Fomes annosus* (Fr.). Cooke. *Ann. appl. Biol.*, **66**, 441-52

Phillips, D.H. and Young, C.W.T. (1976). *Group Dying of Conifers*, Leafl.For.Commn, No. 65, 7pp.

Piearce, G.D. (1972). *Studies on Verticillium wilt in maples*. Ph.D.Thesis, University of Cambridge

Piearce, G.D. and Gibbs, J.N. (1981). *Verticillium wilt of trees and shrubs*. Arboric. Leafl. No. 9, HMSO, London

Pinon, J. (1973). Les rouilles du Peuplier en France. *Eur. J. For. Path.*, **3**, 221-8

Pinon, J. (1984a). Management of diseases of poplars. *Eur. J. For. Path.*, **14**, 415-25

Pinon, J. (1984b). Propriétés biologiques de la toxine d'*Hypoxylon mammatum*, parasite des peupliers de la section Leuce. *Rev. Cyt. Biol. Veg. Botaniste*, **7**, 271-277 (abstract in *Rev. Pl. Path.* No. 1774, 1985)

Pinon, J. (1986a). Situation d'*Hypoxylon mammatum* en Europe. *EPPO Bull.*, **16**, 543-6

Pinon, J. (1986b). Situation de *Melampsora medusae* en Europe. *EPPO Bull.*, **16**, 547-51

Pinon, J. (1986c). Les pathotypes de rouille sur peuplier. *EPPO Bull.*, **16**, 585-8

Pinon, J., van Dam, B.C., Genetet, I. and Kam, M.de (1987). Two pathogenic races of *Melampsora larici-populina* in north-western Europe. *Eur. J. For. Path.*, **17**, 47-53

Piraux, A. (1980). Observations relatives à la maladie de l'écorce du hêtre dans les Ardennes belges. Bilan d'une épidemie. *Ann. Sci. forest*, **37**, 249-356

Pirone, P.P. (1970). *Diseases and Pests of Ornamental Plants*. Ronald Press, New York

Pirone, P.P., Dodge, B.O. and Rickett, H.W. (1960). *Diseases and Pests of Ornamental Plants, 3rd edn*, Ronald Press, New York

Pisi, A., Marani, F. and Bertaccini, A. (1981). Mycoplasma-like organisms associated with elm witches' broom symptoms. *Phytopathologia Mediterranea*, **20**, 189-91

Plasman, A. (1953). Note sur la découverte en Belgique de *Phomopsis pseudotsugae* Wilson sur *Pseudotsuga taxifolia* (Lam). Britt. *Parasitica*, **9**, 1-5

Plassmann, E. (1927). *Untersuchungen uber den Larchenkrebs*, J.Neumann, Neudamm, 88pp.

Plate, H.P. and Schneider, R. (1965). Ein Fall von asthmatiger Allergie, verursacht durch den Pilz *Cryptostroma corticale*. *NachrBl. dt. PflSchutzdienst*, **17**, 100-1

Plavsic, V. (1979). Gljiva *Armillaria mellea* (Vahl). Quel. u cetinarskim kulturama SR Srbije. *Sumarstvo*, **32**, 17-32

Ploaie, P.G., Ionica, M. and Alexe, A. (1987). Ofilirea stejarului o boala cauzata de organisme din grupul micoplasmelor. *Buletinul de Protectia Plantelor*, **1**, 13-21

Pobegailo, A.I., Ladeishchikova, E.I. and Laduykh, L.F. (1980). Vliyanie udobrenii na potentsial'nuyu ustoichivost sosny k kornevoi gubke. *Biologicheskie Nauki*, **11**, 81-6. (abstract in *Rev. Pl. Path.*, **60**, No. 4696, 1981).

Poeteren, N. van (1935). *Verslag over de verhzaamheden van den Plantenziektenkundigen Dienst in het jaar 1934*. Versl. PlZiekt. Dienst Wageningen, 80, 108pp.

Poeteren, N. van (1938). Versl. PlZiekt. Dienst Wageningen, 1936 and 1937; abstract in *Rev. appl. Mycol.*, **18**, 1939, pp. 153-4

Poleac, E., . Ditu, I. and Dumitru, G. (1948). (Etiology and control of *Cercospora microsora*). *Stud. Cerc. Inst. Cerc. For., Bucaresti (Silv. )*, **26**, 235-51; abstract in *For. Abstr.*, **30**, No. 4153

Poleshchuk, Yu.M. and Yakimov, N.I. (1986). [Limiting the spread of *Heterobasidion annosum* in pine stands] *Les. Khoz.*, **1**, 61-2 (abstract in *Rev. Pl. Path.*, **66**, No. 347, 1987).

Polunin, O. (1969). *Flowers of Europe. A Field Guide*. Oxford University Press, Oxford

Pomerleau, R. (1961). History of the Dutch elm disease in the Province of Quebec. *For. Chron.*, **37**, 356-7

Pomerleau, R. (1968). Progression et localisation de l'infection par le *Ceratocystis ulmi* dans l'orme d'Amerique. *Phytopath. Z.*, **63**, 301-27

Ponchet, J. (1986). Rsultats de l'action commune de recherche sur le chancre corticale du cypres. *EPPO Bull.*, **16**, 487-98

Ponchet, J. and Andréoli, C. (1989). Histopathologie du chancre cortical du cyprés à *Seiridium cardinale*. *Eur. J. For. Path.*, **19**, 212-21

Porter, W.A. (1957).*Biological studies on western red cedar needle blight caused by Keithia thujina* Durand. Can. Dep.Agric.,Forest Biol. Div., Interim Rep., 25pp.

Pospisil, J. (1960). Houbovy skudce na Osice *Venturia tremulae* Aderh. *Lesn. Prace*, **38**; abstract in *Rev. appl. Mycol.*, **39**, 1960, p.631

Potlaychuk, V.I. (1953). Vrednaja mikroflora zeludei i ee razvitie v zavisimosti ot uslovi proizvastanija i khranenija. *Bot. Z.*, **38**, 135-142 (abstract in *Rev appl. Mycol.*, **33**, 1954, p.57)

Potter, M.C. (1901-2). On a canker of oak (*Quercus robur*). *Transactions of the English Arboricultural Society*, **5**, 1, 105-12

Power, A.B. and Dodd, R.S. (1984). Early differential susceptibility of juvenile seedlings and more mature stecklings of *Pinus radiata* to *Dothistroma pini*. *N. Z. Jl. For. Sci.*, **14**, 223-8

Prakash, C.S. and Thielges, B.A. (1987). Pathogenic variation in *Melampsora medusae* leaf rust of poplars. *Euphytica*, **36**, 563-70

Pratt, J.E. (1979a). *Fomes annosus* butt rot of Sitka spruce. I. Observations on the development of butt rot in individual trees and in stands. *Forestry*, **52**, 11-29

Pratt,J.E.(1979b). *Fomes annosus* butt rot of Sitka spruce. II. Loss of strength of wood in various categories of rot. *Forestry*, **18**, 31-45

Pratt, J.E. (1979c). *Fomes annosus* butt rot of Sitka spruce. III. Losses in yield and value of timber in diseased trees and stands. *Forestry*, **52**, 113-27

Pratt, J.E. and Greig, B.J.W. (1988). *Heterobasidion annosum*: Development of butt rot following thinning in two young first rotation stands of Norway spruce. *Forestry*, **61**, 339-47

Preece, T.F. (1978). *Watermark Disease of the Cricket Bat Willow*, Leafl. For. Commn, No. 20, 9pp.

Preece, T.F., Wong, W.C. and Adegege, A.O. (1979). Diagnosis of watermark in willows and some characteristics of *Erwinia salicis* (Day). Chester. In *Plant Pathogens* (ed. D.W. Lovelock). Society for Applied Bacteriology Technical Series, 12, 1-17. Academic Press, London.

Preece, T.F. and Wortley, M.D. (1979). *Erwinia salicis* in Cricket Bat Willows: legislation as a means of control. In Ebbels, D.L. and King, J.E. (Eds), *Plant Health: the Scientific Basis for Administrative Control of Plant Diseases and Pests*. Blackwell, Oxford.

Prihoda, A. (1950). Kerovity rust Lipovych sazenic jako nasledik nepadeni houbon *Pyrenochaeta pubescens* Rostr. *Ochr. Rost.*, 23, 366-8

Prihoda, A. (1957). Nakaza zivych smrku vaclavkou. *Les (Bratislava)*, 13, 5

Prokazin, A.E. and Kurakin, B.N. (1983). Provenance of Scots pine seedlings and their resistance to needle cast caused by *Lophodermium pinastri*. *Les. Khoz.* No. 2, 51-3

Prunier, J.P., Luisetti, J., Gardan, L., Germain, E. and Sarraquigne, J. (1976). La bactériose du noisetier (*Xanthomonas corylina*). *Revue. Hort.*, 170, 31-40,

Przybyl, K. (1984a). Development of the fungus *Ceratocystis fimbriata* in shoots of poplar clones with differing resistance. *Eur. J. For. Path.*, 14, 177-83

Przybyl, K. (1984b). Pathological changes and defence responses in poplar tissues caused by *Ceratocystis fimbriata*. *Eur. J. For. Path.*, 14, 183-91

Przybyl, K. (1984c). Diseases of poplar caused by *Ceratocystis fimbriata* Ell. et Halst. I. Isolation of *Ceratocystis fimbriata*, symptoms of the disease and evaluation of resistance of poplar clones resulting from artificial infection. *Arboretum Korn.*, 29, 89-103

Przybyl, K. (1984d). Diseases of poplar caused by *Ceratocystis fimbriata* Ell. et Halst. II. Morphology of the pathogen. *Arboretum Korn.*, 29, 105-18

Przybyl, K. (1988). The response of *Populus* 'NE42' (*P. maximowiczii x P. trichocarpa*) to infection by *Ceratocystis fimbriata* isolates from cacao tree and plane. *Eur. J. For. Path.*, 18, 8-12

Psallidas, P.G. (1987). The problem of bacterial canker of hazelnuts in Greece caused by *Pseudomonas syringae* pv. *avellanae*. *EPPO Bull.*, 17, 257-62

Pupavkin, D.M. (1982). (Rust canker of fir). *Zaschita rastenii*. No. 8, 24 (Abstr. in *Rev. Pl. Path.*, 62, No. 415, 1983).

Raabe, R.D. (1958). The effect of light upon growth of *Armillaria mellea* in culture. *Phytopathology*, 48, 397

Raabe, R.D. (1962). Host list of the root-rot fungus *Armillaria mellea*. *Hilgardia*, 3, 25-88

Rabenhorst, G.L. (1884). *Kryptogamenflora von Deutschlands, Osterreich und der Schweiz*, Vol. 1, p.40

Rack, K. (1957a). Untersuchungen über die Anfalligkeit verschiedener Eichenprovenienzen gegenüber dem Eichenmehltau. *Allg. Forst. u. Jagdztg*, 128, 150-6

Rack, K. (1957b). Versuche zur Bekampfung der Eichenmehltaus. *Forst. u. Holz*, 12, 5-6

Raddi, P. and Fagnani, A. (1978). Miglioramento genetico del pino per la resistenza alla ruggine vesciculosa: controllo ed efficacia della densita di inoculo. *Phytopathologia Mediterranea*, 17, 8-13

Raddi, P. and Fagnani, A. (1981). Blister rust in maritime pine. *Eur. J. For. Path.*, 11, 187-90

Raddi, P. and Panconesi, A. (1984). Pathogenicity of some isolates of *Seiridium* (*Coryneum*) *cardinale*, agent of cypress canker disease. *Eur. J. For. Path.*, 14, 348-54

Ragazzi, A., Fagnani, A., Fedi, I.D. and Mesturino, L. (1987). *Pinus pinaster* e *Pinus sylvestris*: due specie a diverso comportamento verso *Cronartium flaccidum*. *Phytopathologia Mediterranea*, 26, 81-4 (abstract in *Rev. Pl. Path.*, 67, No. 3650, 1988).

Ragazzi, A., Fedi, I.D. and Mesturino, L. (1986). *Cronartium flaccidum* on *Pinus* spp.; relation inoculum concentration to symptom development. *Eur. J. For. Path.*, 16, 16-21

Ragazzi, A. and Mesturino, L. (1987). *Diplodia mutila* in Italia associata al'deperimento della quercia? *Italia for. Mont.*, 42, 264-74

Rahman, M.A. (1982). *Dieback of Pinus contorta caused by Ramichloridium pini in Scotland*. PhD Thesis, University of Aberdeen

Ram Reddy, M.A., Salt, G.A. and Last, F.T. (1964). Growth of *Picea sitchensis* in old forest nurseries. *Ann.appl.Biol.*, 54, 397-414

Ramsbottom, J. (1953). *Mushrooms and Toadstools*, Collins, London

Ramsbottom, J. and Balfour-Brown, F.L. (1951). List of discomycetes reported from the British Isles. *Trans. Br. mycol. Soc.*, 34, 38-137

Rane, K.K. and Tattar, T.A. (1987). Pathogenicity of blue stain fungi associated with

*Dendroctonus terebrans. Pl. Dis.*, **71**, 879-83

Raspopov, P.M. and Petrova, M.V. (1985). The pathogens of needle cast of Scots pine in the Urals and Trans-Urals. *Les. Khoz.*, No. 12, 43-6

Raspopov, P.M. and Petrova, M.V. (1989). [Snow cast of Scots pine and its prevention in nurseries in the Southern Urals]. *Mikol. fitopat.*, **23**, 281-8 (abstract in *Rev. Pl. Path.*, **69**, No. 3270, 1990)

Rayner, A.D.M. (1977). Fungal colonisation of hardwood stumps from natural sources. II Basidiomycetes. *Trans. Br. mycol. Soc.*, **69**, 303-12

Rayner, A.D.M., Bevercombe, G.P., Brown, T.C. and Robinson, A. (1981). Fungal growth in a lattice: a tentative explanation for the shape of diamond-cankers in sycamore. *New Phytol.*, **87**, 383-93

Rayner, R.W. (1959). Root rot of coffee and ring-barking of shade trees. *Kenya Coffee*, **24**, 361-5

Rea, C. (1922). *British Basidiomycetes*. Cambridge University Press Cambridge

Read, D.J. (1967). *Brunchorstia dieback of Corsican pine*. Forest Rec., Lond., No. 61

Redfern, D.B. (1968). The ecology of *Armillaria mellea* in Britain: biological control. *Ann. Bot.*, **32**, 293-300

Redfern, D.B. (1971). In Forest Pathology. *Rep. Forest Res., Lond., 1971*, p.83

Redfern, D.B. (1973). *Sirococcus strobilinus*. *Rep. Forest. Res., Lond., 1973*, pp. 100-101

Redfern, D.B. (1974). In Forest Pathology. *Rep. Forest Res., Lond., 1974*, p.38

Redfern, D.B. (1975). The influence of food base on rhizomorph growth and pathogenicity of *Armillaria mellea* isolates, in G.W. Bruehl (Ed.), *Biology and Control of Soil-borne Plant Pathogens*, American Phytpathological Society, St.Paul, Minnesota, pp. 69-73

Redfern, D.B. (1978). Infection of *Armillaria mellea* and some factors affecting host resistance and the severity of disease. *Forestry*, **51**, 121-35

Redfern, D.B., Gregory, S.C. and Low, J.D. (1973). In Pathology. *Rep. Forest Res., Lond., 1973*, p. 102-3

Redfern, D.B., Gregory, S.C. and Low, J.D. (1974). In Forest Pathology. *Rep. Forest Res., Lond., 1974*, p. 38

Redfern, D.B., Gregory, S.C. and Low, J.D. (1975). In *Rep. For. Res. Lond., 1975*, p. 35.

Redfern, D.B., Gregory, S.C. and Low, J.D. (1976). In Forest Pathology. *Rep. For. Res. Lond., 1976*, p. 35.

Redfern, D.B., Gregory, S.C. and Low, J.D. (1979). In Forest Pathology. *Rep. Forest Res. Lond., 1979*. pp. 33-4

Redfern, D.B., Gregory, S.C. and Low, J.D. (1981). In Pathology. *Rep. Forest Res. Edin., 1981*, pp. 37-38

Redfern, D.B., Gregory, S.C. and Low, J.D. (1982). In Forest Pathology. *Rep. Forest Res. Edin., 1982* p. 28

Redfern, D.B., Gregory, S.C. and Low, J.D. (1986). In *Rep. For. Res. Lond., 1986*, p.36.

Redfern, D.B., Gregory, S.C., MacAskill, G.A. and Pratt, J.E. (1987). In Pathology. *Rep. Forest Res. Edin., 1987*, pp. 42-3

Redfern, D.B., Gregory, S.C. and Pratt, J.E. (1983). In Forest Pathology. *Rep. Forest Res. Edin., 1983*, pp. 31-2

Redfern, D.B., Gregory, S.C. and Pratt, J.E. (1984). In Pathology. *Rep. Forest Res. Edin., 1984*, p. 33

Redfern, D.B., Gregory, S.C. and Pratt, J.E. (1985). In Pathology. *Rep. Forest Res. Edin., 1985*, pp. 33-6

Redfern, D.B. and Low, J.D. (1972). In Forest Pathology. *Rep. Forest Res. Lond., 1972*. pp. 97-98

Redfern, D.B. and Sutton, B. (1981). Canker and dieback of *Ulmus glabra* caused by *Plectophomella concentrica* and its relationship to *P. ulmi*. *Trans. Br. mycol. Soc.*, **77**, 381-90

Redhead, S.A. (1975). The genus *Cristulariella*. *Can. J. Bot.*, **53**, 700-7

Rees, A.A. and Phillips, D.H. (1986). *Detection, presence and control of seed-borne pests and diseases of trees, with special reference to seeds of tropical and sub-tropical pines*. Technical Note No. 28, DANIDA Forest Seed Centre, Humlebaek, Denmark

Regenmortel, M.H.V.van. (1982). *Serology and Immunochemistry of Plant Viruses*. Academic Press, New York

Rehm, H. (1896). *Die Pilze Deutschlands, Oesterreichs und der Schweiz. III Abt.*, Kummer, Leipzig

Reichard, M. and Bolay, A. (1986). La maladie de l'encre du châtaignier dans le canton de Genève. *Revue Suisse de Viticulture, d'Arboriculture et d'Horticulture*, **18**, 243-50

Reid, D.A. (1969). New and interesting British plant diseases. *Trans. Br. mycol. Soc.*, **52**, 19-38

Reid, D.A. (1976). *Inonotus obliquus* (Pers. ex Fr.)Pilat in Britain. *Trans. Br. mycol. Soc.*, **67**, 329-32

Reid, D.A. (1985). An annotated list of some fungi from the Channel Islands, mostly from Jersey. *Trans. Br. mycol. Soc.*, **84**, 709-14

Reid, J. and Cain, R.F. (1960). Additional Diaporthales on twigs from Ontario. *Can. J. Bot.*, **38**, 945-50

Reinfeldt, K. R. (1979). Greinbaum pa *Salix*. *Gartneryrket*, **69**, 188-91 (abstract in *Rev. Pl. Path.*, **60**, No. 3978, 1981)

Reyna, R.N. (1983). Control of blue stain on *Pinus caribaea* var. *hondurensis* Barr.& Golf. Wood in Turrialba, Costa Rica. *Turrialba*, **33**, 387-92

Rhoads, A.S. (1921). Some new or little-known hosts of wood-destroying fungi. *Phytopathology*, **11**, 319-26

Rhumbler, L. (1922). In Nusslin, O. and Rhumbler, L., Die Buchen-Wollschildlaus (*Coccus* [*Cryptococcus*] *fagi*). Bärenspr, in *Forstinsektenkunde*, Paul Paray, Berlin pp. 125-39

Ribeiro, O.K. (1978). *A Source Book of the Genus Phytophthora*, J.Cramer, Lehre

Richter, J. (1988). *Venturia inaequalis* (Cooke) Winter. In Smith *et al.*, 1988 q.v., pp. 385-6

Ridé, M. (1958). Sur l'étiologie du chancre suintant du peuplier. *C. r. Acad. Sci.*, *Paris*, **246**, 2795-8

Ridé, M. (1959). Aspects biologiques de la maladie du chancre suintant du peuplier. *Publ. FAO Int. Poplar Commn Res. Dis. Group*, 4pp.

Ridé, M. (1963). Our present knowledge of bacterial canker on poplar caused by *Aplanobacterium populi*. *Publ. FAO Int. Poplar Commn Res. Dis. Group*, 9pp.

Ridé, M. (1966). Chancre bacterien. Inoculum développement des pousses de l'année et expression des symptomes. Cycle biologique. *Publ. FAO Int. Poplar Comn Res. Dis. Group*, 1p.

Ridé, M. and Ridé, S. (1978). *Xanthomonas populi* Ridé comb. nov.(syn.*Aplanobacter populi* Ridé), specificité, variabilité et absence de relations avec *Erwinia cancerogena* Ur. *Eur. J. For. Path.*, **8**, 310-33

Ridé, M., Ridé, S., Steenackers, M. and Steenackers, V. (1986). Artificial infection of different poplar clones with different geographical isolates of *Xanthomonas populi*. *Mededelingen van de Faculteit Landbouwwetenschappen Rijksuniversiteit Gent*, **51**, 1331-45

Ridé, M. and Viart, M. (1966). Étude de la contamination d'une peupleraie par le chancre bacterien. *Bull. Serv. Cult. & Etudes Peupl. & Saule*, **1-2**, 45-61

Riecken, I. (1985). Versuche zur Verhutung der Rotpustelkrankheit (*Nectria cinnabarina*). *Pflanzenkrankheiten und Pflanzenschutz*, **92**, 516-29

Riecken, I. (1986). Mit Spritzungen gegen die Rotpustelkrankeit? *Deutsche Baumschule*, **38**, 417, 419

Riggenbach, A. (1956). Untersuchungen uber den Eschenkrebs. *Phytopath. Z.*, **27**, 1-40

Rijkaert, C., Tomme, R. van, Steenackers, V., Swings, J., and Ley, J. de (1984). The occurrence of the watermark disease of willows (*Salix*) in Belgium. *Mededelingen van de Faculteit Landbouwwetenschappen Rijksuniversiteit Gent*, **49**, 509-15

Rimpaü, R.H. (1962). Untersuchungen über die gattung *Drepanopeziza* (Kleb.). v.Höhn. *Phytopath. Z.*, **43**, 257-306

Rishbeth, J. (1951a). Observations on the biology of *Fomes annosus*, with particular reference to East Anglian pine plantations. II. Spore production, stump infection, and saprophytic activity in stumps. *Ann. Bot.*, **15**, 1-21

Rishbeth, J. (1951b). Observations on the biology of *Fomes annosus*, with particular reference to East Anglian pine plantations. III. Natural and experimental infection of pines, and some factors affecting severity of the disease. *Ann. Bot.*, **15**, 221-46

Rishbeth, J. (1952). Control of *Fomes annosus*. *Forestry*, **25**, 41-50

Rishbeth, J. (1957). Some further observations on *Fomes annosus* Fr. *Forestry*, **30**, 69-89

Rishbeth, J. (1959a). Stump protection against *Fomes annosus*. I. Treatment with creosote. *Ann. appl. Biol.*, **47**, 519-28

Rishbeth, J. (1959b). Stump protection against *Fomes annosus*. II. Treatments with substances other than creosote. *Ann. appl. Biol.*, **47**, 529-41

Rishbeth, J. (1959c). Dispersal of *Fomes annosus* Fr. and *Peniophora gigantea* (Fr.). Massee. *Trans. Br. mycol. Soc.*, **42**, 243-60

Rishbeth, J. (1963). Stump protection against *Fomes annosus*. III. Inoculation with *Peniophora gigantea*. *Ann. appl. Biol.*, **52**, 63-77

Rishbeth, J. (1964). Stump infection by basidiospores of *Armillaria mellea*. *Trans. Br. mycol. Soc.*, **47**, 460

Rishbeth, J. (1980). *Bacterial wetwood*. Arboriculture Research Note 20/80/PAT. DOE

Arboricultural Advisory and Information Service, Forestry Commission, Farnham.
Rishbeth, J. (1982). Species of *Armillaria* in southern England. *Pl. Path.*, **31**, 9-17
Rishbeth, J. (1983). The importance of honey fungus (*Armillaria*). in urban forestry. *Arboric. J.*, **7**, 217-25
Rishbeth, J. (1985). Infection cycle of Armillaria and host response. *Eur. J. For. Path.*, **15**, 332-41
Rishbeth, J. (1986). Some characteristics of English Armillaria species in culture. *Trans. Br. mycol. Soc.*, **86**, 213-8
Rishbeth, J. (1988). Stump infection by Armillaria in first-rotation conifers. *Eur. J. For. Path.*, **18**, 401-8
Risley, J.A. and Silverborg, S.B. (1958). *Stereum sanguinolentum* on living Norway spruce following pruning. *Phytopathology*, **48**, 337-8
Robak, H. (1946). Tre skogsykdommer som hittil har vaert lite kjent eller piaktet i Norge. *Tidsskr. Skogbr. 1946*, **10-1**, 323-34
Robak, H. (1952a). *Dothichiza pithyophila* (Cda). Petr., the pycnidial stage of a mycelium of the type *Pullularia pullulans* (de B.). Berkh. *Sydowia*, **6**, 361-2
Robak, H. (1952b). *Phomopsis pseudotsugae* Wilson—*Discula pinicola* (Naumov). Petr. as a saprophyte on coniferous woods. *Sydowia*, **6**, 378-82
Robertson, N.F. and Macfarlane, I. (1946). The occurrence of perithecia of the oak mildew in Britain. *Trans. Br. mycol. Soc.*, **29**, 219-20
Rohde, T. (1932). Das Vordringen der Rhabdocline-Schütte in Deutschland. Die folgen des Rhabdocline-Befalls in deutschen Douglasienbeständen. Welche Douglasien sind in Deutschland durch Rhabdocline gefährdet? *Forstarchiv.*, **8**, 247-9
Roll-Hansen, F. (1964). *Scleroderris lagerbergii* Gremmen (*Crumenula abietina*) and girdling of *Pinus sylvestris* L. *Meddr. norske SkogsforsVes.*, **68**, 159-87
Roll-Hansen, F. (1965). *Pucciniastrum areolatum* on *Picea engelmannii*. Identification by spermagonia. Rept. from *Meddr norske Skogfors Ves.*, **20**, 391-7
Roll-Hansen, F. (1966). Some notes on *Microsphaera hypophylla* Nevodovskij. *Meddr. norske SkogsforsVes.*, **78** (21), 19-22
Roll-Hansen, R. (1985). The *Armillaria* species in Europe. A literature review. *Eur. J. For. Path.*, **15**, 22-31
Roll-Hansen, F. (1989). *Phacidium infestans*. A literature review. *Eur J. For. Path.*, **19**, 237-50
Roll-Hansen, F. and Roll-Hansen, H.(1980a). Microorganisms which invade *Picea abies* in seasonal stem wounds 1.General aspects. Hymenomycetes. *Eur. J. For. Path.*, **10**, 321-339
Roll-Hansen, F. and Roll-Hansen, H. (1980b). Urediniospore chains in *Melampsoridium betulinum*. *Eur. J. For. Path.*, **10**, 382-4
Roll-Hansen, F. and Roll-Hansen, H. (1980c). Microorganisms which invade *Picea abies* in seasonal stem wounds. II. Ascomycetes, fungi imperfecti, and bacteria. General discussion, Hymenomycetes included. *Eur. J. For. Path.*, **10**, 396-410
Roll-Hansen, F. and Roll-Hansen, H. (1981). *Melampsoridium* on *Alnus* in Europe. *M. alni* conspecific with *M. betulinum*. *Eur. J. For. Path.*, **11**, 77-87
Romagnesi, H. (1962). *Petit Atlas Des Champignons, 3 vols*, Bordas, Paris and Harraps, London
Romagnesi, H. (1970). Observations sur les Armillariella (1). *Bull. Soc. mycol. Fr.*, **86**, 257-68
Romagnesi, H. (1973). Observations sur les Armillariella (2). *Bull. Soc. mycol. Fr.*, **89**, 195-206
Romagnesi, H. and Marxmüller, H. (1983). Etude complémentaire sur les armillaires annelées. *Bull. Soc. mycol. France*, **99**, 301-24
Rose, D. (1970). Disease resistant willows sought. *Gdnrs' Chron.*, **167**, 40
Rosnev, B. (1983). [Possibilities of limiting the primary infection of Scots pine by *Heterobasidion annosum*]. *Gorskostopanska Nauka*, **20**, 73-80 (abstract in *Rev. Pl. Path.*, **64**, No. 1305, 1985)
Rossman, A.Y., Palm, M.E., and Spielman, L.J. (1987). *A Literature Guide for the Identification of Plant Pathogenic Fungi*. American Phytopathological Society, St Paul, Minnesota
Roth, E.R. and Hepting, G.H. (1954). Eradication and thinning tests for Nectria and Strumella canker control in Maryland. *J. For.*, **52**, 253-6
Rumbold, C. (1936). Three blue staining fungi including two new species associated with bark beetles. *J. agric. Res.*, **52**, 419-37
Rupert, J A. and Leach, J.G. (1942). Willow blight in West Virginia. *Phytopathology*, **32**, 1095-6
Rushton, B.S. (1977). Artificial hybridisation between *Quercus robur* L. and *Quercus petraea* (Matt.). Liebl. *Watsonia*, **11**, 229-36
Russo, M., Martelli, G.P. and Savino, V. (1978). Cherry leaf roll virus in walnut. III. Ultrastructure of infected cells. *Phytopathologia Mediterranea*, **17**, 90-5
Rybak-Mikitiuk, T. (1962). *Acta agrobot.*, **11**, 93-129; abstract in *Rev. appl. Mycol.*, **42**, 1963,

p.158

Rykowski, K. and Sierota, Z. (1983). [Economic effectiveness of the biopreparation 'Pg IBL' for protection against *Heterobasidion annosum*]. O ekonomiczny efektywnosci stosowania biopreparst 'PgIBL' przeciw hubie korzeni.*Las pol.* No. 12, 26-28. (abstract in *Rev. Pl. Path.*, **66**, No. 4499. 1987)

Rykovski, K. and Sierota, Z. (1984). Aspekt ekonomiczny wystepowania huby korzeni w drzewostanach sosnowych na gruntach porolnych. *Sylwan*, **128**, 11-21 (abstract in *Rev. Pl. Path.*, **66**, No. 3517, 1987)

Ryvarden, L. (1976). *The Polyporaceae of north Europe*, Fungiflora, Oslo

Sabet, K.A. (1953). Studies on the bacterial dieback and canker disease of poplar. III. Freezing in relation to the disease. *Ann. appl. Biol.*, **40**, 645-50

Sabet, K.A. and Dowson, W.J. (1952). Studies on the bacterial dieback and canker disease of poplar. I. The disease and its cause. *Ann. appl. Biol.*, **39**, 609-16

Salmon, E.S. (1900). *A Monograph of the Erysyphaceae*. Mem. Torr. Bot. Club IX

Salt, G.A. (1964). Pathology experiments on Sitka spruce seedlings. *Rep Forest Res., Lond., 1963*, pp. 83-7

Salt, G.A. (1965). Pathology experiments on Sitka spruce seedlings. *Rep. Forest Res., Lond., 1964*, pp. 89-95

Salt, G.A. (1967). Pathology experiments on Sitka spruce seedlings. *Rep. Forest Res., Lond., 1966*, pp. 104-8

Salt, G.A. (1974). Etiology and morphology of *Geniculodendron pyriforme* gen. et sp. nov., a pathogen of conifer seeds. *Trans. Br. mycol. Soc.*, **63**, 339-51

Salter, P.J. and Goode, J.E. (1967). *Crop responses to Water at Different Stages of Growth*, Commonwealth Agricultural Bureau

Salter, P.J. and Williams, J.B. (1969). The influence of texture on the moisture characteristics of soil. V. Relationships between particle size, composition and moisture contents at the upper and lower limits of available water. *J. Soil Sci.*, **20**, 126-31

Sanderson, P.G. and Worf, G.L. (1986). *Phomopsis* and *Sirococcus* shoot blights of Colorado blue spruce in Wisconsin. *Pl. Dis.*, **70**, 1159

Santamour, F.S. (1976). Resistance to sycamore anthracnose disease in hybrid *Platanus*. *Pl. Dis. Reptr.*, **60**, 161-2

Saric, A. and Milatovic, I. (1960). Pokus suzbijanja rde Topola u sumskom rasadniku Banova Jaruga. *Sumarski List*, **84**, 290-91; abstract in *For. Abs.*, **22**, 1961, p.424

Sato, K., Ota, N. and Shoji, T. (1955). Influence of MH-30 treatment upon the control of overgrowth of 'sugi' seedlings. Especially on the effects of the frost damage and grey mould control. *J. Jap. For. Soc.*, **37**, 533-7

Sato, K., Shoji, T. and Ota, N. (1959). Studies on the snow moulding of conifer seedlings. I. Grey mould and sclerotial disease. *Bull. Govt Forest Exp. Stn Meguro*, **110**, 1-153

Savill, P.S. and Mather, R.A. (1990). A possible indicator of shake in oak: relationship between flushing dates and vessel sizes. *Forestry*, **63**, 355-62

Savory, J.G., Nash-Wortham, J., Phillips, D.H. and Stewart, D.H. (1970). Control of blue stain in unbarked logs by a fungicide and an insecticide. *Forestry*, **43**, 161-74

Savvin, I.M. (1984). [Interaction of the fungi *Microsphaera alphitoides* and *Septoria quercina*, pathogens of oak seedlings]. (Abstr. in *Rev. Pl. Path.*, **64**, No. 5512, 1985)

Scaramuzzi, G. (1954). Sul seccume delle foglie d'ippocastano. *Ann. sper. agr., N. S.*, **8**, 1256-81

Schaad, N.W.(editor). (1988). *Laboratory Guide for Identification of Plant Pathogenic Bacteria*, 2nd edn, American Phytopathological Society, St Paul, Minnesota

Scheffer, R.J., Elgersma, D.M., Weger, L.A.de and Strobel, G.A. (1989). *Pseudomonas* for biological control of Dutch elm disease I. Labeling, detection and identification of *Pseudomonas* isolates injected into elms; comparison of various methods. *Neth. J. Pl. Path.*, **95**, 281-92

Scheffer, T.C. and Lindgren, R.M. (1940). *Stains of Sapwood and Sapwood Products and Their Control*. Tech.Bull. U.S. Dept.Agric., No. 714, 123pp.

Schell, E. (1922). Diseases of the French chestnut tree particularly the 'ink-malady'. *J. Am. Leather Chem. Ass.*, **17**, 353-9

Schimalski, H.H., Albrecht, H.J. and Kegler, H. (1980). Samenübertragung des Kirschenblattroll-Virus (Cherry leaf roll virus). bei der Hängebirke. *Arch. Phytopathol. u. PflSchutz.*, **16**, 231-2

Schink, B., Ward, J.C. and Zeikus, J.G. (1981). Microbiology of wetwood: importance of pectin degradation and Clostridium species in living trees. *Applied and Environmental Microbiology*, **42**, 526-32

Schipper, M.A.A. and Heybroek, H.M. (1957). Het toetsen van stammen van *Nectria cinnabarina* (Tode). Fr. op levende takken in vitro. *Tijdschr. PlZiekt.*, **63**, 193-4

Schmelzer, K. (1972). Das Kirschenblattroll-virus aus der Birke (Betula pendula Roth.). *Zentbl. Bakt. ParasitKde, Abt. II*, **127**, 10-12

Schmidle, A. (1953). Zur Kenntnis der Biologie und der Pathogenität von Dothichiza populea Sacc. et Briard, dem Erreger eines Rindenbrandes der Pappel. *Phytopath. Z.*, **21**, 189-209

Schneider, A. and Sutra, G. (1969). Les modalités de l'infection de *Populus nigra* L. par *Taphrina populina*. *C. r. hebd. Séanc. Acad. Sci. Paris, Sér. D*, **269**, 1056-9

Schneider, R. (1961). Untersuchungen uber das Auftreten der Guignardia Blattbraune der Rosskastanie (*Aesculus hippocastanum*). in Westdeutschland und ihren Erreger. *Phytopath. Z.*, **42**, 272-8

Schneider, R. and Arx, J.A. von(1966). Zwei neue, als Erreger von Zweisterben nachgewesene Pilze: *Kabatina thujae* n.g., n.sp. *Phytopath. Z.*, **57**, 176-82

Schneider, R. and Paetzholdt, M. (1964). Ascochyta piniperda als Erreger einer Triebsterbens an Blaufichten in Baumschulen. *NachrBl. dt. PflSchutzdienst, Braunschweig*, **16**, 73-5

Schöber, R. and Fröhlich, H.J. (1967). *Der Gahrenberger Larchen Provenierengversuch*. Schr.Forst.Foh.Univ.Gottingen, 37/38, 208pp.

Schoeneweiss, D.F. (1969). Susceptibility of evergreen hosts to the Juniper blight fungus, *Phomopsis juniperovora* under epidemic conditions. *J. Am. Soc. Hort. Sci.*, **94**, 609-11

Schönhar, S. (1952). Untersuchungen über den Erreger des Pappelrindentodes. *Allg. Forstz.*, **49**, 509-12

Schönhar, S. (1953). Untersuchungen über die biologie von *Dothichiza populea* (Erreger des Pappelrindentodes). *Forstw. Zbl.*, **72**, 358-68

Schönhar, S. (1958). Bekampfung der durch *Meria laricis* verursachten Larchenschutte. *Allg. Forstz.*, **13**, 100

Schönhar, S. (1959). Eine nedelkrankung der Fichte. *Allg. Forstz.*, **49**, 1

Schönhar, S. (1965). Schaden an Jungtannen durch Rostpilzbefall. *Allg. Forstz.*, **20**, 120

Schönhar, S. (1980). Untersuchungen über die Uberlebensdauer von *Fomes annosus*-Sporen im Boden. *Forst- u. Jagdztg.*, **151**, 197-9

Schönhar, S. (1981). Untersuchungen über die Bedeutung frischer Stubben für die Ausbreitung von *Fomes annosus* in Fichten-Erstaufforstungen.*Allg. Forst- u. Jagdztg.*, **152**, 217-8

Schönhar, S. (1984). Infektionsversuche an Fichten- und kiefernkeimlingen mit aus faulen Feinwurzeln von Nadelbäumen häufig isolierten Pilzen. *Allg. Forst- u. Jagdztg.*, **155**, 191-2

Schreiner, E.J. (1931a). The rôle of disease in the growing of poplar. *J. For.*, **29**, 79-82

Schreiner, E.J. (1931b). Two species of *Valsa* causing disease in *Populus*. *Am. J. Bot.*, **18**, 1-29

Schreiner, E.J. (1959). *Rating poplars for Melampsora leaf rust infection*. For.Res.notes Northwestern For.Exp.Sta.Upper Darby, Pa., No. 90, pp3.

Schubert, G.M. (1960). Fungi associated with viability losses of sugar pine seed during cold storage. *Proc. Soc. Amer. For., 1960*, pp. 18-21

Schuldt, P.H. (1955). Comparison of anthracnose fungi on oak, sycamore and other trees. *Contr. Boyce Thompson Inst.*, **18**, 85-107

Schultz, M.E. and Parmeter, J.R, Jr. (1990). A canker disease of *Abies concolor* caused by *Nectria fuckeliana*. *Pl. Dis.*, **74**, 178-80.

Schultz, M.G. and Harrap, K.A. (1975). Bacilliform particles associated with vein yellowing of *Laburnum anagyroides*. *Ann. appl. Biol.*, **79**, 247-50

Schütt, P. (1979). Buchen- und Tannensterben, zwei altbekannte Waldkrankheiten von höchster Aktualität. *Mitt. dt. dendrol. Ges.*, **71**, 229-35

Schütt, P. (1985). Das Waldsterben—eine Pilzkrankheit? *Forstwiss. CentBl.*, **104**, 169-77

Schütt, P. and Lang, K.J. (1980). *Buchen-Rindennekrose, Waldschutz-Merkblatt*, No. 1, Paul Paray, Hamburg and Berlin

Schwarz, M.B. (1922). *Das Zweigsterben der Ulmen, Trauerweiden und Pfirsichbaume*, Thesis, University of Utrecht; abstract in *Rev. appl. mycol.*, **2**, 1923, pp. 92-4

Schwerdtfeger, F. (1963). Eine wetterbedingte Baumkrankheit. *Die Umschau in Wissenschaft und Technik.*, **15**, 476-8

Scopes, N. and Stables, L. (editors). (1989). *Pest and Disease Control Handbook*. BCPC, Thornton Heath

Scott, E.S. (1984a). Populations of bacteria in poplar stems. *Eur. J. For. Path.*, **14**, 103-12

Scott, E.S. (1984b). Detection of wetwood in living poplar trees by electrical resistance measurements. *Eur. J. For. Path.*, **14**, 334-9

Seaby, D.A. (1985). Forest tree biology. In *Annual Report on Research and Technical Work of the Department of Agriculture for Northern Ireland, 1984*. Belfast, Northern Ireland

Seemüller, E. (1988a). *Stigmina carpophila* (Lev.) M.B.Ellis. In Smith *et al.*, 1988 q.v., pp. 414-5

Seemüller, E.(1988b). *Blumeriella jaapii* (Rehm) v.Arx. In Smith *et al.*, 1988, q.v., pp. 447-8

Sequeira, M.P. da S. de (1981). O oidio do platano em Portugal. *Agronomia Lusitania*, 40, 299-301 (abstract in *Rev. Pl. Path.*, 60, No. 6668, 1981)

Servazzi, O. (1935). Contributi alla patologia dei pioppi. II. La 'tafrinosi' o 'bolla fogliare' dei pioppi. *Difesa Piante*, 12, 48-62

Shafranskaya, V.N. (1960). Regul'taty ispytanu novykh slozhnovrqonicheshikh preparator protiv boleznei khvoi Listvemnitsy. *Sborni. Rab. les. Khoz. vaes. n. -i Inst. Lesovod Mekhan. les. Khoz. 1960*, 43, 141-53; abstract in *Rev. appl. Mycol.*, 41, 1962, pp. 66-7

Shang, R.Z. and Pei, M.H. (1984). [Study on the leaf rust of Davids European aspen caused by *Melampsora larici-tremulae* Kleb.]. *Journal of North-eastern Forestry Institute, China*, 12, 47-55 (Abstract in *Rev. Pl. Path.*, 65, No. 1536, 1986)

Sharma, A.K. and Cousin, M.T. (1986). Mycoplasma-like organisms (MLOs). associated with witches' broom disease of poplar. *Journal of Phytopathology*, 117, 349-56

Sharma, I.K. and Heather, W.A. (1988). Light and electron microscope studies on *Cladosporium tenuissimum*, mycoparasite on poplar leaf rust. *Trans. Br. mycol. Soc.*, 90, 125-31

Sharma, M.P. and Sharma, R. (1981). *Lophodermium piceae* (Fuck.). Höhn (Phacidiaceae) — a new record from India. *Curr. Sci.*, 50, 725-6

Shavrova, L.A. (1967). K biologii vozbuditelya parshi Osiny. *Mikol. i Fitopatol.*, 1, 321-9

Shaw, C.G.III(1989a). Root disease threat minimal in young stands of western hemlock and Sitka spruce in southeastern Alaska. *Pl. Dis.*, 73, 573-7

Shaw, C.G.III(1989b). *Armillaria ostoyae* associated with mortality of new hosts in Chihuahua, Mexico. *Pl. Dis.*, 73, 775

Shea, K.R. (1960). *Mould fungi on forest tree seed.* For.Res.Note Weyerhaeuser Co.Centralia, Wash., No. 31, 10pp.

Sheppard, L.J. and Cannell, M.G.R. (1987). Frost hardiness of subalpine eucalypts in Britain. *Forestry*, 60, 239-48

Sheridan, J.E. (1981). *Poplar leaf rust and its effect on tree growth and suitability for match splints.* Report, Botany Department, Victoria University of Wellington, No. 18, 25pp.

Sherwood-Pike, M., Stone, J.K. and Carroll, G.C. (1986). *Rhabdocline parkeri*, a ubiquitous foliar endophyte of Douglas-fir. *Can. J. Bot.*, 64, 1849-1855.

Shigo, A.L. (1963). *Beech Bark Disease.* U.S. Dept.Agric.Forest Service For.Pest Leafl., 75, 8pp.

Shigo, A.L. (1964). Organism interactions in the beech bark disease. *Phytopathology*, 54, 263-9

Shigo, A.L. (1977). *Compartmentalisation of Decay in trees*, U.S. Dept.Agric.Agriculture Information Bulletin, No. 405

Shigo, A.L. (1986). *A new tree biology: Facts, photos and philosophies on trees and their problems and proper care.* Shigo and Trees, Associates Durham N.H., USA

Shirnina, L.V. (1986). [Role of saprophytic fungi in the pathogenesis of poplar bark. IV. Antagonistic activity of saprophytic fungi on *Dothichiza populea* Sacc. & Br.] *Mikologiya i Fitopatologiya*, 20, 424-429.(abstract in *Rev. Pl. Path.*, 66, No. 4488, 1987)

Shirnina, L.V. (1987). [Short-term forecasting of the appearance of oak powdery mildew]. *Mikol. Fitopatol.*, 21, 278-81 (abstract in *Rev. Pl. Path.*, 67, No. 4127, 1988)

Shirnina, L.V. and Nachaeva, M.Yu (1986). [Role of saprophytic fungi in the pathogenesis of poplar bark. III. Interaction of epiphytic saprophytic fungi and the pathogen *Dothichiza populea* Sacc. & Br.] *Mikologiya i Fitopatologiya*, 20, 309-12

Shishkina, A.K. (1969). Etiologia usykhania pobegov i list'a Gruzinskogo Duba v rannevesennii period. *Mikol. i Fitopatol.*, 3, 365-7

Shoemaker, R.A. (1967). *Cucurbitaria piceae* and associated Sphaeropsidales parasitic on spruce buds. *Can J. Bot.*, 45, 1243-8

Siepmann, R. (1988). Intersterilitätsgruppen und Klone von *Heterobasidion annosum* Isolaten aus Koniferen-Wurzel und Stammfäulen. *Eur. J. For. Path.*, 18, 93-7

Sierota, Z. (1984). [Survival of *Phlebia gigantea* in Scots pine stands after its use for the biological protection of stumps against *Heterobasidion annosum* ] Ocena przezywalnosci grzyba *Phlebia gigantea* (Fr.). Donk w drzewostanach sosnowych po zabiegu biologicznej ochrony pniakow przed huba koprzeni. *Sylwan*, 128, 29-40 (abstract in *Rev. Pl. Path.*, 66, No. 4500, 1987)

Siggers, P.V. (1944). *The Brown Spot Needle Blight of Pine Seedlings.* U.S. Dept. Agric. Tech. Bull., No. 370

Simonyan, S.A. and Mamikonyan, T.O. (1982). [Diseases of plane trees]. *Zashchita Rastenii*, 8, 23-4]. (abstract in *Rev. Pl. Path.*, 62, No. 406, 1983

Sinclair W.A. and Campana, R.J. (1978). Dutch elm disease, perspectives after 60 years. *Search, Agriculture*, **8** (5), 52pp.

Sinclair, W.A., Lyon, H.H. and Johnson, W.T. (1987). *Diseases of trees and shrubs*. Comstock Publishing Associates, Ithaca and London

Singer, R. (1956). The *Armillariella mellea* complex. *Lloydia*, **19**, 176-87

Singh, S.J. and Heather, W.A. (1982). Temperature sensitivity of quantitative race-cultivar interactions in *Melampsora medusae* Thüm. and *Populus* species. *Eur. J. For. Path.*, **12**, 123-7

Sivanesan, A. (1977). *The Taxonomy and Pathology of Venturia species*. Biblthca. mycol., Lehre, **59**, 1-123

Skilling, D.D. (1974). Control of *Lophodermium* needlecast in Scots pine Christmas tree plantations. *Pl. Dis. Reptr.*, **58**, 853-6

Skilling, D.D. (1977). The development of a more virulent strain of *Scleroderris lagerbergii* in New York State. *Eur. J. For. Path.*, **7**, 297-302

Skilling, D.D. and Nicholls, T.H. (1975). The development of *Lophodermium pinastri* in conifer nurseries and plantations in North America. *Eur. J. For. Path.*, **5**, 193-7

Skilling, D.D., O'Brien, J.T. and Bell, J.A. (1979). *Scleroderris Canker on Northern Conifers*, U.S. Dept.Agric.Forest & Insect Leafl., No. 130

Skilling, D.D. and Waddell, C.D. (1974). Fungicides for the control of Scleroderris canker. *Pl. Dis. Reptr.*, **58**, 1097-1100

Skoric, V. (1938). Jasenov rak i njegov uzrocnik. *Ann. appl. for.*, *Zagreb*, **6**, 66-97

Smalley, E.B. (1962). Prevention of Dutch elm disease by treatments with 2, 3, 6-trichlorophenyl acetic acid.*Phytopathology*, **52**, 1090-1

Smalley, E.B. (1978). Control tactics in research and practice, in W.A.Sinclair and R.J.Campane (Eds), Dutch Elm Disease, Perspectives After 60 Years, *Search, Agriculture*, **8**, 34-9

Smalley, E.B. and Riker, A.J. (1962). *Tropical members of the Ulmaceae resistent to Dutch elm disease*. Univ. Wis. Res. Not., No. 77, 4pp.

Smerlis, E. (1962). Taxonomy and morphology of *Potebniamyces balsamicola* sp.nov. associated with a twig and branch blight of balsam fir in Quebec. *Can. J. Bot.*, **40**, 351-9

Smith, C.O. (1938). Inoculation on conifers with the cypress Coryneum. *Phytopathology*, **28**, 760-2

Smith, I.M. (1979). EPPO: the work of a regional plant protection organisation, with particular reference to phytosanitary regulations. In D.L.Ebbels and J.E.King (Eds), *Plant Health, The Scientific Basis for Administrative Control of Plant Diseases and Pests* pp. 13-22, Blackwell, Oxford

Smith, I.M., Dunez, J., Lelliott, R.A., Phillips, D.H. and Archer, S.A.(editors). (1988). *European Handbook of Plant Diseases*. Blackwell, Oxford

Smith, K.M. (1972). *A Textbook of Plant Virus Diseases*, 3rd edn, Longman, London

Smith, L.D. (1983). Major nutrients influence on *Verticillium dahliae* infections of *Acer saccharum*. *J. Arb.*, **9**, 277-81

Smith, R.S. Jr. and Graham, D. (1973). *Sirococcus* tip dieback of *Pinus* spp. in California forest nurseries. *Pl. Dis. Reptr.*, **57**, 69-77

Smith, R.S. Jr. McCain, A.H., Srago, M., Krohn, R.F. and Perry, D. (1972). Control of *Sirococcus* tip blight of Jeffrey pine seedlings *Pl. Dis. Reptr.*, **56**, 241-2

Smith, W.H. (1981). *Air pollution and forests*. New York Springer Verlag

Smucker, S.J. (1935). Air currents as a possible carrier of *Ceratocystis ulmi*. *Phytopathology*, **25**, 442-3

Smucker, S.J. (1937). Relation of injuries to infection of American elm by *Ceratocystis ulmi*. *Phytopathology*, **27**, 140

Soegaard, B. (1969). Resistance studies in Thuja. *Det. Forstl. Forsogsv Danm.*, No. 245, **31**, 287-396

Solel, Z., Messinger, R., Golan, Y. and Madar, Z. (1983). Coryneum canker of cypress in Israel. *Pl. Dis.*, **67**, 550-1

Somda, B. and Pinon, J. (1981). Ecophysiologie du stade urédien de *Melampsora larici-populina* Kleb. et de *M. allii-populina* Kleb. *Eur. J. For. Path.*, **11**, 243-54

Soutrenon, A. (1986). Observations sur le comportement de provenances de douglas verts et de douglas bleus a une attaque de rhabdocline. *Forets de France*, No. 294, 10-6

Spaulding, P. (1938). A suggested method of converting some heavily Nectria-cankered hardwood stands in northern New England to soft-woods. *J. For.*, **36**, 72

Spaulding, P. (1961). *Foreign diseases of forest trees of the world*. Agric. Handb. USDA, No. 197

Spaulding, P., MacAloney, H.J. and Cline, A.C. (1935). *Stereum sanguinolentum. A dangerous fungus in pruning wounds on northern white pine*. U.S. Dept.Agric.Tech.Note, No. 19

Spiers, A.G. (1978). Effect of light, temperature and relative humidity on germination of urediniospores of, and infection of poplars by, *Melampsora larici-populina and M. medusae*. *N. Z. Jl. Sci.*, **21**, 393-400

Spiers, A.G. (1983a). Host range and pathogenicity studies of *Marssonina brunnea* to poplars. *Eur. J. For. Path.*, **13**, 181-196

Spiers, A.G. (1983b). Host range and pathogenic studies of *Marssonina castagnei* to poplars. *Eur. J. For. Path.*, **13**, 218-27

Spiers, A.G. (1984). Comparative studies of host specificity and symptoms exhibited by poplars infected with *Marssonina brunnea, Marssonina castagnei* and *Marssonina populi*. *Eur. J. For. Path.*, **14**, 202-18

Spiers, A.G. (1988). Studies of *Marssonina castagnei* in Australia. *Eur. J. For. Path.*, **18**, 65-76

Spiers, A.G. and Hopcroft, D.H. (1984). Influence of leaf age, leaf surface and frequency of stomata on the susceptibility of poplar cultivars to *Marssonina brunnea*. *Eur. J. For. Path.*, **14**, 270-82

Spiers, A.G. and Hopcroft, D.H. (1985). Ultrastructural studies of the spermatial and aecial stages of *Melampsora larici-populina* and *Melampsora epitea* on *Larix decidua*. *N. Z. Jl. Bot.*, **23**, 101-16

Spiers, A.G. and Wenham, H.T. (1983a). Poplar seed transmission of *Marssonina brunnea*. *Eur. J. For. Path.*, **13**, 305-14

Spiers, A.G. and Wenham, H.T. (1983b). Fungicidal control of *Marssonina brunnea* on poplar seed. *Eur. J. For. Path.*, **13**, 344-8

Spies, J.L., Knösel, D. and Meier, D. (1985). Anthracnose und Hitzeschäden an Platanen im Stadtgebiet von Hamburg. *NachrBl. dt. PflSchutzdienst. Berl.*, **37**, 17-21

Srivastava, S.L. and Kumari, R. (1983). Powdery mildew of oak from Garhwal. *Indian Phytopathology*, **36**, 552-3

Stahl, W. (1967). An investigation on the Cytospora disease of poplars in Australia. *Pap. 14th I. U. F. R. O. Conf. Sec. 24, Munich, 1967*, pp. 428-44

Staley, J.M. (1965). Decline and mortality of red and scarlet oaks. *Forest Sci.*, **11**, 9-17

Stalpers, J.A. and Vlug, I. (1983). *Confistulina hepatica* new genus new combination the anamorph of *Fistulina hepatica*. *Can. J. Bot.*, **61** (6), 1660-6

Stapp, C. (1961). *Bacterial Plant Pathogens*, Oxford University Press

Steenackers, M. (1988). Breeding poplars for rust resistance; recent advances. *Meded. Landbouwwetenschappen Rijksuniversiteit Gent*, **53**, 417-22

Steenackers, M. (1989). Study of the genetic diversity of the *Xanthomonas populi* (Ridé). Ridé and Ridé population in Belgium. *Meded Landbouwwetenschappen Rijksuniversiteit Gent*, **54**, 611-7

Steinmetz, F.H. and Prince, A.E. (1938). Observations on willow blight in Maine, 1927-1938. *Pl. Dis. Reptr.*, **22**, 282-3

Stenlid, J. (1987). Controlling and predicting the spread of *Heterobasidion annosum* from infected stumps and trees of *Picea abies*. *Scand. J. For. Res.*, **2**, 187-98

Stenlid, J. and Johansson, M. (1987). Infection of roots of Norway spruce (*Picea abies*) by *Heterobasidion annosum* II. Early changes in phenolic content and toxicity. *Eur. J. For. Path.*, **17**, 217-26

Stenlid, J. and Swedjemark, G. (1988). Differential growth of S- and P- isolates of *Heterobasidion annosum* in *Picea abies* and *Pinus sylvestris*. *Trans. Br. mycol. Soc.*, **90**, 209-13

Stenlid, J. and Wästerlund, I. (1986). Estimating the frequency of stem rot in *Picea abies* using an increment borer. *Scand. J. For. Res.*, **1**, 303-8

Stephan, B.K. (1975). Resistance in pine species to *Lophodermium pinastri*, in *Lophodermium in Pines*, pp. 105-112, Reinbek, No. 103

Stephan, B.R. (1985). Resistance of five-needled pines to blister rust. *Allg. Forst.*, **26**, 695-7

Stephan, B.R. and Butin, H. (1980). Krebsartige an *Pinus contorta*-Herkunften. *Eur. J. For. Path.*, **10**, 410-9

Stevens, F.L. (1925). *Plant disease fungi*, Macmillan, New York

Stewart, V.B. (1916). The leaf blotch disease of horse-chestnut, *Phytopathology*, **6**, 5-20

Stipes, J. and Campana, R.J. (1981). *Compendium of Elm Diseases*, American Phytopathological Society

Stott, K.G., Hunter, T., Parfitt, R.I. and Stinchcombe, G.R. (1980). Control of basket willow rust. In *Rep. Long Ashton Res. Sta.*, 1980, 6-7

Strong, F.C. (1936). Maple wilt. *Q. Bull. Mich. agric. Exp. Sta.*, **18**, 225-7

Strouts, R.G. (1970). Coryneum canker of Cupressus. *Pl. Path.*, **19**, 149-50

Strouts, R.G. (1973). Canker of cypresses caused by *Coryneum cardinale* Wag. in Britain. *Eur. J.*

*For. Path.*, **3**, 13-24

Strouts, R.G. (1981a). *Phytophthora diseases of trees and shrubs*. Arboric. Leafl. No. 8, HMSO, London

Strouts, R.G. (1981b). In Forest Pathology. *Rep. Forest Res., Edin., 1981*, p.37

Strouts, R.G. and Patch, D. (1983). The cold winter of 1981-1982. *Rep. Forest Res. Edin., 1983*, pp. 35-6

Strouts, R.G., Reffold, T.C. and Rose, D.R. (1982). In Forest Pathology. *Rep. Forest Res. Edin. 1982*, pp 27-8

Strouts, R.G., Rose, D.R. and Reffold, T.C. (1983). In Forest Pathology. *Rep. Forest Res. Edin. 1983*, p.31

Strouts, R.G., Rose, D.R. and Reffold, T.C. (1984). In Pathology. *Rep. Forest. Res. Edin., 1984*, pp. 32-3

Strouts, R.G., Rose, D.R. and Reffold, T.C. (1985). In Pathology. *Rep. Forest Res. Edin., 1985*, p.33

Strouts, R.G., Rose, D.R. and Reffold, T.C. (1986). In Pathology. *Rep. Forest Res. Edin., 1986*, pp. 35-6

Strouts, R.G., Rose, D.R. and Reffold, T.C. (1987). In Pathology. *Rep. Forest Res. Edin., 1987*. pp. 43-4

Strouts, R.G., Rose, D.R. and Reffold, T.C. (1988). In Pathology. *Rep. Forest Res. Edin., 1988*, p.40

Strouts, R.G., Rose, D.R. and Reffold, T.C. (1989). In Pathology. *Rep. Forest Res., Edin. 1989*, pp. 42-3

Subikova, V., Sutakova, G. and Bojnansky, V. (1985). Etiologia prederavenosti listov javora. *Biologicke Prace*, **31**, 72pp. (abstract in *Rev. Pl. Path.*, **65**, No. 4555, 1986).

Summers, D., Sutherland, J.R. and Woods, T.A.D. (1986). Inland spruce cone rust (*Chrysomyxa pirolata*) control: relation of ferbam application to basidiospore production, rainfall, and cone phenology. *Can. J. For. Res.*, **16**, 360-2

Suske, J. and Acker, G. (1987). Internal hyphae in young symptomless needles of *Picea abies*: electron microscopic and cultural investigation. *Can. J. Bot.*, **65**, 2098-103

Sutherland, J.R., Hopkinson, S.J. and Farris, S.H. (1984). Inland spruce cone rust, *Chrysomyxa pirolata*, in *Pyrola asarifolia* and cones of *Picea glauca*, and morphology of the spore stages. *Can. J. Bot.*, **62**, 2441-7

Sutherland, J.R., Lock, W. and Farris, S.H. (1981). *Sirococcus* blight: a seed-borne disease of container-grown spruce seedlings in Coastal British Columbia forest nurseries. *Can. J. Bot.*, **59** (5), 559-62

Sutherland, J.R., Miller, T. and Salinas Quinard, R.(editors). (1987). *Cone and Seed Diseases of North American Conifers*. Publication, North American Forestry Commission, No. 1, vi plus 77pp.

Sutton, B.C. (1961). British records. 58. *Pollaccia radiosa* (Lib.). Bald. & Cif. *Trans. Br. mycol. Soc.*, **44**, 608-9

Sutton, B.C. (1980). *The Coelomycetes*, Commonwealth Mycological Institute, Kew

Sutton, B.C. and Pirozynski, K.A. (1963). Notes on British microfungi. 1. *Trans. Br. mycol. Soc.*, **46**, 505-22

Swart, W.J., Knox-Davies, P.S. and Wingfield, M.J. (1985). Sphaeropsis sapinea, with special reference to its occurrence on *Pinus* spp. in South Africa. *South African Forestry Journal* No. **135**, 1-8,

Swinburne, T.R. (1975). European canker of apple (*Nectria galligena*). *Rev. Pl. Path.*, **54**, 787-99

Swinburne, T.R., Cartwright, J., Flack, N.J. and Brown, A.E. (1975). The control of apple canker (*Nectria galligena*). in a young orchard with established infections. *Ann. appl. Biol.*, **81**, 61-73

Sylvestre-Guinot, G. (1981). Etude de l'émission des ascospores du *Lachnellula willkommii* (Hartig) Dennis dans L'Est de la France. *Eur. J. For. Path.*, **11**, 275-83

Sylvestre-Guinot, G. (1986). Etude des sites d'infection du *Lachnellula willkommii* (Hartig) Dennis chez *Larix decidua* Miller. *Ann. Sci. For.*, **43**, 199-206

Szanto, I. (1948). A Bükkfa rakja mint éghajlati betegség. *Erdész. Kisérl.*, **48**, 10-31; abstract in *Rev. appl. Mycol.*, **29**, 234, 1950

Szontagh, P. (1981, publ.1983). [Phytopathological evaluation of poplar stands irrigated with waste water]. *Erdeszeti Kutasok*, **74**, 369-77 (abstract in *Rev. Pl. Path.*, No. 5740, 1986).

Takahashi, O. and Saho, H. (1985). Notes on the Japanese rust fungi. IX. Dissemination of *Chrysomyxa abietis* (Wallroth). Unger. *Trans. mycol. Soc. Japan*, **26**, 433-9

Talboys, P.W. and Davies, M.K. (1984). Robinia stem canker. *Rep. E. Malling Res. Stn 1983*, pp. 106-7

Tao, D., Li, P.H., Carter, J.V. and Ostry, M.E. (1984). Relationship of environmental stress and *Cytospora chrysosperma* infection to spring dieback of poplar shoots. *Forest Science*, **30**, 645-51

Taris, B. (1956). Résistance au froid du mycélium de *Dothichiza populea* Sacc. et Briard et de *Cytospora chrysosperma* (Pers.). Fr. *C. r. Acad. Sci., Paris*, **242**, 1648-9

Taris, B. (1959). *Contribution à l'étude des maladies cryptogamiques des rameaux et des jeunes plantes de peupliers*. Publ.FAO Int.Poplar Commn Res.Dis.Group, 13pp.

Taris, B. (1970). Contribution à l'étude du *Taphrina aurea* (Pers.). Fr., agent de la cloque dorée du Peuplier. *Bulletin du Service de Culture et l'Etudes du Peuplier et du Saule*, pp. 55-155

Taris, B. (1980). Un nouveau danger pour la populiculture française: la tavelure (*Pollaccia elegans* Serv.). *C. r. Séanc. Acad. Agric. Fr.*, **66**, 774-8

Tarr, S.A.J. (1972). *The Principles of Plant Pathology*. Macmillan, London

Tattar, T.A. (1978). *Diseases of Shade Trees*, Academic Press, New York

Taylor, C.M.A. and Tabbush, P.M. (1990). *Nitrogen deficiency in Sitka spruce plantations*. For.Commn Bull. 89, HMSO, London

Taylor, O.C. and Eaton, F.M. (1966). Suppression of plant growth by nitrogen dioxide, *Pl. Physiol., Lancaster*, **41**, 132-6

Tehon, L.R. (1935). *A Monographic Rearrangement of Lophodermium*. Illinois biol.Monogr., 13, 151pp.

Terrier, C. (1953). Note sur *Lophodermium macrosporum* (Hartig) Rehm. *Phytopath. Z.*, **20**, 397-404

Thomas, A. (1974). Society autumn foray, Cardiff. *Bull. Br. mycol. Soc.*, **8**, 51-7

Thomas, H.E. (1934). Studies on *Armillaria mellea* (Vahl). Quél. Infection, parasitism, and host resistance. *J. agric. Res.*, **48**, 187-218

Thomas, M.D. (1961). Effects of air pollution on plants, in *Air Pollution, Wld Hlth Org. Monogr. Ser. 46*, 233-78, Columbia University Press, New York

Thomsen, M., Buchwald, N.F. and Hauberg, P.A. (1949). Angreb af *Cryptococcus fagi*, Nectria galligena, *og andre parasiter paa bog i Danmark 1939-43*, Kendrup and Wunsch, Kobenhavn

Thyr, B.D. and Shaw, C.G. (1964). Identity of fungus causing red band disease of pines. *Mycologia*, **56**, 103-9

Tiberi, R., Panconesi, A. and Roversi, P.F. (1988). Ulteriori indagini sul metodo per iniezione nella lotta contro *Corythuca ciliata* (Say). e *Gnomonia platani* (Kleb.). *Redia*, **71**, 227-45 (abstract in *Rev. Pl. Path.*, **69**, No. 1852, 1990)

Tinsley, T.W. (1967). Virus diseases of forest trees. *Rep. Forest Res., Lond.*, *1966*, p.112.

Tippett, J.T. and Shigo, A.L. (1980). Barrier zone anatomy in red pine roots invaded by *Heterobasidion annosum*. *Can. J. For. Res.*, **10**, 224-32

Tippett, J.T. and Shigo, A.L. (1981). Barriers to decay in conifer roots. *Eur. J. For. Path.*, **11**, 51-9

Tomiczek, C. (1985). Fungal diseases in afforestation of high altitude: experiences with Scleroderris and larch canker epidemics. *Proc. 3rd Int Workshop, IUFRO project group Pl. 07-00 Sept. 1984. Switzerland*

Toole, E.R. (1955). *Polyporus hispidus* on southern bottomland oaks. *Phytopathology*, **45**, 177-80

Toole, E.R. (1966). Root rot caused by *Polyporus lucidus*. *Pl. Dis. Reptr.*, **50**, 945-6

Towers, B. (1966). Effect of induced soil moisture stress on the growth of *Fomes annosus* in inoculated loblolly pine stumps. *Pl. Dis. Reptr.*, **50**, 747-9

Towey, J.W., Sweany, H.C. and Huron, W.H. (1932). Severe bronchial asthma apparently due to fungus spores found in maple bark. *J. Am. med. Ass.*, **99**, 453-9

Townrow, J.A. (1954). The biology of *Cryptostroma corticale* and the sooty bark disease of sycamore. *Rep Forest Res., Lond.*, *1953*, pp. 118-120

Townsend, A.M., Schreiber, L.R., Hall, T.J. and Bentz, S.E. (1990). Variation in response of Norway maple cultivars to *Verticillium dahliae*. *Pl. Dis.*, **74**, 44-6

Trench, T.N., Baxter, A.P. and Churchill, H. (1987). Report of *Melampsora medusae* on *Populus deltoides* in Southern Africa. *Pl. Dis.*, **71**, 761

Treshow, M. and Harward, M. (1965). Preliminary investigations in the incidence of canker and decline in Utah aspen stands. *Proc. Utah Acad. Sci.*, **42**, 196-200

True, R.P. and Tryon, E.H. (1956). Oak stem cankers initiated in the drought year 1953. *Phytopathology*, **46**, 617-21

Truter, S.J. (1947). Een voorlopig onderzoek naar de insterving van *Alnus glutinosa* (L.). Gaertner. Thesis, University of Utrecht; abstract in *Rev. appl. Mycol.*, p.576, 1947

Tsao, P.H. and Ocana, G. (1969). Selective isolation of species of Phytophthora from natural soils on an improved antibiotic medium. *Nature, Lond.*, **223**, 636-8

Tsyplakova, O.D. (1967). O vzaimosvyazi fitonsidnykh svoistv Topolei i ikh ustoichivosti k tsitosporozu. *Biol. Nauk.*, **10**, 139-41

Tubeuf, C.von (1930). Gnomonia pseudoplatani n.sp., die Ursache der Riesenflecken auf den Blättern des Bergahorns (*Acer pseudoplatanus*). *Z. PflKrankh. PflPath. PflSchutz.*, **40**, 364-75

Tubeuf, C.v. (1936). Tuberkulose, Krebs und Rindengrind der Eschen- (*Fraxinus*) Arten und die sie veranlassenden Bakterien, Nectria-pilze und Borkenkfer. *Z. PflKrankh.*, **46**, 449-83

Tubeuf, K.F. (1895). *Pflanzenkrankheiten durch kryptogame Parasiten verursacht.* Springer, Berlin

Turchetti, T. and Chelazzi, G. (1984). Possible role of slugs as vectors of the chestnut blight fungus. *Eur. J. For. Path.*, **14**, 125-7

Turchetti, T. and Panconesi, A. (1982). Osservazioni preliminari sull' antagonismo di alcune specie di Bacillus verso *Ceratocystis fimbriata* (Ell. and Halst.). Davidson f. *platani* Walter. *Riv. Patol. veg.*, Padova, **18**, 71-6

Turnau, K. and Czerwonka, M. (1988). Scanning ultrastructural ontogeny of cleistothecia in the powdery mildew *Microsphaera alphitoides*. *Acta Mycologica (1986, publ. 1988).*, **22**, 223-6

Turner, J.A. and Fox, R.T.V. (1988). Prospects for the chemical control of Armillaria species. In *Brighton Crop Protction Conference. Pests and Diseases 1988. Vol. 1.* British Crop Protection Council, Thornton Heath

Twarowska, I. (1968). Gospodarcze znaczenie, biologia i podstawy zwalczania grzyba *Pollaccia saliciperda* (All. et Tub.). v. Arx. *Prace Inst. Bad. Lésn.*, No. 351/353, pp. 3-54

Twyman, E.S. (1946). Notes on the dieback of oak caused by *Colpoma quercinum* (Fr.). Wallr. *Trans. Br. mycol. Soc.*, **29**, 234-41

Upadhyang, R.K. (1986). Effect of zineb and copper oxychloride on leaf inhabiting microfungi of *Eucalyptus globulus*. *Pesticides*, **20**, 34-9

Urbasch, I. (1989). *Pestalotia funerea* Desm. Untersuchungen zur biologie und Vorkommenshäufigkeit an Thuja. *NachrBl. dt. PflSchutzdienst.*, **41**, 33-5

Uri, J. (1948). Het parasitsme van *Nectria cinnabarina* (Tode). Fr. *Tijdschr. PlZiekt.*, **54**, 29-73

Urosevic, B. (1957). Mykoflora skladovanych zaludu. *Prace vyzk. Ust. lesn. CSR*, **13**, 149-200

Urosevic, B. (1958). Mumifikace semen nasich listnacu. *Lesn. prace*, **37**, 320-4

Urosevic, B. (1961). Mykoflora zaludu v obdobi dozravani, sberu a skladovani. *Prace vyzk. Ust. lesn. CSR*, **21**, 81-203

Urosevic, B. (1979). K otazce patogenity mykoflory semen jenlicnanu. *Lesnictvi*, **25**, 325-38. (abstract in *Rev. Pl. Path.*, **61**, No. 416, 1982)

Urosevic, B. (1987). *Tracheomycotic diseases in oak.* Communicationes Instituti Forestalis Cechosloveniae (1983, publ. 1987), **13**, 85-100

Urosevic, B. and Jancarik, V. (1957). [Some important diseases of oak saplings in our nurseries]. *Prace vyzk. Ust. lesn. C. S. R.*; abstract in *Rev. appl. Mycol.*, **37**, 423, 1958

Urquijo Landaluze, P. (1942). La enfermedad de la 'tinta' del castano y sn tratamiento. *Agricultura Madr.*, **11**, 54-6

Urquijo Landaluze, P. (1944). Aspectos de la obtencion de hebridos resistentes a la enfermedad del castano. *Bol. Pat. veg. Ent. agric., Madr.*, **13**, 447-62

Urquijo Landaluze, P. (1949). Accion de las sales de cobre el hongo *Phytophthora cinnamomi* y difusion de aquellas en la tierra. *Bol. Pat. veg. Ent. agric., Madr.*, **16**, 295-310

Urquijo Landaluze, P. (1963). Nuevos aspectos de la produccion de castanos resistentes a la enfermedad de la tinta'. *Bol. Pat. veg. Ent. agric., Madr.*, **26**, 163-79

Uscuplic, M. (1981). Infekcioni period *Lophodermium seditiosum* Min., Stal. and Mill. i mogucnosti njegovog suzbijanja u rasadnicima. *Zastita Bilja*, **32**, 375-82 (abstract in *Rev. Pl. Path.*, **61**, No. 3719, 1982)

Uscuplic, M. (1983). [New investigations on sweet chestnut blight]. *Zastita Bilja*, **34**, 317-28 (abstract in *Rev. Pl. Path.*, **63**, No. 3041, 1984

Vaartaja, O. (1964). Chemical treatment of seedbeds to control nursery diseases. *Bot. Rev.*, **30**, 1-91

Vajna, L. (1986). Branch canker and dieback of sessile oak (*Quercus petraea*) in Hungary caused by *Diplodia mutila*. I. Identification of the pathogen. *Eur. J. For. Path.*, **16**, 223-9

Valdivieso, J.A. and Luisi, N. (1987). El cancro del cipres (*Seiridium cardinale* (Wag.) Sutt. & Gibs.). en Chile. *Fitopatologia*, **22**, 79-84

Valentine, F.A., Carlson, K.D., Westfall.R.D. and Manion, P.D. (1981). Testing Verticillium wilt resistance in urban Norway maples. *J. Arb.*, **7**, 317-25

Valentine, H.T. (1983). An approach to modelling the consequences of beech mortality from beech bark disease. In Houston, D.R. and Wainhouse, D. (1983), q.v.134-7

Vamos, R. (1954). A Fenyocsemete dolese. *Erdo*, **3**, 34-40

Van Arsdel, E.P., Riker, A.J. and Patton, R.F. (1956). The effects of temperature and moisture on the spread of white pine blister rust. *Phytopathology*, **46**, 307-18

van der Kamp.B.J. (1968). *Peridermium pini* (Pers.). Lev. and the resin-top disease of Scots pine. I.A review of literature. *Forestry*, **41**, 189-98

van der Kamp, B.J. (1970). *Peridermium pini* (Pers.)Lev. and the resin-top disease of Scots pine. III. Infection and lesion development. *Forestry*, **43**, 73-88

van der Meiden, H.A. (1964). Virus bijpopulier. *Ned. Bosb. Tijdschr.*, **36**, 269-75

van der Meiden, H.A. and van Vloten, M. (1958). Roest en schorsbrand als bedreiging van de teelt van populier. *Ned. Bosb. Tijdschr.*, **30**, 261-73

van Vloten, H. (1944). Is verrijking van de mycoflora mogelijk? (Nar aanleiding van de populierenroest). *Tijdschr. PlZiekt.*, **1**, 49-62

Van Vloten, H. and Gremmen, J. (1953). Studies in the discomycete genera *Crumenula* De Not. and *Cenangium* Fr. *Acta Bot. Neerl.*, **2**, 226-41

Van Zanen, G.C.N. (1988). *Sparassis laminosa* new record versus *Sparassis crispa. Coolia*, **31** (4), 93-5

Vanderwalle, R. (1970). Station de Phytopathologie. *Rapport d'activité Centre de Recherches agronomiques de l'Etat, Gembloux, 1969*, 35-44.

Vedernikov, N.M. (1985). [Predicting needle-cast and snow blight in nurseries] *Les. Khoz.* No. 4, 63-5 (abstract in *Rev. Pl. Path.*, **65**, No. 4124, 1986)

Vedernikov, N.M. (1986). [Short-term forecasting of dates of spraying pine against snow blight in nurseries] *Mikol. Fitopatol.*, **20**, 53-6 (abstract in *Rev. Pl. Path.*, **65**, No. 6212, 1986)

Vedernikov, N.M. and Fedorova, N.S. (1986). [The technique of integrated control of diseases of conifers in nurseries] *Les. Khoz.*, No. 3, 59-61 (abstract in *Rev. Pl. Path.*, **66**, No. 2081, 1987).

Vegh, I. (1984). Identification en France de *Rosellinia herpotrichioides* Hepting et Davidson sur *Picea excelsa* Link ornamental. *Revue Hort.*, **247**, 35-37.

Veldeman, R. and Welvaert, W. (1960). Schorsbrand bij populier. *Meded. LandbHogesch. Gent*, **25**, 1107-1115; abstract in *For. Abs.*, **23**, No. 808, 1962

Velenovsky, J. (1920). Ceske houby. II. S cetnymi obrazy. Praha. *(quoted by Roll-Hansen, 1985)*

Venn, K.O. (1986). Threats created by import of timber and chips into Norway. *EPPO Bull.*, **16**, 457-9

Vergara Castillo, C. (1953). Un aporte al estudio de campo y de laboratorio del hongo *Nectria galligena* Bres. *Agricultura Tecnica, Santiago*, **13**, 62-85

Verrall, A.F. (1937). *Variation in Fomes igniarius (L.) Gill.* U.S. Dept.Agric.Tech.Bull. Minn.agric.Exp.Sta. No. 117

Viennot-Bourgin, G. (1949). *Les Champignons Parasites des Plantes Cultivées*, Masson, Paris

Viennot-Bourgin,G.(1956). *Mildious, Oidiums, Caries, Charbons, Rouilles des Plantes de France*. Encyclopedie Mycologique Vol. XXVI. Lechevalier, Paris

Viennot-Bourgin, G. (1981). A propos du Coryneum des cyprès. *C. r. Séanc. Acad. Agric. Fr.*, **67**, 266-72

Viennot-Bourgin, G. and Taris, B. (1957). Les maladies cryptogamiques des peupliers (Etat des travaux realises en France). *6th Int. Pop. Cong. Rep. & Comm. Fr.*, pp. 75-90, Paris

Vigouroux, A. (1986). Les maladies du platane, avec référence particulière au chancre coloré; situation actuelle en France.*EPPO Bull.*, **16**, 527-32

Viney, R. (1970). L'oidium du chêne. *Revue for. Fr.*, **22**, 365-9

Vliet, J.I.van (1931). Esschenkankers en hun bouw. Thesis, Univ. Utrecht. Abstract in *Rev. appl. Mycol.*, **11**, 12-13, 1932

Vloten, H. van (1930). Aantasting van *Pseudotsuga taxifolia* Britton (Douglasspar) door *Rhabdocline pseudotsugae* Sydow en Chermes cooleyi Gillette. *Nederl. BoschbTijdschr.*, **3**, 283-98

Vloten, H.van (1943). Verschillen in virulentie bij *Nectria cinnabarina* (Tode). Fr. *Tijdschr. PlZiekt.*, **49**, 164-71

Vloten, H.van (1952). Evidence of host-parasite relations by experiments with *Phomopsis pseudotsugae* Wilson. *Scott. For.*, **6**, 38-46

Voglino, P. (1931). Il nerume delle Castagne. *La Difesa delle Piante*, **8**, 1-4

Vucinic, Z. (1977). *Marssonina brunnea (Ell. et Everh.).* P. *Magn.*—prouzrokovac smede pjegavosti lisca topole. *Poljoprivreda i Sumarstvo*, **23**, 13-24

Vuillemin, P. (1893). Les Hypostomacées. *Comptes rend.*, **122**, 543

Wagener, W.W. (1928). Coryneum canker of cypress. *Science, N. S.*, **47**, 1745, p.584

Wagener, W.W. (1939). The canker of Cupressus induced by *Coryneum cardinale* n.sp. *J. agric. Res.*, **58**, 1-46

Wagener, W.W. (1948). Diseases of cypress. *El. Aliso*, **1**, 255-321

Wagn, O. (1987). Smitteforsog med rodfordoeveren, *Fomes annosus* (Fr.). Cooke, i loetroeer. II. Afsluttende opgorelse. *Tidsskr. Pl.*, **91**, 173-81

Wagner, K. (1927). Erkrankungen unserer Gehölze *Rhytisma acerinum* (Runzelschorf). *Gartenflora*, **76**, 81-2

Wakefield, E.M. and Bisby, G.R. (1941). List of hyphomycetes recorded for Britain. *Trans. Br. mycol. Soc.*, **25**, 49-126

Wakefield, E.M. and Dennis, R.W.G. (1981). *Common British Fungi*. Saiga Publishing, Hindhead

Waldie, J.S.L. (1930). An oak seedling disease caused by *Rosellinia quercina*.*Forestry*, **4**, 1-6

Wall, R.E. and Magasi, L.P. (1976). Environmental factors affecting Sirococcus shoot blight of black spruce. *Can. J. For. Res.*, **6**, 448-52

Walla, J.A. and Stack, R.W. (1980). Dip treatment for control of blackstem on Populus cuttings. *Pl. Dis.*, **64**, 1092-5

Waller, S. (1952). The Wanstead fungus disease of sycamore. *Essex Nat.*, **29**, 9-13

Wallis, G.W. (1960). Survey of *Fomes annosus* in East Anglian pine plantations. *Forestry*, **33**, 203-14

Wan, X.C. and Wang, A.L. (1987). [Different resistance of six clones of Aegeiros poplars to *Marssonina populi*]. *Journal of Nanjing Forestry University*, No. 1, 35-41 (Abstract in *Rev. Pl. Path.*, **69**, No. 1264, 1990)

Warcup, J.H. (1951). The effect of partial sterilisation on the occurrence of fungi in the soil. *Rep. Forest Res., Lond., 1950*, pp. 107-10

Warcup, J.H. (1952). Effect of partial sterilisation by steam or formalin on damping-off of Sitka spruce. *Trans. Br. mycol. Soc.*, **35**, 248-62

Warcup, J.H. (1953). Effect of partial sterilisation by steam or formalin on damping-off of Sitka spruce in an old forest nursery. *Rep. Forest Res. Lond., 1952*, p.108

Wardlaw, T.J. and Palzer, C. (1985). Stem disease in nursery seedlings caused by *Phytophthora cactorum, P. citricola* and *Pythium anandrum*. *Australian Plant Pathology*, **14**, 57-9

Waterhouse, G.M. (1963). *Key to the species of Phytophthora de Bary*, Mycol. Pap., No. 92

Waterhouse, G.M. (1970). *The Genus Phytophthora de Bary*, Mycol. Pap., No. 122

Waterman, A.M. (1945). Tip blight of species of *Abies* caused by a new species of *Rehmiellopsis*. *J. agric. Res.*, **70**, 315-37

Waterman, A.M. (1947). *Rhizosphaera kalkhoffii* associated with a needle cast of *Picea pungens*.*Phytopathology*, **37**, 507-11

Waterman, A.M. (1957). Canker and dieback of poplars caused by *Dothichiza populea*. *For. Sci.*, **3**, 175-83

Waterman, A.M. and Marshall, R.P. (1947). A new species of *Cristulariella* associated with a leaf spot of maple. *Mycologia*, **39**, 690-8

Watling, R., Kile, G.A. and Gregory, N.M. (1982). The genus *Armillaria*—nomenclature, typification, the identity of *Armillaria mellea* and species differentiation. *Trans. Br. mycol. Soc.*, **78**, 271-85

Watson, A.R. and Millar, C.A. (1971). Hypodermataceous needle-inhabiting fungi in pines in Scotland. *Trans Proc. Bot. Soc. Edinb.*, **41**, 250

Watson, H. (1928). Notes on attack by *Rhizoctonia crocorum* on Sitka spruce (*Picea sitchensis*). *Scott. For. J.*, **42**, 58-61

Watson, H. (1933). Diseases attacking common silver fir (*Abies pectinata*). *Scott. For.*, **47**, 71-2

Webber, J.F. and Hedger, J.N. (1986). Comparison of interactions between *Ceratocystis ulmi* and elm bark saprobes in vitro and in vivo. *Trans. Br. mycol. Soc.*, **86**, 93-101

Weber, G.F. (1941). *Leaf blister of oaks*. Pr. Bull. Fla. agric. Exp. Sta., No. 558, 2pp.

Webster, J. (1977). *Introduction to Fungi*, 2nd edn, Cambridge University Press, Cambridge

Wedgeworth, H.H. (1926). Leaf blister of oak. *Q. Bull. State Pl. Board Mississippi*, **6**, 10-2

Wehmeyer, L.E. (1932). The British species of the genus *Diaporthe* Nits. and its segregates. *Trans. Br. mycol. Soc.*, **17**, 237-95

Weijman, A.C.M. and Hoog, G.S. (1975). On the subdivision of the genus *Ceratocystis. Antonie van Leewenhoek*, **41**, 353-60

Weinstein, L.H. and Alscher-Herman, R. (1982). Physiological responses of plants to fluorine. In *Effects of gaseous air pollution in agriculture and horticulture* (edited by Unsworth, M.H. and Ormrod, D.P.). London, UK Butterworths.139-167

Weir, J.R. (1915). Observations on *Rhizina inflata*. *J. agric. Res.*, **4**, 93

Weir, J.R. and Hubert, E.E. (1918). Notes on the overwintering of forest tree rusts. *Phytopathology*, **8**, 55-9

Weisberger, H. (1969). Untersuchungen über *Pollaccia radiosa*, den erreger der TriebspitzenKrankheit an Pappeln der Sektion Leuce Duby. *Phytopath. Z.*, **66**, 50-68

Wells, J.M. and Payne, J.A. (1980). Mycoflora and market quality of chestnuts treated with hot water to control the chestnut weevil. *Pl. Dis.*, **64**, 999-1001

Went, J.C. (1948). Verslag van de onderzoekingen over de iepenziekte en andere boomziekten, 1974. *Meded. Iepenziektecomite, Baarn*, **43**, 15pp.

Went, J.C. (1949). Verslag over 1948. *Meded. iepenziektecomite, Baarn*, **44**, 15pp.

Werner, A. (1987). Responses of in vitro grown pine seedlings to infection by four strains of *Heterobasidion annosum*. *Eur. J. For. Path.*, **17**, 93-101

Werner, A. and Siwecki, R. (1978). Histological studies of the infection process by *Dothichiza populea* Sacc. et Briard in susceptible and resistant poplar clones. *Eur J. For. Path.*, **8**, 217-26

Wettstein, W. and Donaubauer, E. (1958). *Survey of Dothichiza populea*. Austria. Docum 14th Sess.stand.exec.Comm.Int.Poplar Cómm.Rome 1958, No. FAO/CIP/16-A, 1958, 6pp.

Whalley, A.J.S. and Watling, C. (1982). Distribution of *Daldinia concentrica* in the British Isles. *Trans. Br. mycol. Soc.*, **78** (1), 47-54

Whitbread, R. (1967). Bacterial canker of poplars in Britain. 1. The cause of the disease and the role of leaf scars in infection. *Ann. appl. Biol.*, **59**, 123-31

Whittle, A.M. (1977). Mycoflora of cones and seeds of *Pinus sylvestris*. *Trans. Br. mycol. Soc.*, **69**, 47-57

Wicker, E.F. (1965). A Phomopsis canker on western larch. *Pl. Dis. Reptr.*, **49**, 102-5

Wicker, E.F. (1981). Natural control of white pine blister rust by *Tuberculina maxima*. Stakman-Craigie Symposium on rust diseases. *Phytopathology*, **71**, 997-1000

Wiehe, P.O. (1952). The spread of *Armillaria mellea* (Fr.). Quel. in Tung orchards. *E. Afr. agric. J.*, **18**, 67-72

Wiejak, K. (1960). Observacje nad wysepowaniem zgorzeli pedow Wikliny powodowanej przez *Physalospora miyabeana* Fukushi. *Biul. Inst. Ochr. Ros., Poznan*, **9**, 205-13

Wilhelm, S. and Ferguson, J. (1953). Soil fumigation against *Verticillium albo-atrum*. *Phytopathology*, **43**, 593-5

Wilkins, W.H. (1934). Studies in the genus *Ustulina* with special reference to parasitism. I.Introduction, survey of the previous literature and host index. *Trans. Br. mycol. Soc.*, **18**, 320-46

Wilkins, W.H. (1936). Studies in the genus *Ustulina*.II.A disease of common lime (*Tilia vulgaris*) caused by *Ustulina*. *Trans. Br. mycol. Soc.*, **20**, 133-56

Wilkins, W.H. (1939a). Studies in the genus *Ustulina*.IV. Conidia, germination and infection. *Trans. Br. mycol. Soc.*, **23**, 65-85

Wilkins, W.H. (1939b). Studies in the genus *Ustulina*. V.A disease of elm caused by *Ustulina*. *Trans. Br. mycol. Soc.*, **23**, 171-85

Wilkins, W.H. (1943). Studies in the genus *Ustulina*.VI. A brief account of heartrot of beech caused by *Ustulina*. *Trans. Br. mycol. Soc.*, **26**, 169-70

Williams, D.J. and Moser, B.C. (1975). Critical level of airborne sea salt inducing foliar injury to bean. *Hort. Sci.*, **10**, 615-6

Wilson, C.L. (1965). *Ceratocystis ulmi* in elm wood.*Phytopathology*, **55**, 477

Wilson, K.W. (1981). *Removal of tree stumps*. Arboric. Leafl.7, HMSO, London

Wilson, M. (1920). A new species of *Phomopsis* parasitic on the Douglas fir. *Trans. Bot. Soc., Edin.*, **28**, 47-9

Wilson, M. (1921). A newly-recorded disease on Japanese larch. *Trans. R. Scott. arboric. Soc.*, **35**, 73-4

Wilson, M. (1922). Studies in the pathology of young trees and seedlings 1.The Rosellinia disease of the spruce.*Trans. R. Scott. Arboric. Soc.*, **36**, 226-35

Wilson, M. (1925). *The Phomopsis Disease of Conifers*, Bull. For.Commn, Lond., No. 6, 34pp.

Wilson, M. (1937). The occurrence of *Keithia tsugae* in Scotland. *Scott. For.*, **51**, 46-7

Wilson, M. and Hahn, G.G. (1928). The identity of *Phoma pitya* Sacc., *Phoma abietina* Hart. and their relation to *Phomopsis pseudotsugae* Wilson. *Trans. Br. mycol. Soc.*, **13**, 261-78

Wilson, M. and Henderson, D.H. (1966). *British Rust Fungi*, Cambridge University Press, Cambridge

Wilson, M. and Macdonald, J. (1924). A new disease of the silver firs in Scotland. *Trans. R. Scott. arboric. Soc.*, **38**, 114-6

Wilson, M. and Waldie, J.S.L. (1926). *Rhizosphaera kalkhoffii* Bubak as a cause of defoliation of conifers. *Trans. R. Scott. arboric. soc.*, **40**, 34-6

Wilson, M. and Waldie, J.S.L. (1927). An oak leaf disease caused by *Sclerotinia candolleana* (Lév.) Fuckel. *Ann. appl. Biol.*, **14**, 193-6

Wiltshire, S.P. (1921). Studies on the apple canker fungus. 1. Leaf scar infection. *Ann. appl. Biol.*, **8**, 182

Winter, C. (1897). *Die Pilze Deutschlands, Österreichs und der Schweiz*. 2 Abt., Kummer, Leipzig

Winter, G. (1887). *Die Pilze Deutschlands, Oesterreichs und der Schweiz*. II Abt., Kummer, Leipzig

Witcher, W. and Lane, C.L. (1980). Annosus root rot in slash pine plantations in the sandhill section of South Carolina. *Pl. Dis.*, **64**, 398-9

Wittman,W.(1980). *Rhytisma acerinum* (Pers.) Fr. = *Melasmia acerina* Lev. *Pflanzenarzt*, **33**, 89

Woeste, U. (1956). Anatomische Untersuchungen über die Infektionswege einiger Wurzelpilze. *Phytopath. Z.*, **26**, 225-72

Wong, W.C., Nash, T.H. and Preece, T.F. (1974). A field survey of watermark disease of cricket bat willow in Essex and observations on some of the probable sources of the disease. *Pl. Path.*, **23**, 25-9

Wong, W.C. and Preece, T.F. (1973). Infection of cricket bat willow (*Salix alba* var. *coerulea* Sm.). by *Erwinia salicis* (Day). Chester detected in the field by the use of a specific antiserum. *Pl. Path.* 95-7

Wong, W.C. and Preece, T.F. (1978a). *Erwinia salicis* in cricket bat willows: histology and histochemistry of infected wood. *Physiol. Pl. Path.*, **12**, 321-32

Wong, W.C. and Preece, T.F. (1978b). *Erwinia salicis* in cricket bat willows: peroxidase, polyphenoloxidase, ß-glucosidase, pectinolytic and cellulolytic enzyme activity in diseased wood. *Physiol. Pl. Path.*, **12**, 333-47

Wong, W.C. and Preece, T.F. (1978c). *Erwinia salicis* in cricket bat willows: phenolic constituents in healthy and diseased willows. *Physiol. Pl. Path.*, **12**, 349-57

Woo, J.Y. and Partridge, A.D. (1969). The life history and cytology of *Rhytisma punctatum* on bigleaf maple. *Mycologia*, **61**, 1085-95

Woodgate-Jones, P. and Hunter, T. (1983). Integrated control of *Chondrostereum purpureum* in plum by treatment of pruning wounds. *J. Hortic. Sci.*, **58** (4), 491-496

Woodward, R.C., Waldie, J.S.L. and Steven, H.M. (1929). Oak mildew and its control in forest nurseries. *Forestry*, **3**, 38-56

Wormald, H. (1930). Bacterial 'blight' of walnuts in Britain. *Ann. appl. Biol.*, **17**, 59-70

Wormald, H. (1937). Leaf blotch of hawthorn. *Gdnrs' Chron.*, **102**, 47

Wormald, H. (1955). *Diseases of Fruits and Hops*, Crosby Lockwood, London

Worrall, J. (1983). Cytospora canker of poplars and willows. *California Plant Pathology*, No. 64, 4-5

Worrall, J.J. and Parmeter, J.R., Jr. (1982). Wetwood formation as a host response in white fir. *Eur. J. For. Path.*, **12**, 432-41

Worrall, J.J. and Parmeter, J.R., Jr. (1983). Inhibition of wood decay fungi by wetwood of white fir. *Phytopathology*, **73**, 1140-5

Worrall, J.J., Parmeter, J.R., Jr. and Cobb, F.W., Jr. (1983). Host specialization of *Heterobasidion annosum*. *Phytopathology*, **73**, 304-7

Wright, W.R. (1960). Storage decays of domestically grown chestnuts. *Pl. Dis. Reptr.*, **44**, 820-5

Wuhlisch, G.von and Stephan, B.R. (1986). Entwicklung der Nachtkommenschaften frei abgeblühten Schleswig-Holsteinischer Plusbäume von *Pinus sylvestris* L. bis zum Alter von 33 Jahren. *Allg. Forst. - u. Jagdztg.*, **157**, 112-24

Wulf, A.(1988). *Pleuroceras pseudoplatani* (v.Tubeuf) Monod, Erreger einer Blattbraune an Bergahorn (*Acer pseudoplatanus* L.) *NachrBl. dt PflSchutzdienst, Berl.*, **40**, 65-70

Xenopoulos, S. and Diamandis, S. (1985). A distribution map for *Seiridium cardinale* causing cypress canker disease in Greece. *Eur. J. For. Path.*, **15**, 223-6

Xiang, Y.Y., Xi, Z.K. and Zhang, H.L. (1984). [A study on the properties of poplar mosaic virus]. *Scientia Silvae Sinicae*, **20**, 441-6 (abstract in *Rev. Pl. Path.*, **65**, No. 5182, 1986)

Yatsenko-Khmelevky, A.A. (1938). Sur l'échauffure du bois de l'hêtre. *C. r. Acad. Sci. U. R. S. S.*, **18**, 207-12

Yde-Anderson, A. (1961). Om angreb af *Polyporus schweinitzii* Fr. i naletraebevorkinger. *Dansk Skogforen. Tidsskr.*, **46**

Yde-Andersen, A. (1963). Plantedod i naletraekulturer som folge af kvasbraending. *Dansk Skovforen. Tidsskr.*, **48**, 112-23

Yokota, S. (1983). Etiological and pathological studies on Scleroderris canker in Hokkaido, Japan. *Bull. For. Prod. Res. Inst.*, No. 321, 89-116

York, H., Wean, R.E. and Childs, T.W. (1936). Some results of investigations on *Polyporus schweinitzii* Fr. *Science, N. S.*, **84**, 160-1

Young, C.W.T. (1965). *Death of Pedunculate Oak and Variations in Annual Radial Increment Related to Climate*, Forest Rec.Lond., No. 55, 15pp.

Young, C.W.T. (1978). *Sooty Bark Disease of Sycamore*, Arboric. Leafl., No. 3, 8pp., HMSO, London

Young, C.W.T. (1984). Revised by Lonsdale, D. *External signs of decay in trees*, Arboric.Leafl., No. 1. HMSO, London

Young, C.W.T. and Strouts, R.G. (1971). In Forest Pathology. *Rep. Forest Res., Lond., 1971*, 77-84

Young, C.W.T. and Strouts, R.G. (1973). In Pathology. *Rep. Forest Res., Lond., 1973*, p.101

Young, C.W.T. and Strouts, R.G. (1974). *Rep. Forest Res., Lond., 1974*, 37-8

Young, C.W.T. and Strouts, R.G. (1975). In Forest Pathology. *Rep. Forest Res., Lond., 1975*, pp. 34-5

Young, C.W.T. and Strouts, R.G. (1976). In Forest Pathology. *Rep. Forest Res., Lond., 1976*, pp. 34-5

Young, C.W.T. and Strouts, R.G. (1977). In Forest Pathology. *Rep. Forest Res., Lond., 1977*, pp. 31-2

Young, C.W.T. and Strouts, R.G. (1978). In Forest Pathology. *Rep. Forest Res., Lond., 1978*, pp. 33-4

Young, C.W.T. and Strouts, R.G. (1979). In Forest Pathology. *Rep. Forest Res., Edin., 1979*, pp. 32-3

Yu, Y.N. and Lai, Y.Q. (1982). [Taxonomic studies on the genus *Microsphaera* of China, IV. New and known species of *Microsphaera* on the family Fagaceae]. *Journal of the North-Eastern Forestry Institute, China*, No. 4, 24-36. (Abstract in *Rev. Pl. Path.*, **65**, No. 1530, 1986)

Yuan, J.Y. (1982). [Studies on Valsa kunzei Nits. Growth characteristics, pathogenicity and its morphological identification]. *Journal of North-Eastern Forestry Institute, China*, **1**, 13-9. (abstract in *Rev. Pl. Path.*, **65**, No. 1545, 1986).

Zahorovska, E. (1985). [Germination of the conidia of the fungus *Microsphaera alphitoides*]. *Acta Facultatis Rerum naturalium Universitatis Comenianae Botanica*, **32**, 77-84. Abstract in *Rev. Pl. Path.*, **66**, No. 3043, 1987

Zalesky, H. (1968). Penetration and initial establishment of *Nectria galligena* in aspen and peachleaf willow. *Can. J. Bot.*, **46**, 57-60

Zbinden, W. (1986). Pflanzenschultz bei Walmissbäumen. *Schweiz. Z. Obst Weinbau*, **122**, 228-3

Zeller, S.M. (1926). Observations on infection of apple and prune roots by *Armillaria mellea*. *Phytopathology*, **16**, 479-84

Zeller, S.M. (1935). Some miscellaneous fungi of the Pacific Northwest. *Mycologia*, **27**, 449-66

Zeller, S.M. and Owens, C.E. (1921). European canker on the Pacific Slope. *Phytopathology*, **11**, 464-8

Zeller, W. (1979). Resistance and resistance breeding in ornamentals. *EPPO Bull.*, **9**, 35-44

Zentmeyer, G.A. (1961). Chemotaxis of zoospores for root exudates. *Science*, **133**, 1595-6

Zhang, F.J. and Fang, K.J. (1984). [A preliminary report of experimental control of *Cronartium flaccidum* on *Pinus sylvestris* with pine tar oil]. *Forest Science and Technology (Linye Keji Tongxum)*, No. 5, 25 (abstract in *Rev. Pl. Path.* No. 1170, 1987).

Zhang, J.J., Zhou, J. and Xiang, W.N. (1988). [Studies in the types of *Agrobacterium tumefasciens* from poplars in China and their sensitivity to agrocins]. *Acta Microbiologica Sinica (Weishengwa Xuebas).*, **28**, 12-18 (abstract in *Rev. Pl. Path.*, **68**, No. 3951, 1959)

Zhao, J.Z., Xue, M.X. and Wang, L.F. (1983). [A study on the big spot type canker of poplars. II. Biological characteristics of the pathogen]. *Forest Science andTechnology (Linye Keiji Tongxun)*. No. 10, 24-27 (abstract in *Rev. Pl. Path.*, **64**, No. 5137, 1985)

Zhong, Z.K. (1982). [Study on poplar canker caused by *Dothichiza populea*]. *Journal of the North-Eastern Forestry Institute, China*, No. 2, 63-69 (abstract in *Rev. Pl. Path.*, **65**, No. 1535, 1986)

Zhu, Z.C. Zou, K.M. and Xu, G.G. (1984). [Fungicide smokes for curing forest diseases]. *J. Nanjing Inst. For.*, **3**, 68-79

Ziller, W.G. (1974). *The Tree Rusts of Western Canada*. Canadian Forest Service Publication No. 1329. Department of the Environment, British Columbia

Zinkernagel, C. and Flückiger, W. (1987). Uber das Auftreten von *Leucocytospora kunzei* bei *Pinus griffithii* in Parkanlagen der Stadt Basel. *Eur. J. For. Path.*, **17**, 441-3

Zogg, H. (1943). Untersuchungen über die Gattung *Hysterographium* Corda insbesondere über *Hysterographium fraxini* (Pers.). de Not. *Phytopath. Z.*, **14**, 310-84

Zollfrank, V., Sautter, C. and Hock, B. (1987). Fluorescence immunohistochemical detection of *Armillaria* and *Heterobasidion* in Norway spruce. *Eur. J. For. Path.*, **17**, 230-7

Zweep, P. van der and Kam, M. de (1982). The occurrence of *Erwinia salicis*, the cause of watermark disease, in the phyllosphere of *Salix alba*. *Eur. J. For. Path.*, **12**, 257-61

Zycha, H. (1951a). Das Rindensterben der Buche. *Phytopath. Z.*, **17**, 444-61

Zycha, H. (1951b). Das Buchensterben Ursache und Prognose. *SchReihe forstl. Fak. Univ. Gottingen*, **2**, 72-8

Zycha, H. (1955). Eine Krebserkrankung der Sitke-Fichte (*Picea sitchensis* [Bong.] Carr.). *Forstwiss. ZentBl.*, **74**, 293-304

Zycha, H. (1959a). Stand unserer Kenntnis vam Rindensterben der Buche. *Allg. Forstz.*, **14**, 786-9

Zycha, H. (1959b). Zur frage der Infektion beim Larchenkrebs. *Phytopath Z.*, **37**, 61-74

Zycha, H. (1961/2). Trieb- und Nadelschäden an jungen Lärchen im Bestand. Reprinted from *Allg. Forstz.*, **27/28**, 2pp.

Zycha, H. (1965). Die Marssonina-Krankheit der Pappel. *Forstwiss. Cbl.*, **7/8**, pp. 254-9

# Glossary

Only a short, simple, mainly mycological glossary is provided here, as most of the terms used in this book are already explained in the text, and can be found readily in either a good general dictionary or a dictionary of biology, in which expanded definitions of the terms included here can be found.

ACERVULUS   A more or less saucer-shaped fruit body forming conidia (asexual spores) and embedded in the host tissues

AECIOSPORE   A spore formed in an *aecium*. Also called an *aecidiospore*

AECIUM (pl. AECIA)   One of the spore-producing bodies found in the rust fungi (Uredinales), usually cup-like or blister-like in form, with a distinct wall. Also called an *aecidium* (pl. *aecidia*)

ANAMORPH   The form of a fungus which produces the asexual spores. Also called the *imperfect state* or *stage*

APOTHECIUM (pl. APOTHECIA)   A usually disc-like or saucer-like fruit body found in the Discomycetes, one of the groups of the Ascomycotina

ASCOCARP   A general term for the perfect (sexual) fruit body in the Ascomycotina

ASCOMYCOTINA   The Ascomycetes, a main group of the fungi, the perfect spores of which are formed in *asci*

ASCOSPORES   The spores formed in *asci*

ASCUS (pl. ASCI)   The cell in which the ascospores of the Ascomycotina are formed

BASIDIOMYCOTINA   The Basidiomycetes, a main group of the fungi, the perfect (sexual) spores of which are formed on *basidia*

BASIDIOSPORES   The spores formed on *basidia*

BASIDIUM (pl. BASIDIA)   The cell on which the basidiospores of the Basidomycotina are formed

CAEOMA (pl. CAEOMATA)   An *aecium* lacking a distinct wall

CLEISTOCARP, CLEISTOTHECIUM (pl. CLEISTOTHECIA)   An ascal fruit body with no specialised opening for the release of spores

DEUTEROMYCOTINA   An artificial assemblage of fungi. Most either are forms with no perfect (sexual) stages or are the imperfect (asexual) forms of ascomycetes

FUNGI IMPERFECTI   Deuteromycotina

HYMENIUM   The layer of a fruit body bearing the spores

HYPHA (pl. HYPHAE)   One of the filaments which form the *mycelium* in the fungi

HYSTEROTHECIUM (pl. HYSTEROTHECIA)   A long, narrow ascocarp

MASTIGOMYCOTINA   One of the main groups of fungi, in which the hyphae are generally non-septate, the fruit bodies are simple, and the sex cells are motile. cf. *Zygomycotina*

MYCELIUM (pl. MYCELIA)   The hyphal mass forming the body of a fungus

OIDIUM (pl. OIDIA)   Term used for the usually barrel-shaped asexual spores of the mildews, and for superficially similar spores formed by the breaking up of the hyphae of, for example, some Basidiomycotina

PERFECT STATE   The state in a fungal life cycle in which spores such as basidiospores or ascospores are formed following nuclear fusion or by parthenogenesis

PERITHECIUM (pl. PERITHECIA)   A subglobose or flask-shaped ascocarp found in the Pyrenomycetes, one of the groups of the Ascomycotina. A term often used to include *pseudothecium*

PHYCOMYCETES   The 'lower fungi', now separated into the Mastigomycotina and the Zygomycotina

PSEUDOTHECIUM (pl. PSEUDOTHECIA)   A form of *perithecium* in which the asci are borne in unwalled spaces in a fungal stroma

PYCNIDIUM (pl. PYCNIDIA)   A subglobose or flask-like fruit body in the Deuteromycotina. It superficially resembles a perithecium but contains asexual spores

RHIZOMORPH   A cord-like or bootlace-like body made up of fungal hyphae

SCLEROTIUM (pl. SCLEROTIA)   A generally rounded and often hardened mass of fungal hyphae on which no spores are borne. It may act as a resting stage and later give rise to a fruit body

SEPTATE (of fungal hyphae)   Divided by cross-walls

SPERMATIUM (pl. SPERMATIA)   A sex cell formed in some fungi (for example, the rusts and some ascomycetes)

SPERMAGONIUM (pl. SPERMAGONIA)   A walled body in which spermatia are produced. Also spelled *spermogonium*

STROMA (pl. STROMATA)   A mass of fungal hyphae on or in which spores are borne

TELEOMORPH   The form of a fungus which produces the sexual spores. Also called the *perfect state* or *stage*, q.v.

TELIOSPORE   The 'resting spore' or 'winter spore' produced in a *telium*. Also called a *teleutospore*

TELIUM (pl. TELIA)   One of the spore stages of the rusts (Uredinales) in which *teliospores* ('resting spores', 'winter spores') are produced. Also called a *teleutosorus*

UREDINIOSPORE   The 'summer spore' produced in a *uredinium*

UREDINIUM (pl. UREDINIA)   One of the spore stages of the rusts (Uredinales), in which urediniospores ('summer spores') are produced. Also called a *uredosorus*

VECTOR   An organism, usually an insect or other small animal, which carries the living agents of disease from one host to another

ZYGOMYCOTINA   One of the main groups of fungi, in which the hyphae are generally non-septate, the fruit bodies are simple, and the sex cells are non-motile. cf. *Mastigomycotina*

# Index

Pages with black and white figures are indexed in *italics*. Terms such as 'fungi' (which are described throughout the book under their individual names) are indexed only where they appear in the General Introduction (Chapter 1). Under the various host trees, general references to each host genus appear first, followed by those to individual species. In the text we sometimes refer to host names used by the authors of quoted papers. What appear to be current host names, with some common synonyms, can be found by using this index.

Printed in the United States
By Bookmasters